Lecture Notes in Artificial Intell

Edited by J. G. Carbonell and J. Siekmann

Subseries of Lecture Notes in Computer Science

Sergei O. Kuznetsov Stefan Schmidt (Eds.)

Formal Concept Analysis

5th International Conference, ICFCA 2007
Clermont-Ferrand, France, February 12-16, 2007
Proceedings

Springer

Series Editors

Jaime G. Carbonell, Carnegie Mellon University, Pittsburgh, PA, USA
Jörg Siekmann, University of Saarland, Saarbrücken, Germany

Volume Editors

Sergei O. Kuznetsov
Higher School of Economics
Department of Applied Mathematics
105679 Kirpichnaya 33/5 Moscow, Russia
E-mail: skuznetsov@hse.ru

Stefan Schmidt
Technische Universität Dresden
Institut fuer Algebra
01062 Dresden, Germany
E-mail: Stefan.Schmidt@tu-dresden.de

Library of Congress Control Number: 2007920052

CR Subject Classification (1998): I.2, G.2.1-2, F.4.1-2, D.2.4, H.3

LNCS Sublibrary: SL 7 – Artificial Intelligence

ISSN 0302-9743
ISBN-10 3-540-70828-6 Springer Berlin Heidelberg New York
ISBN-13 978-3-540-70828-5 Springer Berlin Heidelberg New York

Springer is a part of Springer Science+Business Media

springer.com

Printed in Germany

Typesetting: Camera-ready by author, data conversion by Scientific Publishing Services, Chennai, India
Printed on acid-free paper SPIN: 12018079 06/3142 5 4 3 2 1 0

Preface

This volume contains the Proceedings of ICFCA 2007, the Fifth International Conference on Formal Concept Analysis. The ICFCA conference series intends to serve as a distinguished forum for state-of-the-art research from foundational to applied lattice theory and related fields, all of which involve methods and techniques in the realm of FCA. We believe this year's conference provided a continuation of the high standard of its predecessors.

Formal concept analysis ranges from restructuring general algebra and lattice theory, on the one hand, to providing high-level conceptual analysis and knowledge processing techniques in various applied disciplines, on the other hand. This is due to the fact that FCA allows transforming relational data into implications and dependencies and, vice versa, exploring plausible hypotheses, suggested implications, or expected dependencies against data. This all is connected with the investigation of hierarchical aspects of information—based on a mathematical formalization of *conceptual hierarchy*. The field of FCA has been developing way beyond its original scope and attracts a still-growing number of sophisticated research scholars, who vary from theoreticians to more practical problem solvers. This volume reflects the diversity between fundamental methods and applied techniques, explored by an enthusiastic research community. Here, we only mention research areas such as data visualization, information retrieval, machine learning, data analysis and knowledge management.

ICFCA 2007 covered practical data analysis and problem solving as well as foundational progress of the field. The algorithmic side was in balance with theoretical discoveries. All regular papers in this volume were refereed by independent domain experts, although the Program Chairs are responsible for the final decision on publications in this volume. To assure a high-quality standard, the close involvement of the Program Committee and the Editorial Board in any decision making was crucial.

The General Conference Chair of ICFCA 2007, held in Clermont-Ferrand, France, was Lhouari Nourine. The success of the conference was a result of all involved, in particular, the General Conference Chair together with the Conference Organizing Committee, and the Editorial Board in adjunction with the Program Committee—to all of whom we express our warmest thanks. The activity of the first Program Co-chair in preparing the volume was supported by the DFG project "Concepts and Models" (COMO).

February 2007

Sergei O. Kuznetsov
Stefan E. Schmidt

Organization

Executive Committee

Organizing Conference Chair

Lhouari Nourine	LIMOS, Université Blaise Pascal, France

Conference Organization Committee

Pierre Colomb	LIMOS, Université Blaise Pascal, France
Alain Gély	LIMOS, Université d'Auvergne, France
Raoul Medina	LIMOS, Université Blaise Pascal, France
Olivier Raynaud	LIMOS, Université Blaise Pascal, France
Yoan Renaud	LIMOS, Université Blaise Pascal, France

Program and Conference Proceedings

Program Chairs

Sergei O. Kuznetsov	Higher School of Economics, Moscow, Russia
Stefan E. Schmidt	Technische Universität Dresden, Germany

Editorial Board

Peter Eklund	University of Wollongong, Australia
Bernhard Ganter	Technische Universität Dresden, Germany
Robert Godin	Université du Québec à Montréal (UQAM), Canada
Sergei O. Kuznetsov	Higher School of Economics, Moscow, Russia
Rokia Missaoui	Université du Québec en Outaouais (UQO), Gatineau, Canada
Uta Priss	Napier University, Edinburgh, United Kingdom
Stefan E. Schmidt	Technische Universität Dresden, Germany
Gregor Snelting	University of Passau, Germany
Gerd Stumme	University of Kassel, Germany
Rudolf Wille	Technische Universität Darmstadt, Germany
Karl Erich Wolff	University of Applied Sciences, Darmstadt, Germany

Program Committee

Radim Belohlavek	Palacky University of Olomouc, Czech Republic
Claudio Carpineto	Fondazione Ugo Bordoni, Rome, Italy
Richard J. Cole	University of Queensland, Brisbane, Australia
Paul Compton	University of New South Wales, Sydney, Australia

Frithjof Dau	Technische Universität Darmstadt, Germany
Vincent Duquenne	Université Pierre et Marie Curie, Paris, France
Sébastien Ferré	Université de Renne, France
Alain Gély	LIMOS, Université d'Auvergne, France
Wolfgang Hesse	Universität Marburg, Germany
Andreas Hotho	Kassel University, Germany
Bjoern Koester	Technische Universität Dresden, Germany
Marzena Kryszkiewicz	Warsaw University of Technology, Warsaw, Poland
Léonard Kwuida	Universität Bern, Switzerland
Wilfried Lex	Universität Clausthal, Germany
Chirstian Lindig	Saarland University, Saarbrücken, Germany
Raoul Medina	LIMOS, Université Blaise Pascal, France
Engelbert Mephu Nguifo	IUT de Lens - Université d'Artois, France
Rokia Missaoui	Université du Québec en Outaouais (UQO), Gatineau, Canada
Lhouari Nourine	LIMOS, Université Blaise Pascal, France
Sergei Obiedkov	Higher School of Economics, Moscow, Russia
Jean-Marc Petit	LIRIS, INSA-Lyon, France
Alex Pogel	New Mexico State University, Las Cruces, USA
Sandor Radeleczki	University of Miskolc, Hungary
Stefan E. Schmidt	Technische Universität Dresden, Germany
Jürg Schmid	Universität Bern, Switzerland
Bernd Schröder	Louisiana Tech University, Ruston, USA
Selma Strahringer	University of Applied Sciences, Cologne, Germany
Petko Valtchev	DIRO, Université de Montréal, Canada
Mohammed J. Zaki	Rensselaer Polytechnic Institute, New York, USA

Sponsoring Institutions

Université Blaise Pascal
UFR ST, Université Blaise Pascal
LIMOS, Laboratoire d'informatique, de modélisation et d'optimisation des systèmes
Département de mathématique et informatique de l'Université Blaise Pascal
ISIMA, Institut supérieur d'informatique de modélisation et de leurs applications
Ecole doctorale SPI, Université Blaise Pascal
Le Conseil Général du Puy de Dome
Conseil régional d'Auvergne
Ville de Clermont-Ferrand
Clermont-communauté

Table of Contents

Relational Galois Connections

Bernhard Ganter

Institut für Algebra
Technische Universität Dresden

Abstract. Galois connections can be defined for lattices and for ordered sets. We discuss a rather wide generalisation, which was introduced by Weiqun Xia and has been reinvented under different names: Relational Galois connections between relations. It turns out that the generalised notion is of importance for the original one and can be utilised, e.g., for computing Galois connections.

The present paper may be understood as an attempt to bring together ideas of Wille [15], Xia [16], Domenach and Leclerc [3], and others and to suggest a unifying language.

1 Galois Connections Between Relations

It is usual to define a **Galois connection** between two complete lattices $(L_1, \leq_1)$ and $(L_2, \leq_2)$ as a pair (φ, ψ) of mappings

$$\varphi : L_1 \to L_2, \quad \psi : L_2 \to L_1$$

satisfying for all $x \in L_1$ and all $y \in L_2$

$$x \leq_1 \psi(y) \iff y \leq_2 \varphi(x).$$

It is well known that if this condition is satisfied, both mappings φ and ψ are order reversing and that their compositions $\varphi \circ \psi$ and $\psi \circ \varphi$ are closure operators on L_2 and L_1, respectively, with dually isomorphic lattices of closed sets (see, e.g., [4]).

Not so obvious is how to *construct* Galois connections for given lattices $(L_1, \leq_1)$ and $(L_2, \leq_2)$. We shall address this question later. For the case that both lattices are power set lattices, the answer was given by Birkhoff [2]: Let G and M be sets, let $I \subseteq G \times M$ be some relation. Define

$$A^I := \{m \in M \mid g\, I\, m \text{ for all } g \in A\} \qquad \text{if } A \subseteq G$$

and

$$B^I := \{g \in G \mid g\, I\, m \text{ for all } m \in B\} \qquad \text{if } B \subseteq M.$$

Then

$$\varphi(X) := X^I \quad \text{for } X \subseteq G \qquad \text{and} \qquad \psi(Y) := Y^I \quad \text{for } Y \subseteq M$$

S.O. Kuznetsov and S. Schmidt (Eds.): ICFCA 2007, LNAI 4390, pp. 1–17, 2007.

defines a Galois connection between the power set lattice of G and the power set lattice of M. The set

$$\mathfrak{B}(G, M, I) := \{(A, B) \mid A \subseteq G, B \subseteq M, A^I = B, A = B^I\},$$

ordered by

$$(A_1, B_1) \leq (A_2, B_2) :\iff A_1 \subseteq A_2 \iff B_2 \subseteq B_1$$

is a complete lattice, called the **concept lattice**[1]. The sets of the form A^I, $A \subseteq G$, are called the **intents** of (G, M, I), and those of the form B^I, $B \subseteq M$ are the **extents**. These are the closed sets of the two closure operators.

The notion of a Galois connection can be generalised to ordered sets and, even further, to arbitrary binary relations $I \subseteq G \times M$, $J \subseteq H \times N$, as in the next definition:

Definition 1. *A* **Galois connection**[2] *between* (G, M, I) *and* (H, N, J) *is a pair* (φ, ψ) *of mappings*

$$\varphi : G \to N, \quad \psi : H \to M$$

satisfying

$$g\ I\ \psi(h) \iff h\ J\ \varphi(g).$$

This definition is symmetric: if (φ, ψ) is a Galois connection between (G, M, I) and (H, N, J), then (ψ, φ) is a Galois connection between (H, N, J) and (G, M, I). This corresponds to the original, *contravariant* definition of Galois connections. Some authors consider also the *covariant* version, which allows for composition of Galois connections. This is achieved when (H, N, J) is replaced by the *dual context* (N, H, J^{-1}). These mappings are closely related to *infomorphisms* and to *Chu morphisms*, see Section 6 for more.

One might argue that this definition deviates from the original one for lattices or ordered sets. But it is only a natural generalisation. For two ordered sets $(P, \leq_1)$ and $(Q, \leq_2)$ the condition that (φ, ψ) is a Galois connection between $(P, P, \leq_1)$ and $(Q, Q, \leq_2)$ is

$$x \leq_1 \psi(y) \iff y \leq_2 \varphi(x),$$

as usual.

We may generalise even further, replacing the pair of mappings by a pair of relations $\Phi \subseteq G \times N$ and $\Psi \subseteq H \times M$. The natural condition then is that

$$g\ I\ h^{\Psi} \iff h\ J\ g^{\Phi}$$

holds for all $g \in G$ and all $h \in H$. We call this the (relational) **Galois condition**. However, this condition by itself turns out to be not strong enough. We therefore define

---

[1] Our notation is that of Formal Concept Analysis [5], where (G, M, I) is called a **formal context**. Other authors use names like **classification** [1], **Chu-space** [11], etc.

[2] Called **context–Galois connection** by Xia [16].

Definition 2. *A* **relational Galois connection**[3] *between* (G, M, I) *and* (H, N, J) *is a pair* (Φ, Ψ) *of relations*

$$\Phi \subseteq G \times N \quad \text{and} \quad \Psi \subseteq H \times M$$

satisfying

1. $g\, I\, h^{\Psi} \iff h\, J\, g^{\Phi}$ *for all* $g \in G, h \in H$ *(the Galois condition),*
2. Φ *is the largest relation satisfying the Galois condition for the given* Ψ*, and conversely.*

	M	N
G	I	Φ
H	Ψ	J

Fig. 1. The Galois condition requires that the left hand part h^{Ψ} of a row from the lower part is contained in a row g^{I} of the upper part iff the right hand side g^{Φ} of the latter is contained in the former, in h^{J}

Proposition 1. *The second condition of Definition 2 can be reformulated as follows:*

2'. g^{Φ} *is an intent of* (H, N, J) *and* h^{Ψ} *is an intent of* (G, M, I).

Proof. Fix $\Psi \subseteq H \times M$. A relation Φ satisfies the direction "$\Rightarrow$" of the Galois condition iff

$$g^{\Phi} \subseteq \{h \in H \mid g\, I\, h^{\Psi}\}^{J}.$$

The implication "$\Leftarrow$" is equivalent to

$$g^{\Phi J} \subseteq \{h \in H \mid g\, I\, h^{\Psi}\},$$

which implies

$$g^{\Phi} \subseteq \{h \in H \mid g\, I\, h^{\Psi}\}^{J} = g^{\Phi JJ}.$$

Thus from Φ we obtain another relation satisfying the Galois condition for Ψ by replacing each g^{Φ} by its closure $g^{\Phi JJ}$. This then is the largest possible choice. The dual argument works for Ψ. □

The next proposition shows that this is in fact a further generalisation.

[3] Called **essential Galois bond** by Xia [16]. Xia's condition (iii) is implied.

Proposition 2. *If* (φ, ψ) *is a Galois connection between* (G, M, I) *and* (H, N, J) *then* (Φ, Ψ)*, given by*

$$g^{\Phi} := \varphi(g)^{JJ} \text{ for } g \in G, \quad h^{\Psi} := \psi(h)^{II} \text{ for } h \in H,$$

is a relational Galois connection.

The proof of Proposition 1 shows that we get the second condition essentially "for free":

Proposition 3. *Whenever* $R \subseteq G \times N$ *and* $S \subseteq H \times M$ *are relations with*

$$g\ I\ h^S \iff h\ J\ g^R$$

then there is a unique relational Galois connection (Φ, Ψ) *such that*

$$g\ I\ h^S \iff g\ I\ h^{\Psi} \iff h\ J\ g^{\Phi} \iff h\ J\ g^R.$$

Proof. Define Φ and Ψ as follows: for $g \in G$ let $g^{\Phi} := (g^R)^{JJ}$, and for $h \in H$ let $h^{\Psi} := (h^S)^{II}$. □

The next observation shows that our "radical" generalisation of Galois connections leads back to the original definition:

Lemma 1. *If* (ϕ, ψ) *is a Galois connection between concepts lattices* $\underline{\mathfrak{B}}(G, M, I)$ *and* $\underline{\mathfrak{B}}(H, N, J)$*, then*

$$g^{\Phi} = Y :\iff \phi(g^{II}, g^I) = (Y^J, Y)$$

and

$$h^{\Psi} = X :\iff \psi(h^{JJ}, h^J) = (X^I, X)$$

defines a relational Galois connection between (G, M, I) *and* (H, N, J).

Conversely we obtain from each relational Galois connection (Φ, Ψ) *between* (G, M, I) *and* (H, N, J) *a Galois connection* (ϕ, ψ) *between the concept lattices* $\underline{\mathfrak{B}}(G, M, I)$ *and* $\underline{\mathfrak{B}}(H, N, J)$ *by*

$$\phi(X, X^I) := (X^{\Phi J}, X^{\Phi}) \quad \text{and} \quad \psi(Y, Y^J) := (Y^{\Psi I}, X^{\Psi}).$$

The two constructions are inverse to each other.

Proof. The proof is straightforward. □

The following proposition is no surprise:

Proposition 4. *In a relational Galois connection, the two parts determine each other. More precisely, if* (Φ_1, Ψ) *and* (Φ_2, Ψ) *are relational Galois connections between* (G, M, I) *and* (H, N, J)*, then* $\Phi_1 = \Phi_2$*, and dually.*

Proof. For each $g \in G$, we have

$$(g^{\Phi_1})^J = \{h \in H \mid g\ I\ h^{\Psi}\} = (g^{\Phi_2})^J$$

and therefore $g^{\Phi_1} = g^{\Phi_2}$. □

2 G-Relations and Dual Bonds

We have seen in Proposition 4 that it suffices to study only one of the two parts of a relational Galois connection. So let us call $\Phi \subseteq G \times N$ a **G-relation** from (G, M, I) to (H, N, J) if there is some Ψ such that (Φ, Ψ) is a relational Galois connection (this term was coined by Xia [16] following Shmuely's notion of a G-ideal [12,13]). A G-relation is, loosely spoken, one half of a relational Galois connection.

Proposition 5. *A relation $\Phi \subseteq G \times N$ is a G-relation from (G, M, I) to (H, N, J) if and only if it satisfies*

1. *for each $g \in G$, g^Φ is an intent of (H, N, J),*
2. *for each $h \in H$, $\{g \in G \mid h\ J\ g^\Phi\}$ is an extent of (G, M, I).*

Proof. If (Φ, Ψ) is a relational Galois connection then the two conditions obviously have to be satisfied. Conversely, assume 1.) and 2.) and define Ψ by

$$h^\Psi := \{g \in G \mid h\ J\ g^\Phi\}^I.$$

Clearly this is an intent and we have

$$\begin{aligned} g\ I\ h^\Psi &\iff g \in (h^\Psi)^I \\ &\iff g \in \{x \in G \mid h\ J\ x^\Phi\}^{II} \\ &\iff g \in \{x \in G \mid h\ J\ x^\Phi\} \\ &\iff h\ J\ g^\Phi. \end{aligned}$$ □

Note that the set of relation pairs satisfying the Galois condition is closed under (arbitrary) unions:

$$g\ I\ h^{\Psi_1} \iff h\ J\ g^{\Phi_1} \quad \text{and} \quad g\ I\ h^{\Psi_2} \iff h\ J\ g^{\Phi_2}$$

together imply

$$g\ I\ h^{\Psi_1 \cup \Psi_2} \iff h\ J\ g^{\Phi_1 \cup \Phi_2}.$$

Therefore, the relations $\Phi \subseteq G \times N$ for which there is some $\Psi \subseteq H \times M$ satisfying the Galois condition, ordered by inclusion, form a complete lattice. Moreover, Proposition 3 guarantees that for each such relation there is a smallest G-relation containing it (its **G-closure**[4]). As a consequence, we obtain

Theorem 1 (Xia [16]). *The G-relations from (G, M, I) to (H, N, J), ordered by set inclusion, form a complete lattice. The supremum of G-relations is the G-closure of their union.*

This lattice is isomorphic to the complete lattice of Galois connections between the corresponding concept lattices $\mathfrak{B}(G, M, I)$ and $\mathfrak{B}(H, N, J)$.

An example is given in Figure 4.

Another encoding of a relational Galois connection in only one relation is analogous to Birkhoff's construction:

[4] Note, however, that G-relations are not necessarily closed under intersections.

Definition 3. *A dual bond[5] from (G, M, I) to (H, N, J) is a relation $R \subseteq G \times H$ for which it holds that*

- *for every $g \in G$, g^R is an extent of (H, N, J) and*
- *for every $h \in H$, h^R is an extent of (G, M, I).*

Theorem 2. *From each relational Galois connection (Φ, Ψ) between (G, M, I) and (H, N, J) we obtain a dual bond $R \subseteq G \times H$ by means of*

$$g \, R \, h \quad :\Longleftrightarrow \quad g \, I \, h^\Psi \quad (\Longleftrightarrow h \, J \, g^\Phi).$$

Conversely, if R is a dual bond between (G, M, I) and (H, N, J), then

$$g^\Phi := g^{RJ} \quad \text{for } g \in G,$$
$$h^\Psi := h^{RI} \quad \text{for } h \in H$$

defines a relational Galois connection. The two constructions are inverse to each other[6].

Proof. If R is defined as above from a relational Galois connection (Φ, Ψ) we get

$$g^R = g^{\Phi J} \quad \text{and} \quad h^R = h^{\Psi I} \quad \text{for } g \in G, h \in H.$$

Then obviously R is a dual bond. Conversely, if R is a dual bond, then $g^R = g^{RJJ}$ and thus

$$h \, J \, g^\Phi \iff h \in g^{\Phi J} = g^{RJJ} = g^R \iff g \, R \, h,$$

and analogously $g \, I \, h^\Psi \iff g \, R \, h$. This shows that (Φ, Ψ) is a relational Galois connection. $\qquad\square$

As a corollary, we obtain a generalised version of Birkhoff's construction:

Corollary 1 ([5], Theorem 53). *For every dual bond $R \subseteq G \times H$,*

$$\phi(X, X^I) := (X^R, X^{RJ}), \quad \psi(Y, Y^J) := (Y^R, Y^{RI})$$

defines a Galois connection between $\underline{\mathfrak{B}}(G, M, I)$ and $\underline{\mathfrak{B}}(H, N, J)$. Conversely, for every Galois connection (ϕ, ψ),

$$R := \{(g, h) \mid (g^{II}, g^I) \le \psi(h^{JJ}, h^J)\} = \{(g, h) \mid (h^{JJ}, h^J) \le \phi(g^{II}, g^I)\}$$

is a dual bond. The two constructions are inverse to each other.

Theorem 2 has several nice consequences. First of all, it shows that relational Galois connections lead to closure operators:

---

[5] Called *biclosed relation* by Domenach and Leclerc [3].
[6] And induce a *dual* isomorphism between the lattice of all relational Galois connections and the lattice of all dual bonds.

Proposition 6. *If (Φ, Ψ) is a relational Galois connection between (G, M, I) and (H, N, J), and if R is the corresponding dual bond, then the mappings*

$$A \mapsto A^{RR} \quad \text{for } A \subseteq G$$
$$C \mapsto C^{RR} \quad \text{for } C \subseteq H$$

are closure operators with dually isomorphic closure systems.

Note that the closed sets of these closure operators are just the extents of the images of the two mappings φ and ψ in Lemma 1.

Secondly, Theorem 2 shows us a simple way to construct the lattice of all (relational) Galois connections. This comes from the fact that the dual bonds form a closure system with a simple closure operator:

Proposition 7. *A relation $R \subseteq G \times H$ is a dual bond from (G, M, I) to (H, N, J) if and only if*

$$A \times \{h\} \subseteq R \Rightarrow A^{II} \times \{h\} \subseteq R \quad \text{and} \quad \{g\} \times B \subseteq R \Rightarrow \{g\} \times B^{JJ} \subseteq R.$$

Proof. This is immediate from Definition 3. □

Proposition 7 explains how to find for a given relation $S \subseteq G \times H$ the smallest dual bond R containing S: We have to form the closure of S with respect to the conditions of the proposition. This was elaborated by Domenach and Leclerc [3].

3 Simultaneous Implications

The possibility to describe Galois connections via dual bonds shows that there is yet another possible view. To define a dual bond, it is not necessary to know the two formal contexts (G, M, I) and (H, N, J), it suffices to know the induced closure operators $X \mapsto X^{II}$ and $Y \mapsto Y^{JJ}$. The condition on a relation R for being a dual bond the is that for each $g \in G$ the set g^R must be closed with respect to the closure operator $Y \mapsto Y^{JJ}$, and dually for each $h \in H$ the set h^R must be closed under $X \mapsto X^{II}$.

This can be generalised. An important theme in Formal Concept Analysis is that of *attribute implications*, i.e., expressions of the form $A \to B$, where A and B are subsets of the attribute set M. It is easy to see that from each family $\mathcal{F}$ of such implications one obtains a closure operator on M. It associates to each subset X of M the smallest set containing X that is closed under all implications from $\mathcal{F}$. When a formal context is to be constructed in which these implications hold, it is necessary and sufficient that all objects intents are closed under this closure operator.

Dually we can infer conditions on the attribute intents when given *object implications* have to be respected. The considerations that we have made above now allow to characterise formal contexts compatible with given attribute implications and object implications simultaneously.

Consider the following setting: Let G and M be sets, let $\mathcal{F}_1$ be a set of implications on G and let $\mathcal{F}_2$ be a set of implications on M. Which relations $I \subseteq G \times M$ have the property that in the formal context (G, M, I) all implications from $\mathcal{F}_1$ are valid object implications and all implications from $\mathcal{F}_2$ are valid attribute implications?

Each of the two implication sets induces a closure operator that associates to every subset its *implicational closure*. Let us denote these operators by $A \mapsto \widehat{A}$ and $B \mapsto \breve{B}$, respectively. Then the problem translates to the following: Find those relations $I \subseteq G \times M$ that satisfy

- for each $A \subseteq G$, $\widehat{A} \subseteq A^{II}$, and
- for each $B \subseteq M$, $\breve{B} \subseteq B^{II}$.

If A is closed then this implies $A \subseteq \widehat{A} \subseteq A^{II} = A$, i.e., $A = \widehat{A}$. The condition therefore is equivalent to

- Every extent of (G, M, I) must be closed under $A \mapsto \widehat{A}$ and
- every intent must be closed under $B \mapsto \breve{B}$.

In other words:

Proposition 8. *The formal contexts (G, M, I) compatible with given implication sets on G and on M are precisely those for which I is a dual bond between the closure operators generated by these implications.*

4 Computing Galois Connections

Dual bonds are closed under intersection and thus form a closure system. The most elegant way of describing the family of all dual bonds between two given contexts would be to find a formal context whose extents are precisely the dual bonds. There is a natural candidate:

Definition 4. *The direct product of formal contexts (G, M, I) and (H, N, J) is the context*

$$(G \times H, M \times N, \nabla),$$

given by

$$(g, h)\nabla(m, n) :\iff (gIm \text{ or } hJn).$$

Proposition 9. *All extents of the direct product are dual bonds.*

Proof. It suffices to prove this for attribute extents. We have

$$(m, n)^{\nabla} = m^I \times H \cup G \times n^I.$$

Abbreviating $R := (m, n)^{\nabla}$, we find that $R \subseteq G \times H$ is a relation such that

- $g^R = H$ or $g^R = n^J$ if $g \in G$ and
- $h^R = G$ or $h^R = m^I$ if $h \in H$.

In any case the result is an extent of (H, N, J) or of (G, M, I), respectively. □

The concept lattice of a direct product is the *tensor product* of the concept lattices, and experience shows that the tensor product tends to be rather large compared to its factors. We may therefore expect that there are usually *many* (relational) Galois connections between two formal contexts, respectively between their concept lattices. The converse of Proposition 9 does not always hold: there are dual bonds that are not extents of the direct product. An example is given in Figure 3, which displays the dual bonds from the formal context $\mathbb{N}_3 := (\{1,2,3\}, \{1,2,3\}, =)$ to itself (see Figure 2). So we consider the example

$$(G, M, I) := \mathbb{N}_3 =: (H, N, J).$$

We find 50 dual bonds, but the tensor product has only 44 elements, see Figure 3. In fact, the equality relation is an example of a dual bond which is not an extent of the direct product.

It seems somewhat mysterious that "most" of the dual bonds are extents of the direct product, but some are not. This has been studied by Krötzsch, Hitzler, and Zhang [7] and by Krötzsch and Malik [8].

	1	2	3
1	×		
2		×	
3			×

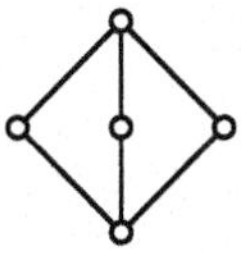

Fig. 2. The formal context $\mathbb{N}_3$ and its concept lattice $\mathbf{M}_3$

Corollary 2. *The tensor product is dually isomorphic to a complete supremum-subsemilattice of the lattice of all Galois connections.*

We do not know of a natural context construction for the closure system of all dual bonds. Nevertheless it is easy to compute all dual bonds between two given (finite) formal contexts (and thereby all relational Galois connections, and all Galois connections between the corresponding concept lattices as well): From Proposition 7 we know what the closure operator for the closure system of dual bonds is, and there are simple and fast algorithms for generating all closed sets of a given closure operator (see [5]).

5 Bonds

A **bond** from (G, M, I) to (H, N, J) is a dual bond from (G, M, I) to (N, H, J^{-1}). Definition 3 thus gives:

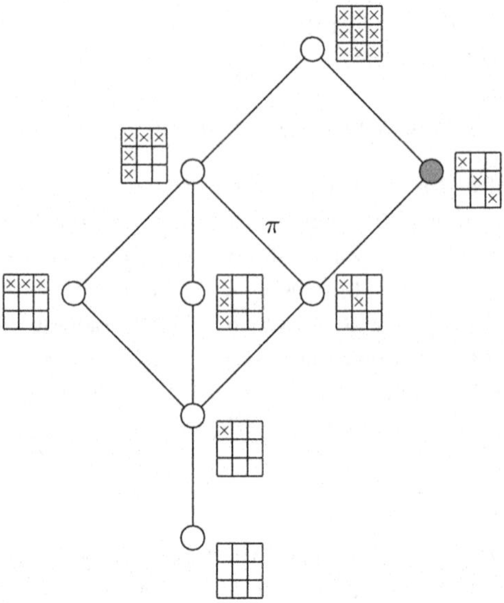

Fig. 3. The 50 dual bonds from $\mathbb{N}_3$ to $\mathbb{N}_3$. This *orbifold diagram* is folded "modulo automorphisms": Only one representative is given for each isomorphism class. The label π represents a permutation that is needed to make the labelling correct, see [17] for details. The six dual bonds represented by the shaded vertex are not extents of the direct product of the contexts.

Definition 5. *A* **bond** *from* (G, M, I) *to* (H, N, J) *is a relation* $R \subseteq G \times N$ *for which it holds that*

- *for every* $g \in G$, g^R *is an intent of* (H, N, J) *and*
- *for every* $n \in N$, n^R *is an extent of* (G, M, I).

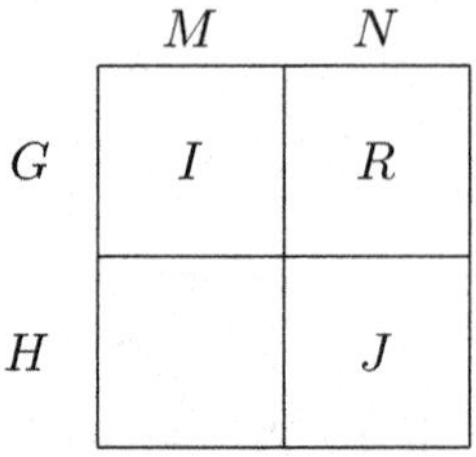

Bonds were introduced by R. Wille [14] to describe subdirect products of concept lattices, see [5] for an introduction. Subdirect products are certain complete sublattices of the direct product, and since direct products correspond to *context sums* and complete sublattices to *closed relations*, the following is natural:

Proposition 10. $R \subseteq G \times N$ *is a bond from* (G, M, I) *to* (H, N, J) *if and only if*

$$I \cup R \cup J \cup (H \times M)$$

is a closed relation of the context sum

$$(G, M, I) + (H, N, J)$$

(assuming $G \cap H = \emptyset = M \cap N$*).*

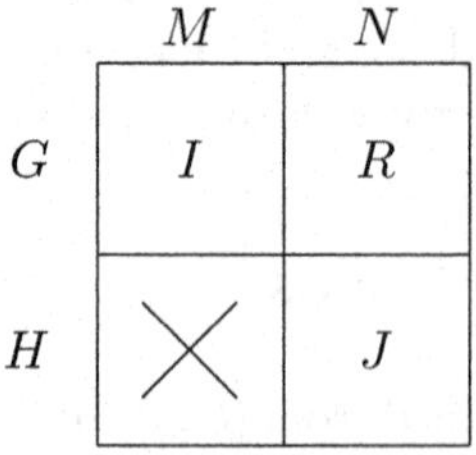

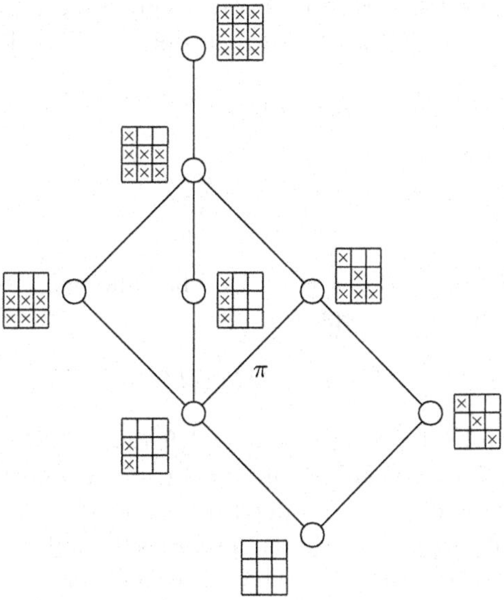

Fig. 4. The 50 Galois connections from $\mathbb{M}_3$ to $\mathbb{M}_3$ form a lattice which is isomorphic to the lattice of all G-relations from $\mathbb{N}_3$ to $\mathbb{N}_3$, and dually isomorphic to the lattice in Figure 3. As in Figure 3 the diagram is an orbifold diagram.

Proof. The concepts of the sum are precisely the pairs of the form

$$(A_1 \cup A_2, B_1 \cup B_2),$$

where $(A_1, B_1) \in \underline{\mathfrak{B}}(G, M, I)$ and $(A_1, B_2) \in \underline{\mathfrak{B}}(H, N, J)$. Now let (U, V) be a concept of (G, N, R). Then $(U \cup V^J, V \cup U^J)$ is a concept of

$$(G \cup H, M \cup N, I \cup R \cup J \cup (H \times M)).$$

In order for this to be a concept of the sum, it is necessary that U is an extent of (G, M, I) and that V is an intent of (H, N, J). In other words, R must be a bond.

But conversely, if R is a bond, and if (A, B) is a concept of

$$(G \cup H, M \cup N, I \cup R \cup J \cup (H \times M)),$$

then let $U := A \cap G$ and $V := B \cap N$. Since

$$U = (B \cap M)^I \cap V^R,$$

U must be an extent of (G, M, I) and $B \cap M = U^I$. Dually we get that V is an intent of (H, N, J) and $V^J = A \cap H$. Thus (A, B) is a concept of the sum $(G, M, I) + (H, N, J)$. □

According to the proposition, the bonds correspond to certain sublattices of the direct product, and it can be said to which ones. Let us denote the smallest element of $\underline{\mathfrak{B}}(G, M, I)$ by

$$\mathbf{0}_1 := (M^I, M),$$

the largest element of $\underline{\mathfrak{B}}(H, N, J)$ by

$$\mathbf{1}_2 := (H, H^J),$$

and similarly $\mathbf{0}_2 := (N^J, N)$, $\mathbf{1}_1 := (G, G^I)$. Then the pair $(\mathbf{0}_1, \mathbf{1}_2)$ of the direct product corresponds to the concept

$$(H \cup M^I, M \cup H^J)$$

of the sum context and roughly to the lower left quarter in the graphics shown with Proposition 10. This element is contained in the complete sublattice $\underline{\mathfrak{B}}(\mathbb{K})$ of the direct product, which was obtained from the bond R in Proposition 10. According to Proposition 82 of [5], this sublattice is actually a subdirect product. This leads to the following theorem, which I learnt from Kaarli, Kuchmei and Schmidt [6], who generously attribute it to Shmuely [12], although it is not stated there explicitly.

Theorem 3 ([6]). *The complete sublattices of the direct product*

$$\underline{\mathfrak{B}}(G, M, I) \times \underline{\mathfrak{B}}(H, N, J)$$

corresponding to the closed relations associated with bonds, as described in Proposition 10, are precisely those that contain the sublattice

$$[(\mathbf{0}_1, \mathbf{0}_2), (\mathbf{0}_1, \mathbf{1}_2)] \cup [(\mathbf{0}_1, \mathbf{1}_2), (\mathbf{1}_1, \mathbf{1}_2)].$$

Some of the relations discussed in the mathematical foundations of Formal Concept Analysis are actually **self-bonds**, i.e., bonds from (G, M, I) to (G, M, I). These include *closed relations* (associated with complete sublattices) and *block relations* (associated with complete tolerance relations). For example, any closed relation $C \subseteq I$ automatically also is a self-bond. From Proposition 10 we know that C then induces a closed relation in the context sum of (G, M, I). To both of these closed relations corresponds a complete sublattice, the first being a sublattice of $\underline{\mathfrak{B}}(G, M, I)$, the second of $\underline{\mathfrak{B}}(G, M, I) \times \underline{\mathfrak{B}}(G, M, I)$. We describe this in the easiest case, when $C := I$.

Proposition 11. *The concept lattice of the formal context*

$$\begin{array}{c|c} (G, M, I) & (G, M, I) \\ \hline (G, M, G \times M) & (G, M, I) \end{array}$$

is isomorphic to the order relation of $\underline{\mathfrak{B}}(G, M, I)$, with the component-wise order.

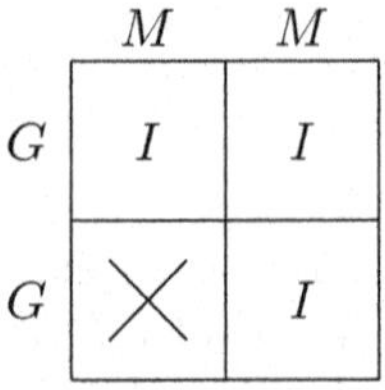

Before we give a proof, some words of explanation are needed. The context in this proposition is the same as in Proposition 10, except that (G, M, I) and (H, N, J) the same and $R = C$. The double occurrence of G (and of M) is meant as an abbreviation for using two disjoint copies.

By Proposition 10, the incidence relation of this context is a closed relation of the context sum $(G, M, I) + (G, M, I)$. Its concept lattice therefore is isomorphic to the direct square of $\underline{\mathfrak{B}}(G, M, I)$. The elements of the direct square are the pairs of elements of $\underline{\mathfrak{B}}(G, M, I)$, with the component-wise order. The *order relation* of $\underline{\mathfrak{B}}(G, M, I)$ is also a set of pairs, and is in fact a complete sublattice of the square. The proposition gives the corresponding closed relation.

Proof. Let (G_1, M_1, I_1), (G_2, M_2, I_2) be disjoint copies of (G, M, I), let $I_{12} \subseteq G \times M$ and let $i : G_1 \to M_2$, $J : M_2 \to M_1$ be bijections such that

$$g_1 I_1 j(m_2) \iff g_1 I_{12} m_2 \iff i(g_1) I_2 m_2.$$

Then for all $A_1 \subseteq G_1$, $B_2 \subseteq M_2$ we get

$$A_1^{I_{12}} = j^{-1}(A_1^{I_1}) \quad \text{and} \quad B_2^{I_{12}} = i^{-1}(B_2^{I_2}).$$

The concepts of the direct sum $(G_1, M_1, I_1) + (G_2, M_2, I_2)$ are precisely the pairs $(A_1 \cup A_2, B_1 \cup B_2)$, where (A_1, B_1) and (A_2, B_2) are concepts of (G_1, M_1, I_1), (G_2, M_2, I_2), respectively. Such a pair is a concept of the closed subrelation

$$\begin{array}{c|c} I_1 & I_{12} \\ \hline \times & I_2 \end{array}$$

iff

$$A_1 = B_1^{I_1} \cap B_2^{I_{12}}, \quad A_2 = B_2^{I_2}, \quad B_1 = A_1^{I_1}, \quad \text{and} \quad B_2 = A_1^{I_{12}} \cap A_2^{I_2}.$$

This simplifies to

$$A_1 = B_1^{I_1} \cap i^{-1}(A_2), \quad B_2 = A_2^{I_2} \cap j^{-1}(B_1),$$

and further to

$$A_1 = A_1 \cap i^{-1}(A_2), \quad B_2 = B_2 \cap j^{-1}(B_1),$$

which is satisfied iff (A_1, B_1) is a subconcept of $(i^{-1}(A_2), j(B_2))$. □

Observe that the formal context in Proposition 11 is the context product of (G, M, I) with $\begin{array}{|c|c|}\hline & \\ \hline \times & \\ \hline\end{array}$, the standard context of the three-element chain. Thus the concept lattice of that context is isomorphic to the tensor product of $\underline{\mathfrak{B}}(G, M, I)$ with the three element chain.

Corollary 3. *The order relation of a complete lattice* $\mathbf{V}$*, considered as a sublattice of* $\mathbf{V} \times \mathbf{V}$*, is isomorphic to the tensor product of* $\mathbf{V}$ *with the three element chain.*

In particular, the free distributive lattice with $n + 1$ generators is isomorphic to the order relation of the free distributive lattice with n generators, since free distributive lattices are tensor powers of the three element chain. This is, however, not surprising and can easily be obtained by a direct argument.

6 Concatenating Bonds

A reason why some authors prefer a *covariant* version of Galois connections is that such mappings can be concatenated. This is of particular importance if a category theoretic point of view is taken. This was recently broadly elaborated by Mori [9], who also studies the category of *Chu Spaces* [10]. A *Chu map* between two formal contexts (G, M, I) and (H, N, J) is a pair (f, g) of mappings $f : G \to H$, $g : N \to M$, satisfying

$$f(g)\ J\ n \iff g\ I\ g(n)$$

for all $g \in G, n \in N$. His *Chu correspondences* correspond to what we call relational Galois connections here. The category of Chu correspondences, including the functors that naturally arise there, is described in his work.

Covariant Galois connections correspond to bonds as Galois connections correspond to dual bonds (see Theorem 2 and Corollary 1 above). Before we come back to this, we give an elementary result on reducing a formal context. The rôle of the following proposition is to motivate a "relation product" for relations $R \subseteq G \times N$ and $S \subseteq H \times M$ between formal contexts (G, M, I) and (H, N, J), defined by

$$R \circ_I S := \{B^S \times A^R \mid (A, B) \in \mathfrak{B}(G, M, I)\}.$$

A situation where this product plays a rôle is that of a formal context with *reducible* objects and attributes. Recall from [5] that a subcontext is called **dense** if every object and every attribute outside this subcontext is reducible. To a certain degree, the structure "outside" a dense subcontext is determined by the dense subcontext. The next proposition makes this precise.

Lemma 2. *Let G, H, M, N be sets with $G \cap H = \emptyset = M \cap N$, and let I, R, S, T be relations with $I \subseteq G \times M$, $R \subseteq G \times N$, $S \subseteq H \times M$, $T \subseteq H \times N$ (see Figure 5). Then (G, M, I) is a dense subcontext of*

$$\mathbb{K} := (G \dot\cup H, M \dot\cup N, I \cup R \cup S \cup T)$$

if and only if the following conditions are all satisfied:

1. *For each $n \in N$ is n^R an extent of (G, M, I),*
2. *for each $h \in H$ is h^S an intent of (G, M, I),*
3. *$T = R \circ_I S := \{B^S \times A^R \mid (A, B) \in \mathfrak{B}(G, M, I)\}$.*

Condition 3 can equivalently be replaced by each of the following:

4. *$T = \{(h, n) \mid n^{RI} \subseteq h^S\}$.*
5. *$T = \{(h, n) \mid h^{SI} \subseteq n^R\}$.*

Proof. Conditions *1.* and *2.* are obviously necessary for (G, M, I) to be a dense subcontext. To see that *3.* also is necessary, pick an arbitrary $h \in H$ and let

$$A_h := h^{SI} \cap h^{TR}.$$

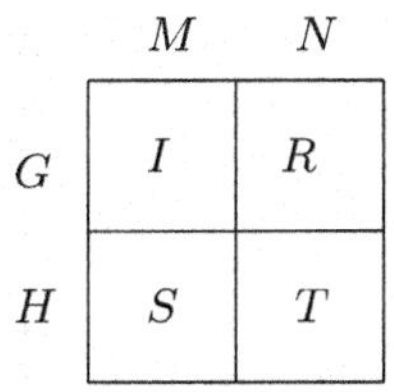

Fig. 5. With reference to Lemma 2

Both h^{SI} and h^{TR} are extents of (G, M, I) (the latter according to *1.*), and therefore A_h is the extent of some concept $(A_h, A_h^I) \in \underline{\mathfrak{B}}(G, M, I)$. When (G, M, I) is dense, then $A_h' = h'$, which is short for $A_h^I = h^S$ and $A_h^R = h^T$. From $h \in A_h^{IS}$ we then get

$$\{h\} \times h^T \subseteq A_h^{IS} \times A_h^R,$$

which proves that T is contained in the right hand side of *3.*, i.e., in $R \circ_I S$.

For the other inclusion let $(X, Y) \in \underline{\mathfrak{B}}(G, M, I)$ such that $(h, n) \in Y^S \times X^R$. Then $h \in Y^S$ and thus $Y \subseteq h^S = A_h^I$, which implies that $A_h \subseteq X$ and therefore $X^R \subseteq A_h^R$. This together with $n \in X^R$ forces $n \in A_h^R = h^T$, i.e., $(h, n) \in T$.

To see that *3.* is sufficient when *1.* and *2.* are given, we must infer from *3.* that each $h \in H$ is reducible. So let $Y := h^S$, $X := Y^I$. Then Y is an intent, $h \in Y^S$ and from *3.* we get $h^T \supseteq X^R$. Let $n \in h^T \setminus X^R$. Then there is some $(U, V) \in \underline{\mathfrak{B}}(G, M, I)$ such that $h \in V^S$ and $n \in U^R$. But $h \in V^S$ yields $V \subseteq h^S = Y$ and consequently $U = V^I \supseteq Y^I = X$. We get $X \subseteq U$ and conclude $U^R \subseteq X^R$, which is a contradiction to $n \in U^R$, $n \notin X^R$.

To prove that *4.* is equivalent to *3.*, assume first that $(h, n) \in B^S \times A^R$ for some concept $(A, B) \in \underline{\mathfrak{B}}(G, M, I)$. The $B \subseteq h^S$ and $A \subseteq n^R$, thus $B = A^I \supseteq n^{RI}$, which implies $n^{RI} \subseteq h^S$. Conversely, if $n^{RI} \subseteq h^S$, then $A := n^R$ is an extent with corresponding concept intent $B := A^I = n^{RI}$. We get $h^S \subseteq B$ and thus $(h, n) \in B^S \times A^R$. The proof for *5.* is analogous. □

The main reason why we are interested in this relation product $\circ_I$ is that in the case of bonds it corresponds to the product of the morphisms. This has already been described in [5]. We repeat the result here for completeness.

Proposition 12 (see Corollary 112 of [5]). *From every bond $R \subseteq G \times N$ between contexts (G, M, I) and (H, N, J) we obtain a pair (φ_R, ψ_R) of mappings*

$$\varphi_R : \underline{\mathfrak{B}}(G, M, I) \to \underline{\mathfrak{B}}(H, N, J), \quad \psi_R : \underline{\mathfrak{B}}(H, N, J) \to \underline{\mathfrak{B}}(G, M, I),$$

such that φ_R is a $\bigvee$-morphism and ψ_R is a $\bigwedge$-morphism residual to φ_R by

$$\varphi_R(A, A^I) := (A^{RJ}, A^R) \quad \psi_R(B^J, B) := (B^R, B^{RI}).$$

Every $\bigvee$-morphism is obtained in this way from exactly one bond, in fact,

$$\begin{aligned} R &= \{(g, n) \mid \varphi\gamma g \leq \mu n\} \\ &= \{(g, n) \mid \gamma g \leq \psi\mu n\}. \end{aligned}$$

□

I_1	J_{12}	J_{13}
	I_2	J_{23}
		I_3

Fig. 6. With reference to Proposition 13

Proposition 13 (see Proposition 113 of [5]). *Let* (G_1, M_1, I_1), (G_2, M_2, I_2), *and* (G_3, M_3, I_3) *be contexts, and let (see Figure 6)*

- J_{12} *be a bond from* (G_1, M_1, I_1) *to* (G_2, M_2, I_2),
- J_{23} *be a bond from* (G_2, M_2, I_2) *to* (G_3, M_3, I_3), *and*
- J_{13} *be a bond from* (G_1, M_1, I_1) *to* (G_3, M_3, I_3).

Moreover, let $(\varphi_{J_{12}}, \psi_{J_{12}})$, $(\varphi_{J_{23}}, \psi_{J_{23}})$, *and* $(\varphi_{J_{13}}, \psi_{J_{13}})$ *be the morphism pairs corresponding to these bonds according to Proposition 12. Then*

$$\varphi_{J_{13}} = \varphi_{J_{23}} \circ \varphi_{J_{12}} \quad \text{and} \quad \psi_{J_{13}} = \psi_{J_{12}} \circ \psi_{J_{23}}$$

if and only if

$$J_{13} = J_{12} \circ_{I_2} J_{23}.$$

Proof. (g, m) is in the bond corresponding to $\varphi_{J_{23}} \circ \varphi_{J_{12}}$ iff $\varphi_{23} \circ \varphi_{12} \gamma g \leq \mu n$. The intent of $\varphi_{23} \circ \varphi_{12} \gamma g$ is $g^{J_{12} I_2 J_{23}}$. Therefore the condition is that $n \in g^{J_{12} I_2 J_{23}}$, or, equivalently, that $g^{J_{12} I_2} \subseteq n^{J_{23}}$. This is just condition *5.* of Lemma 2. □

Acknowledgements

The first version of this paper was inspired by discussions during a stay with Prof. T. Tsujishita at Hokkaido University in 2001. Most of its results are based on the work of Weiqun Xia [16]. Further ideas resulted from fruitful discussions with Grit Malik, Hideo Mori, and Stefan E. Schmidt.

References

1. J. Barwise and J. Seligman, *Information Flow – The Logic of Distributed Systems.* Cambridge University Press, 1997.
2. G. Birkhoff, *Lattice Theory.* First edition. Amer. Math. Soc. Coll. Publ. 25 Providence, R.I., 1940.
3. F. Domenach, B. Leclerc, *Biclosed Binary Relations and Galois Connections.* Order, 18:1, 2001.
4. M. Erne, J. Koslowski, A. Melton, G.E. Strecker *A primer on Galois connections.* In: A. R. Todd, editor, *Papers on general topology and applications.* Madison, WI, 1991.

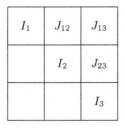

5. B. Ganter and R. Wille, *Formal Concept Analysis – Mathematical Foundations.* Springer Verlag 1999.
6. K. Kaarli, V. Kuchmei and S.E. Schmidt, *Sublattices of the direct product.* 2006. To appear.
7. M. Krötzsch, P. Hitzler, and GQ Zhang. *Morphisms in context.* In: F. Dau, M.-L. Mugnier, G. Stumme (editors) *Conceptual Structures: Common Semantics for Sharing Knowledge.* Proceedings of ICCS 2005. Springer LNAI 3596, 2005.
8. M. Krötzsch and G. Malik, *The tensor product as a lattice of regular Galois connections.* In: Rokia Missaoui, Jürg Schmid (eds.): *Formal concept Analysis.* Proceedings of ICFCA06. Springer LNAI 3874, 2006.
9. H. Mori, *Chu correspondences.* 2006. To appear.
10. H. Mori, *Functorial properties of concept lattices.* 2006. To appear.
11. V. Pratt. *Chu spaces: Automata with quantum aspects.* In: Proc. PhysComp '94, Dallas, IEEE, 1994.
12. Z. Shmuely, *The structure of Galois connections,* Pac. J. Math. 54(1974), 209–225.
13. Z. Shmuely, *The tensor product of distributive lattices.* Alg. Universalis 9(1979), 281–296.
14. R. Wille. *Sur la fusion des contextes individuels.* Math. Sci. Hum. 85 (1984), pp. 57–71.
15. W. Wille. *Subdirect product constructions of concept lattices.* Discrete Mathematics, 63 (1987), 305–313.
16. W. Xia, *Morphismen als formale Begriffe – Darstellung und Erzeugung.* Verlag Shaker 1993.
17. M. Zickwolff. *Darstellung symmetrischer Strukturen durch Transversale.* In: D. Dorninger, G. Eigenthaler, H.K. Kaiser, and W.B. Müller, editors. Contributions to general algebra, volume 7. Hölder-Pichler-Tempsky, Vienna, 1991.

Semantology as Basis for Conceptual Knowledge Processing

Peter Eklund and Rudolf Wille

The University of Wollongong,
School of Economics and Information Systems
peklund@uow.edu.au
Technische Universität Darmstadt,
Fachbereich Mathematik
wille@mathematik.tu-darmstadt.de

Abstract. *Semantology* has been introduced as the theory of semantic structures and their connections which, in particular, covers the methodology of activating semantic structures for *representing conceptual knowledge*. It is the main aim of this paper to explain and demonstrate that semantic structures are in fact basic for *conceptual knowledge processing* which comprises activities such as representing, infering, acquiring, and communicating conceptual knowledge.

Contents

1 Conceptual Knowledge Representation: An Example

In this paper we assume that *Conceptual Knowledge* is based on the so-called "main functions of human thought: concept, judgment, and conclusion" (cf. [Ka88], p.6). Therefore we can benefit from the traditional philosophical logic with its basic doctrine of concepts, doctrine of judgments, and doctrine of conclusions. Our assumption allows us to understand *Conceptual Knowledge Representation* as the presentation of semantic structures that makes possible to recover (at least partly) the original knowledge (cf.[GW06]). Conceptual knowledge representations have been elaborated in related work by *methods of Formal Concept Analysis and Contextual Logic* (cf. [Wi06]). How those representations may be formed shall be demonstrated in this introductory section by an example.

The example originated in a research project of political scientists in the late 1980's. This project was concerned with analysing *international regimes and their relationships* on the basis of empirical material gained by different case studies (see [Ko89]). An evaluation of the empirical material is presented in the

S.O. Kuznetsov and S. Schmidt (Eds.): ICFCA 2007, LNAI 4390, pp. 18–38, 2007.

Regimes	1-1	1-2	1-3	1-4	1-5	1-6	1-7	1-8	1-9	2-1-1	2-1-2	2-2	2-3	2-4	2-5	2-6	2-7	2-8	2-9	2-10	2-11	3-1	3-2	3-3
Cocom	h	cy,f	w	ai	n	h	i	m	e	r	ia	n	oi	e	so,st	k	p	cc	b	h	e	ip	h	h
Antarctic Regime	h	cn,cy	w	ri	n	l	i	r	u	g	ie,g	n	oi	u	so,st	d	p	ct	u	h	e,a	ip	l	m
Nonproliferation	h	i,cy	s,w	ri	w	l	i,ni	m,r,s	u	g	ie,g	c	oi	u	so,st,io	d	f	cc	u	h	e,a	ip	h	h
Conflict Settling	e	i	s	ri	n	l	ni	s	e	r	ie,ew	c	mi	e	st	d	f	cc	u	l	a	d	l	l
Human Rights	e	i	p	v	n	l	ni	r	e	r	ie,ew	c	oi	u	so,st	d	p	cc	u	m	a	ia	m	m
Economy & Technology	e	i	w	ai	w	l	i	m,r	u	r	ie,ew	c	mi	e	so,st	k	f	cc	u	l	a	d	m	m
Free Movement	e	i	p	v	w	l	ni	r	e	r	ie,ew	n	oi	u	st	d	p	cc	b	h	a	ip	h	m
Journalism	e	i	p	v	w	l	i	r	e	r	ie,ew	n	oi	u	st	d	p	cc	b	m	a	ia	h	m
KVAE	e	i,cn	s	ai	n	l	ni	s	e	r	ie,ew	c	mi	u	st	d	f	cc	b	h	e,a	ip	h	m
Nonintervention	h	f	s	ri	n	l	ni	s	e	g	ie,g		mi	e	st	k	p	cc		l	wo	d	l	l
Rhine	e	i	w	ai	w	h	i	m	e	r	ia	c	oi	u	so,st	k	p	cc	b	h	e,a	ip	h	h
North Sea	e	cn	w	ai	w	h	i	m	e	r	ia	c	mi	e	so,st	k	p	cc	b	m	wo	ia	m	h
Mediteranian Sea	e	i,cn	w	ai	w	l	i	m	e	r	ie,g	c	mi	e	so,st	k	p	cc		l	e,a	ia	l	l
East Sea	e	cn,cy	w	ai	w	l	i	m	u	r	ie,ew	c	oi	u	so,st	k	p	cc	b	h	e,a	ip	h	h
Air	e	cn,cy,f	w	ai	w	l	i,ni	m	u	g	ie,g	n	oi	u	so,st	k	p	cc	b	h	e,a	ip	h	h
Ozone	h	cn,cy,f	w	ai	w	l	i,ni	m	u	g	ie,g	n	oi	u	so,st	k	p	cc	b	h	e,a	ip	h	h
Data	h	cn,cy,f	w	ai	w	h	i	m	e	g	ie,g	c	oi	u	so,st,io	k	p	cc	u	l	e,a	ia	l	l
Obligations	h	cy,f	w	ai	w	l	i	m	u	g	ie,g	c	mi	u	so,st,io	k	p	cc	b	h	e,a	ip	h	h

Fig. 1. Evaluations of international regimes

1.1 *power structure*
e : egalitarian
h : hegemonial
1.2 *inst. environment*
i : institution
cn : commission
cy : country
f : forum
1.3 *field of distribution*
s : security
w : welfare
p : power
1.4 *object of conflict*
v : values
ai : absolute interests
ri : relative interests
1.5 *net of politics*
n : narrow
w : wide
1.6 *transnationality*
h : high
l : low
1.7 *imputability*
i : imputable
ni : not imputable
1.8 *allocation modus*
m : market
r : regulation
s : state activity
1.9 *conincidence actors*
e : equal
u : unequal
2.1.1 *catchment area*
g : global
r : regional
2.1.2 *catchment area systemic*
ia : intrasystemic west-west
ie : intersystemic
ew : east-west
g : global
2.2 *extent*
c : complex
n : narrow
2.3 *origin*
oi : one-sided interest
mi : mutual interest
2.4 *impact of distribution*
e : equal
u : unequal
2.5 *level of action*
so : society
st : state
io : international organisation
2.6 *conincidence*
c : consence
d : dissence
2.7 *purpose/function*
p : purpose
f : function
2.8 *coherence*
cc : concurrent
ct : contradictory
2.9 *balance*
b : balanced
u : unbalanced
2.10 *degree of arrangement*
h : high
m : middle
l : low
2.11 *organizational stabilization*
e : execution
a : advancement
wo : without organisation
3.1 *degree of development*
d : declaratory
ia : instructive to action
ip : implemented
3.2 *effectivity*
h : high
m : middle
l : low
3.3 *durability*
h : high
m : middle
l : low

Fig. 2. Legend for the evaluations of international regimes in Fig. 1

data table shown in Fig. 1. Those data tables may be viewed as elementary representations of conceptual knowledge. The data table of our example, for instance, presents a basic semantic structure which relates objects (international regimes) with attributes (e.g. power structure) and attribute values (e.g. egalitarian, hegemonial). Richer semantic structures can be obtained by elaborating *conceptual structures* inherent in the data. Such conceptual structures can be determined by mathematical methods of *Formal Concept Analysis* [GW99] which shall be briefly sketched here to make this paper more self-contained (cf. [VWW91]).

Formal Concept Analysis is based on the notion of a *formal context* which is defined as a set structure (G, M, I) consisting of sets G and M together with a binary relation I between G and M; the elements of G and M are called *formal objects* and *formal attributes*, respectively, and gIm is read: *the formal object g has the formal attribute m*. A *formal concept* of the formal context (G, M, I) is defined as a pair (A, B) with $A \subseteq G$, $B \subseteq M$, $A = \{g \in G \mid gIm \text{ for all } m \in B\}$, and $B = \{m \in M \mid gIm \text{ for all } g \in A\}$; A and B are called the *extent* and the *intent* of the formal concept (A, B), respectively. The hierarchical relation *subconcept-superconcept* - expressed by sentences as "the formal concept (A_1, B_1) is a subconcept of the formal concept (A_2, B_2)" - is modelled by the definition:

$$(A_1, B_1) \leq (A_2, B_2) :\Longleftrightarrow A_1 \subseteq A_2 \quad (\Longleftrightarrow B_1 \supseteq B_2)$$

The set of all formal concepts of (G, M, I) with this order relation is a complete lattice, called the *concept lattice* of the formal context (G, M, I) and denoted by $\mathfrak{B}(G, M, I)$. Formal contexts are usually represented by cross-tables as in Fig. 3 and concept lattices by line diagrams as in Fig. 4, Fig. 5, and Fig. 6 (for more detailed information see [GW99]).

As with most empirical data, the table in Fig. 1 is not in the form of a cross-table from which hierarchical structures of formal concepts could be directly derived. Fig. 1 presents a *many-valued context* which, in general, is defined as a set structure (G, M, W, I) where G, M, and W are sets together with a ternary relation $I \subseteq G \times M \times W$ such that $(g, m, w_1) \in I$ and $(g, m, w_2) \in I$ always imply $w_1 = w_2$; $(g, m, w) \in I$ is read: *the object g has the value w for the attribute m.* In the many-valued context of Fig. 1, for instance, the object *antarctic regime* has the value *hegemonial* for the attribute *power structure*. To obtain formal concepts, the many-valued context has to be transformed into a formal context as follows: The set of potential values of each (many-valued) attribute $m \in M$ is interpreted as the object set G_m of a *conceptual scale* (G_m, M_m, I_m) which, in Formal Concept Analysis, is understood as a formal context with a clear conceptual structure reflecting some meaning. With those chosen scales, one derives the formal context $(G, \bigcup_{m \in M} M_m, J)$ where $gJn :\Leftrightarrow wI_m n$ for $(g, m, w) \in I$ and $n \in M_m$ (cf. [GW99]). The derived context of the many-valued context in Fig 1 is shown in Fig. 3; some concept lattices of the chosen scales can be seen in Fig. 4.

A comprehensive semantic structure for the data table in Fig. 1 is given by the concept lattice of the formal context in Fig. 3. Unfortunately, that concept lattice consists of 1,535 formal concepts. Therefore a presentation of the full lattice as a human-readable diagram is difficult and barely within reach. But, for almost all questions posed by the researchers, it was enough to examine much smaller concept lattices of suitable subcontexts of the formal context in Fig. 3. Let us demonstrate this in two cases (cf. [VWW91]):

	Cocom	Antarctic Regime	Nonproliferation	Conflict Settling	Human Rights	Economy&Technology	Free Movement	Journalism	KVAE	Nonintervention	Rhine	North Sea	Mediterranean Sea	Baltic Sea	Air	Ozone	Data	Obligations
hegemonial structure	×	×	×							×						×	×	×
egalitarian structure				×	×	×	×	×	×		×	×	×	×	×			
institution			×	×	×	×	×	×	×		×		×					
commission		×										×	×	×	×	×	×	×
country	×	×	×											×	×	×	×	
forum	×									×					×	×	×	×
security			×	×					×	×								
welfare	×	×	×			×					×	×	×	×	×	×	×	×
power					×		×	×										
value conflict					×		×	×										
conflicts of interests	×	×	×	×		×			×	×	×	×	×	×	×	×	×	×
absolute interests	×				×				×		×	×	×	×	×	×	×	×
relative interests		×	×	×						×								
narrow net	×	×		×	×				×	×								
wide net			×			×	×	×			×	×	×	×	×	×	×	×
transnationality: high	×										×	×					×	×
transnationality: low		×	×	×	×	×	×	×	×	×			×	×	×	×		
imputable	×	×	×			×		×			×	×	×	×	×	×	×	×
not imputable			×	×	×		×		×	×					×	×		
allocation: market	×		×			×					×	×	×	×	×	×	×	×
allocation: regulation		×	×		×	×	×	×										
allocation: state activity			×	×					×	×								
allocation: equal	×			×	×		×	×	×	×	×	×	×				×	
allocation: unequal		×	×			×								×	×	×		×
catchment area: global		×	×							×					×	×	×	×
catchment area: regional	×			×	×	×	×	×	×		×	×	×	×				
intrasystemic: west-west	×										×	×						
intersystemic		×	×	×	×	×	×	×	×	×			×	×	×	×	×	×
east-west				×	×	×	×	×	×					×				
global		×	×							×			×		×	×	×	×
extent: complex			×	×	×	×			×		×	×	×	×			×	×
extent: narrow	×	×					×	×							×	×		
one-sided interest	×	×	×		×		×	×			×			×	×	×	×	
mutual interest				×		×			×	×		×	×					×
distribution: equal	×			×		×				×		×	×					
distribution: unequal		×	×		×		×	×	×		×			×	×	×	×	×
level of action: society	×	×	×		×	×					×	×	×	×	×	×	×	×
level of action: state	×	×	×	×	×	×	×	×	×	×	×	×	×	×	×	×	×	×
level of action: intern. organisation			×														×	×
consent	×					×				×	×	×	×	×	×	×	×	×
dissent		×	×	×	×		×	×	×									
purpose	×	×			×		×	×		×	×	×	×	×	×	×	×	×
function			×	×		×			×									
concurrent	×		×	×	×	×	×	×	×	×	×	×	×	×	×	×	×	×
contradictory		×																
balanced	×						×	×	×		×	×		×	×	×		×
unbalanced		×	×	×	×	×												
degree of arrangement: high	×	×	×				×		×		×			×	×	×		×
degree of arrangement: middle-high	×	×	×		×		×	×	×		×	×		×	×	×		×
degree of arrangement: low-middle				×	×	×		×		×		×	×				×	
degree of arrangement: low				×		×				×			×				×	
stabilization: execution	×	×	×						×		×		×	×	×	×	×	×
stabilization: advancement		×	×	×	×	×	×	×	×		×		×	×	×	×	×	×
stabilization: without organisation										×		×						
declaratory				×		×				×								
instructive to action					×			×				×	×				×	
implemented	×	×	×				×		×		×			×	×	×		×
effectivity: high	×		×				×	×	×		×			×	×	×		×
effectivity: middle-high	×		×		×	×	×	×	×		×	×		×	×	×		×
effectivity: low-middle		×		×	×	×				×		×	×				×	
effectivity: low		×		×						×			×				×	
durability: high	×		×								×	×		×	×	×		×
durability: middle-high	×	×	×		×	×	×	×	×		×	×		×	×	×		×
durability: low-middle		×		×	×	×	×	×	×	×			×				×	
durability: low				×						×			×				×	

Fig. 3. Derived context of the many-valued context in Fig. 1

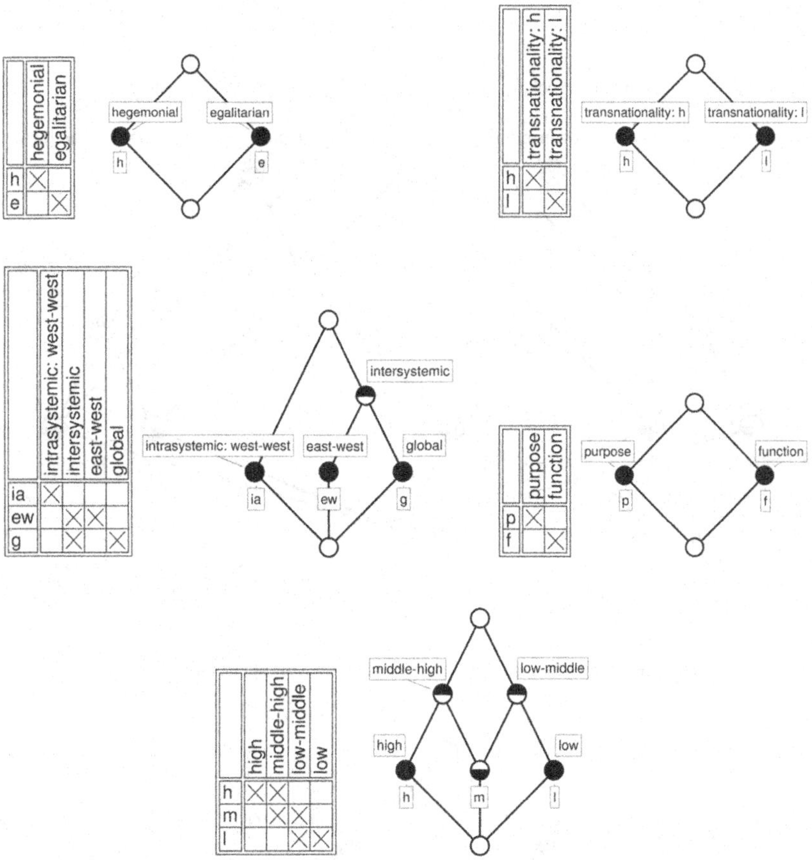

Fig. 4. Some conceptual scales for the many-valued context in Fig. 1

First, we consider the seemingly plausible hypothesis: *Intense regimes are found mostly under hegemonial structures*. Since a previous analysis of the data has shown that the (many-valued) attribute *degree of arrangement* is a good indicator for the intensity of a regime, the researchers could concentrate on the conceptual scales *degree of arrangement* and *power structure*, the concept lattices of which are presented in Fig. 5 (let us recall that, in a line diagram representing a concept lattice, there is an ascending path of line segments from a circle labelled by an object name to a circle labelled by an attribute name if and only if the object has the attribute). The line diagram makes apparent that there are more regimes with egalitarian structure than with hegemonial structure, especially among the regimes of middle or high degree of arrangement. This was interpreted by the researchers in the way that the data do not support the hypothesis.

Secondly, a concept lattice is used to examine the following hypothesis: *In a setting of dense interactions international regimes support the enactment of*

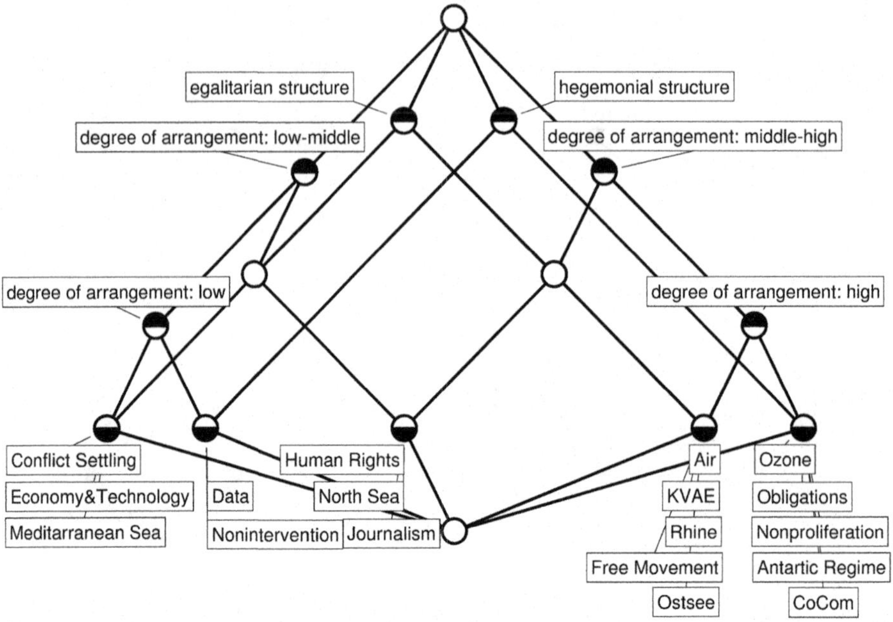

Fig. 5. Regime intensity and power structure

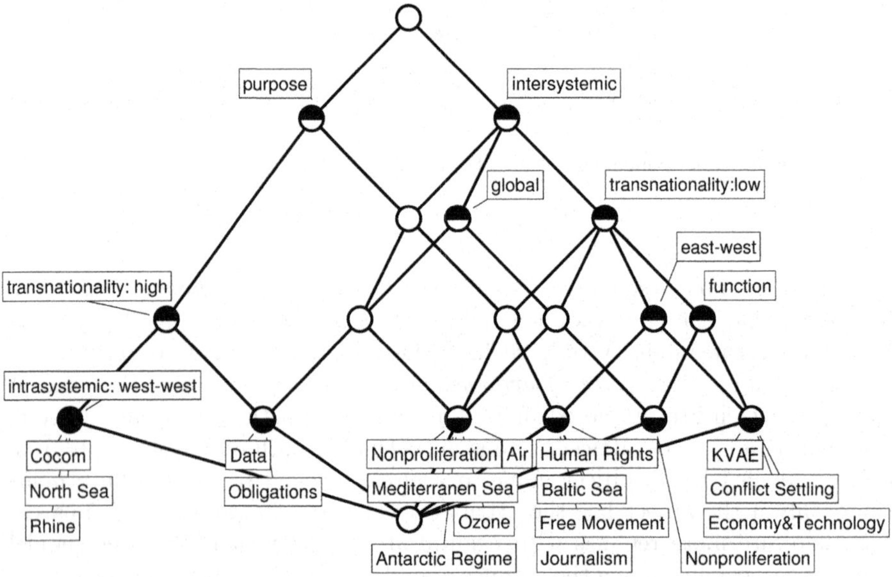

Fig. 6. Relative importance of purpose and function of regimes

a specific agreement (purpose), whereas in a situation/setting charaterized by a low level of inter- and transnational relations regimes have the function of providing convergent actors' expectations to enable them to come to international agreements. To test this hypothesis, the researchers selected the subcontext determined by the attributes *transnationality: high*, *transnationality: low*, *intrasystemic: west-west*, *intersystemic*, *east-west*, *global*, *purpose*, and *function*. Its concept lattice presented in Fig. 6 shows that all regimes between western industrial nations have a high density of transnational references and are directed by their object or aim. For instance, the regime *Rhine* is active because the nations at the river Rhine wish to clear the water. On the other hand, regimes which are directed by their function (e.g. to establish convergent actors' expectations) have only few transnational references and the involved countries belong to different systems. Such conceptual analysis brought the researchers to the conclusion that the data do not contradict the hypothesis, but they do also not fully support it; on this case more data are needed.

2 Semantology as Theory of Semantic Structures

According to our introductory assumption, *concepts* are the basic units of representing and processing Conceptual Knowledge. This corresponds well to the concept understanding in Piaget's structure-genetic approach to developmental psychology (cf. [Pi70]) where concepts are considered as *cognitive structures* (see also [Se01]). Changing from Piaget's epistomological view to the semantical view of knowledge represention makes it appropriate to replace the term "cognitive structure" by the term *"semantic structure"* for describing concepts and their meaningful combinations in knowledge representation. The general theory of semantic structures and their connections has been named *"Semantology"* in [GW06].

Concept hierarchies represented as concept lattices have proven to be useful and well communicable semantic structures. For example, the *Research Group on Concept Analysis* of the TU Darmstadt presented the so-called TOSCANA sofware at the *CeBIT '93* in Hannover/Germany and, in particular, showed results of the collaboration with the political scientists about international regimes as discussed in Section 1. Once the Hessian Science Minister came to our presentation at the fair and saw also the diagram of Fig. 5. He immediately understood the relationships and was very surprised that the hegemonial regimes are not dominant under the intense regimes; when he left, he even turned back to point with his finger to the circle representing the egalitarian intense regimes and said: "Unbelievable!".

The *meaning of semantic structures* in Conceptual Knowledge Representation and Processing can be analysed on at least three levels. This shall be briefly scatched:

First, there is the meaning on the *concrete level* on which the considered conceptual knowledge originates: this is usually the semantics belonging to the sciences whose language and understanding are used to describe that knowledge. In the example of Section 1, the meaning on the concrete level is given by the semantics of the political sciences, in particular of the researchers analysing international regimes. For instance, the meaning of all word concepts in Fig. 1 is that on which the experts for international regimes research agree.

Second, there is the meaning on the general *philosophic-logical level* on which the semantics is highly abstracted from the semantics of the concrete level, but is still related to actual realities. It is the semantics of the traditional philosophical logic based on the main functions of human thought: concept, judgment, and conclusion. For instance, the line diagram in Fig.5 represents a concept hierarchy in the sense of philosophical logic which understands concepts to be constituted by their extensions and intensions.

Third, there is the meaning on the *mathematical level* on which the semantics is strongly restricted the purely abstract: like numbers, ideal geometric figures and, since the twentieth century, set structures (and their generalizations). This very rigid semantics makes possible the high consensus about the validity of mathematical results, of which the semantic structures may also benefit. How philosophic-logical concept hierarchies can be successfully mathematized by concept lattices has already been explained and demonstrated in Section 1.

Emphasizing the three levels of semantics for semantic structures has been inspired by *Peirce's classification of sciences* in which *Mathematics* is viewed as the most abstract science studying hypotheses exclusively and dealing only with potential realities, *Philosophy* is considered as the most abstract science dealing with actual realities, while all *other sciences* are more concrete in dealing with special types of actual realities (cf.[GW06]).

According to Peirce, mathematicians are "gradually uncovering a great Cosmos of Forms, a world of potential being" ([Pe92], p.120). Peirce valued this growing cosmos of mathematical forms so far-reaching that for him all *deductive reasoning* will finally become *mathematical reasoning*. This underlines that appropriate mathematizations of semantic structures and their relationships are very important for understanding and handling those structures. For creating *adequate mathematizations*, elaborated semantic structures on the philosophical level may function as a useful bridge to reach well abstracted mathematical structures from given concrete structures. For instance, conceptual structures inherent in the data table about international regimes in Fig. 1 could be made mathematically explicit in labelled line diagrams of concept lattices as, for instance, in Fig. 4, Fig. 5, and Fig. 6 because the philosophic-logical understanding of concepts constituted by their extension and intension could be naturally turned into the mathematical definition of formal concepts of formal contexts, which form concept lattices.

An extremely close *connection between the concrete level and the mathematical level* is that between a concrete cross-table as the one in Fig. 3 and the formal context represented by that cross-table. But even more valuable is the close connection between the concrete cross-table and the labelled concept lattice of the corresponding formal context, because the concrete cross-table can be completely reconstructed from that concept lattice, i.e., the labelled concept lattice represents the full concept hierarchy inherent in the considered cross-table. A concrete data table - such as the one in Fig. 1 - corresponds uniquely to the many-valued context represented by that data table. But there is not a unique concept lattice derived from the many-valued context. Only after choosing for each (many-valued) attribute a meaningful conceptual scale as explained in Section 1, a uniquely determined formal context is derived. Thus, the connection between the concrete level and the mathematical level is in this case a consequence of the choice of meaningful scales. Despite the dependence on those choices, the connection between the two levels is still transparent for human discourse.

Less transparent is when data are mathematically reduced to what is thought to be the *essentials*, which usually does not allow to reconstruct the original data back from the essentials. Such opaques endanger a successful development of substantial semantic structures in human thought and should therefore be critically judged. It is important that semantic structures and their connections can be tested with respect to their meaningfulness. Measurement theory is a discipline which has convincingly investigated the meaningfulness of applications of mathematical models in a wide range of fields (see [KLST71]).

3 Semantological Methods for Conceptual Knowledge Processing

Semantic structures are not only fundamental for representing conceptual knowledge, they more generally play a basic role in processing conceptual knowledge. The main reason for this is that human knowledge is formed in a continuous process of thinking, arguing, recognizing, and communicating. *Conceptual Knowledge Processing* shall therefore be understood as the general discipline which investigates activities such as representing, reasoning, acquiring, and communicating conceptual knowledge (cf. [Wi94], [Wi97]).

Semantology as general theory of semantic structures and their connections has also to develop and to maintain the *methodology* of activating and applying semantic structures in Conceptual Knowledge Processing. Since concepts are the basic semantic structures in Conceptual Knowledge, *semantological methods* for Conceptual Knowledge Processing can be based on the analysis of concepts and concept hierarchies according to the threefold semantics discussed in Section 2. For instance, the derivation of the data table in Fig. 1 represented as cross-table in Fig. 3 was performed by applying the semantological method *"Conceptual Scaling"* [GW89]. From the semantological viewpoint, the data table in Fig. 1 represents a semantic structure consisting of object concepts, attribute concepts, and attribute value concepts such that each pair of object concept and attribute concept uniquely determines an attribute value concept. Turning this semantic structure into the semantic structure represented by the cross-table in Fig. 3 is guided by so-called conceptual scales which are viewed as semantic structures of the same type as the one in Fig. 3, namely they consist of related pairs of an object concept and an attribute concept represented in the corresponding table by a cross as shown in Fig. 3 and Fig. 4.

Another semantological method is *"Determining a Concept Hierarchy"*. This method derives, for instance, from the semantic structure which is represented by the sub-table in Fig. 3 consisting of the columns with the headings *transnationality: high*, *transnationality: low*, *intrasystemic: west-west*, *intersystemic*, *east-west*, *global*, *purpose*, and *function*, the semantic structure represented by the labelled line diagram in Fig. 6. The method "Determining a Concept Hierarchy" is one of six basic semantological methods for Conceptual Knowledge Representation discussed in [GW06].

In [Wi06], 38 methods of Conceptual Knowledge Processing are presented, all of them are of semantological nature. With the focus

- on *Conceptual Knowledge Representation*, nineteen methods are discussed under the class headings: 1. Conceptual Knowledge Representation, 2. Determination of Concepts and Contexts, 3. Conceptual Scaling, 4. Conceptual Classification,
- on *Conceptual Knowledge Inference*, ten methods are discussed under the class headings: 5. Analysis of Concept Hierarchies, 6. Aggregation of Concept Hierarchies, 7. Conceptual Identification, 8. Conceptual Knowledge Inferences,
- on *Conceptual Knowledge Acquisition*, five methods are discussed under the class headings: 9. Conceptual Knowledge Acquisition, 10. Conceptual Knowledge Retrieval,
- on *Conceptual Knowledge Communication*, four methods are discussed under the class headings: 11. Conceptual Theory Building, 12. Contextual Logic.

Applications of methods of conceptual knowledge representation have already been presented in Section 1. Therefore the rest of this section shall be used to give also an idea how to apply methods concerning infering, acquiring, and comunicating conceptual knowledge.

Conceptual Knowledge Inference is usually activated on already represented knowledge. How this is done shall be demonstrated by an exhibit which was designed for the Symmetry Exhibition at the Mathildenhöhe Darmstadt in 1986 (see [Ga86], p.130). The interactive exhibit was based on a computer presentation of the concept lattice representing the 17 symmetry types of two-dimensional patterns und their interrelationships. An example of such a pattern is shown in Fig. 7. The 17 symmetry types are represented in the concept lattice by the object concepts which are indicated in the line diagram of Fig. 8 by the circles having a black lower half; the possible symmetry transformation are represented in the diagram by circles having a black upper half (cf. [Wi00b], p.363).

At the exhibition visitors were invited to determine the symmetry type of two-dimensional patterns by using the lattice presentation on the computer screen. For the pattern example in Fig. 7 this could be done as follows: The user might first recognize that a black point in the pattern is the center of a 90°-rotation which transforms the (infinitely extended) pattern onto itself. Therefore the user marks "rota90" in the list of symmetry transformations at the computer screen. This has the effect that all circles turn into grey except the circles labelled by "p4", "p4m", and "p4g", respectively, and the circle which is linked with those

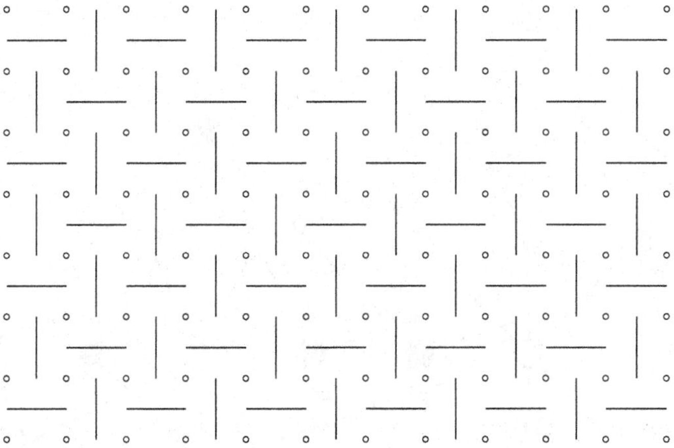

Fig. 7. What is the symmetry type of the shown two-dimensional pattern?

three circles; furthermore, in the list of symmetry transformations, all those transformations are erased which do not apply in patterns having the attribute "rota90" (for instance "rota120"). Such inferential support might help the user to find out that the center of a 90°-rotation of the pattern lies on the axis of a glide-reflection, i.e. the pattern has the attribute "rota90ongli". Marking this attribute in the present list of symmetry transformations gives the result that the pattern is of type "p4g" and has all those symmetry transformations which are then shown in the final transformation list. The line diagram in Fig. 8 shows that the object concept of "p4g" equals the attribute concept of "rota90ongli", which even visualizes the attribute implication:

$$\{\text{"rota90ongli"}\} \Longrightarrow \{\text{"rota90offref"}, \text{"rota180onref"}, \text{"4glideref"}, \text{"2reflect"}\}.$$

Conceptual Knowledge Acquisition, first outlined in [Wi89] as a methodology of Formal Concept Analysis, shall here only be demonstrated by an application of the manifoldly used *"attribute exploration"* (method M9.1 in [Wi06], see also [GW99], p.85). Our application deals with the question: "How can adjectives be used to judge upon musical compositions?" For allowing a transparent explanation of the method, we restrict ourselves to the five adjectives "lively", "sprightly", "rhythmizing", "fast", and "playful" (a more extensive treatment of the above question can be found in [WW06]). Now, the underlying *knowledge universe* for the attribute exploration is fixed as the formal context having as its formal objects all compositions of classical music and as attributes the five listed adjectives; the object-attribute-relation consists of those pairs which combine a composition with an adjective applying to that composition. The universal knowledge acquired by the attribute exploration consists of all *attribute implications* valid in the fixed universal context.

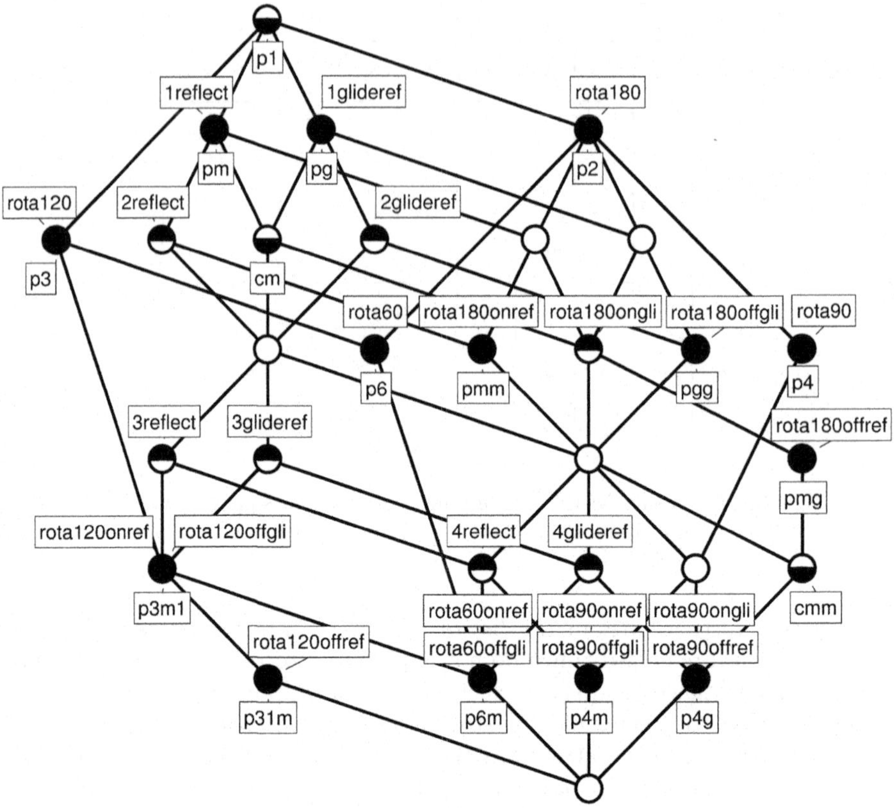

Fig. 8. Concept lattice of the 17 symmetry types of two-dimensional patterns

The exploration procedure may start with declaring some *pre-knowledge* which in our example shall be the information that Bartok's Concert for orchestra has as attributes "lively", "sprightly", "rhythmizing", and "playful" but not "fast" and Bach's 3rd Brandenburg concerto, 3rd movement, has as attributes "lively", "sprightly", "fast", and "playful" but not "rhythmizing". This information is represented in a 2×5-protocol context. Then the *exploration program* presents attribute implications to experts who have to judge whether the implications are valid in the universal context or not; in the case of "not", the experts have to name a musical counterexample to the presented implication. For our example, the sequence of questions and answers developed into the following list:

1. Are all compositions "lively", "spritely", "playful"? No!
   Counterexample: Beethoven: String quartett Op. 131, final movement
2. Is "playful" $\Longrightarrow$ "lively", "sprightly" valid? No!
   Counterexample: Bizet: Suite arlesienne
3. Is "fast", "playful" $\Longrightarrow$ "lively", "sprightly" valid? No!
   Counterexample: Ligeti: Continuum

4. Is "fast", "playful" $\Longrightarrow$ "lively" valid? Yes!
5. Is "sprightly $\Longrightarrow$ "lively", "playful" valid? Yes!
6. Is "lively" $\Longrightarrow$ "playful" valid? No!
   Counterexample: Bach: WTP 1: prelude c minor
7. Is "lively", "rhythmizing" $\Longrightarrow$ "sprightly", "playful" valid? No!
   Counterexample: Mahler: 9th symphony, 2nd movement (Ländler)
8. Is "lively", rhythmizing, "fast" $\Longrightarrow$ "sprightly" valid? Yes!
9. Is "lively", rhythmizing, "fast" $\Longrightarrow$ "sprightly", "playful" valid? No!
   Counterexample: Beethoven: Moonlight sonata, 3rd movement

In this procedure, each counterexample is instantly represented in the growing protocol context by marking the attributes which apply to the chosen counterexample. This leads finally to the formal context in Fig. 9 whose attribute implications are the same as the attribute implications of the universal context. Thus, the concept lattice in Fig. 10 is isomorphic to the concept lattice of the universal context which therefore has the Duquenne-Guigues-Basis: 1. "fast", "playful" $\Longrightarrow$ "lively", 2. "sprightly $\Longrightarrow$ "lively", "playful", 3. "lively", rhythmizing, "fast" $\Longrightarrow$ "sprightly". The performed attribute exploration based on an algorithm created by B. Ganter [Ga86b] was executed with P. Burmeister's program *"ConImp"* [Bu03].

	lively	sprightly	rhythmizing	fast	playful
Bartok: Concert for orchestra	X	X	X		X
Bach: 3rd Brandenburg Concerto, 3rd movement	X	X		X	X
Beethoven: Stringquartett Op. 131, final movement			X	X	
Bizet: Suite arlesienne			X		X
Ligeti: Continuum	X			X	X
Bach: WTP 1: prelude c minor	X			X	
Mahler: 9th symphony, 2nd movement (Ländler)	X		X		
Beethoven: Moonlight sonata, 3rd movement	X		X	X	

Fig. 9. Formal context derived from an attribute exploration

Conceptual Knowledge Communication is necessary in all performances of Conceptual Knowledge Processing in which creative human thought plays a constitutive role in its process (cf. [SWW01]). We shall only demonstrate this here by an application of the method *"Theory Building with TOSCANA"* (M11.2 in [Wi06]). The method has been substantially used to support a dissertation about *"Simplicity. Reconstruction of a conceptual landscape in the esthetics of music of the 18th century"*. The methodological foundation for this application was elaborated in [MW99]. The empirical collection of objects was given by 270 historical documents which were made accessible by a normed vocabulary of more than 400 text attributes. Those text attributes were used to form more general attributes for the conceptual scales of the approached TOSCANA-system. By repeatedly examining

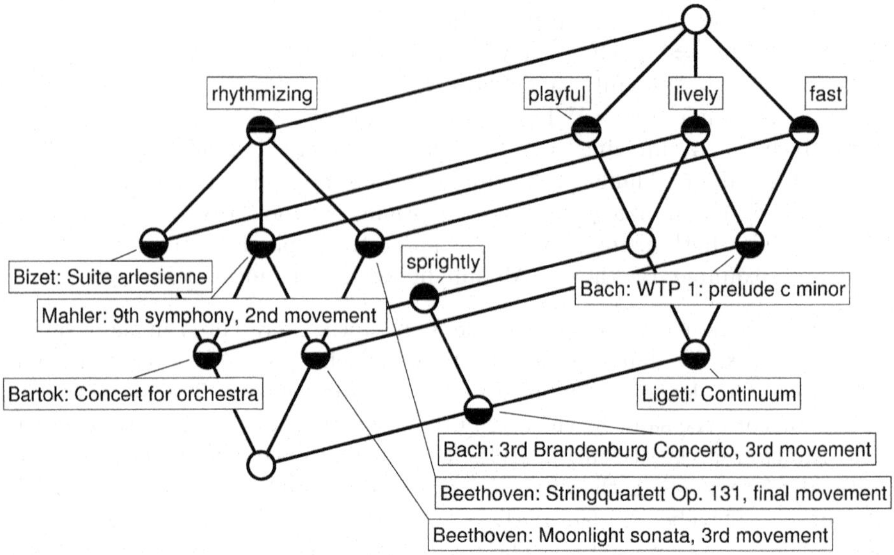

Fig. 10. Concept lattice of the formal context in Fig. 9

aggregations of scales and their concept lattices many improvements were made so that, finally, a well-founded TOSCANA-system was established. All this has been considered as a convincing process of theory building which could successfully support musicological research (see [Ma00] and [Wi05a], p.23ff.).

4 Semantological Software

In order to support *semantological methods* for Conceptual Knowledge Processing, new software paradigms need to be developed and evaluated. For more than twenty years this idea has led to various software frameworks that support one or multiple semantological methods, in particular, the 38 methods presented in [Wi06]. In this section we survey some of the software tools that support semantological methods. The approach we take in the presentation is broadly chronological and, where appropriate, we tie the software to its Conceptual Knowledge Processing methods (shortly *CKP-methods*) identified in [GW06] and [Wi06]. The best (and most complete) survey coverage for Conceptual Knowledge Processing can be found in the PhD thesis of T. Tilley [TT04]. However, our purpose is to identify examples of software that support semantological CKP-methods.

In the early 1980th, when *Formal Concept Analysis* started to be developed in the Research Group on General Algebra at Darmstadt University of Technology, personal computers became available at universities. This had the effect that several members of the Darmstadt group (B. Ganter, M. Skorsky, F. Vogt etc.) wrote useful *PC-programs* for basic procedures of Formal Concept Analysis. Here we only emphasize P. Burmeister's pioneering program (written in PASCAL) which is still available under the name "ConImp" in Windows (DOS) and Linux (see [Bu03]).

ConImp is text-based and has no capability for graphics and lattice rendering, the program can however be used to manipulate formal contexts and compute concept listings from which a line diagram can be drawn. Furthermore, the *Duquenne-Guigues-Basis* of the attribute implications of a finite context can be computed by ConImp. Subsequently, even in 3-valued contexts with the values "true", "false", and "unknown", ConImp can be used for *attribute exploration* to determine the implicational knowledge about a given conceptual universe by a question and answering dialogue as far as possible with the missing parts clearly indicated. In these ways, ConImp is able to support CKP-methods classified in Section 3 under the classes 1., 2., 3., 4., 5., 7., 8., and 9. ConImp remains a progenitor for the community in terms of its capacity to deliver an understanding of Conceptual Knowledge Processing though it is still useful, in particular as a demonstrator of the early willingness of the CKP research community to build practical tools for CKP problem solving and analysis. In the semantological sense ConImp is the most complete and influential program in Conceptual Knowledge Processing. It is therefore interesting to note that it has never been re-engineered for *modern operating environments.* This task remains as a challenge to the global CKP community.

Also in the 1980's, V. Duquenne started in Paris the development of software for rendering concept lattices which is available under the name *"GLAD: General Lattice Analysis and Design"* [Du05]. This software is written in FORTRAN with the lattice rendering in the Hewlett Packard Graphics Language (HPGL). While Duquenne's program broadly supports the CKP-method of *representing a concept lattice by a labelled line diagram*, it could not be said to be an analysis framework since it does not support the modern notions of software re-usability nor did it create a broad user base. The fact that GLAD was not fully reported until 2005 reinforces its minor impact to the broader Conceptual Knowledge Processing community. We therefore consider GLAD an early proof of concept for more extensible CKP software frameworks for drawing concept lattices that subsequently followed or were developed in parallel.

The visual rendering of concept lattices by line diagrams has widespread appeal - for a survey of this issue see [EDB04] - and therefore systems based on the *TOSCANA software* (see [KSVW94], [VW95]) have enjoyed greater impact on Conceptual Knowledge Processing than any others. Most importantly, the idea of a development and analysis methodology of *TOSCANA-systems* (see [EKSW00], [EGSW00]); in [BH05] described as *Conceptual Knowledge Systems*) has emerged due to TOSCANA's widespread use. The methodology (or workflow) for system development using the TOSCANA tools allowed the CKP modelling to scale beyond the original research and to develop TOSCANA to a much broader audience.

A TOSCANA-system incorporates a *multi-phase development process* in which two software tools - TOSCANA and ANACONDA - interact with a relational database and an intermediate format called a "conceptual systems file" [VWW91]. The process starts with the data to be analysed being stored in a relational database. Knowledge from a domain expert is than used to create queries in the form of conceptual scales using the ANACONDA program. The scales capture expert knowledge and are stored in the conceptual systems file. As an intelligent schema, the conceptual systems file contains queries against the relational database and allows to determine logical subcontexts as well as the conceptual hierarchy. ANACONDA not only helps to build logical scales, which provide a library of views of the data for the conceptual knowledge system, but is also a toolkit for creating and editing contexts and line diagrams representing scales. The workflow of ANACONDA aids the knowledge engineering process and it can therefore be understood as an analysis and design tool.

Description of line diagrams - as well as conceptual scales - are stored in the conceptual file format. TOSCANA takes as its input the associated conceptual file format and the relational database. In the methodology and workflow of TOSCANA-systems, TOSCANA becomes the conceptual systems browser. While TOSCANA users cannot create scales, the existing library of scales created by ANACONDA can be used to form nested line diagrams. TOSCANA is the conceptual browser, tying together (or realising) the relational data via its interaction with the conceptual systems file as an intelligent schema.

There are three versions of TOSCANA based on the *Formal Concept Analysis C^{++}-library* of F. Vogt [Vo96]. For several years an open-source Java version - *ToscanaJ* [BH05] - has also been available. The ToscanaJ scale and context editor are given by the programs *Siena* and *Elba*. These tools exhibit subtle variations on the initial workflow of TOSCANA-system development: including varying degrees of interoperability with ODBC, ANACONDA, ConImp, and another tool called "Cernato".

Cernato, a program developed by P. Becker and Navicon, is CKP software worthy of mention because it departs from Conceptual Knowledge Systems workflow of TOSCANA-systems: effectively combining both ANACONDA and TOSCANA functionality into a single software framework. In Cernato the context is presented in the form of a spreadsheet and line diagrams are constructed incrementally. Scales are called "views" and nested line diagrams are not supported. The unconstrained nature of Cernato is both its most important asset and its most significant problem. Cernato is difficult for novices and can easily generate unwieldy line diagrams. Our *experiments teaching Conceptual Knowledge Processing* using Cernato at the University of Queensland proved that Cernato could not easily be used or understood by 4th year computer science students who had elementary training in the methods of Conceptual Knowledge Processing. Cernato is therefore a tool more suited to CKP experts: the reverse of its design intention.

A similar argument runs for *ConExp* [Ye00], another Java-based open-source software development for Conceptual Knowledge Processing. Like Cernato, ConExp combines context creation, lattice line diagrams design, and data realisation into a single tool. Similar usability concerns are apparent for ConExp as for Cernato however. On the positive side, ConExp implements the *largest set of operations* from the foundational book [GW99], including calculation of association rules and the Duquenne-Guigues-Basis of implications. The program is therefore modern CKP software that most closely matches the functionality of ConImp. Unfortunately, the graphical representation of concept lattices is not as satisfying as in Cernato and ANACONDA. Neither ConExp nor Cernato have been reported in any significant way in the CKP literature even though they are widely used.

Experiences with Cernato and ConExp show the separation of TOSCANA and ANACONDA is a decompositional feature that reinforces the importance of the management of complexity of *semantological methods in Conceptual Knowledge Processing.* In the TOSCANA-system methodology, the management of the context and the user-defined rendering of the line diagram as a conceptual scale are the domain of ANACONDA (or Siena and Elba when using ToscanaJ). The realization of the scale line diagrams with data, or nested line diagrams - multiple scales, one or more embedded within another - become the responsibility of the TOSCANA-program. Thus there is a principle separation of intent as design (in ANACONDA) with extent as the realization of design (in TOSCANA). This results in the *TOSCANA-systems development methodology* reinforcing the underlying semantological methods in Conceptual Knowledge Processing.

The systems we have surveyed to this point may give support to the first eleven classes of semantological methods for Conceptual Knowledge Processing. In each TOSCANA-system, and to various degrees, it is possible to: 1. Represent conceptual knowledge in formal contexts and line diagrams of concept lattices; 2. Determine concepts and contexts; 3. Insert and combine conceptual scales; 4. Conceptually classify objects; 5. Analyse concept hierarchies; 6. Aggregate concept hierarchies; 7. Identify concepts and concept patterns; 8. Determine conceptual inferences; 9. Aquire conceptual information and knowledge; 10. Retrieve conceptual information and knowledge; 11. Perform conceptual theory building (cf. [Wi06]).

5 An Outlook

Higher level semantic structures in Conceptual Knowledge Processing which are based on Contextual Logic [Wi00a] remain the realm of research software and have yet to find a broad and general audience as have software systems such as ConImp, ConExp, or TOSCANA. *Contextual Logic* has been developed as a mathematization of the philosophical doctrines of concepts, judgments, and conclusions. The mathematization of concepts follows the approach of Formal Concept Analysis [GW99], and the mathematization of judgments uses, in addition, the Theory of Conceptual Graphs [So84]. The resulting *formal judgments* are composed by formal concepts which are either concepts of a basic formal context or are concepts of a k-ary relational context. Those contexts are taken from a so-called *power context family* (for more details see [Wi00a]).

B. Groh, the author of TOSCANA 3.0, has demonstrated in [Gr02] the importance of power context families in Conceptual Knowledge Processing by elaborating further the approach offered in [EGSW00]. Groh created a *representation of relational data* through realising a TOSCANA-system, adapted from the source code of TOSCANA 3.0, for representing and searching airline flights in Austria. Representations of knowledge - flight data - encoded in the TOSCANA-system, which has been extended by its ability to accommodate power context families, could be used to vizualize route-finding within in the airline timetable. The results are rendered as a significantly reduced form of line diagram equivalent to a directed graph. This rendering is well suited to the application while abstracting some of the complexity of the underlying concept lattice as a representation of the information space.

However convincing, Groh's work serves only as a proof a concept of the value of power context families. The real-world application of power context families to Conceptual Knowledge Processing are as likely as not to be found within the realm of *computational linguistics* and in learning relational structures to be found in the description of prototypical texts in document classification domains for precision-based *information retrieval.* This work has yet to be realised but is a promising area of research and application for power context families.

Description Logics have become a popular subset of First-Order Logic that have decidable tableau theorem provers and are sound and complete. Description Logics have also become the de facto standard for representing knowledge in the *Semantic Web* via W3C recommendations such as OWL. F. Dau and P. W. Eklund [DE06] have explored whether several existing well-known *diagrammatic reasoning systems* (including conceptual graphs) are compatible with Description Logics, which have no diagrammatic form. The main emphasis of this work is the evaluation of spider and constraint diagrams as compared to conceptual graphs can supplement the popularity of Description Logic by providing a principle way of performing diagrammatic reasoning on the Semantic Web.

Concept graphs (CKP-method class "12. Contextual Logic" in [Wi06]), as expressions of formal judgments based on knowledge represented by an underlying power context family, form the actual underlying mathematical basis for Conceptual Knowledge Processing in general. Therefore those structures should be explored further and worked off by CKP-methods to increase real-world applications of *Formal Concept Analysis* and *Contextual Logic.*

References

[BH05] P. Becker, J. Hereth Correia: The ToscanaJ Suite for implementing conceptual information systems. In: [GSW05], 324–348.

[Bu03] P. Burmeiter: Formal Concept Analysis with ConImp: Introduction to the basic features. http://www.mathematik.tu-darmstadt.de/ Burmeister. 2003.

[DE06] F. Dau, P. W. Eklund: Towards a diagrammatic reasoning system for Description Logic. Preprint (submitted to *Visual Languages and Computing*)

[Du05] V. Duquenne: GLAD: A program for general lattice analysis and design. Preprint, 2005.

[EDB04] P. W. Eklund, J. Ducrou, P. Prawn: Concept lattices for information visualisation: Can novices read line-diagrams? In: P. W. Eklund (ed.): *Concept lattices.* LNAI **2961**. Springer, Heidelberg 2004, 57–73.

[EGSW00] P. W. Eklund, B. Groh, G. Stumme, R. Wille: A contextual-logic extension of TOSCANA. In: B. Ganter, G. Mineau (eds.): *Conceptual structures: logical, linguistic and computational issues.* LNAI **1867**. Springer, Heidelberg 2000, 453-467.

[EKSW00] D. Eschenfelder, W. Kollewe, M. Skorsky, R. Wille: Ein Erkundungssystem zum Baurecht: Methoden der Entwicklung eines TOSCANA-Systems. In: [SW00], 254–272.

[Ga86] B. Ganter: Der Ornamentcomputer. In: G. Mazzola (ed.): *Symmetrie - Band3 - Spiel, Natur und Wissenschaft.* E. Roether Verlag, Darmstadt 1986, 130.

[Ga86b] B. Ganter: Algorithmen zur Formalen Begriffsanalyse. In: B. Ganter, R. Wille, K. E. Wolff (eds.): *Beiträge zur Begriffsanalyse.* BI-Wissenschaftsverlag, Mannheim 1986, 241–254.

[GSW05] B. Ganter, G. Stumme, R. Wille (eds.): *Formal Concept Analysis: foundations and applications.* State-of-the-Art Survey. LNAI **3626**. Springer, Heidelberg 2005.

[GW89] B. Ganter, R. Wille: Conceptual scaling. In: F. Roberts (ed.): *Applications of combinatorics and graph theory in the biological and social sciences.* Springer, New York 1989, 139–167.

[GW99] B. Ganter, R. Wille: *Formal Concept Analysis: mathematical foundations.* Springer, Heidelberg 1999.

[GW06] P. Gehring, R. Wille: Semantology: basic methods for knowledge representations. In: H. Schärfe, Pascal Hitzler, Peter Øhrstrøm (eds.): *Conceptual structures: inspiration and application.* LNAI **4068**. Springer, Heidelberg 2006, 215–228.

[Gr02] B. Groh: *A contextual logic framework based on relational power context families.* PhD thesis. The University of Queensland, Brisbane 2002.

[Ka88] I. Kant: Logic. Dover, Mineola 1988.

[Ko89] B. Kohler-Koch: Zur Empirie und Theorie internationaler Regime. In: B. Kohler-Koch (ed.): *Regime in den internationalen Beziehungen.* Nomos, Baden-Baden 1989, 15–85.

[KV00] B. Kohler-Koch, F. Vogt: Normen- und regelgeleitete internationale Kooperationen - Formale Begriffsanalyse in der Politikwissenschaft. In: [SW00], 325–340.

[KSVW94] W. Kollewe, M. Skorsky, F. Vogt, R. Wille: TOSCANA - ein Werkzeug zur begrifflichen Analyse und Erkundung von Daten. In: [WZ94], 267–288.

[KLST71] D. Krantz, D. Luce, P. Suppes, A. Tversky: *Foundation of measurement.* vol.1 (1971), vol.2 (1989), vol.3 (1990). Academic Press

[Ma00] K. Mackensen: *Simplizität. Genese und Wandel einer musikästhetischen Kategorie des 18. Jahrhunderts.* Bärenreiter, Kassel 2000.

[MW99] K. Mackensen, U. Wille: Qualitative text analysis supported by conceptual data systems. *Quality & Quantity* **33** (1999), 135–156.

[Pe92] Ch. S. Peirce: *Reasoning and the logic of things.* Edited by K. L. Ketner; with an introduction by K. L. Ketner and H. Putnam. Havard University Press, Cambridge 1992.

[Pi70] J. Piaget: *Genetic Epistomology.* Columbia University Press, New York 1970.

[Se01] Th. B. Seiler: *Begreifen und Verstehen. Ein Buch ber Begriffe und Bedeutungen.* Verlag Allgemeine Wissenschaft, Mühltal 2001.

[So84] J. F. Sowa: *Conceptual structures: information processing in mind and machine.* Adison-Wesley, Reading 1984.

[SWW01] S. Strahringer, R. Wille, U. Wille: Mathematical support of empirical theory building. In: H. S. Delugach, G. Stumme (eds.): *Conceptual structures: broadening the base.* LNAI **2120**. Springer, Heidelberg 2001, 169–186.

[SW00] G. Stumme, R. Wille (Hrsg.): *Begriffliche Wissensverarbeitung: Methoden und Anwendungen.* Springer, Heidelberg 2000.

[TT04] T. Tiley: *Formal Concept Analysis applications to requirements engineering and design.* PhD thesis. The University of Queensland, Brisbane 2004.

[Vo96] F. Vogt: *Formale Begriffsanalyse mit* C^{++}. Springer, Heidelberg 1996.

[VWW91] F. Vogt, C. Wachter, R. Wille: Data analysis based on a conceptual file. In: H.-H. Bock, P. Ihm (eds.): *Classification, data analysis, and knowledge organization.* Springer, Heidelberg 1991, 131–140.

[VW95] F. Vogt, R. Wille: TOSCANA – A graphical tool for analyzing and exploring data. In: R. Tamassia, I. G. Tollis (eds.): *Graph drawing '94.* LNCS **894**. Springer, Heidelberg 1995, 226–233.

[Wi89] R. Wille: Knowledge Acquisition by methods of Formal Concept Analysis. In: E. Diday (ed.): *Data analysis, learning symbolic and numeric knowledge.* Nova Sience Publishers, New York 1989, 365–380.

[Wi94] R. Wille: Plädoyer für eine philosophische Grundlegung der Begrifflichen Wissensverarbeitung. In: [WZ94], 11–25.

[Wi97] R. Wille: Conceptual landscapes of knowledge: a pragmatic paradigm for knowledge processing. In: G. Mineau, A. Fall (eds.): *Proceedings of the International Symposium on Knowledge Representation, Use, and Storage Efficiency.* Simon Fraser University, Vancouver 1997, 2–13; reprinted in: W. Gaul, H. Locarek-Junge (Eds.): *Classification in the Information Age.* Springer, Heidelberg 1999, 344–356.

[Wi00a] R. Wille: Contextual Logic summary. In: G. Stumme (ed.): *Working with conceptual structures. Contributions to ICCS 2000.* Shaker-Verlag, Aachen 2000, 265–276.

[Wi00b] R. Wille: Begriffliche Wissensverarbeitung: Theorie und Praxis. Informatik Spektrum **23** (2000), 357–369.

[Wi05a] R. Wille: Formal Concept Analysis as mathematical theory of concepts and concept hierarchies. In: [GSW05], 1–33.

[Wi05b] R. Wille: Conceptual Knowledge Processing in the field of economics. In: [GSW05], 226–249.

[Wi06] R. Wille: Methods of Conceptual Knowledge Processing. In: R. Missaoui, J. Schmid (eds.): *Formal Concept Analysis. ICFCA 2006.* LNAI . Springer, Heidelberg 2006, 1–29.

[WW06] R. Wille, R. Wille-Henning: Beurteilung von Musikstücken durch Adjektive. In: K. Proost, E. Richter (Hrsg.): *Von Intenionalität zur Bedeutung konventionalisierter Zeichen. Festschrift für Gisela Harras zum 65. Geburtstag.* Narr, Tübingen 2006, 453–475.

[WZ94] R. Wille, M. Zickwolff: *Begriffliche Wissensverarbeitung: Grundfragen und Aufgaben.* B.I.-Wissenschaftsverlag, Mannheim 1994.

[Ye00] S. Yevtushenko: Systems of data analysis "Concept Explorer". In: *Proceedings of the sevens national conference on Artificial Intelligence KII-2000,* Russia 2000, 127–134. (in Russian)

A New and Useful Syntactic Restriction on Rule Semantics for Tabular Datasets

Marie Agier[1,2] and Jean-Marc Petit[3]

[1] DIAGNOGENE, Aurillac
[2] LIMOS, UMR 6158 CNRS, Univ. Clermont-Ferrand II
[3] LIRIS, UMR 5205 CNRS, INSA Lyon France

Abstract. Different rule semantics have been successively defined in many contexts such as implications in artificial intelligence, functional dependencies in databases or association rules in data mining. We are interested in defining on tabular datasets a class of rule semantics for which Armstrong's axioms are sound and complete, so-called *well-formed semantics*. The main contribution of this paper is to show that an *equivalence* does exist between some syntactic restrictions on the natural definition of a given semantics and the fact that this semantics is well-formed. From a practical point of view, this equivalence allows to prove easily whether or not a new semantics is well-formed. We also point out the relationship between our generic definition of rule satisfaction and the underlying data mining problem, i.e. given a well-formed semantics and a tabular dataset, discover a cover of rules satisfied in this dataset. This work takes its roots from a bioinformatics application, the discovery of gene regulatory networks from gene expression data.

1 Introduction

The notion of *rules* or *implications* is very popular and appears in different flavors in databases, data mining or artificial intelligence communities. Famous examples of rules are functional dependencies [1], implications [2] or association rules [3]. As such, a simple remark can be done on such rules: their syntax is the same but their semantics widely differs.

In this paper, we consider rules to be defined on *tabular datasets*. Basically, a tabular dataset is equivalent to a *relation* over a set U of distinguished attributes in databases terminology [4]. In this setting, a *rule* is an expression of the shape $X \rightarrow Y$ i.e. "X implies Y" with $X, Y \subseteq U$. The *semantics* of a rule $X \rightarrow Y$ over U is the *meaning*, the *sense* one wants to give to this rule: Given a relation r, a rule $X \rightarrow Y$ is said to be *satisfied* in r with the semantics s, noted $r \models_s X \rightarrow Y$ if the semantics of the rule is true (or valid) in r.

From our analysis of existing rule semantics, we identify two main components to specify a rule semantics: the subsets of the relation on which the rule applies and the predicates occurring in the "if... then..." part of the rule. By the way, a natural and "generic" definition of rule semantics can be elaborated in order to be able not only to capture most of existing semantics already known on tabular datasets, but also to devise new semantics specific to some application domains.

S.O. Kuznetsov and S. Schmidt (Eds.): ICFCA 2007, LNAI 4390, pp. 39–50, 2007.

We also chose to focus on those semantics verifying Armstrong's axioms, i.e. semantics for rules on which Armstrong's axioms are sound and complete, so-called "well-formed semantics". For functional dependencies and implications, this result is known for a long time but more surprisingly, many other semantics also fit into this framework [2]. Roughly speaking, our aim is to define syntactical boundaries of well-formed semantics. Practical interests are for instance that some form of *reasoning* can be done on rules (e.g. implication problem in linear time [5]). Moreover, it is also possible to work on "small" *covers* of rules [6, 7, 8] and to use data mining techniques specific to the considered cover, but applicable to *every* well-formed semantics.

Paper contribution. The contribution of this paper is to show that an *equivalence* does exist between some syntactic restrictions on the natural definition of a given semantics and the fact that this semantics is *well-formed.*

From a practical point of view, this equivalence allows to prove easily whether or not a new semantics is well-formed: So far, for a given semantics, a proof of the soundness and the completeness of the Armstrong's axioms for this semantics should be given. Now, it is just enough to show that this semantics complies with the proposed syntactic restrictions.

We also point out a relationship between our generic definition of rule satisfaction and the underlying data mining problem, i.e. given a well-formed semantics and a relation, discover a cover of rules satisfied in this relation. More precisely, we show how a base of the closure system for any well-formed semantics can be computed from a dataset.

Application. This work takes its roots from a bioinformatics application, the discovery of gene regulatory networks from gene expression data. The challenge is to find out relationships between genes that reflect observations of how expression level of each gene affects those of others. The conjecture that association rules could be a model for the discovery of gene regulatory networks has been partially validated, see for example [9, 10, 11]. Nevertheless, we believe that many different kinds of rules between genes could be useful with respect to some biological objectives and the restricted setting of association rules could be not enough to cope with this diversity. Therefore, the main application of this paper is to offer a framework in which biologists may define their "own customized semantics" for rules with regard to their requirements. It is worth noting that other application domains could benefit from the propositions made in this paper.

Paper organization. We give in Section 2 the motivations of our proposition with examples of rule semantics. In Section 3, we propose a natural definition of a semantics using some syntactic restrictions. In Section 4, we further restrict the syntax and give the main result of this paper then we point out the relationships between our proposition and the underlying data mining problem. In Section 5, we give the related contributions of this work and finally, we conclude and give some perspectives in Section 6.

2 Motivating Examples

We give in the sequel three examples of semantics for tabular datasets, some of them in the context of gene expression data.

Table 1. A running example

r	g_1	g_2	g_3	g_4	g_5	g_6	*time*
t_1	1.7	1.5	1.2	-0.3	1.4	1.6	0
t_2	1.8	-0.7	1.3	0.8	-0.1	1.7	1
t_3	-1.8	0.4	1.7	1.8	0.6	-0.4	2
t_4	-1.7	-1.4	0.9	0.5	-1.8	-0.2	3
t_5	0.0	1.9	-1.9	1.7	1.6	-0.5	4

To illustrate the following examples, let us consider a relation r made of 5 tuples over a set U of 6 attributes, given in Table 1. This relation represents a gene expression dataset, each attribute being a gene with real numbers as domain.

Example 1. Let r be a relation over U and $X, Y \subseteq U$. Let s_1 be a semantics studying for example the levels of gene expression, i.e. the under- or over-expression of the genes. Given a user-supplied threshold ε, s_1 can be defined as follows:

$r \models_{s_1} X \rightarrow Y$ if and only if $\forall t \in r$, if $\forall A \in X, t[A] \geq \varepsilon$ then $\forall A \in Y, t[A] \geq \varepsilon$.

For example, with $\varepsilon = 1.0$ i.e. for over-expressed genes, the rule $g_1 \rightarrow g_3$ is satisfied in the relation given in Table 1 since $\forall t \in r$ when $t[g_1] \geq 1.0$ then $t[g_3] \geq 1.0$.

This semantics is not too far from association rules, except that no explicit discretization phase does exist and minimum support threshold is not taken into account with s_1.

Example 2. We consider here that an *order* exists among tuples: In the context of gene expression data, the first experience represents the state of a cell at the moment 0, then after injection of a drug, the cell is analyzed six hours later to give the second experience etc. until the fifth experience 30 hours later. This process allows to show the impact of a drug on gene expression of the cell in the time. In that case, the time can be represented as an external attribute as depicted in Table 1.

Let s_2 be a semantics studying for example the *evolution in time* of gene expression levels. Given a user-supplied threshold ε, s_2 can be defined as follows:

$r \models_{s_2} X \rightarrow Y$ if and only if $\forall t_i, t_{i+1} \in r$ such that t_i and t_{i+1} are two consecutive tuples, if $\forall A \in X, t_{i+1}[A] - t_i[A] \geq \varepsilon$ then $\forall A \in Y, t_{i+1}[A] - t_i[A] \geq \varepsilon$.

For example, with $\varepsilon = 1.0$, the rule $g_2 \rightarrow g_4$ is satisfied in the relation given in Table 1 since $\forall t_i, t_{i+1} \in r$ when $t_{i+1}[g_2] - t_i[g_2] \geq 1.0$ (i.e. for t_2/t_3 and t_4/t_5) then $t_{i+1}[g_4] - t_i[g_4] \geq 1.0$.

Example 3. Now we are interested in a semantics, called s_d, studying the Euclidean distance between gene expression profiles. Given a user-supplied threshold ε, s_d can be defined as follows:

$r \models_{s_d} X \rightarrow Y$ if and only if $\forall t_i, t_j \in r$, if $d(t_i[X], t_j[X]) \geq \varepsilon$ then $d(t_i[Y], t_j[Y]) \geq \varepsilon$.

where d is a function computing the Euclidean distance between two tuples on a set of attributes.

These three examples show that with the same syntax, a rule may have quite different meanings and from the same dataset, several semantics can be defined and hopefully interesting for the application domain.

3 A Generic and Natural Definition of a Semantics

In the sequel, we are going to explicit the common underlying structures of rule semantics.

For a given dataset, the nature of the data being analyzed clearly influences the definition of a semantics. Association rules and implications require binary data while functional dependencies can be defined on arbitrary attribute domains. Furthermore, external information may also be available and useful to define semantics. In example 2, a $time$ attribute have been added to define an order among tuples (see Table 1).

Moreover, the two important components in the definition of a rule semantics can be identified:

- The subsets of the relation on which a rule applies. We can for example study tuples one by one (e.g. implication or semantics given in example 1), we can also do a pairwise comparison of tuples (e.g. functional dependencies or semantics given in example 3) or compare the tuple i with the tuple $i+1$ (e.g. example 2) whenever an order exists among tuples.
- The predicates occurring in the "if $Pred_1$ is true then $Pred_2$ is true" part of the rule. Note that these two predicates can be the same. For example, for functional dependencies, the predicates can be formulated as: $[\ \forall A \in X, t_1[A] = t_2[A]\ ]$, where t_1, t_2 are two tuples, and X is a subset of attributes. These predicates really give the meaning of the semantics.

A "generic" definition of a semantics based on this analysis is described in the sequel.

Given a relation r over U and two subset of attributes $X, Y \subseteq U$, the *satisfaction* of a rule $X \rightarrow Y$ in r for a semantics s, noted $r \models_s X \rightarrow Y$, can be defined in a general way as follows:

Definition 1. *Let $X, Y \subseteq U$ and r be a relation over U. The* satisfaction *of the rule $X \rightarrow Y$ in r for a semantics s, noted $r \models_s X \rightarrow Y$, is defined by:*
$r \models_s X \rightarrow Y$ if and only if $\forall r' \subseteq r$ verifying $c(r')$, if $Pred_1(X, r')$ is true then $Pred_2(Y, r')$ is true where:

- *$c(r')$ specifies a constraint which has to be verified by $r' \subseteq r$.*
- *$Pred_1(X, r')$ (resp. $Pred_2(Y, r')$) is a predicate specifying a condition on X (resp. Y) over r'.*

A semantics is thus characterized by a constraint c and two predicates $Pred_1$ and $Pred_2$ defined for a subset of attributes. The constraint c is an expression specifying the condition that the subset r' of r must verify, to be considered in the predicates (e.g. $[\ |r'| = 2\ ]$ for functional dependencies). The predicates $Pred_1$ and $Pred_2$ are expressions which have to be defined on X (or Y) and r' **only**. Neither other attributes nor other subsets of r are allowed in their definition. In other words, they must be defined on $\pi_X(r')$ (or $\pi_Y(r')$).

In the sequel, we shall say that a semantics s *complies* with definition 1 if s can be syntactically expressed within the setting of definition 1. The class $\mathcal{C}$ will denote the class of semantics complying with definition 1.

3.1 Definition of the Constraint and the Predicates

We give some examples and define the permitted constraints and predicates.

Example 4. The three semantics s_1, s_2 and s_d presented in examples 1, 2 and 3 can be redefined to comply with the definition 1 as follows:

1. Semantics s_1:
   - $c(r') = [\,|r'| = 1\,]$.
   - $Pred_1(X, \{t\}) = Pred_2(X, \{t\}) =$ $[\,\forall A \in X, t[A] \geq \varepsilon\,]$.
2. Semantics s_2:
   - $c(r') = [\,|r'| = 2$ and for $r' = \{t_i, t_j\}, t_j[time] = t_i[time] + 1\,]$.
   - $Pred_1(X, \{t_i, t_j\}) = Pred_2(X, \{t_i, t_j\}) = [\,\forall A \in X, t_j[A] - t_i[A] \geq \varepsilon\,]$.
3. Semantics s_d:
   - $c(r') = [\,|r'| = 2\,]$.
   - $Pred_1(X, \{t_i, t_j\}) = Pred_2(X, \{t_i, t_j\}) =$ $[\,d(t_i[X], t_j[X]) \geq \varepsilon\,]$.

More formally, the constraint c is a `condition` defined on the relation r' and the set of attributes U whereas the predicates $Pred_1$ (resp. $Pred_2$) are `conditions` defined on the relation r' and the set of attributes X (resp. Y).

`Conditions` are defined inductively as follows:

A `simple condition` over a set of attributes X and a relation r, is an expression of the form: `<term>` θ `<term>`, where:

- θ is a comparison operator: $=, <, >, \leq, \geq, \neq$.
- `<term>` is one the following (with $A, B \in X$, $Y \subseteq X$ and $t \in r$):
  - A value in $\pi_X(r)$: $t[A], t[B], \ldots$
  - A constant: $a, b, 8, \varepsilon, null, \ldots$
  - A function: $fct(r, X), fct(r, A), fct(t, Y), \ldots$ e.g $d(t[A], t[B])$ or $|r|$.

A `condition` over X and r, is an expression composed of one or more `simple conditions` over X and r, using the logical connectives: AND, OR, NOT, () and the variables $A, B, Y, \ldots$ (resp t) are introduced using $\forall$ and $\exists$ quantifiers over X (resp. r).

3.2 Removing Dummy Semantics

Some semantics may give rise to rules always (resp. never) satisfied in any dataset. Therefore, these semantics are useless in practice and should be safely removed. We shall say that these semantics are *pathological* semantics, their definition is given below.

Definition 2. *Let s be a semantics characterized by a constraint c and two predicates $Pred_1$ and $Pred_2$.*

The semantics s is said to be a pathological semantics *if for any relation r over U, one of the following conditions is true:*

- $\nexists r' \subseteq r$ *verifying* $c(r')$.
- $\forall X, Y \subseteq U$ *and* $\forall r' \subseteq r$ *verifying* $c(r')$, $Pred_1(X, r')$ *is true and* $Pred_2(Y, r')$ *is false.*
- $\forall X, Y \subseteq U$ *and* $\forall r' \subseteq r$ *verifying* $c(r')$, $Pred_1(X, r')$ *false or* $Pred_2(Y, r')$ *true.*

Example 5. Let s the semantics partially defined as follows:

- $c(r') = [\, |r'| = 1 \text{ and } \forall A \in U, t[A] < 0 \,]$.
- $Pred_1(X, \{t\}) = [\, \forall A \in X, t[A] > 0 \,]$.

s is clearly a pathological semantics since $Pred_1$ is always false whatever the definition of $Pred_2$. For this semantics, all the rules are satisfied in any relation.

In the sequel, C_P will denote the class of pathological semantics of $\mathcal{C}$ and C_X will denote non pathological semantics of $\mathcal{C}$ i.e. $C_X = \mathcal{C} \setminus C_P$.

3.3 Well-Formed Semantics

A general framework can be borrowed from theoretical investigations performed over functional dependencies and Armstrong's axioms [12, 13]. This framework allows to resume interesting properties defined for functional dependencies like reasoning on rules and generating covers for rules. To be sure that a semantics fulfills this framework, the notion of well-formed semantics can be defined as follows:

Definition 3. *A semantics s is* well-formed *if Armstrong's axioms are sound and complete for s.*

Let F be a set of rules, recall that the notation $F \vdash X \rightarrow Y$ means that a proof of $X \rightarrow Y$ can be obtained using Armstrong's axiom system from F. Moreover, given a semantics s, the notation $F \models_s X \rightarrow Y$ means that for all relations r over U, if $r \models_s F$ then $r \models_s X \rightarrow Y$. In other words, for any well-formed semantics s, $\vdash$ and $\models_s$ coincide.

4 More Syntactic Restrictions

We have given the class C_X of semantics based on a natural and "generic" definition (see definition 1) of a rule semantics, able to capture a wide class of semantics on tabular datasets.

Moreover, since we are interested in well-formed semantics, a basic question comes in mind: "Is there an equivalence between the class of well-formed semantics and the class C_X?"

Not surprisingly, the answer turns out to be negative as shown in the following counterexample.

Example 6. Let s be a semantics defined as follows:

- $c(r') = [\,|r'| = 1\,]$.
- $Pred_1(X, \{t\}) = [\,\forall A \in X, t[A] = 1.0\,]$.
- $Pred_2(X, \{t\}) = [\,\forall A \in X, t[A] = 2.0\,]$.

Clearly, the reflexivity axiom is not sound for this semantics.

Therefore, the idea is to find out further syntactic restrictions on rule satisfaction definition in order to ensure the well-formedness of the semantics. In other words, given a new semantics, we would not have to prove anything to be sure that Armstrong's axioms apply: It would be just enough to show that the semantics complies with these syntactic restrictions.

We propose a new definition below based on the definition 1.

Definition 4. *Let $X, Y \subseteq U$ and r be a relation over U. The* satisfaction *of the rule $X \rightarrow Y$ in r for a semantics s, noted $r \models_s X \rightarrow Y$, is defined by:*
$r \models_s X \rightarrow Y$ if and only if $\forall r' \subseteq r$ verifying $c(r')$, if $\forall A \in X$, $Pred(A, r')$ is true then $\forall A \in Y$, $Pred(A, r')$ is true where:

- *$c(r')$ specifies a constraint which has to be verified over $r' \subseteq r$.*
- *$Pred(A, r')$ is a predicate specifying a condition on A over r'.*

The first item of this new semantic definition does not change with regard to the definition 1. Nevertheless, the difference is twofold:

- Firstly, the two predicates $Pred_1$ and $Pred_2$ are equivalent.
- Secondly, a restriction is posed on the predicate: Now, it must be satisfied for each *single attribute* $A \in X$ instead of being satisfied for the *subset of attributes* X.

Since two syntactically different definitions of predicates could be equivalent, we need to precisely define what "equivalent" means. Intuitively, two predicates are equivalent if for any relation r and for any subset of attributes $X \subseteq U$, $Pred_1$ and $Pred_2$ are satisfied for X on the same subsets of r. Here is the formal definition:

Definition 5. *Let s be a semantics characterized by a constraint c and two predicates $Pred_1$ and $Pred_2$. The two predicates $Pred_1$ and $Pred_2$ are said to be* equivalent, *denoted by $Pred_1 \equiv Pred_2$, if and only if for any relation r and for any subset of attributes $X \subseteq U$, we have:*
$\{r' \subseteq r \mid r'$ verifying $c(r')$ and $Pred_1(X, r')$ is true$\} = \{r' \subseteq r \mid r'$ verifying $c(r')$ and $Pred_2(X, r')$ is true$\}$.

In the sequel, we shall note by C_A the class of rule semantics complying with definition 4, C_A being a subset of C_X.

Recall that pathological semantics for which all the rules are always true (or never) are not considered in C_X.

4.1 Main Result

The main result of the paper gives an equivalence between well-formed semantics and semantics complying with definition 4 and is stated as follows:

Theorem 1. *Let $s \in C_X$ be a rule semantics. The semantics s is well-formed if and only if $s \in C_A$.*

Proof. Let $s \in C_X$ be a rule semantics. We have first to prove that if $s \in C_A$ then s is well-formed and secondly that if s is well-formed then $s \in C_A$ or equivalently that if $s \notin C_A$ then s is not well-formed.

Lemma 1. *Let $s \in C_X$ be a rule semantics. If $s \in C_A$ then s is well-formed.*

Proof. This proof borrows the classical proof of soundness and completeness for functional dependencies [14]. The only technical difficulty is to use the constraint c in order to avoid to explicitly handle the couple of tuples on which functional dependencies are defined.

The (only if) part of the proof is somewhat surprising since it tells us that any well-formed semantics complies with definition 4:

Lemma 2. *Let $s \in C_X$ be a rule semantics. If the semantics s is well-formed, then $s \in C_A$.*

Proof. We have to show that if s is well-formed, then $s \in C_A$ or equivalently, if $s \notin C_A$ (i.e. s does not comply with definition 4) then s is not well-formed.

Suppose that s does not comply with definition 4, two cases are possible: Either the two predicates are not equivalent or they are equivalent with the following restriction: it does not exist an equivalent predicate which can be formulated as a condition on each single attribute.

Let us consider the first case, i.e. $Pred_1$ and $Pred_2$ are not equivalent: In that case, since $s \in \mathcal{C}$, it does exist a relation r and a subset of attributes $Y \subseteq U$ such that $\{r' \subseteq r \mid r'$ verifying $c(r')$ and $Pred_1(Y, r')$ is true$\} \neq \{r' \subseteq r \mid r'$ verifying $c(r')$ and $Pred_2(Y, r')$ is true$\}$.

Two cases are thus possible:

- $\exists r' \subseteq r$ verifying $c(r')$ such that $Pred_1(Y, r')$ is true and $Pred_2(Y, r')$ is false: Let us assume that s is well-formed. By reflexivity, we have $\forall X \subseteq Y, r \models_s Y \rightarrow X$ and thus $r \models_s Y \rightarrow Y$ i.e. $\forall r' \subseteq r$ verifying $c(r')$, if $Pred_1(Y, r')$ is true then $Pred_2(Y, r')$ is true which is a contradiction.
- $\exists r' \subseteq r$ verifying $c(r')$ such that $Pred_1(Y, r')$ is false and $Pred_2(Y, r')$ is true: Without loss of generality, let us assume there exists $X \in U \setminus Y$ and $Z \in U \setminus Y$ such that $Pred_1(X, r')$ is true and $Pred_2(Z, r')$ is false, as depicted in Table 2 (clearly, such a relation r always exists when s is in C_X, otherwize s would be in C_P). Thus, we have $r' \models_s X \rightarrow Y$ and $r' \models_s Y \rightarrow Z$.

  Assume now that s is well-formed. By transitivity, we should have $r' \models_s X \rightarrow Z$, which is false and leads to a contradiction.

Table 2. Example

r	X	Y	Z	...
...	...	...	...	...
$r'\{$	P_1 true	P_1 false and P_2 true	P_2 false	...
...	...	...	...	...

Finally, we have shown that if $Pred_1$ and $Pred_2$ are not equivalent, then s is not-well-formed.

Now, let us consider the second case, i.e. $Pred_1 \equiv Pred_2$ but they are not equivalent to the predicate $Pred'(Y) = [\forall A \in Y, Pred_1(A)]$: In that case, it does exist a relation r and a subset of attributes $Y \subseteq U$ such that $\{r' \subseteq r \mid r'$ verifying $c(r')$ and $Pred_1(Y, r')$ is true$\} \neq \{r' \subseteq r \mid r'$ verifying $c(r')$ and $\forall A \in Y, Pred_1(A, r')$ is true$\}$. We can show that reflexivity and transitivity axioms are not sound in r'. The proof is similar to the previous one and is omitted.

4.2 Usefulness of These Syntactic Restrictions

We shall point out the interest of this result through several examples.

Example 7. The semantics s_1 and s_2 (see example 4) clearly comply with the definition 4, i.e. $s_1, s_2 \in C_A$. From theorem 1, the following result holds:

Corollary 1. *The semantics s_1 and s_2 are* well-formed.

Example 8. For the semantics s_d, we have the following result:

Corollary 2. *The semantics s_d is not* well-formed.

Proof. We have to show that the semantics s_d does not comply with definition 4. The semantics s_d complies with definition 1, but the predicate $Pred(X, \{t_i, t_j\}) = [\, d(t_i[X], t_j[X]) \geq \varepsilon \,]$ is not equivalent to the predicate $Pred'(X, \{t_i, t_j\}) = [\, \forall A \in X, d(t_i[A], t_j[A]) \geq \varepsilon \,]$. Indeed, let us consider the following counterexample where $Pred(X, r)$ is true and $Pred'(X, r)$ is false:

Let r be the relation described in Table 3 and let $\varepsilon = 4$.

Table 3. Counterexample

r	A	B
t_1	2	10
t_2	5	6

We can see that $d(t_1[AB], t_2[AB]) = 5$ i.e. $Pred(AB, r)$ is true whereas $d(t_1[A], t_2[A]) = 3$ i.e. $Pred'(AB, r)$ is false . This counterexample shows that $s \notin C_A$ and thus, by theorem 1, that s is not well-formed.

4.3 Relationships with Data Mining

In the setting of this paper, we may question about the underlying discovery problem for any well-formed semantics i.e. for semantics $s \in C_A$. Rule generation for a semantics $s \in C_X \setminus C_A$ is also interesting but out of the scope of this paper.

First, recall that a one-to-one correspondence does exist between a set of rules and a closure system (A closure system C on U is such that $U \in C$ and $\forall X, Y \in C$, $X \cap Y \in C$) [2]. Second, given a relation r and a well-formed semantics s, we can always define a closure system with respect to the set of satisfied rules in the relation r for the semantics s.

Two main techniques do exist to compute a cover of rules:

- Those which enumerate the closure system to generate for example a minimum cover [7], generally used for association rules generation [15].
- Those which avoid the enumeration of the closure system to generate the canonical cover, generally used for the inference of functional dependencies [12, 13].

In both cases, the first step is to compute a *base* of the closure system from the dataset. The base is obviously specific to the considered semantics and can be formulated as follows:

Definition 6. *Let r be a relation over U and s a given well-formed semantics. Let $B_s(r)$ the set defined as follows:*

$$B_s(r) = \bigcup_{r' \subseteq r \mid c(r')} \{A \in U \mid Pred(A, r') \text{ is true}\}.$$

We have the following result which extends in our context a well-known result obtained in the setting of functional dependencies [16]:

Proposition 1. *$B_s(r)$ is a* base *of the closure system with respect to the set $F_s(r)$ of satisfied rules in r for the semantics s.*

Proof. Let us recall that a sub-family B of a closure system C is a *base* if $Inf \subseteq B \subseteq C$ where Inf is the set of meet-irreducible sets. The proof is omitted.

We note that the base $B_s(r)$ is defined at the level of single attributes $A \in U$, which shows the necessity that the semantics complies with the definition 4.

From a data mining point of view, the computation of $B_s(r)$ is a crucial step since data accesses are only performed here.

5 Related Works and Discussion

To the best of our knowledge, we are not aware of related works in the literature dealing with syntactical characterizations with regard to some inference systems. Nevertheless, related contributions have been done on association rules in data mining [3, 17] and functional dependencies in databases [13, 18]. Rule mining often results in a huge amount of rules and as a consequence, rules turn out to be useless for experts. This is the well-known post-processing step in a KDD process. To address this problem, different lines of research have been proposed. Firstly, rules may be filtered out a priori, based on user-defined templates of rules [19, 20]. Secondly, inference rules or inference systems have been proposed to reduce the number of rules given to the experts [17, 21, 22, 23]. Thirdly, many quality measures have been developed to select only the most interesting rules [3, 24] with regard to these measures.

It is worth noting that we have not integrated quality and error measures in this paper. *Error measures* like for example confidence defined for association rules or error indications defined for functional dependencies [25], allow to append approximate rules to exact rules, i.e. those rules which are almost satisfied. *Quality measures* like support, dependency or informative rate [3, 24], allow at contrary to limit the number of rules and possibly to sort out the obtained rules. These measures are very interesting since they allow to give to the experts the rules which seem to be the most surprising, the most interesting with regard to the chosen statistical criteria.

These measures can be generally applied to a wide class of semantics. Nevertheless, they do not belong to what we believe to be the core definition of a semantics for rules. Moreover, we do not want to define as much semantics as there are measures. In a data mining context, error and quality measures can be integrated a posteriori to sort and to qualify the rules.

6 Conclusion

In this paper, we have pointed out that an equivalence does exist between some syntactic restrictions on the natural definition of a given semantics and the fact that this semantics is well-formed, i.e. Armstrong's axioms are sound and complete for this semantics. From a practical point of view, this equivalence allows to prove easily whether or not a given semantics is well-formed. We have also pointed out the relationship between our generic definition of rule satisfaction and the underlying data mining problem.

For gene expression data, this work brings some foundations to build new well-formed semantics which best fit into biologists' requirements. In the context of an ongoing bioinformatics project, a convenient graphical user interface has been developed to facilitate the discovery process of biologists. We chose to integrate it as a module into an existing open-source system devoted to microarray data analysis: MeV developed by The Institute for Genomic Research (TIGR) [26].

References

[1] Armstrong, W.: Dependency structures of data base relationships. In: Proc. of the IFIP Congress 1974. (1974) 580–583

[2] Ganter, B., Wille, R.: Formal Concept Analysis. Springer-Verlag (1999)

[3] Agrawal, R., Imielinski, T., Swami, A.: Mining association rules between sets of items in large databases. In: ACM SIGMOD International Conference on Management of Data, Washington D.C., ACM Press (1993) 207–216

[4] Abiteboul, S., Hull, R., Vianu, V.: Fondements des bases de données. Addison Wesley (2000)

[5] Beeri, C., Berstein, P.: Computational problems related to the design of normal form relation schemes. ACM TODS **4**(1) (1979) 30–59
[6] Maier, D.: Minimum covers in the relational database model. JACM **27**(4) (1980) 664–674
[7] Guigues, J.L., Duquenne, V.: Familles minimales d'implications informatives résultant d'un tableau de données binaires. Math. Sci. Humaines **24**(95) (1986) 5–18
[8] Gottlob, G., Libkin, L.: Investigations on Armstrong relations, dependency inference, and excluded functional dependencies. Acta Cybernetica **9**(4) (1990) 385–402
[9] Icev, A., Ruiz, C., Ryder, E.: Distance-enhanced association rules for gene expression. In: BIOKDD'03, in conjunction with ACM SIGKDD, Washington, DC, USA. (2003)
[10] Creighton, C., Hanash, S.: Mining gene expression databases for association rules. Bioinformatics **19** (2003) 79–86
[11] Cong, G., Tung, A., Xu, X., Pan, F., Yang, J.: Farmer: Finding interesting rule groups in microarray datasets. In: Proc. of the ACM SIGMOD. (2004) 143–154
[12] Mannila, H., Räihä, K.J.: Algorithms for inferring functional dependencies from relations. DKE **12**(1) (1994) 83–99
[13] Demetrovics, J., Thi, V.: Some remarks on generating Armstrong and inferring functional dependencies relation. Acta Cybernetica **12**(2) (1995) 167–180
[14] Ullman, J.: Principles of Database Systems. Second edn. Computer Science Press (1982)
[15] Zaki, M.: Generating non-redundant association rules. In: Proceedings of the sixth ACM SIGKDD international conference on Knowledge discovery and data mining, ACM Press (2000) 34–43
[16] Beeri, C., Dowd, M., Fagin, R., Statman, R.: On the structure of Armstrong relations for functional dependencies. JACM **31**(1) (1984) 30–46
[17] Bastide, Y., Pasquier, N., Taouil, R., Stumme, G., Lakhal, L.: Mining minimal non-redundant association rules using frequent closed itemsets. In: Computational Logic, UK. Volume 1861 of LNCS., Springer (2000) 972–986
[18] Huhtala, Y., Krkkinen, J., Porkka, P., Toivonen, H.: Efficient discovery of functional and approximate dependencies using partitions. In: Proc. of the 14th IEEE ICDE. (1998) 392–401
[19] Klemettinen, M., Mannila, H., Ronkainen, P., Toivonen, H., Verkamo, A.: Finding Interesting Rules from Large Sets of Discovered Association Rules. In: Proc. of the 3^{td} CIKM. (1994) 401–407
[20] Baralis, E., Psaila, G.: Designing templates for mining association rules. J. Intell. Inf. Syst. **9**(1) (1997) 7–32
[21] Cristofor, L., Simovici, D.: Generating an informative cover for association rules. In: Proceedings of the 2002 IEEE ICDM, Japan, IEEE Computer Society (2002) 597–600
[22] Li, G., Hamilton, H.: Basic association rules. In: Proceedings of the Fourth SIAM International Conference on Data Mining, Lake Buena Vista, Florida, USA, SIAM (2004)
[23] Luong, V.: The representative basis for association rules. In: IEEE ICDM'01. (2001) 639–640
[24] Tan, P.N., Kumar, V., Srivastava, J.: Selecting the right objective measure for association analysis. Information Systems **29**(4) (2004) 293–313
[25] Kivinen, J., Mannila, H.: Approximate inference of functional dependencies from relations. TCS **149**(1) (1995) 129–149
[26] Saeed, A., Sharov, V., White, J., Li, J., Liang, W., Bhagabati, N., Braisted, J., Klapa, M., Currier, T., Thiagarajan, M., Sturn, A., Snuffin, M., Rezantsev, A., Popov, D., Ryltsov, A., Kostukovich, E., Borisovsky, I., Liu, Z., Vinsavich, A., Trush, V., Quackenbush, J.: TM4: a free, open-source system for microarray data management and analysis. Biotechniques **34**(2) (2003) 374–378

A Proposal for Combining Formal Concept Analysis and Description Logics for Mining Relational Data

Mohamed Hacene Rouane[1], Marianne Huchard[2], Amedeo Napoli[3], and Petko Valtchev[1,4]

[1] DIRO, Université de Montréal, C.P. 6128, Succ. CV, Montréal, Canada, H3C 3J7
[2] LIRMM, 161, rue Ada, F-34392 Montpellier Cedex 5
[3] LORIA, B.P. 239, F-54506 Vandœuvre-lès-Nancy
[4] Dépt. informatique, UQAM, C.P. 8888, Succ. CV, Montréal, Canada H3C 3P8

Abstract. Recent advances in data and knowledge engineering have emphasized the need for formal concept analysis (FCA) tools taking into account structured data. There are a few adaptations of the classical FCA methodology for handling contexts holding on complex data formats, e.g. graph-based or relational data. In this paper, relational concept analysis (RCA) is proposed, as an adaptation of FCA for analyzing objects described both by binary and relational attributes. The RCA process takes as input a collection of contexts and of inter-context relations, and yields a set of lattices, one per context, whose concepts are linked by relations. Moreover, a way of representing the concepts and relations extracted with RCA is proposed in the framework of a description logic. The RCA process has been implemented within the GALICIA platform, offering new and efficient tools for knowledge and software engineering.

1 Introduction

Formal concept analysis (FCA) has been successfully applied to a range of knowledge engineering problems [22,24]. Nevertheless, FCA methods and tools aimed at directly processing data for producing knowledge units represented within a knowledge representation language –based on description logics (DL) [1] such as OWL DL– are still under study. One key difficulty lies in the presence and management of relational attributes or links in the data, such as "spouse", "reference", and "part-of". For example, a target group for a marketing campaign may be the class of "spouses of Master Gold credit card holders", that involves both binary and relational attributes.

Current FCA methods and tools have no capabilities for taking into account relational attributes. This is a rather hard problem to solve, since relational attributes introduce dependencies and even cycles between the data items. A standard way for producing DL-like concept descriptions from a formal context including binary and relational attributes remains to be designed. Accordingly, one of the objectives of this paper is to present a methodology for taking into

S.O. Kuznetsov and S. Schmidt (Eds.): ICFCA 2007, LNAI 4390, pp. 51–65, 2007.

account relational attributes within FCA, leading to what could be called "relational concept analysis" or RCA.

The introduction of relational information, e.g. relational attributes, in the data formats for FCA has been studied for almost a decade now, leading to three main categories of research lines: (i) the relational attributes remain within the formal objects [10,11,12], (ii) relational attributes are considered as first-class citizens and organized into an independent lattice, separated from the standard concept lattice [14] (just like relation types are represented within the conceptual graph formalism [18]), (iii) relations between concepts are established independently from concept construction, on a manual or semi-automated basis [15]. Although these three approaches successfully deal with relational attributes for solving a specific task, they are still not general enough and do not allow to combine and process binary and relational attributes as object descriptors at the concept formation step. Such a need arises in various practical situations, for example in model engineering for software development or in ontology learning from data.

A first introduction of relational concept analysis (RCA) has been proposed in [9]. The data structure on which is based the relational concept analysis process is called a "relational context family" (RCF): it is composed of a collection of contexts and inter-context relations, the latter being binary relations between pairs of object sets lying in two different contexts. The objective is to build a set of lattices whose concepts are related by relational attributes, similar to DL roles or to UML associations. In addition, there are needs for associating restrictions with relational attributes for describing specific characteristics. RCA has been initially motivated by an application on the engineering of UML static models (see [5]) with an emphasis on expressiveness and algorithmic aspects. Meanwhile, the need for processing complex data such as relational data has become an important problem, especially in the field of knowledge discovery in databases [7], and calls for a formalization of RCA.

In this paper, we propose a global and declarative description of the relational structure within the RCA approach, based on a set of lattices resulting from the processing of the contexts that are successively considered. One feature of the relational structure is that an object lying in the extent of an RCA concept can be connected with another object lying in the extent of another RCA concept, through a set of relational attributes or links. The inter-concept links can be nested leading to a relational structure of an arbitrary depth. An auxiliary graph structure is defined for covering these inter-object links.

Moreover, as experiences with UML model analysis reveal, the complexity of the final concept descriptions calls for a knowledge representation formalism, for managing and taking into account the semantics of the inter-concept links, e.g. classifying links and checking their consistency. In the second part of the paper, it is shown how concepts and relations from RCA can be mapped into a knowledge base (KB) represented within a DL of the $\mathcal{ALE}$ family (more precisely $\mathcal{FL}^-\mathcal{E}$). The connection between the structure of the original data mapped into the ABox of the KB (set of individuals) and the RCA concepts stored in the TBox of the KB (set of concepts) is also studied.

The paper starts with a recall of basic notions from FCA (section 2) and from DL (section 3). Then, the RCA framework is presented in section 4. Section 5 describes the translation of the set of relational concepts into a knowledge base represented within a knowledge representation language. Finally, related work and a discussion are proposed in section 6.

Table 1. The formal context $\mathcal{K}_{papers}$ (in a transposed form for space minimization). The descriptors of papers are: requirement analysis (ra), architectural design (ad), detailed design (dd), and software maintenance (sm).

	(1) Fun95	(2) God93	(3) God95	(4) God98	(5) Huc99	(6) Huc02	(7) Kro94	(8) Kui00	(9) Leb99	(10) Lin95	(11) Lin97	(12) Sah97	(13) Sif97	(14) Sne96	(15) Sne98C	(16) Sne98R	(17) Sne99	(18) Sne00S	(19) Sne00U	(20) Str99	(21) Til03S	(22) Til03T	(23) Ton99	(24) Tone01	(25) Van98
ra																					x				
ad																					x	x			
dd	x	x	x	x	x	x	x	x	x	x	x	x	x	x	x	x	x	x	x	x		x	x	x	x
sm	x	x	x	x	x	x	x	x	x	x	x	x	x	x	x	x	x	x	x	x			x	x	x

2 The Basics of FCA

FCA is the process of abstracting conceptual descriptions from a set of individuals described by attributes [8]. Formally, a *context* $\mathcal{K}$ associates a set of objects (O) to a set of attributes (A) through an incidence relation $I \subseteq O \times A$. An example of formal context, namely $\mathcal{K}_{papers}$, is depicted in table 1, where O is a set of scientific publications on the applications of FCA in software engineering, and A the set of ISO software engineering activities (this example is adapted from [20]). Two operators, both denoted by $'$, connect the powerset of objects, 2^O and the powerset of attributes 2^A as follows:

$$' : 2^O \rightarrow 2^A,\ X' = \{a \in A \mid \forall o \in X, oIa\}$$

The operator $'$ is dually defined on attributes. The pair of $'$ operators induces a Galois connection between 2^O and 2^A [3]. The composition operators $''$ are closure operators and induce two families of closed sets, respectively $\mathcal{C}^o \subseteq 2^O$ and $\mathcal{C}^a \subseteq 2^A$. These two sets, provided with set-inclusion order, form two complete lattices (anti-isomorphic by $'$). A pair (X, Y) where $X \in 2^O$, $Y \in 2^A$, $X = Y'$, and $Y = X'$, is a *(formal) concept*, with X as *extent* and Y as *intent*. The set $\mathcal{C}_\mathcal{K}$ of all concepts extracted from $\mathcal{K}$ ordered by extent inclusion forms a complete lattice, $\mathcal{L}_\mathcal{K} = \langle \mathcal{C}_\mathcal{K}, \leq_\mathcal{K} \rangle$, called the *concept lattice* of the context (or the *Galois lattice* of the binary relation I). The lattice of $\mathcal{K}_{papers}$ associated with the formal context $\mathcal{K}_{papers}$ is drawn on the left-hand side of Figure 1.

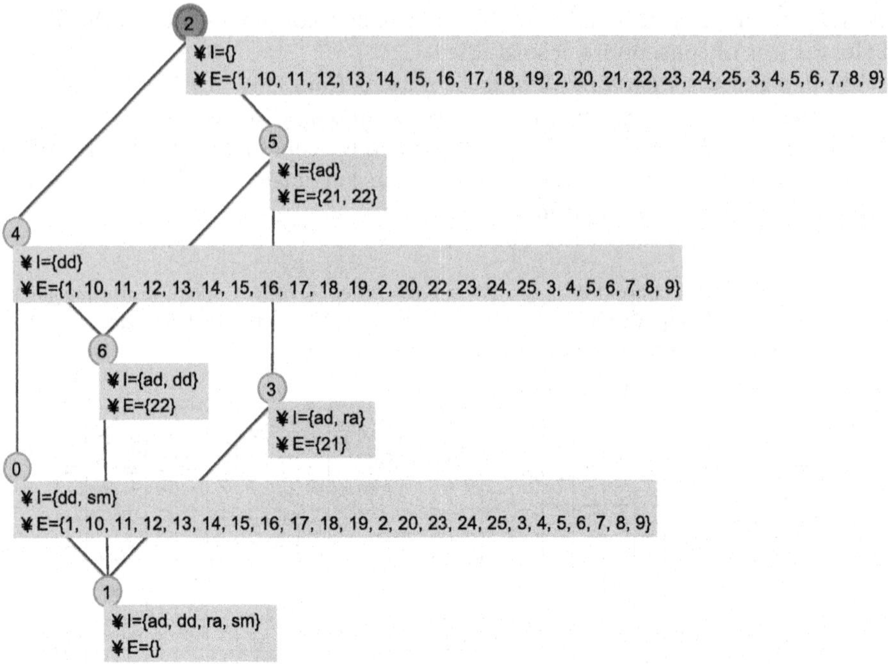

Fig. 1. The initial lattice $\mathcal{L}^0_{papers}$

The lattice represents in an exhaustive way the sharing of structures among objects: two objects are in the extent of a concept iff they share at least one attribute, meaning, in mathematical terms, that concept extents may be seen as the intersection of the extents of the attributes from their respective intents. As it will be shown in section 4, RCA relies on a similar principle, extended to attributes that summarize inter-object links.

Conceptual scaling deals with non-binary data descriptions [8], organized into a *many-valued context* $\mathcal{K} = (O, A, V, J)$, where J is a ternary relation between objects in O, attributes in A, and values in V. The scaling replaces a many-valued attribute with a set of binary attributes, each one representing a value that the many-valued attribute holds. In RCA, as explained farther, scaling is used to propagate structure sharing along inter-object links. Actually, shared structure among referred objects, i.e. concepts of the underlying context, is propagated to the referring objects, where it induces further sharing and hence new concepts.

3 The Basics of Description Logics

Description logics (DL) are formalisms for knowledge representation based on concepts, roles, and individuals [1]. A concept represents a set of individuals while a role determines a binary relationship between concepts. Concepts and roles are designed according to a syntax and a semantics, as in any logic-based formalism. The subsumption relation is a partial ordering relation, used for declaring (and detecting) specialization relations between concepts and roles, and for organizing concepts and roles within a hierarchy. Instance and concept classification are basic reasoning mechanisms: the former for finding the concepts an individual is an instance of, the latter for searching for the most specific subsumers (ascendants) and the most general subsumees (descendants) of a concept in the concept hierarchy.

The representation of concepts and roles in DL and in FCA is considered with different points of view: (i) the FCA approach is "inductive" and mainly interested in building concepts from a formal context (starting from individuals), (ii) the DL approach is "deductive" and mainly interested in designing concepts and inferring subsumption and instantiation relations for reasoning purposes (starting from concepts). Accordingly, FCA and DL can play complementary roles in understanding and managing complex data and knowledge units. Attempts for the integration of both approaches may be found in, e.g. [13,2,17].

From a practical point of view, a KB in DL consists of a set of concept and role descriptions (respectively comparable to unary and binary predicates in first order logic), that may be primitive or defined (the concept hierarchy is also called a TBox). "Primitive concepts" are ground descriptions that are used for forming more complex descriptions, the "defined concepts", by means of a set of constructors, such as conjunction ($\sqcap$), universal value quantification ($\forall$), existential value quantification ($\exists$), disjunction ($\sqcup$), negation ($\neg$), etc. While a primitive concept can be considered as an atom of the KB, a defined concept is described by a set of conceptual expressions –role introductions– that can be regarded as a set of necessary and sufficient conditions for detecting that an individual is an instance of the concept, allowing for classification-based reasoning.

The choice of a set of constructors has a direct influence on the complexity of reasoning, i.e. classification and instantiation (see, e.g. [6]). In the following, a simple representation language, called $\mathcal{FL}^{-}\mathcal{E}$, is considered, based on the constructors $\mathsf{C} \sqcap \mathsf{D}$, $\forall \mathsf{r.C}$, $\exists \mathsf{r.C}$, $\top$ (the top concept whose extension is the set of all individuals), $\bot$ (the bottom concept whose extension is the empty set), and $\equiv$ for the concept definition (no negation in this language).

For an illustration, consider the paper dataset in Fig. 2. This dataset could be represented in DL using the primitive concepts AboutDetailedDesign, AboutMaintenance, AboutArchitecture, and AboutRequirements, whereas a primitive role cite could be added to express bibliographic references. In the resulting representation, the concept of the papers about detailed design citing at least one paper on maintenance –denoted by ADD– is defined as a conjunction of the primitive concept AboutDetailedDesign and an existential role restriction on cite:

$$\mathsf{ADD} \equiv \mathsf{AboutDetailedDesign} \sqcap \exists \mathsf{cite.AboutMaintenance}$$

The semantics of the descriptions in a KB is defined by means of an *interpretation*, i.e. a pair $\mathcal{I} = (\Delta^{\mathcal{I}}, .^{\mathcal{I}})$, where $\Delta^{\mathcal{I}}$ is a set of individuals called the *interpretation domain*, and $.^{\mathcal{I}}$ is the *interpretation function*. The $.^{\mathcal{I}}$ function maps a concept description to a subset of $\Delta^{\mathcal{I}}$ and a role description to a subset of $\Delta^{\mathcal{I}} \times \Delta^{\mathcal{I}}$. Moreover, each constructor from the language translates into a specific set-theoretic operation on the interpretations of its argument expressions (e.g., $\sqcap$ translates into $\cap$). For example, the interpretation of the concept ADD with respect to a domain based on the paper context (in Table 1) and the cite relation (Fig. 2) is:

$$\{1, 3, 4, 5, 6, 8, 9, 10, 11, 12, 13, 14, 15, 16, 17, 18, 19, 20, 23, 24, 25\}$$

The subsumption relationship between concepts is the basis of the main inferential service provided by a DL reasoner: a concept C is subsumed by a concept D, denoted $C \sqsubseteq D$, iff $C^{\mathcal{I}} \subseteq D^{\mathcal{I}}$ for every interpretation $\mathcal{I}$. Subsumption is a pre-ordering relation, that can be considered as a partial ordering up to an equivalence, two concepts being equivalent as soon as the first subsumes the second, and reciprocally, i.e. they have the same set of instances.

A concept definition $A \equiv C$ assigns a concept description C to a concept name A (as in the definition of the ADD concept above). An assertion stores facts about actual individuals. Assertions are of two kinds, namely concept and role instantiation, e.g. AboutDetailedDesign(12), AboutMaintenance(13), and cite(12,13) are assertions representing facts in the current example on scientific papers. In this way, a KB is composed of a TBox $\mathcal{T}$ storing concept introductions, and an ABox $\mathcal{A}$ storing assertions about individuals.

It can be noticed that the descriptions of DL concept and the intents of formal concepts in FCA can be both considered as conjunctions of descriptors that act as predicates on individuals or objects. This observation calls for the introduction of relation between objects in FCA, similar to the relations materialized by roles between individuals in DL. Accordingly, a mapping could be obtained between concept descriptions produced by FCA and DL-based descriptions, leading to FCA as a method for building an ontology from data [19,4,16]. This is the purpose of the next section.

4 Relational Concept Analysis

Relational Concept Analysis (RCA) is an original approach for extracting formal concepts from sets of data described by binary and relational attributes. In this section, the formal background of RCA is introduced and detailed.

4.1 Data Model

In RCA, data are organized within a structure composed of a set of contexts $\mathbf{K} = \{\mathcal{K}_i\}$ and of a set of binary relations $\mathbf{R} = \{r_k\}$, where $r_k \subseteq O_i \times O_j$, O_i and O_j being sets of objects (respectively in $\mathcal{K}_i$ and $\mathcal{K}_j$). The structure $(\mathbf{K}, \mathbf{R})$ is called a *relational context family* (RCF) and can be compared to a relational database schema, including both classes of individuals and classes of relations. In our running example, $\mathbf{K} = \{K_{papers}\}$ and $\mathbf{R} = \{cite\}$. The following definition introduced in [5] gives a formal account of RCF.

Definition 1. *A relational context family* $\mathcal{R}$ *is a pair* $(\mathbf{K}, \mathbf{R})$*, where* $\mathbf{K}$ *is a set of contexts* $\mathcal{K}_i = (O_i, A_i, I_i)$*,* $\mathbf{R}$ *is a set of relations* $r_k \subseteq O_i \times O_j$*, where* O_i *and* O_j *are the object sets of the formal contexts* $\mathcal{K}_i$ *and* $\mathcal{K}_j$*.*

A relation $r_k \subseteq O_i \times O_j$ has a *domain* and a *range*, where :

- $\mathbf{O} = \{O_i / O_i \in \mathcal{K}_i = (O_i, A_i, I_i), \mathcal{K}_i \in \mathbf{K}\}$
- $r_k : O_i \rightarrow 2^{O_j}$
- $dom : \mathbf{R} \rightarrow \mathbf{O}$ and $dom(r_k) = O_i$
- $ran : \mathbf{R} \rightarrow \mathbf{O}$ and $ran(r) = O_j$,
- $rel : \mathbf{K} \rightarrow 2^{\mathbf{R}}$ and $rel(\mathcal{K}_i) = \{r_k | dom(r_k) = O_i\}$.

The instances of a relation r_k, say $r_k(o_i, o_j)$, where $o_i \in O_i$ and $o_j \in O_j$, are called *links*. For example, the figure 2 shows the binary table of the `cite` relation on the paper example, thus the set of links that are considered in the following. The links can be "scaled" in order to be included as binary attributes in a formal original context, through a mechanism called *relational scaling* and explained in the following sections.

Contexts, objects, attributes, and relational attributes, are considered to bear a *unique name*, within a name space holding for the whole RCF. Contexts are uniquely determined by their set of objects, that constitute the only component remaining invariant during the relational analysis process. In the same way, formal concepts are uniquely determined by their extents, that are invariant during relational scaling, just as object sets for contexts. The addition of new attributes in a formal context leads to an expansion of the underlying lattice: the augmented lattice contains the original lattice as a join-sub-semi-lattice [23]. Moreover, an order embedding of the original lattice into the augmented lattice may be set on and allows to separate invariant formal concepts, that are lying in both lattices, and *new* formal concepts in the augmented lattice, having no counterpart in the original lattice.

4.2 The Scaling of Relations

Our goal is to consistently introduce abstractions similar to the role restrictions $\forall$r.C or $\exists$r.C in DL into FCA representation and analysis frameworks. Role restrictions provide summaries of object links, e.g. $\forall$r.C expresses that for all the objects satisfying the restriction, their "r links" point to instances of C.

The same could be done on formal objects with links, provided that a suitable set of concepts to use as relational "restrictors" is given beforehand. The most natural, although not the unique, way of forming such a concept set is to take the formal concepts over the range context of a relation. In other words, given a relation r such that $dom(r) = O_i$ and $ran(r) = O_j$, the target set of concepts will correspond to *a* concept lattice[1] of the context underlying O_j. For each of the

[1] As we shall demonstrate in the following paragraphs, several such lattices will be composed depending on how much relational information is inserted into the initial context $\mathcal{K}_j$.

	(1) Fun95	(2) God93	(3) God95	(4) God98	(5) Huc99	(6) Huc02	(7) Kro94	(8) Kui00	(9) Leb99	(10) Lin95	(11) Lin97	(12) Sah97	(13) Sif97	(14) Sne96	(15) Sne98C	(16) Sne98R	(17) Sne99	(18) Sne00S	(19) Sne00U	(20) Str99	(21) Til03S	(22) Til03T	(23) Ton99	(24) Tone01	(25) Van98
(1) Fun95							x																		
(2) God93																									
(3) God95		x																							
(4) God98		x	x				x																		
(5) Huc99		x																							
(6) Huc02		x														x			x						
(7) Kro94																									
(8) Kui00											x		x	x	x	x		x							x
(9) Leb99		x																							
(10) Lin95							x																		
(11) Lin97	x						x							x											
(12) Sah97			x								x		x												
(13) Sif97											x	x		x											
(14) Sne96							x			x															
(15) Sne98C	x						x			x	x		x	x											
(16) Sne98R		x					x				x		x	x	x										
(17) Sne99	x	x		x			x				x		x	x	x	x									
(18) Sne00S							x				x		x	x		x	x								x
(19) Sne00U		x		x			x				x			x	x	x									
(20) Str99														x	x	x									
(21) Til03S											x					x									
(22) Til03T																									
(23) Ton99							x				x		x	x											
(24) Tone01							x				x		x	x			x								
(25) Van98											x		x		x	x									

Fig. 2. The table of citations between papers of the formal context $\mathcal{K}_{papers}$

available concepts c (in the lattice), an attribute will be assigned corresponding to the type of restriction to be enforced. So far, only universal and existential restrictions have been defined, although others could equally apply. They have been designed as separate modes of scaling, called *encoding schemes*. Thus, the *narrow* scheme assigns the attribute corresponding to a pair (r, c) to all the objects $o \in O_i$ whose set of r links, $r(o)$ is strictly included in the extent of c. By contrast, the *wide* scheme only requires that the two sets have a non-empty intersection. In the following, c refers to a concept in the lattice while C refers to a DL concept.

In mathematical terms, given $\mathcal{K}_i = (O_i, A_i, I_i)$, the scaling of $\mathcal{K}_i$ for a relation $r \in rel(\mathcal{K}_i)$ such that $ran(r) = O_j$ with respect to the lattice $\mathcal{L}_j$ yields an extension of A_i and I_i, but keeps O_i unchanged. The attributes added to A_i are of the form $r : c$, as made precise in the definition below:

Definition 2. *Given a relation $r \in rel(\mathcal{K}_i)$ and a lattice $\mathcal{L}_j$ on $\mathcal{K}_j$ with $O_j = ran(r)$, the narrow scaling operator $sc_{\times}^{(r,\mathcal{L}_j)} : \mathbf{K} \rightarrow \mathbf{K}$ is defined as follows:*

$$sc_{\times}^{(r,\mathcal{L}_j)}(\mathcal{K}_i) = (O_i^{(r,\mathcal{L}_j)}, A_i^{(r,\mathcal{L}_j)}, I_i^{(r,\mathcal{L}_j)})$$

where $O_i^{(r,\mathcal{L}_j)} = O_i$, $A_i^{(r,\mathcal{L}_j)} = A_i \cup \{r : c | c \in \mathcal{L}_j\}$, and
$I_i^{(r,\mathcal{L}_j)} = I_i \cup \{(o, r : c) | o \in O_i, c \in \mathcal{L}_j, r(o) \neq \emptyset, r(o) \subseteq extent(c)\}$

The *wide scaling operator* $sc_{+}^{(r,\mathcal{L}_j)}$ is defined in a similar way, the only difference lying in the incidence relation for $sc_{+}^{(r,\mathcal{L}_j)}(\mathcal{K}_i)$:

$$I_i \cup \{(o, r : c) | o \in O_i, c \in \mathcal{L}_j, r(o) \cap extent(c) \neq \emptyset\}$$

For example, suppose that the context $\mathcal{K}_{papers}$ has to be scaled on the cite relation with respect to the lattice given in Fig. 1. In this lattice, the concept denoted by 0 groups all papers on both detailed design and maintenance (attributes dd and sm). The extent of 0 actually comprises all papers but 21 and 22. In the scaled context, all objects from the $\mathcal{K}_{papers}$ context citing at least one paper from 0 will be assigned the attribute *cite* : 0 (equivalent to the DL expression ∃cite.C0, where C0 represents the concept 0 in the lattice). Besides, in the final lattice, shown in Fig. 3, these papers constitute the extent of the concept denoted by 8.

The complete relational scaling of a context $\mathcal{K}_i$ is the scaling of all the relations in $rel(\mathcal{K}_i)$. Considering a context $\mathcal{K}_i$, the relation set $rel(\mathcal{K}_i) = \{r_l\}_{l=1..p_i}$, and $\mathcal{L}_{j_l}$ the lattice associated to the context of the range object set for r_l for each l in $[1, p_i]$, the result of the scaling of $\mathcal{K}_i$ on all pairs $(r_l, \mathcal{L}_{j_l})$ is denoted by:

$$\mathcal{K}_i^{rel} = sc_{\times}^{(r_1,\mathcal{L}_{j1})}(sc_{\times}^{(r_2,\mathcal{L}_{j2})}(\ldots sc_{\times}^{(r_{k_i},\mathcal{L}_{jp_i})}(\mathcal{K}_i)))$$

Thus, when all the contexts of a RCF are scaled for lattice construction, a scaled version of a context $\mathcal{K}_j$ may possibly be no longer consistent with a prior scaling $sc_{\times}^{(r,\mathcal{L}_j)}$. When there is no loop within relations in the whole RCF, the inconsistent situation may be avoided by properly ordering contexts and the associated lattice construction tasks. A general method for constructing the lattices of a RCF is presented below.

4.3 Lattice Construction

The building of a lattice associated with a RCF is an iterative process that alternates pure lattice construction and expansion of the contexts through relational scaling. Let us consider a relational context family $\mathcal{R} = (\mathbf{K}, \mathbf{R})$, where $\mathbf{K}$ and $\mathbf{R}$ are finite, i.e. the number of contexts and the number of relations are finite. The process starts with a preliminary lattice construction for each context, exclusively focusing on the available binary attributes and omitting relations. This first construction provides the necessary basis for relational scaling that is applied on every relation of the RCF at the next iteration. Then, a new step of lattice construction is carried on, followed by a new scaling step, and so on. The construction process ends whenever a lattice construction does not introduce any new concept on the entire set of lattices. This means that a fixed point for the RCF scaling operator has been met, the computation ends up and the process terminates.

More formally, the evolution of the content of each context $\mathcal{K}_i$ is captured in a series of context contents $\mathcal{K}_i^p$. The initial element is the input context $\mathcal{K}_i^0 = (O_i, A_i^0, I_i^0)$. Each member of the series is obtained from the previous one by complete relational scaling: $\mathcal{K}_i^{p+1} = (\mathcal{K}_i^p)^{rel_p}$, where the $.^{rel_p}$ operator denotes the fact that the scaling is computed with respect to the context contents at the step p of the process. Finally, $\mathcal{K}_i^\infty$ is the fixed point of the series which is a context whose size is "non-decreasing". Indeed, during the scaling, the size of concept set of a context cannot decrease as the introduction of new (relational) attributes may only augment or leave unchanged that concept set.

Furthermore, to express the global evolution of the context set, a composite operator is defined for $\mathbf{K}^n$, that is considered as a vector of contexts. The operator, denoted $.^{\mathbf{rel}_p^*}$, denotes the application of $.^{rel_p}$ to all contexts in $\mathbf{K}$. Here again, a series $\mathbf{K}^n$ can be defined by $\mathbf{K}^0 = \mathbf{K}$, $\mathbf{K}^{p+1} = (\mathbf{K}^p)^{\mathbf{rel}_p^*}$ for all $p \geq 0$. The resulting series has an upper bound since all component series are upper bounded. The series is non-decreasing as well, and thus has a limit. This limit, denoted by $\mathbf{K}^\infty$, is the element where scaling produces no more concepts in any of the contexts. In the paper example, the final lattice $\mathcal{L}_{papers}^\infty$ is given in Fig. 3.

It can be noticed that the intents of the concepts on this figure have been simplified in the sense that, given a concept c, the presence in the intent of c of two relational attributes $cite : c_1$ and$cite : c_2$ such that $c_2 \leq c_1$ makes the expression $cite : c_1$ redundant and implies its removal. Accordingly, while all the objects of the concept 13 satisfy $cite : 9$, this last attribute is no longer in the intent of 13: $cite : 9$ does not bring any additional knowledge w.r.t. $cite : 13$ that is already in the intent of intent of 13.

From a pure computational standpoint, the fixed point is reached whenever at two subsequent steps all pairs of the corresponding lattices remain isomorphic.

5 Mapping the RCA Constructs to a DL KB

RCA provides relational descriptions that can be fully exploited by means of a representation formalism supporting reasoning, i.e. classification, instance and consistency checking. As argued above, the DL family is a natural choice for a formalism such as RCA. More specifically, we need constructors for the conjunction of the concepts (concept intents are basically conjunctions), universal role quantification (or value restriction, $\forall \mathsf{r}.\mathsf{C}$), and existential role quantification ($\exists \mathsf{r}.\mathsf{C}$) The closest description language is therefore $\mathcal{FL}^-\mathcal{E}$ (actually a subset of $\mathcal{ALE}$). Furthermore, an appropriate translation mechanism is necessary to ensure that the semantics of the expressions output by RCA, i.e. concept intents, is preserved. Thus, we assume a bijective mapping α that associates a DL construct to each RCF entity. The individual translation rules supporting the mapping are described below.

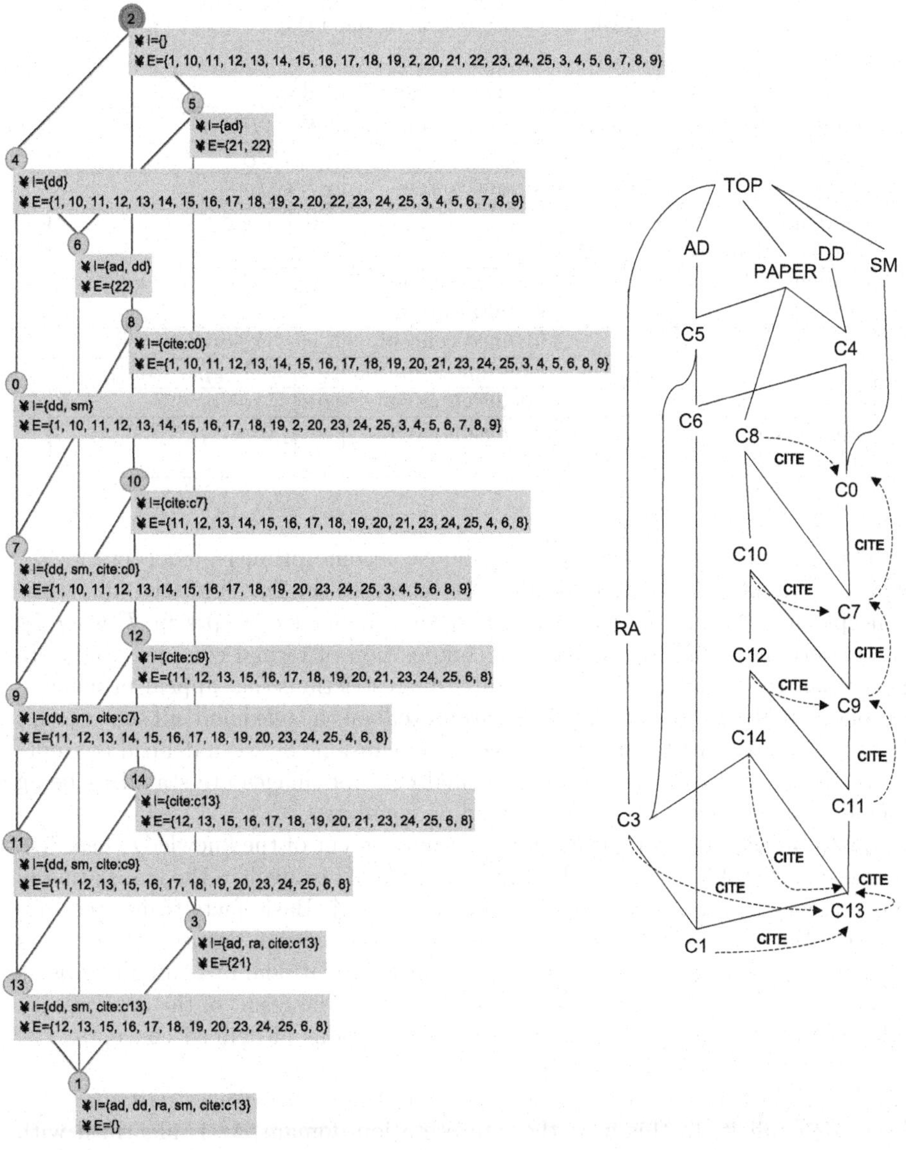

Fig. 3. Left: The final relational lattice $\mathcal{L}^{\infty}_{papers}$. **Right:** The corresponding TBox.

A knowledge base $\mathcal{KB}$ in $\mathcal{FL}^{-}\mathcal{E}$ is a pair $(\mathcal{T}(\mathcal{T}_C, \mathcal{T}_R), \mathcal{A})$. A knowledge base can be designed on top of a RCF $\mathcal{R}$ and the corresponding set of final lattices $\mathbf{L} = \{\mathcal{L}_i\}$ in the following way. First, the TBox $(\mathcal{T}(\mathcal{T}_C, \mathcal{T}_R)$ hosts the translations of the symbols for context, attribute, relation (relational attribute), and concept names from $\mathcal{R}$ and $\mathbf{L}$. The respective rules are presented in Table 2:

Table 2. Construction of the TBox

Source RCA entity	Notation	Target DL element	Rule
context	$\mathcal{K} \in \mathbf{K}$	primitive concept	$\alpha[\mathcal{K}] \in \mathcal{T}_C$
atomic attribute	$a \in A_i$	primitive concept	$\alpha[a] \in \mathcal{T}_C$
relation	$r \in \mathbf{R}$	primitive role	$\alpha[r] \in \mathcal{T}_R$
relational attribute (narrow)	$(r : c) \in A_i^\infty$	value restriction	$\forall\alpha[r].\alpha[c] \in \mathcal{T}$
relational attribute (wide)	$(r : c) \in A_i^\infty$	existential role quantification	$\exists\alpha[r].\alpha[c] \in \mathcal{T}$
concept	$c \in \mathcal{L}_i^\infty$, $c \neq \top_{\mathcal{L}_i^\infty}$	defined concept	$\alpha[c] \in definitions(\mathcal{T})$
		concept definition	$(\alpha[c] \equiv \sqcap_{a_j \in int(c)} \alpha[a_j]) \in \mathcal{T}$
		inclusion axiom	$(\alpha[c] \sqsubseteq \alpha[\mathcal{K}_i]) \in \mathcal{T}$
sub-concept link	$c_1 \leq_{\mathcal{L}_i^\infty} c_2$	inclusion axiom	$(\alpha[c_1] \sqsubseteq \alpha[c_2]) \in \mathcal{T}$

This means that contexts and attributes become primitive concepts in $\mathcal{T}_C$, while relations (relational attributes) become roles in $\mathcal{T}_R$. Similarly, all formal concepts in $\mathcal{L}_i^\infty$ but the top concept become defined concepts in $\mathcal{T}$ whereby the underlying definitions reflect the composition of formal concept intents. In particular, relational attributes generate role restrictions that appear in the definitions. Finally, all the inclusions ($\sqsubseteq$) are stated: on the one hand, all sub-concept links in the final lattices are translated into inclusions between defined concepts in $\mathcal{T}$, while, on the other hand, all formal concept images are stated as being included in the image of the corresponding contexts.

Based on the above translation, the actual content of the knowledge base, i.e. the set of ground facts in the $\mathcal{A}$ ABox is created. This involves the translation of objects which become individuals (constants) in $\mathcal{A}$. Basic facts translated into instantiations for primitive concepts include the incidence between objects and contexts, and the initial incidence facts from I_i^0. Relational links among objects become instances of roles, i.e. pairs of individuals. The result of the analysis, i.e. all concept extents, are translated into instantiations of defined concepts. The respective rules are listed in Table 3.

The connection between the data part in $\mathcal{A}$ and the schema part in $\mathcal{T}$ can be made explicit. In this way, the interpretation domain $\Delta^\mathcal{I}$ is identified with the ABox $\mathcal{A}$, meaning that concept descriptions are interpreted in terms of individuals in the ABox, and role descriptions in terms of pairs of individuals. Moreover, the formal concept extents in $\mathcal{L}_i^\infty$ have been explicitly translated into facts of the ABox. A question remains whether all relations in the data have been expressed in the $\mathcal{FL}^-\mathcal{E}$ knowledge base $(\mathcal{T}, \mathcal{A})$. The answer is that under some very reasonable hypotheses, the set of all possible semantics, i.e. the actual object sets that can be described by a formula of $\mathcal{FL}^-\mathcal{E}$ using only concept and role names in $\mathcal{T}(\mathcal{T}_C, \mathcal{T}_R)$, are already in the TBox $\mathcal{T}$ constructed in the above fashion.

Table 3. Construction of the ABox

Source RCA entity	Notation	Target DL element	Rule
objects	$o \in O_i$	individual	$\alpha[o] \in \mathit{individuals}(\mathcal{A})$
		primitive concept instantiation	$\alpha[\mathcal{K}_i](\alpha[o]) \in \mathcal{A}$
object-to-attribute incidence	$(o,a) \in I_j^0$	primitive concept instantiation	$\alpha[a](\alpha[o]) \in \mathcal{A}$
relational link	$r(o_1, o_2), r_i \in \mathbf{R}$	role instance	$\alpha[r](\alpha[o_1], \alpha[o_2]) \in \mathcal{A}$
object-to-concept incidence	$c \in \mathcal{L}_i^\infty, o \in \mathit{ext}(c)$	defined concept instantiation	$\alpha[c](\alpha[o]) \in \mathcal{A}$

This can be stated in the following property. Given an arbitrary description D in $\mathcal{FL}^-\mathcal{E}$, with $\mathcal{T}(\mathcal{T}_C, \mathcal{T}_R)$, there exists a concept C in $\mathcal{T}(\mathcal{T}_C, \mathcal{T}_R)$ such that D and C are semantically equivalent for the model provided by the ABox, i.e. $D^\mathcal{A} \equiv C^\mathcal{A}$.

6 Conclusion

Several research works hold on the extension of the classical binary context model of FCA to the handling of more complex contexts including relational descriptions, while preserving the closure properties of formal concepts. For example, in the approach based on *power context families* in [14], inter-object links are processed as high-order objects and formal concepts are composed these links. This approach offers a uniform way of processing n-ary links between objects, without giving details on the way of representing the concepts in the framework of a knowledge representation language.

Graph-based descriptions are considered in [10,11,12]. These papers propose efficient extensions of the FCA machinery to complex data formats described as graphs, e.g. chemical compound models, conceptual graphs, RDF triples. As above, the resulting methods are not intended to provide DL concept descriptions by means of FCA, contrasting the objectives of the present approach.

In this paper, we have proposed "relational concept analysis", an extension of FCA for the representation and manipulation of relational data. Concepts and relations extracted with RCA techniques can then be represented in the framework of description logics, allowing reasoning and problem-solving. The RCA approach has been implemented in the GALICIA platform[2] [21], and is operational for small datasets. Indeed, scalability is still an open issue, since the size of lattices grows rapidly w.r.t. the growth of relations between contexts. Various techniques for preventing combinatorial explosion can be used, e.g. iceberg lattices or Galois sub-hierarchies. Algorithmic issues are one of the current primary concern: further efficient progress can be realized based on a technique for iterative lattice

[2] `http://sourceforge.net/projects/galicia/`

design and maintenance (replacing design from scratch). Moreover, a challenging research track includes as well the coupling of GALICIA with a DL reasoner for knowledge representation and problem-solving purposes, and for ontology and software engineering.

Acknowledgments

This work was partially supported by the first author's FQRNT fellowship and the last author's NSERC Discovery and FQRNT Team grants, as well as by a FQRNT–INRIA collaborative grant hold by the last two authors.

References

1. F. Baader, D. Calvanese, D. McGuinness, D. Nardi, and P. Patel-Schneider, editors. *The Description Logic Handbook.* Cambridge University Press, Cambridge, UK, 2003.
2. F. Baader and B. Sertkaya. Applying formal concept analysis to description logics. In P. Eklund, editor, *Second International Conference on Formal Concept Analysis (ICFCA 2004), Sydney, Australia*, pages 261–286. LNAI 2961, Springer, Berlin, 2004.
3. M. Barbut and B. Monjardet. *Ordre et classification – Algèbre et combinatoire.* Hachette, Paris, 1970.
4. P. Cimiano, A. Hotho, G. Stumme, and J. Tane. Conceptual Knowledge Processing with Formal Concept Analysis and Ontologies. In P.W. Eklund, editor, *Concept Lattices, Second International Conference on Formal Concept Analysis (ICFCA 2004), Sydney, Australia*, pages 189–207, Berlin, 2004. LNCS 2961, Springer.
5. M. Dao, M. Huchard, M. Rouane Hacene, C. Roume, and P. Valtchev. Improving Generalization Level in UML Models: Iterative Cross Generalization in Practice. In K.E. Wolff, H.D. Pfeiffer, and H.S. Delugach, editors, *Conceptual Structures at Work: Proceedings of the 12th International Conference on Conceptual Structures, ICCS 2004, Huntsville, AL*, pages 346–360, Berlin, 2004. LNAI 3127, Springer.
6. F.M. Donini, M. Lenzerini, D. Nardi, and W. Nutt. The complexity of concept languages. *Information and Computation*, 134(1):1–58, 1997.
7. S. Dzeroski and N. Lavrac, editors. *Relational Data Mining.* Springer, Berlin, 2001.
8. B. Ganter and R. Wille. *Formal Concept Analysis.* Springer, Berlin, 1999.
9. M. Huchard, C. Roume, and P. Valtchev. When concepts point at other concepts: the case of UML diagram reconstruction. In V. Duquenne, B. Ganter, M. Liquiere, E. Mephu-Nguifo, and G. Stumme, editors, *Proceedings of the Workshop on Formal Concept Analysis for Knowledge Discovery in Databases (FCAKDD at ECAI-02)*, pages 32–43, 2002.
10. S.O. Kuznetsov. Learning of simple conceptual graphs from positive and negative examples. In *Proceedings of the Third European Conference on Principles of Data Mining and Knowledge Discovery (PKDD '99), Prague, Czech Republic*, pages 384–391, Berlin, 1999. LNCS 1704, Springer, Berlin.
11. S.O. Kuznetsov. Machine learning and formal concept analysis. In P.W. Eklund, editor, *Concept Lattices, Second International Conference on Formal Concept Analysis (ICFCA 2004), Sydney, Australia*, pages 287–312, Berlin, 2004. LNCS 2961, Springer.

12. M. Liquiere and J. Sallantin. Structural Machine Learning with Galois Lattice and Graphs. In J.W. Shavlik, editor, *Proceedings of the Fifteenth International Conference on Machine Learning (ICML 1998), Madison, Wisconson*, pages 305–313. Morgan Kaufmann, 1998.
13. S. Prediger and G. Stumme. Theory-driven logical scaling: conceptual information systems meet description logics. In E. Franconi and M. Kifer, editors, *Proceedings of the 6th International Workshop on Knowledge Representation meets Databases (KRDB'99), Linköping Sweden*, 1999.
14. S. Prediger and R. Wille. The lattice of concept graphs of a relationally scaled context. In W.M. Tepfenhart and W.R. Cyre, editors, *Proceedings of the 7th International Conference on Conceptual Structures (ICCS'99), Blacksburg, Virginia*, pages 401–414, Berlin, 1999. LNCS 1640, Springer.
15. U. Priss. *Relational Concept Analysis: Semantic Structures in Dictionaries and Lexical Databases.* PhD thesis, Aachen University, 1996.
16. T.T. Quan, S.C. Hui, A.C.M. Fong, and T.H. Cao. Automatic generation of ontology for scholarly semantic web. In S.A. McIlraith, D. Plexousakis, and F. Van Harmelen, editors, *International Conference on Semantic Web, ISWC 2004, Hiroshima, Japan*, pages 726–740, Berlin, 2004. LNCS 3298, Springer.
17. S. Rudolph. Exploring relational structures via fle. In K.E. Wolff, H.D. Pfeiffer, and H.S. Delugach, editors, *Conceptual Structures at Work: 12th International Conference on Conceptual Structures Proceedings (ICCS 2004), Huntsville, AL*, pages 261–286. LNCS 3127, Springer, Berlin, 2004.
18. J.F. Sowa, editor. *Principles of Semantic Networks: Explorations in the Representation of Knowledge.* Morgan Kaufmann Publishers, Inc., San Mateo, California, 1991.
19. G. Stumme and A. Maedche. FCA-MERGE: Bottom-up merging of ontologies. In *Proceedings of IJCAI'01, Seattle (WA)*, pages 225–234, 2001.
20. T. Tilley, R. Cole, P. Becker, and P. Eklund. A survey of formal concept analysis support for software engineering activities. In *Proceedings of the First International Conference on Formal Concept Analysis, Darmstadt, Germany*, Berlin, 2003. Springer Verlag.
21. P. Valtchev, D. Grosser, C. Roume, and M.H. Rouane. Galicia: an open platform for lattices. In A. de Moor, W. Lex, and B. Ganter, editors, *Contributions to the 11th International Conference on Conceptual Structures (ICCS'03), Dresden, Germany*, pages 241–254. Shaker Verlag, 2003.
22. P. Valtchev, R. Missaoui, and R. Godin. Formal concept analysis for knowledge discovery and data mining: The new challenges. In P.W. Eklund, editor, *Concept Lattices, Second International Conference on Formal Concept Analysis (ICFCA 2004), Sydney, Australia*, pages 352–371, Berlin, 2004. LNCS 2961, Springer.
23. P. Valtchev, R. Missaoui, and P. Lebrun. A partition-based approach towards constructing galois (concept) lattices. *Discrete Mathematics*, 256(3):801–829, 2002.
24. R. Wille. Knowledge acquisition by methods of formal concept analysis. In *Proceedings of the conference on Data analysis, learning symbolic and numeric knowledge*, pages 365–380. Nova Science Publishers, Inc., 1989.

Computing Intensions of Digital Library Collections

Carlo Meghini[1] and Nicolas Spyratos[2]

[1] Consiglio Nazionale delle Ricerche, Istituto della Scienza e delle Tecnologie della Informazione, Pisa, Italy
meghini@isti.cnr.it
[2] Université Paris-Sud, Laboratoire de Recherche en Informatique, Orsay Cedex, France
spyratos@lri.fr

Abstract. We model a Digital Library as a formal context in which objects are documents and attributes are terms describing documents contents. A formal concept is very close to the notion of a collection: the concept extent is the extension of the collection; the concept intent consists of a set of terms, the collection intension. The collection intension can be viewed as a simple conjunctive query which evaluates precisely to the extension. However, for certain collections no concept may exist, in which case the concept that best approximates the extension must be used. In so doing, we may end up with a too imprecise concept, in case too many documents denoted by the intension are outside the extension. We then look for a more precise intension by exploring 3 different query languages: conjunctive queries with negation; disjunctions of negation-free conjunctive queries; and disjunctions of conjunctive queries with negation. We show that a precise description can always be found in one of these languages for any set of documents. However, when disjunction is introduced, uniqueness of the solution is lost. In order to deal with this problem, we define a preferential criterion on queries, based on the conciseness of their expression. We then show that minimal queries are hard to find in the last 2 of the three languages above.

1 Introduction

In a Digital Library (DL for short), collections [14,16,1] are sets of documents defined to facilitate the tasks of various DL actors, ranging from content providers for whom physical collections are provided, to users, for whom logical collections are provided. The latter kind of collections typically helps the user in carrying out information access. For discovery, the user requires a "place" where to accumulate the discovered documents, similar to the shopping cart of an e-commerce Web site. This concept is commonly known as *static* collection [20,2]. Static collections are also useful in other tasks, such as cooperative work, where they play the role of a shared information space within a community. A classical example of static collection is the *book-mark* (or *favorites*) of a Web browser. Users may also associate a description of their "view" of the DL to a collection, and access

S.O. Kuznetsov and S. Schmidt (Eds.): ICFCA 2007, LNAI 4390, pp. 66–81, 2007.

the collection whenever they need to explore this view. This concept is commonly captured by so-called *dynamic* collections [4,5,3]. Dynamic collections are not the only way users have in order to know at once the changes in the DL that may be of interest to them. Publish/subscribe (pub-sub for short) mechanisms are another way of achieving the same goal, but with a different modality: while in dynamic collection users are *active,* in the sense that they act by accessing collections, in pub-sub users are *passive,* in the sense that the system intercepts changes in the DL which may be of interest for users, and notifies them. This distinction is also known as *pull* vs. *push* access mode.

We argue that the notions of static and dynamic collections are two sides of the same coin, and propose a general notion of collection, which generalizes both. According to this notion, collections have an extension and an intension, very much like classes in object models or predicates in predicate logics. We then solve a basic problem, arising upon collection creation: the determination of the intension of a collection based on a given extension.

The paper is organized as follows: Sections 2 to 5 introduce our model of a DL, illustrating the most relevant concepts. Section 6 states in precise terms the problem we address. Sections 7 to 10 present different solutions to the problem, by examining different description languages for expressing collection intensions.

2 Terms

The basic ingredient of descriptions are *terms.* A term denotes a set of documents. As such, it may be a keyword describing the content of documents (such as *nuclear waste disposal* or *database*), or their type (*image*); or may be thought of as an attribute value (for instance, *creator*= *"CM"*). For generality, we do not impose any syntax on terms and treat them just as symbols making up a finite, non-empty set T, which is a proper subset of a countable domain $\mathcal{T}$, $\mathsf{T} \subset \mathcal{T}$, always containing the special term *true*, standing for truth.

Terms are arranged in a taxonomy, that is a binary relation $\leq_\mathsf{T}$ on T, reflexive and transitive, having *true* as the greatest element, that is

$$\forall \mathsf{t} \in \mathsf{T},\ \mathsf{t} \leq \mathit{true} \text{ and } \mathit{true} \leq \mathsf{t} \text{ implies } \mathsf{t} = \mathit{true}.$$

Based on $\leq_\mathsf{T}$, we define $\equiv_\mathsf{T}$ as follows: for any two terms $\mathsf{t}_1, \mathsf{t}_2 \in \mathsf{T}$,

$$\mathsf{t}_1 \equiv_\mathsf{T} \mathsf{t}_2 \text{ if and only if } \mathsf{t}_1 \leq \mathsf{t}_2 \text{ and } \mathsf{t}_2 \leq \mathsf{t}_1.$$

It is easy to see that $\equiv_\mathsf{T}$ is an equivalence relation. Let T_e be the set of equivalence classes induced by $\equiv_\mathsf{T}$, *i.e.*

$$\mathsf{T}_e = \{\ [\mathsf{t}] \mid \mathsf{t} \in \mathsf{T}\}.$$

Clearly, $[\mathit{true}] = \{\mathit{true}\}$. Furthermore, let us extend $\leq_\mathsf{T}$ to T_e as follows:

$$[\mathsf{t}_1] \leq_\mathsf{T} [\mathsf{t}_2] \text{ iff } \mathsf{t}_1 \leq_\mathsf{T} \mathsf{t}_2.$$

$(\mathsf{T}_e, \leq_\mathsf{T})$ is now a partial order, in which equivalent terms have been collapsed into the same equivalence class, having as greatest element $[true]$. To simplify notation, we will consider these equivalence classes as terms, therefore using the symbol T in place of T_e, and understand $\leq_\mathsf{T}$ as a partial order.

For any two terms $\mathsf{t}_1, \mathsf{t}_2 \in \mathsf{T}$, if $\mathsf{t}_1 \leq \mathsf{t}_2$ we say that t_1 *is a specialization (or sub-term) of* t_2, or that t_2 *is a generalization (or super-term) of* t_1.

3 The Description Directory

Description are used to annotate the documents of a digital library, which for the present purposes we just represent as a finite, non-empty subset D. The relation between documents and terms is stored in the *description directory,* which is a relation r from documents to terms, $\mathsf{r} \subseteq \mathsf{D} \times \mathsf{T}$, such that $(\mathsf{d}, \mathsf{t}) \in \mathsf{r}$ means that d is described (or indexed) by term t. We impose on r two requirements:

- it must be total, that is $dom(\mathsf{r}) = \mathsf{D}$. This is not a serious limitation for the users, because if no term qualifies as a satisfactory descriptor of a document, the term *true* can, and indeed should, be used;
- a document cannot be indexed by $\leq$-related terms:

$$\forall \mathsf{d} \in \mathsf{D},\ \mathsf{t}_1, \mathsf{t}_2 \in \mathsf{r}(\mathsf{d}) \text{ implies } \mathsf{t}_1 \parallel \mathsf{t}_2. \tag{1}$$

  This second constraint requires to select independent terms when indexing a document, which we think is not a serious limitation. The constraint also interacts with the previous one by imposing that if *true* is used for describing a document, then no other term can be used to describe that document, which is consistent with the usage of *true* postulated above.

From r we define two functions which will turn out very useful in the sequel:

- the *index,* a function $index : \mathsf{D} \rightarrow \mathcal{P}(\mathsf{T})$, giving the terms which a document is indexed by: $\forall \mathsf{d} \in \mathsf{D},\ index(\mathsf{d}) = \{\mathsf{t} \in \mathsf{T} \mid (\mathsf{d}, \mathsf{t}) \in \mathsf{r}\}$.
- the *extension,* a function $termext : \mathsf{T} \rightarrow \mathcal{P}(\mathsf{D})$, giving the documents which a term describes: $\forall \mathsf{t} \in \mathsf{T},\ termext(\mathsf{t}) = \{\mathsf{d} \in \mathsf{D} \mid (\mathsf{d}, \mathsf{t}) \in \mathsf{r}\}$

Constraint (1) just says that $index(\mathsf{d})$ consists of incomparable terms, for all documents $\mathsf{d} \in \mathsf{D}$.

4 General Descriptions

In general, a description is a propositional formula over the alphabet T, built out of the connectives $\neg$ (negation), $\wedge$ (conjunction) and $\vee$ (disjunction). We will denote the set of such formulas as $\mathcal{L}_T$, or simply $\mathcal{L}$ when there is no danger of ambiguity.

Descriptions denote sets of documents. This is captured by the function *ans*, named after the fact that a typical usage of descriptions is for querying a DL. *ans* is inductively defined as follows, where $\mathsf{t}, \mathsf{t}' \in \mathsf{T}$ and $\mathsf{q}, \mathsf{q}_1, \mathsf{q}_2 \in \mathcal{L}$:

$$\begin{aligned}
ans(\mathsf{t}) &= \bigcup\{termext(\mathsf{t}') \mid \mathsf{t}' \leq \mathsf{t}\} \\
ans(true) &= \mathsf{D} \\
ans(\neg \mathsf{q}) &= \mathsf{D} \setminus ans(\mathsf{q}) \\
ans(\mathsf{q}_1 \wedge \mathsf{q}_2) &= ans(\mathsf{q}_1) \cap ans(\mathsf{q}_2) \\
ans(\mathsf{q}_1 \vee \mathsf{q}_2) &= ans(\mathsf{q}_1) \cup ans(\mathsf{q}_2)
\end{aligned}$$

In the course of our study, we will need to consider several sub-languages of $\mathcal{L}$, corresponding to different types of descriptions. The simplest descriptions are conjunctions of incomparable terms. We will call these descriptions *simple queries,* and denote their set as $\mathcal{L}_S$. In fact, document descriptions can be regarded as simple queries given by the conjunction of the terms which describe the document. That is, assuming that the description of a document d is given by $index(\mathsf{d}) = \{\mathsf{t}_1, \mathsf{t}_2, \ldots, \mathsf{t}_n\}$ and recalling that (1) sanctions the incomparability of the terms $\mathsf{t}_1, \mathsf{t}_2, \ldots, \mathsf{t}_n$, we may, and in fact will assume that:

$$index(\mathsf{d}) = (\mathsf{t}_1 \wedge \mathsf{t}_2 \wedge \ldots \wedge \mathsf{t}_n) \in \mathcal{L}_S$$

Other important classes of descriptions will be introduced in due course.

5 Collections

A collection is a set of documents that make up a significant whole from an application point of view. We model collections as objects belonging to a finite, non-empty set C. The membership of documents into collections is stored in the *classification directory,* which is a relation e from documents to collections, $\mathsf{e} \subseteq \mathsf{D} \times \mathsf{C}$, such that $(\mathsf{d}, \mathsf{c}) \in \mathsf{e}$ means that d is a member of (or belongs to) collection c. In a DL it is usually required that every document belongs to at least one collection: $dom(\mathsf{e}) = \mathsf{D}$.

In order to best serve its purposes, a collection must have both an *extension* and an *intension,* very much like predicates in predicate logics. The extension of a collection is the set of objects that are members of the collection at a given point in time. It can then be defined as the total function $collext : \mathsf{C} \rightarrow \mathcal{P}(\mathsf{D})$ given by: $\forall \mathsf{c} \in \mathsf{C},\ collext(\mathsf{c}) = \{\mathsf{d} \in \mathsf{D} \mid (\mathsf{d}, \mathsf{c}) \in \mathsf{e}\}$. The intension of a collection is a description of the meaning of the collection, that is the peculiar property that the members of the collection collectively possess and that distinguishes the collection from other collections. This should not be confused with the so-called collection metadata (such as the owner or the creation date of the collection), which represent properties of collections required for administration purposes. Formally, the intension of a collection is a description, and is associated to its collection by the total function $collint : \mathsf{C} \rightarrow \mathcal{L}$. The question arises how these two notions should be related. An obvious requirement is that the set of documents belonging to the collection must *agree* with the collection intension. This can be expressed by requiring that the collection intension, when used as a query, should retrieve *at least* the documents in the collection extension. Formally:

$$\forall \mathsf{c} \in \mathsf{C},\ collext(\mathsf{c}) \subseteq ans(collint(\mathsf{c})). \tag{2}$$

As a consequence of this last requirement, we obtain two very useful properties of collections, namely: for any given query $\mathsf{q} \in \mathcal{L}$ and collection $\mathsf{c} \in \mathsf{C}$: (1) if $collint(\mathsf{c}) \wedge \mathsf{q}$ is unsatisfiable, then no document in (the extension of) c satisfies the query, that is: $ans(\mathsf{q}) \cap collext(\mathsf{c}) = \emptyset$. (2) if $collint(\mathsf{c})$ subsumes q, then all documents in (the extension of) c satisfy the query, that is: $collext(\mathsf{c}) \subseteq ans(\mathsf{q})$.

For a given collection $\mathsf{c} \in \mathsf{C}$, we define the *precision* of the collection intension, $prec(\mathsf{c})$, the set of documents denoted by $collint(\mathsf{c})$ which are not members of the collection:

$$prec(\mathsf{c}) = ans(collint(\mathsf{c})) \setminus collext(\mathsf{c})$$

If $prec(\mathsf{c}) = \emptyset$ we say that the collection is *precise,* and *imprecise* otherwise. Clearly, a collection is precise if and only if $collext(\mathsf{c}) = ans(collint(\mathsf{c}))$. More generally, we say that a description α is precise with respect to a set of documents X just in case $X = ans(\alpha)$.

6 The Problem

The problem we want to address in this study is the following: given a DL and a subset X of the documents in it, to find a description $\alpha \in \mathcal{L}$ such that $X \subseteq ans(\alpha)$. This problem typically arises when a user has a set of documents and wants to create a collection having those documents as extension. The documents in question may have been gathered by the user through one or more discoveries, or may have been brought to the user attention by an expert, or may have been notified to him by the system as the result of the user registration to a publish-subscribe mechanism. These are just a few scenarios, in all of which the user likes the documents he has and wants to persist their set in the DL by creating a collection which holds them. To this end, an intension must be defined which satisfies the constraint (2), whence the problem.

Let us define as *conjunctive queries* the descriptions of the form:

$$\bigwedge_{1 \leq j \leq n} l_j \quad (n \geq 1)$$

where each l_j is a *literal,* that is is either a term $t \in \mathsf{T}$, in which case it is called a *positive* literal, or its negation $\neg t$ (negative literal), such that:

- a term and its negation do not occur: for no different indexes $i, j \in [1, n]$, $l_i = \mathsf{t}$ and $l_j = \neg \mathsf{t}$, for some $\mathsf{t} \in \mathsf{T}$.
- literals are pairwise incomparable: two literals are incomparable if they are either both positive or both negative, and the terms occurring in them are incomparable. Let $\mathcal{L}_C$ be the set of conjunctive queries.

A typical conjunctive query is the *description* of a document $\mathsf{d} \in \mathsf{D}$, $\delta(\mathsf{d})$, given by the conjunction of the terms describing the document with the negation of the terms *not* describing the document:

$$\delta(\mathsf{d}) = \bigwedge\{\mathsf{t} \mid \mathsf{t} \in index(\mathsf{d})\} \wedge \bigwedge\{\neg \mathsf{t}' \mid \mathsf{t}' \notin index(\mathsf{d})\}$$

It is easy to see that $\delta(\mathsf{d})$ is more specific (*i.e.*, it is subsumed by) the index of d, $index(\mathsf{d})$; moreover, $\{\mathsf{d}\} \subseteq ans(\delta(\mathsf{d}))$ and a document $\mathsf{d}' \in ans(\delta(\mathsf{d}))$ just in case d' has exactly the same index as d, that is $index(\mathsf{d}) = index(\mathsf{d}')$. We assume this is not the case, *i.e.* all document indexes are different. This is not a serious limitation, since documents with the same index can be treated as a class, of which only one representative is considered.

Now *DNFS queries* are descriptions of the form:

$$\bigvee_{1 \leq i \leq m} D_i \quad (m \geq 1)$$

where each D_i is called a *disjunct* and is a conjunctive query. DNFS queries make up the language $\mathcal{L}_D$.

Evidently, any set of documents X has a trivial description in $\mathcal{L}_D$, given by:

$$\bigvee \{\delta(\mathsf{d}) \mid \mathsf{d} \in X\}$$

which is as precise as a description of X can be in the DL. However, this description is not very interesting: apart from being as large as X itself, it just replicates the index of every document in X, offering no additional information. A more satisfactory formulation of our problem is therefore: given a set of documents X, can we find a description of X which is *better* than the trivial one?

7 An Easy Solution: Formal Concepts

A simple query would certainly be a better description for X than the trivial one. Simple queries have a minimal logical structure (no negation, no disjunction) and therefore convey their meaning in a simple and intuitive way. So we reduce our problem to the following ones:

1. does X have a description in $\mathcal{L}_S$?
2. how precise can it be?

An answer to both questions comes from Formal Concept Analysis (FCA) [11,10,12]. The *formal context* of a DL is the triple $\mathcal{K} = (\mathsf{D}, \mathsf{T}, \mathsf{x})$, where:

$$(\mathsf{d}, \mathsf{t}) \in \mathsf{x} \text{ iff } \exists \mathsf{t}' \leq \mathsf{t} : (\mathsf{d}, \mathsf{t}') \in \mathsf{r}$$

The relation x, called the *incidence* of the context, extends r by taking into account the term taxonomy according to its intuitive meaning: it assigns a term t to a document d just in case d is described by a term t' that is more specific than t. Since t is more specific than itself (*i.e.*, $\leq$ is reflexive) we have that $\mathsf{r} \subseteq \mathsf{x}$. In particular, $\mathsf{r} = \mathsf{x}$ if no term is a sub-term of another term.

As an example, let us consider the DL whose formal context is shown in Figure 1 left in tabular form. In this DL, term D is a sub-term of C and in fact any document described by D is also described by C, and all documents are described by *true*.

	A	B	C	D	E	F	$true$
1	×		×	×	×		×
2		×	×				×
3	×		×	×		×	×
4		×	×	×	×	×	×
5	×	×			×	×	×

Fig. 1. A Formal Context

A *formal concept* in $\mathcal{K}$ is a pair (D, T), where: (1) D, the *extent* of the concept, is a set of documents: $D \subseteq \mathsf{D}$; (2) T, the *intent* of the concept, is a set of terms: $T \subseteq \mathsf{T}$; and (3) T are the terms describing all documents in D and, vice-versa, D are all the documents described by the terms in T. Formally, (D, T) is a concept if and only if $D = \psi(T)$ and $T = \varphi(D)$, where

$$\psi(T) = \bigcap\{\varepsilon(\mathsf{t}) \mid \mathsf{t} \in T\} \text{ for all } T \subseteq \mathsf{T}$$

$$\varphi(D) = \bigcap\{\iota(\mathsf{d}) \mid \mathsf{d} \in D\} \text{ for all } D \subseteq \mathsf{D}$$

$$\varepsilon(\mathsf{t}) = \{\mathsf{d} \in \mathsf{D} \mid (\mathsf{d}, \mathsf{t}) \in \mathsf{x}\} \text{ for all } \mathsf{t} \in \mathsf{T}$$

$$\iota(\mathsf{d}) = \{\mathsf{t} \in \mathsf{T} \mid (\mathsf{d}, \mathsf{t}) \in \mathsf{x}\} \text{ for all } \mathsf{d} \in \mathsf{D}$$

In the formal context shown in Figure 1, $(\{1, 3, 4\}, \{C, D, true\})$ is a concept, while $(\{1, 3\}, \{A, D\})$ is not.

Lemma 1. *For all sets of terms* $Y \subseteq \mathsf{T}$, $\psi(Y) = ans(\bigwedge Y)$.

Since in a concept (D, T), we have that $D = \psi(T)$, the previous Lemma tells us that $D = ans(\bigwedge T)$, that is the extent of a concept is the answer to the intent of the concept, seen as a conjunction of terms.

7.1 Solving the Problem for $\mathcal{L}_S$

It should be evident that a formal concept strongly is a precise collection: the concept extent mirrors the extension of the collection; the concept intent consists of a set of terms, which can be viewed as a simple query which evaluates precisely to the extent. However, for our purposes concept intents tend to be redundant. For instance, given the concept $(\{4, 5\}, \{B, E, F, true\})$, there are simpler queries than $(B \wedge E \wedge F \wedge true)$ which return $\{4, 5\}$, for instance $(B \wedge E)$, $(B \wedge F)$ and $(E \wedge F)$. Part of the problem is that $true$ is the most general term, thus decidedly useless in queries other than $true$ itself. However the problem is more general since none of B, E and F is $\leq$-comparable with the others, yet one of the 3 is clearly redundant.

A term $t \in T$ is *redundant* in a set of terms $T \subseteq \mathsf{T}$, iff for all documents outside $ans(T)$, $d \in \mathsf{D} \setminus ans(T)$, whenever t does not describe d, $(d, t) \notin \mathsf{x}$, then there exists another term t' in T, such that t' does not describe d, $(d, t') \notin \mathsf{x}$. Now it is very simple to check that t is redundant in T iff $ans(T) = ans(T \setminus \{t\})$.

That is, a term is redundant in a set if it can be removed without altering the denotation of that set. Given a set of terms T, a simplification function σ is any function that iterates through the elements of T removing the ones that are found redundant. Notice that if $t \leq t'$ then t' is redundant whenever it co-occurs with t, so by eliminating redundant terms we implicitly eliminate comparable terms. The order in which non-comparable, redundant terms are considered is very important to determine the result. Indeed, the one of the terms B, E, F which is considered first is always redundant, while the remaining 2 are not. So, $\sigma(\{B, E, F, true\})$ may be anyone of $\{E, F\}$, $\{B, F\}$, or $\{B, E\}$. This is not relevant for our study, so we will leave it unspecified.

Proposition 1. *A set of documents $X \subseteq \mathsf{D}$ has a unique precise description in $\mathcal{L}_S$ if and only if X is the extent of a concept in $\mathcal{K}$.*

Proof: ($\leftarrow$) Suppose (X, Y) is a concept in $\mathcal{K}$. By definition, $X = \psi(Y)$ and by the previous Lemma $X = ans(\bigwedge Y)$. Now $(\bigwedge Y)$ may not be in $\mathcal{L}_S$ since some terms in Y may not be incomparable. Now $\sigma(Y)$ consists of incomparable terms and $X = ans(\bigwedge \sigma(Y))$ therefore $(\bigwedge \sigma(Y))$ is an $\mathcal{L}_S$ precise description for X. Since no two concepts can have the same extent, $(\bigwedge \sigma(Y))$ is also unique.
($\rightarrow$) We must prove that (X, Y) is a concept in $\mathcal{K}$ for some set of terms $Y \subseteq \mathsf{T}$. Since X has a precise description in $\mathcal{L}_S$, there exists a set of incomparable terms $T \subseteq \mathsf{T}$, such that $X = ans(\bigwedge T)$. Let $Y = \bigcap\{\iota(\mathsf{d}) \mid \mathsf{d} \in X\}$. By construction, $Y = \varphi(X)$ therefore it remains to be proved that $X = \psi(Y)$. By the previous Lemma, this is the same as proving that $X = ans(Y)$. We do this in 2 steps. (1) $ans(Y) \subseteq X$. By construction, $T \subseteq Y$, hence $\psi(Y) \subseteq \psi(T)$. By applying twice the Lemma, we have $\psi(Y) = ans(Y)$ and $\psi(T) = ans(T) = X$, and therefore we have $ans(Y) \subseteq X$. (2) $X \subseteq ans(Y)$. Now, by construction, for all $x \in X$ and $y \in Y$, $(x, y) \in \mathsf{x}$, hence $x \in \varepsilon(y)$ hence $x \in \bigcap\{\varepsilon(y) \mid y \in Y\}$ hence $x \in \psi(Y) = ans(Y)$. Then $X \subseteq ans(Y)$. □

Now, coupled with the well-known result of FCA that, for all set of documents $X \subseteq \mathsf{D}$, $(\psi(\varphi(X)), \varphi(X))$ is the concept with the smallest extent containing X, this Proposition allows us to answer the questions posed at the beginning of this Section, as follows: all sets X of documents have a description in $\mathcal{L}_S$, which we call the *simple description* of X and denote as $\delta_S(X)$, given by:

$$\delta_S(X) = \sigma(\varphi(X)).$$

The precision of $\delta_S(X)$ is given by:

$$\psi(\varphi(X)) \setminus X$$

The most precise $\mathcal{L}_S$ description for $\{1, 2\}$ is therefore $\sigma(\{C, true\}) = \{C\}$, whose precision is $\{3, 4\}$.

In some cases, the precision may be too large a set for the user, who might therefore be looking for a more precise description. To this end, one of two routes may be followed: the extension relaxation route, in which the user gives up some of the documents in X, or the intension relaxation route, in which the user

	A	B	C	D	E	F	$true$	$\neg A$	$\neg B$	$\neg C$	$\neg D$	$\neg E$	$\neg F$	$false$
1	×		×	×	×		×		×				×	
2		×	×				×	×			×	×	×	
3	×		×	×		×	×		×			×		
4		×	×	×	×	×	×	×						
5	×	×			×	×	×			×	×			

Fig. 2. The augmentation of a formal context

accepts a more complex description than a simple query. We have investigated the former route in [17], so we now concentrate on the latter.

The description language can be made more expressive than $\mathcal{L}_S$ in two different ways: by adding negation of single terms, in which case we end in $\mathcal{L}_C$, or by adding disjunction, in which case we end into a subset of $\mathcal{L}_D$, consisting of disjunctions of simple queries. We will consider each of these two languages in the sequel.

8 Conjunctive Queries

FCA can be very useful also if we admit negation in descriptions. In order to see how, we extend the notion of context to include negated terms. These have been already informally introduced in Section 6. We now give them a more precise mathematical status.

Let $\neg$ be a bijection from T to $\mathsf{T}_\neg$, a subset of $\mathcal{T}$ disjoint from T. For simplicity, we will write $\neg\mathsf{t}$ in place of $\neg(\mathsf{t})$ to indicate the negation of any attribute $\mathsf{t} \in \mathsf{T}$. For clarity, we will denote as $false$ the term $\neg true$. If $T \subseteq \mathsf{T}$ is a set of terms, $\neg(T)$ is the set of the negation of each term in T, *i.e.* $\neg(T) = \{\neg\mathsf{t} \mid \mathsf{t} \in T\}$.

The *augmented* formal context of a DL is the triple $\mathcal{K}_\neg = (\mathsf{D}, \mathsf{T} \cup \mathsf{T}_\neg, \mathsf{x}_\neg)$, where:

$$\mathsf{x}_\neg = \mathsf{x} \cup \{(\mathsf{d}, \neg\mathsf{t}) \mid (\mathsf{d}, \mathsf{t}) \notin \mathsf{x}\}.$$

In practice, the augmentation of a formal context introduces negated terms, whose extensions are the complement of the extensions of the corresponding non-negated terms. We will use $\neg$ as a subscript to indicate that we refer to the augmented context, *e.g.*$\varphi_\neg$ is the correspondent of φ in the augmented context.

The augmentation of the formal context shown in Figure 1 is given in Figure 2. It can be easily seen that augmentation induces a total, one-to-one homomorphism from the concepts of a context to those of the augmentation. In general this is not an isomorphism, as the augmentation may have more concepts. In addition, concept intents may be larger in the augmented context, as they may include negated terms. So in moving from a context to its augmentation we are able to describe more sets of documents. Now, by equating intents of augmented concepts with conjunctive queries, we can state the following Proposition.

Proposition 2. *A set of documents $X \subseteq \mathsf{D}$ has a precise description in $\mathcal{L}_C$ if and only if X is the extent of a concept in $\mathcal{K}_\neg$.*

The proof of this proposition is identical to that of Proposition 1. The most precise $\mathcal{L}_C$ description for a given set of documents X, $\delta_C(X)$ is therefore

$$\delta_C(X) = \sigma(\varphi_\neg(X))$$

and its precision is given by:

$$\psi_\neg(\varphi_\neg(X)) \setminus X. \tag{3}$$

Let us find the most precise $\mathcal{L}_C$ description for the set $\{1, 2\}$ in our running example. We recall that the most precise $\mathcal{L}_S$ description for $\{1, 2\}$ is $\{C\}$, whose precision is $\{3, 4\}$. Now it turns out that this set has a precise $\mathcal{L}_C$ description, since $\psi_\neg(\varphi_\neg(\{1, 2\})) = \psi_\neg(\{C, true, \neg F\}) = \{1, 2\}$. The sought description is given by $\sigma(\varphi_\neg(\{1, 2\})) = \sigma(\{C, true, \neg F\}) = \{\neg F\}$.

We conclude by observing that the set (3) can be computed without computing the augmented context, of course. In fact, it can be verified that, for all sets of terms T and sets of documents D :

$$\psi(T) = \{d \in \mathsf{D} \mid (d, t) \in \mathsf{x} \text{ for all } t \in T \text{ and } (d, t) \notin \mathsf{x} \text{ for all } \neg t \in T\}$$
$$\varphi(D) = \{t \in \mathsf{T} \mid (d, t) \in \mathsf{x} \text{ for all } d \in D\} \cup \{\neg t \mid (d, t) \notin \mathsf{x} \text{ for all } d \in D\}$$

9 Introducing Disjunction

Let $\mathcal{L}_U$ be the sub-language of $\mathcal{L}$ consisting of disjunctions of simple queries, which we call *disjunctive queries* for brevity.

Disjunctive queries can describe many more sets of documents, since disjunction allows to "accumulate" simple queries at will. So, the first question that naturally arises is whether all sets of documents have a precise description in $\mathcal{L}_U$. The answer, perhaps surprisingly, is "no," as the following Proposition shows. Let C^e and C^i denote the extent and the intent of concept C, respectively.

Proposition 3. *A set of documents $X \subseteq \mathsf{D}$ has a precise description in $\mathcal{L}_U$ if and only if $\gamma(\mathsf{d})^e \subseteq X$ for all $\mathsf{d} \in X$.*

Proof: ($\rightarrow$) Let β be the query

$$\beta = \bigvee\{\bigwedge \sigma(\gamma(\mathsf{d})^i) \mid \mathsf{d} \in X\}$$

By definition of ans, we have that:

$$ans(\beta) = \bigcup\{ans(\bigwedge \sigma(\gamma(\mathsf{d})^i)) \mid \mathsf{d} \in X\}.$$

By definition of σ, $ans(\bigwedge Y) = ans(\bigwedge \sigma(Y))$, for all sets of documents Y, therefore:

$$ans(\beta) = \bigcup\{ans(\bigwedge \gamma(\mathsf{d})^i) \mid \mathsf{d} \in X\}.$$

From Lemma 1 we have that $\gamma(\mathsf{d})^e = ans(\bigwedge \gamma(\mathsf{d})^i)$, therefore

$$ans(\beta) = \bigcup\{\gamma(\mathsf{d})^e \mid \mathsf{d} \in X\}$$

By the hypothesis it follows that $ans(\beta) \subseteq X$*. By construction,* $\mathsf{d} \in \gamma(\mathsf{d})^e$*, hence* $X \subseteq ans(\beta)$*. Therefore* $ans(\beta) = X$*, and* β *is a precise description for* X*.*
($\leftarrow$*) We prove that if for some document* $\mathsf{d} \in \mathsf{D}$ $\gamma(\mathsf{d})^e \not\subseteq X$*, then* X *has no precise* $\mathcal{L}_U$ *description. Let* $\mathsf{d}' \notin X$ *and* $\mathsf{d}' \in \gamma(\mathsf{d})^e$*. Then,* $\mathsf{d}' \in C^e$ *for each super-concept of* $\gamma(\mathsf{d})$*. But there is no concept extent containing* d *other than those of the super-concepts of* $\gamma(\mathsf{d})$*. It follows that any description containing* d *also contain* d'*, thus* X *has no precise* $\mathcal{L}_U$ *description.* □

In order to exemplify this last proposition, let us consider again the formal context shown in Figure 1. In this context, $\gamma(2)^e = \{2, 4\}$. This is a consequence of the fact that $\varepsilon(2) \subseteq \varepsilon(4)$ and implies that all concepts having 2 in their extents (*i.e.*$\gamma(2)$ and its super-concepts) also have 4 in their extents, therefore any set of documents containing 2 but not 4 does not have a precise $\mathcal{L}_U$ description. However, the power of disjunction is not to be underestimated, because while $\mathcal{L}_S$ and $\mathcal{L}_C$ precise descriptions are unique, a set of documents X may have more than one precise $\mathcal{L}_U$ description. This is due to the fact that X may be covered by concept extents in more than one way. Let us consider for instance the set $\{2, 3, 4, 5\}$ in our running example. This set has a precise $\mathcal{L}_U$ description, since it satisfies the condition established by the last Proposition, namely $\gamma(2)^e \subseteq \{2, 3, 4, 5\}$ and the same holds for $\gamma(3)^e$, $\gamma(4)^e$ and $\gamma(5)^e$. According to the proof of the last Proposition,

$$\beta = (B \wedge C) \vee (A \wedge D \wedge F) \vee (D \wedge E \wedge F) \vee (A \wedge E \wedge F)$$

is a precise description of $\{2, 3, 4, 5\}$. However, since $\mu(B) = (\{2, 4, 5\}, \{B\})$ and $\gamma(3) = (\{3\}, \{A, C, D, F\})$, also $B \vee (A \wedge D \wedge F)$ is a precise description of $\{2, 3, 4, 5\}$. This latter description is intuitively preferable over the former, since it denotes the same set but it is much shorter. Indeed, every disjunct of the latter description is a subset of a disjunct of the former description; this means that the former description may have more as well as larger disjuncts (set-theoretically speaking), however both of these can be pruned to obtain an equivalent but shorter description.

In order to capture formally this preference criterion, we define a relation between disjunctive queries. To this end and for the sake of simplicity, we will regard simple queries as sets of terms. Given two disjunctive queries $\alpha = D_1 \vee \ldots \vee D_m$ and $\beta = E_1 \vee \ldots \vee E_n$, α *is preferred over* β, $\alpha \sqsubseteq \beta$, if and only if $ans(\alpha) = ans(\beta)$ and for every disjunct D_i in α there exists a disjunct E_j in β such that $D_i \subseteq E_j$. $\sqsubseteq$ is reflexive and transitive, thus $(\mathcal{L}_U, \sqsubseteq)$ is a pre-order. A description is said to be *minimal* if it is a minimal element of $(\mathcal{L}_U, \sqsubseteq)$, that is no description is preferred over it. We then set out to find minimal descriptions. FCA proves very helpful to this end. In order to show how, we must first introduce the notions of candidate concept and minimum set cover.

- Given a set of documents $X \subseteq \mathsf{D}$, a *candidate concept* for X is a concept C such that $C^e \subset X$ and no super-concept D of C exists such that $D^e \subset X$.

- Given a collection $\mathcal{C}$ of subsets of a finite set S, a *set cover* for S is a subset $\mathcal{C}' \subseteq \mathcal{C}$ such that every element in S belongs to at least one member of $\mathcal{C}'$. A set cover is *minimum* if no set cover exists with a smaller cardinality.

As it can be proved: For all sets of documents $X \subseteq \mathsf{D}$,

1. if $X = \psi(\varphi(X))$, then $\sigma(\varphi(X))$ is the only minimal $\mathcal{L}_U$ description of X;
2. if $X \subset \psi(\varphi(X))$, then a $\mathcal{L}_U$ description $D_1 \vee \ldots \vee D_n$ of X is a precise minimal $\mathcal{L}_U$ description for X iff, for all $1 \leq j \leq n$, $D_j = \sigma(C_j^i)$ where C_j is a candidate concept and $C_1^e, \ldots, C_n^e$ is a minimum set cover for X amongst the extents of all candidate concepts for X.

From a computational point of view, the above characterization of minimal precise descriptions does not look particularly good, since these are equated to minimum set covers, whose computation is strongly suspected to be intractable [13]. The question arises whether there exists an equivalent characterization that is more amenable to computation. Unfortunately, the answer is negative. The next Proposition shows that MINIMUM SET COVER can be reduced to the computation of a minimal description, thus giving a lower bound for the latter problem.

Proposition 4. *Computing a minimal $\mathcal{L}_U$ description is NP-hard.*

Proof: We reduce MINIMUM SET COVER to our problem. Given an instance of MINIMUM SET COVER, that is a collection $\mathcal{C}$ of subsets of a set S, we define the formal context (D, T, i) as follows:

- *$D = S \cup \{o\}$ where o is any object not in S;*
- *T has one term t_i for each element C_i of $\mathcal{C}$, plus an extra term u which is any object not in S.*
- *i is defined as follows:*
  - *for all $s \in S$, if $s \in C_i$ then $(s, t_i) \in i$;*
  - *$(o, u) \in i$;*
  - *nothing else is in i.*

It can be proved that each minimum set cover corresponds to a precise, minimal $\mathcal{L}_U$ description for S and vice-versa. □

Candidate concepts play a key role in computing minimal, precise $\mathcal{L}_U$ descriptions, since each of such descriptions is obtained by combining the extents of those concepts so as to form a minimum set cover for X. An efficient way to compute candidate concepts is therefore fundamental. Iterating through all concept extents and retaining the maximal subsets of X is certainly a way of doing it, but not necessarily an efficient one, since a context may have an exponential number of concepts (in the size of the context). Fortunately, there is a more efficient method. It can be easily checked that, for all sets of documents X, the extents of the candidate concepts for X are given by:

$$\max_{t \in \mathsf{T}}\{Y = (\varepsilon(t) \cap X) \mid Y = \psi(\varphi(Y))\} \tag{4}$$

procedure $c3$ (X : set of **document id**)
1.   **begin**
2. $\mathcal{A}_X \leftarrow \emptyset$
3.   **for each** term t in T **do**
4.   **begin**
5. $Y \leftarrow \varepsilon(t) \cap X$
6.   **if** $\nexists\, Z \in \mathcal{A}_X$ such that $Y \subseteq Z^e$ **and** $Y = \psi(\varphi(Y))$ **then**
7.   **begin**
8.   **for each** concept $V \in \mathcal{A}_X$ such that $V^e \subset Y$ **do** $\mathcal{A}_X \leftarrow \mathcal{A}_X \setminus V$
9. $\mathcal{A}_X \leftarrow \mathcal{A}_X \cup (Y, \varphi(Y))$
10.   **end**
11.   **end**
12. **return** $\mathcal{A}_X$
13. **end**

Fig. 3. The $c3$ procedure

where maximality is with respect to set-containment. Clearly, every member of this set is the extent of a concept, a subset of X, and a maximal one. Notice that if $X = \psi(\varphi(X))$, X is the only member of this set.

It follows that the set of candidate concepts of X, $\mathcal{A}_X$, can be computed efficiently by iterating through the terms, as the procedure $c3$ (Figure 3) does. For each term, $c3$ computes in Y the overlapping between the extension of the term and X. If there already is in $\mathcal{A}_X$ a concept with an equal or larger extent that Y, then Y needs no longer to be considered because, even though it turns out to be a concept extent, it will not be maximal. Otherwise, if Y is the extent of a concept, that is $Y = \psi(\varphi(Y))$, then it may be the extent of a candidate concept, so it is added to $\mathcal{A}_X$ after removing from it the concepts with a smaller extent. Thus, when all terms have been examined, $\mathcal{A}_X$ contains the concepts whose extents are all the members of the set (4).

Let us consider again the set $\{2, 3, 4, 5\}$ for which we wish to find a minimal, precise $\mathcal{L}_U$ description in our running example. By running $c3$ on the context, we have the results shown in Table 1. For each term, the Table shows the overlap of the term extension with X, if this is a concept extent, the intent is shown next, and in the last column whether or not the concept is candidate. There turns out to be only 2 candidate concepts, so there is only one minimum set cover for X that can be constructed with the extents of these 2 concepts, therefore the only minimal, precise $\mathcal{L}_U$ for X is:

$$(\bigwedge \sigma(\{B, true\})) \vee (\bigwedge \sigma(\{F, true\})) = B \vee F$$

In this example, the minimum set cover problem has no impact, due to the toy size of the example. In real cases, however, candidate concepts can be as many as the terms, and an approximation technique may have to be used in order to avoid long computations. In alternative, an incomplete method may be chosen, returning a non-minimal description.

Table 1. Run of $c3$ with $X = \{2, 3, 4, 5\}$

t	$\varepsilon(t) \cap X$	*intent*	*candidate*
A	$\{3, 5\}$	$\{A, F, true\}$	no
B	$\{2, 4, 5\}$	$\{B, true\}$	yes
C	$\{2, 3, 4\}$	no	
D	$\{3, 4\}$	$\{C, D, F, true\}$	no
E	$\{4, 5\}$	$\{E, F, true\}$	no
F	$\{3, 4, 5\}$	$\{F, true\}$	yes

9.1 Imprecise $\mathcal{L}_U$ Descriptions

An imprecise $\mathcal{L}_U$ description might be desirable in case a precise one does not exist or is not satisfactory, for instance because too long. Here the problem is: to find the minimal description amongst the descriptions having minimal imprecision. This problem has a unique solution which we have already seen: $\sigma(\varphi(X))$. This is due to the fact that $(\psi(\varphi(X)), \varphi(X))$ is the smallest concept whose extent includes X. Thus, $(\psi(\varphi(X))$ is the only concept extent with minimal imprecision. In our example, if we do not like the description $(B \vee F)$, our best alternative in $\mathcal{L}_U$ is $\sigma(\varphi(X)) = true$.

10 DNFS Descriptions

We conclude this study by considering DNFS descriptions, that is formulas in $\mathcal{L}_D$. As we have already observed in Section 6, a set of documents X has always a precise DNFS description, but from the results of the last Section, we know that there may be more such descriptions. However, since the definition of minimality devised for $\mathcal{L}_U$ descriptions carries over $\mathcal{L}_D$ descriptions, the same technique can be applied. In order to illustrate, let us consider the document set $\{1, 2, 3\}$. Table 2 shows the results of running $c3$ on this set, similarly to Table 1. The extents of the 3 candidate concepts identified by $c3$ allow us to construct two minimal, precise $\mathcal{L}_D$ descriptions for the given set of documents, namely:

$$(\bigwedge \sigma(\{A, \neg B, C, D, true\})) \vee (\bigwedge(\sigma(\{C, \neg E, true\})) = \neg B \vee \neg E$$

$$(\bigwedge \sigma(\{A, \neg B, C, D, true\})) \vee (\bigwedge(\sigma(\{C, \neg F, true\})) = \neg B \vee \neg F$$

11 Related Work

The use of FCA in information system is not new (for a survey, see *e.g.* [19]). The structuring of information that FCA supports has inspired work on browsing [15,6], clustering [7], and ranking [9,18]. A basic drawback of these approaches is that they require the computation of the whole concept lattice, whose size may be exponential in that of the context, as it is well-known. An integrated

Table 2. Run of $c3$ on the augmented context with $X = \{1, 2, 3\}$

t	$\varepsilon(t) \cap X$	*intent*	*candidate*
A	$\{1,3\}$	$\{A, \neg B, C, D, true\}$	yes
B	$\{2\}$	$\{\neg A, B, C, \neg D, \neg E, \neg F, true\}$	no
C	$\{1,2,3\}$	no	
D	$\{1,3\}$	already considered	no
E	$\{1\}$	non-maximal	no
F	$\{3\}$	non-maximal	no
$\neg A$	$\{2\}$	already considered	no
$\neg B$	$\{1,3\}$	already considered	no
$\neg C$	$\{\}$	non-maximal	no
$\neg D$	$\{2\}$	already considered	no
$\neg E$	$\{2,3\}$	$\{C, \neg E, true\}$	yes
$\neg F$	$\{1,2\}$	$\{C, \neg F, true\}$	yes

approach to browsing and querying that uses only part of the lattice, and thus can be computed efficiently, is presented in [8], and extended to include user preferences in [17]. The usage of FCA for computing predicates describing sets of objects is novel, and complements the results of above mentioned approaches on the relationship between queries and concepts.

12 Conclusions

Thanks to the elementary notions of FCA, we have been able to solve a basic problem arising in DL collection management: the determination of a description for a given set of documents. We plan to expand the results obtained in this paper in 2 directions:

- by considering collection updates, in terms of insertion and removal of single documents from a collection extension; and
- by considering extensive usage of collection intensions for query processing, alluded to in Section 5. In fact, by introducing collection intensions we can reduce query processing in a DL to answering queries based on views, a problem that has been intensely studied in the database area in the last decade.

We also plan to set up experiments which would validate from a practical point of view the results obtained in this paper.

References

1. Donna Bergmark. Collection Synthesis. In *Proceeding of the second ACM/IEEE-CS Joint Conference on Digital Libraries*, pages 253–262. ACM Press, 2002.
2. David C. Blair. The challenge of commercial document retrieval, Part II: a strategy for document searching based on identifiable document partitions. *Information Processing and Management*, 38:293–304, 2002.

3. Leonardo Candela. *Virtual Digital Libraries.* PhD thesis, Information Engineering Department, University of Pisa, 2006.
4. Leonardo Candela, Donatella Castelli, and Pasquale Pagano. A Service for Supporting Virtual Views of Large Heterogeneous Digital Libraries. In Traugott Koch and Ingeborg Sølvberg, editors, *7th European Conference on Research and Advanced Technology for Digital Libraries, ECDL 2003*, LNCS vol. 2769, pages 362–373, Trondheim, Norway, August 2003.
5. Leonardo Candela and Umberto Straccia. The Personalized, Collaborative Digital Library Environment CYCLADES and its Collections Management. In Jamie Callan, Fabio Crestani, and Mark Sanderson, editors, *Multimedia Distributed Information Retrieval*, LNCS vol. 2924, pages 156–172, 2004.
6. C. Carpineto and G. Romano. Information retrieval through hybrid navigation of lattice representations. *International Journal of Human-Computer Studies*, 45(5):553–578, 1996.
7. C. Carpineto and G. Romano. A lattice conceptual clustering system and its application to browsing retrieval. *Machine Learning*, 24(2):95–122, 1996.
8. C. Carpineto and G. Romano. Effective reformulation of boolean queries with concept lattices. In *Proceedings of International Conference on Flexible Query Answering Systems*, LNAI vol. 1495, pages 83–94, Roskilde, Denmark, May 1998.
9. C. Carpineto and G. Romano. Order-theoretical ranking. *Journal of American Society for Information Science*, 51(7):587–601, 2000.
10. B.A. Davey and H.A. Priestley. *Introduction to lattices and order*, chapter 3. Cambridge, second edition, 2002.
11. B. Ganter and R. Wille. Applied lattice theory: Formal concept analysis. `http://www.math.tu.dresden.de/~ganter/psfiles/concept.ps`
12. Bernhard Ganter and Rudolf Wille. *Formal Concept Analysis: Mathematical Foundations.* Springer Verlag, 1st edition, 1999.
13. Michael R. Garey and David S. Johnson. *Computers and Intractability, A Guide to the Theory of NP-Completeness.* W.H. Freeman and Company, New York, 1979.
14. Gary Geisler, Sarah Giersch, David McArthur, and Marty McClelland. Creating Virtual Collections in Digital Libraries: Benefits and Implementation Issues. In *Proceedings of the second ACM/IEEE-CS Joint Conference on Digital Libraries*, pages 210–218. ACM Press, 2002.
15. R. Godin, J. Gecsei, and C. Pichet. Design of a browsing interface for information retrieval. In *Proceedings of SIGIR89, the Twelfth Annual International ACM Conference on Research and Development in Information Retrieval*, pages 32–39, Cambridge, MA, 1989.
16. Carl Lagoze and David Fielding. Defining Collections in Distributed Digital Libraries. *D-Lib Magazine*, November 1998.
17. Carlo Meghini and Nicolas Spyratos. Preference-based query tuning through refinement/enlargement in a formal context. In J. Dix and S. Hegner, editors, *Proceedings of FoIKS 2006, the fourth Int. Symp. on Foundations of Information and Knowledge Systems*, LNCS vol. 3861, pages 278–293, Budapest, February 2006.
18. Uta Priss. Lattice-based information retrieval. *Knowledge Organization*, 27(3):132–142, 2000.
19. Uta Priss. Formal concept analysis in information science. *Annual Review of Information Science and Technology*, 40:521–543, 2006.
20. Ian H. Witten, David Bainbridge, and Stefan J. Boddie. Power to the People: End-user Building of Digital Library Collections. In *Proceedings of the first ACM/IEEE-CS joint conference on Digital libraries*, pages 94–103. ACM Press, 2001.

Custom Asymmetric Page Split Generalized Index Search Trees and Formal Concept Analysis

Ben Martin and Peter Eklund

School of Computer Science and Software Engineering
The University of Wollongong
Northfields Avenue, Wollongong, NSW 2522, Australia
monkeyiq@users.sourceforge.net
peklund@uow.edu.au

Abstract. This paper investigates the scalability of applying Formal Concept Analysis to large data sets. In particular we present enhancements based on an existing spatial data structure, the RD-Tree, to better support both specific use with Formal Concept Analysis as well as generic multidimensional applications. Our experiments are motivated by the application of Formal Concept Analysis to a virtual filesystem [11,20,16]. In particular the libferris [1] Semantic File System.

1 Introduction: Information Retrieval and Formal Concept Analysis

In previous work we have shown that the application of spatial indexing to Formal Concept Analysis (FCA) can vastly improve query times [19]. Subsequent research was directed toward improving the spatial indexing techniques themselves [18]. This paper improves upon [18] by applying FCA to improve the spatial indexing structure.

The primary goal of this paper is to improve the efficiency of FCA on large formal contexts. Two subgoals can be seen – the improvement of the spatial indexing structure independent of the data it is indexing and improvements that rely on both the spatial indexing structure and the fact that FCA is being applied using that spatial index. An example of the latter would be the spatial index relying on knowledge from the FCA application in order to employ specialized compression as explained in Section 4.

FCA [10] is a well understood technique of data analysis. FCA takes as input a binary relation I between two sets normally referred to as the object set O and attribute set A and produces a set of "Formal Concepts" which are a minimal representation of the natural clustering of the input relation I. Formal concepts are hereafter referred to simply as concepts. A concept is a pair $(X \subseteq O, Y \subseteq A)$ such that X cannot be enlarged without reducing $|Y|$ and vice versa. The application of FCA to non-binary relations, such as a table in a relational database,

S.O. Kuznetsov and S. Schmidt (Eds.): ICFCA 2007, LNAI 4390, pp. 82–97, 2007.

can be achieved by first transforming or "scaling" the input data into a binary relation [10,21].

For a concept (X, Y), X is called the extent and is the set of all objects that have all of the attributes in Y, similarly Y is called the intent and is the set of all attributes possessed in common by all the objects in X. As the number of attributes in Y increases, the concept becomes more specific, i.e. a specialization ordering is defined over the concepts of a formal context by:

$$(X_1, Y_1) \leq (X_2, Y_2) :\Leftrightarrow Y_2 \subseteq Y_1$$

This ordering is a concept lattice which is normally presented as a Hasse diagram with special labeling rules [10].

A common approach to document and information retrieval using FCA is to convert associations between many-valued attributes and objects into binary associations between the same objects O and new attributes A. For example, modeling a filesystem the files would form the object set O. A many-valued attribute showing a file's size as numeric data may be converted into three attributes: `small`, `medium`, `large` which are then associated with the same set of files O. The binary relation I between $o \in O$ and $a \in A = \{$small, medium, large$\}$ is formed by asserting oIa depending on the level of the numeric size value of file o. The binary relation I is referred to as a formal context in FCA.

This is the approach adopted in the ZIT-library application developed by Rock and Wille [23] as well as the Conceptual Email Manager [6]. The approach is mostly applied to static document collections (such as news classifieds) as in the program RFCA [5] but also to dynamic collections (such as email) as in MAIL-SLEUTH [2] and files in the Logical File System (LISFS) [20]. In all but the latter two the document collection and full-text keyword index are static. Thus, the FCA interface consists of a mechanism for dynamically deriving binary attributes from a static full-text index. Many-valued contexts are used to materialize formal contexts in which objects are document identifiers.

A specialized form of information retrieval system is a virtual file system [11,20,16]. The idea of using FCA to generate a virtual filesystem was first proposed by using a logical generalization of FCA [8,7] and in more recent work using an inverted file index and generating the lattice closure as required by merging inverted lists [20]. In a virtual file system scalability becomes a critical concern because such a system deals with potentially millions of documents and hundreds/thousands of attributes [19,17].

It has been found that spatial indexing structures can greatly reduce typical query times in FCA [19] – we will discuss the type of query we consider in the next section. This has prompted research into improving the existing spatial indexing structures to better support FCA. In many cases the improvements needed by FCA are also applicable to general purpose multidimensional queries. As such, our empirical testing includes application to generic data mining input as well as specific application to FCA. The core focus of the paper remains on improving RD-Trees with the express purpose of improved FCA performance.

2 Indexing and Scalable Knowledge Processing with Formal Concept Analysis

Typical FCA queries seek all index entries which either (a) exactly match a given key or (b) are a superset of the given key. As an example of (b) consider FCA on animal species: one concept might contain the attributes {has-tail, has-fur}, to find the objects which match this concept we want to know all known objects which have *at least* these attributes but may include other attributes as well. Both of these common queries can be vastly aided with spatial indexing [15]. Note that even exact match queries present problems for conventional B-Tree indexes due to attribute ordering in index creation [19].

Other computationally intensive tasks in FCA including the calculation of the set of concepts for a given context are most efficiently handled using Data Mining algorithms [24,12].

The spatial indexing structure most suited to aiding FCA is the RD-Tree. The RD-Tree is derivation of the R-Tree spatial index. Typical secondary storage terminology is assumed: tree nodes are generally size matched to disk blocks [9], nodes at the bottom of a tree are leaf nodes, nodes leading to leaf nodes are internal nodes. The terms tree node and page are used interchangeably where no ambiguity arises.

The R-Tree [13] is a data structure that was created to allow spatial objects to be indexed effectively. Keys in an R-Tree are n-dimensional bounding boxes.

The internal nodes in an R-Tree structure contain entries of the form; `(bounding n-dimensional box, page pointer)`, where pages in the subtree reached by page pointers are within the given bounding n-dimensional box (see Fig. 1). This transitive containment relation is the heart of the R-Tree. R-Trees are not limited to 2 or 3 dimensional data but typically use page sizes allowing branching factors much closer to B-Trees than shown in the example.

Searching for a spatial object in the R-Tree starts at the root node and considers all children whose bounding box contains the query object. Searching for the query object in Fig. 1 begins at the root node (R) – the left node (C1) has a bounding box not containing the query object so only the right child (C2) is followed. In turn, the new left node (C2.1) contains the query object and will be followed whereas (C2.2) is not. At the lowest level (the children of C2.1) many nodes may contain the query object and these are followed to retrieve tuples in the base table.

When data is added to an R-Tree eventually a leaf node will become overfull. When a node is overfull it is split into two nodes: the original node and a new node. Entries are then redistributed between the original node and the new node. The creation of the new node necessitates a new entry in the original node's parent linking the new node into the tree structure. By adding an entry into the parent node the parent itself may become overfull and thus the process of splitting nodes will continue up the tree to the root while overfull nodes still exist.

When an index page becomes overfull and needs to be split into two pages, the R-Tree first selects the two elements who's bounding box are furtherest apart as keys to be placed into the parent. The remaining elements are then distributed

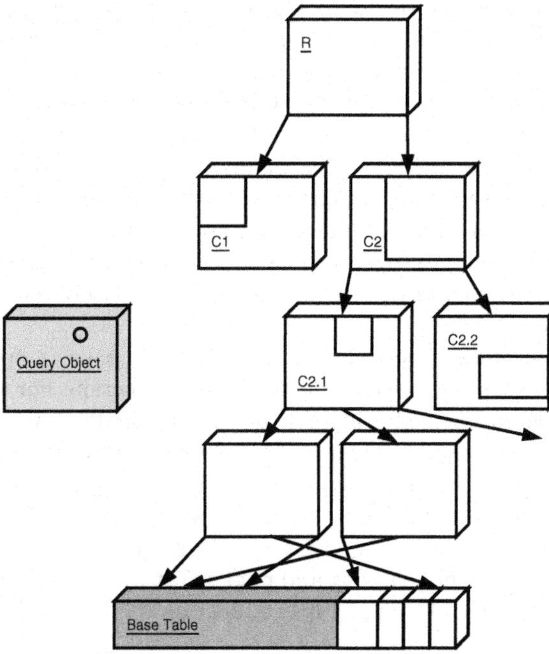

Fig. 1. An example R-Tree with a query object on the left. Each node has a bounding box which fully contains all objects in its child nodes. An implementation stores the bounding box for each child in the parent node. Note the example is limited to 2 dimensional space with a low branching factor for presentation purposes.

as children of one of these new parent keys. A child is distributed into the node to which it causes the least expansion of that node's bounding box. This is the classic Guttman page split [13].

The RD-Tree [15] operates similarly to the R-Tree by treating input as an n-dimensional binary spatial area. The R-Tree notion of containment is replaced by set inclusion and the bounding n-dimensional box replaced by a bounding set. The union of a collection of sets forms the bounding set. The bounding set of a child is thus defined as the union of all the elements in the child. The bounding set defined in this way preserves the "containment" notion of the R-Tree during search as a subset relation. When seeking an element which might be in a child it is sufficient to test if the sought element is a subset of the bounding set for the child to know if that subtree should be considered. The standard RD-Tree page splitting is based on the spatial R-Tree index structure's page splitting with a generalization of Guttman to sets.

A framework for building secondary storage search trees was recently introduced as the Generalized Index Search Tree (GIST) [14,3]. A GIST abstracts the core operations of a tree index structure into a small well defined collection of functions. Both the R-Tree and RD-Tree can be considered as specific GISTs. The page splitting and propagation of page splits toward the tree root as described above directly transfers into a GIST. A major advantage of using GIST is the ability to extend existing tree algorithms or make modifications to various operations to improve performance for specific data sources.

The process of adding an entry to a GIST can be considered as two parts: finding the appropriate leaf node in the index, and adding an entry to that leaf. Finding the leaf node occurs in a similar manner to normal search. The difference between search and insert being when a node has multiple children which could contain the new entry a `penalty` function is used to decide which child node to insert the entry into. An entry can only be inserted into one leaf node. Typically `penalty` will select the child node which is the most closely related to the new entry to aid future searches.

When a node becomes overfull in a GIST a new page is created and the redistribution of keys is delegated to the `picksplit` function. For the R-Tree and RD-Tree the above mentioned Guttman page split would form the `picksplit` function. This function decides for each of the entries in the overfull node whether to store that entry into the original node or new node and provides the updated and new keys for the parent node.

It can be seen from the above description that the functions which will decide the shape of a GIST are the `penalty` and `picksplit` functions. We now consider customizations of the RD-Tree `picksplit` function.

3 Asymmetric Page Split

We now focus on improving the `picksplit` function of the RD-Tree.

There is a major distinction between the standard R-Tree index keys and the RD-Tree index keys: the R-Tree index key is based around n-dimensional bounding boxes whereas the RD-Tree is based around a "bounding set". For a given page the bounding set is simply the union of all the keys in the page. The bounding set faithfully serves the same purpose as bounding box from which it was derived; both can be treated in a similar manner as a "container" in which all keys of a child page must reside.

Any n-dimensional bounding box can always be represented as $2 \times n$ coordinates: those coordinates in n-space on two opposite sides of the bounding box. A bounding set has no such fixed representation and can require an arbitrary number of elements to represent it (up to the set cardinality). This distinction is very important, an RD-Tree based index structure needs to focus on keeping its keys small, particularly those closer to the root of the tree. Both a voluminous bounding box and a bounding set with many elements are less effective in limiting the amount of a GIST that must be searched. However, a large bounding set will also consume more of a page, limiting the branching factor of the tree.

For an index on the same information, a tree with a lower branching factor will be a deeper tree [9]. A limited branching factor index will also require more internal nodes. The efficient caching of internal nodes in a computer's RAM is critical to index performance [9]. Having more internal nodes will decrease the effectiveness of such RAM caches.

Another critical factor in the selection of small bounding sets is that once a bounding set is selected for a parent that bounding set has great difficulty becoming smaller. Effective propagation of such bounding set reductions toward the root of the GIST generally takes so long as to effectively not occur. Further compounding this issue is that poorly chosen bounding sets at the leaf node level will eventually promote overly large bounding sets towards the root of the GIST.

Minimization of the number of elements in any bounding set should be a priority given that variation in size of the bounding set effects the overall tree branching factor. At times a `picksplit` function should favor making one of the new bounding sets slightly larger if it means a that the other bounding set can become substantially smaller.

If the page split is too asymmetric then one of the resulting pages may contain only a single element. Such an index structure may lead to many leaf pages being drastically under full and degrade overall performance. To counter this situation a minimum page fill ratio can be selected. There is a balance between maintaining this minimum ratio and the extension of the bounding set required to do so.

Two methods to achieving an asymmetric page split are considered; a custom distribution function to completely replace the Guttman function [18] and the application of the Guttman distribution followed by a redistribution using FCA.

The first distribution function first appeared in [18]. It is presented in this paper as a good heuristic driven implementation for comparative benchmarking with the FCA driven distribution function.

3.1 Complete Replacement of Guttman

This customized asymmetric `picksplit` algorithm pre-allocates elements to pages where they will not expand the page key, tries to minimize the expansion to one of the bounding sets, incrementally takes into account any expansion of bounding sets while distributing elements and attempts to leniently maintain a minimum page fill. The basic algorithm is shown in Fig. 2.

Note that the complexity of the core algorithm after the initial left and right keys are selected and before post split shuffling is linear in the number of keys to distribute. The shuffle process can range from linear to k^2 where k is the number of keys. There are areas of the algorithm which invite variation. The major one being how best to handle a drastically asymmetric page split.

3.2 Guttman Distribution Followed by FCA

In this method the standard Guttman generalization [15] is applied followed by the use of FCA to improve the cardinality of one of the bounding sets.

The algorithm is shown in Fig. 3. Abstractly the algorithm is mainly concerned with selecting which keys from the source page to move to the target page based on information from the concept lattice of the source page. As the final step of

1. Using the standard RD-Tree method select the initial left L and right R sets as new parent keys.
2. For all sets yet to be distributed, preallocate any set which is a non strict subset of either parent key $\{L, R\}$.
3. For all unallocated keys
   (a) Test if the current key is a subset of either updated parent key, if so then allocate that key to the child page of the respective parent key.
   (b) Attempt to minimize expansion of either parent key, when expansion has to occur prefer to expand the right parent's bounding set.
   (c) If both parents would have to expand the same amount to cater for a key then distribute as per the normal RD-Tree method.
4. If either page is drastically under full shuffle keys into it from the other page. A page is under full if it is less than 10% utilized. Do not expand the under full page's bounding set by more than x elements. For our testing $x = 1$ unless specified.

Fig. 2. Pseudo code for asymmetric page split. Pre-allocation will require a traversal over all keys to be distributed and set union with each key and L and R. The next step is the central part of the algorithm and only loops over keys once. The central distribution will require set unions with each key and both L and R. These cannot be cached from the values computed during pre-allocation because L and R are incrementally expanded during this phase. The final shuffle phase potentially touches most of the keys to be distributed.

updating the extent sizes for the THeap set is a computationally intensive task it is made optional and leads to two implementations for later benchmarking.

Shown in Fig. 4 is the an example concept lattice for a page after the Guttman algorithm has been applied. Notice that the concept nodes are colored depending on the number of keys which exactly match their concept or any downward transitively connected concept. One can immediately see that concepts with intent "c" and "d" are less strongly connected to the page. In the following concepts will be identified by their intent, for example, from the above we shall simply say concept "c" and "d". In particular there is a low overlap between "c" and "a" or "b".

Considering Fig. 4, assuming that the fixed cutoff of 40% allowed 14 of the 36 keys to be moved, the algorithm in Fig. 3 would first move the key matching the "d" concept from the source to the target page. Following this movement of 1 key the next smallest movement would be for concept "c". The movement this time is for 5 keys making a running total of 6 keys moved. By elimination, the next candidate would be "a" or "b". This is where the updating of THeap makes a difference. If THeap is not updated then it is luck if "a" or "b" are selected because their initial counts are identical. If "a" is selected then the algorithm will terminate rather than move 12 more keys making a total of $18 > 14$. If "b" is selected (which is guaranteed when updating THeap) then the 8 keys matching "b" are moved. Note that there are only 8 matches because the previous move of "c" also moved the keys in common with "b and c".

1. Apply the standard Guttman algorithm to obtain the initial two page distribution.
2. Calculate the size of the bounding sets of both pages. The page with the smaller bounding set is the source page and the other the target page.
3. Calculate the maximum number of keys to move k from as the number of keys in the source page multiplied by a cutoff percentage x.
4. For the source page:
   (a) Find the Intent (T_I) of the top formal concept of the page.
   (b) Find the lower covers of (T_I) and sort them by their extent size. Store this in THeap.
   (c) Initialize CumulativeKeysMoved = 0
   (d) While less keys than k have been moved:
      i. select the next lower cover of (T_I) that has yet to be considered working from the lower cover with the smallest extent to the largest.
      ii. Set CumulativeKeysMoved = CumulativeKeysMoved + number of keys in (T_I).
      iii. If CumulativeKeysMoved $> k$ then exit
      iv. Move the keys in (T_I) to the target page
      v. *optionally* Remove (T_I) from THeap and recalculate the extents in THeap and resort it by extent size.
5. If the sum of the size of the final bounding sets for the source and target page are larger than the initial sum of the size of the bounding sets then ignore the results of FCA. Otherwise apply the asymmetric page split.

Fig. 3. Pseudo code for asymmetric page split using Guttman and an FCA post process to achieve superior bounding set sizes

4 Customized Key Compression

If the input bounding sets are generated from a known structure then compression can take advantage of that structure. The use of FCA makes available both full and partial implication information. Consider the application of FCA to a numeric input using a linear ordinal scale: given a set of objects O such that each $o \in O$ has a numeric value $v \in V$ associated with it such a scale will generate a set of (binary) formal attributes $a \in A$ associated with the set O. For example, if the object set was formed using the planets and V was the number of moons of the planet then perhaps $A = \{some, few, many\}$ where planets which have few moons also have some moons by implication.

Such ordinal scaling is typical in many applications of FCA [10].

If we are representing a bounding set as a bitset and the first 63 bits are the result of such a linear ordinal scale then there is a direct implication between a bit and its predecessors. If there is an implication between bits then compression is possible.

Instead of using 63 bits to store this scale we can compress this to just 6 bits by storing only the position of the highest set bit in the first 63 bits. Some examples are shown in Fig. 5.

If one considers a GIST created using two ordinal scales then the bounding set can always be compressed to just two integers. It is much more difficult to take advantage of cross implications between the two ordinal scales.

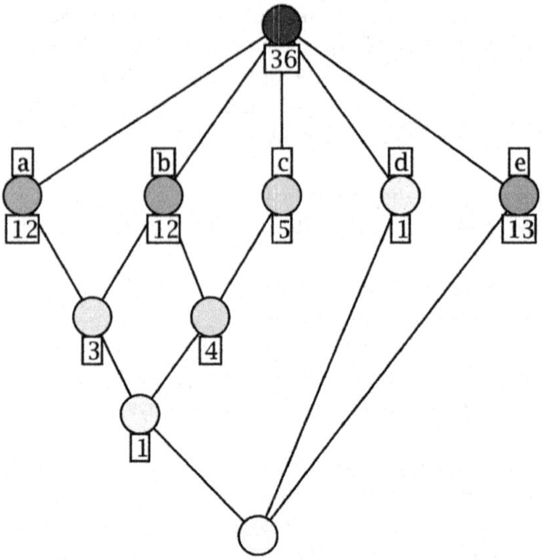

Fig. 4. Concept lattice for the source page after Guttman's algorithm has been applied to obtain the initial distribution. The letters above the nodes indicate which attributes are introduced at that concept. When an attribute is introduced at concept x all concepts connected below concept x in the diagram also have that attribute. The numbers below the nodes indicate how many keys match that concept or any connected below it. For example, there is one key with attributes $\{d\}$, four keys with at least $\{b, c\}$ and one with $\{a, b, c\}$.

Original set	compressed	stored bitset	physical size (bits)
$\{1, 2, 3, 4, 5\}$	N	11111000...0	63
$\{1, 2, 3, 4, 5\}$	Y	000101	6
$\{1, 2, 3, ...35, 36\}$	N	111000...001100...0	63
$\{1, 2, 3, ...35, 36\}$	Y	100100	6

Fig. 5. Compression of a bitset representing a linear scale

If the compression used by the GIST is to change then expensive updates to the index would be required. Potentially, every bounding set which is compressed would have to be loaded, decompressed, recompressed and saved again. Thus any implications between two ordinal scales which are to be factored into the GIST compression would need to be asserted by a domain expert.

The storage of a single integer for each ordinal scale in no way changes the mechanics of the GIST. Due to the storage of two integers it may be tempting to think that the tree is in some way more closely related to the B-Tree [9]. This is not the case, the compression is simply an implementation artifact to enable more bounding sets to be stored in internal nodes.

5 Performance Analysis

The benchmark system is an AMD XP running at 2.4GHz with 1Gb of RAM. The database is stored on a single 160Gb 7200RPM PATA disk. The implementations use the PostgreSQL GIST system. Testing was performed on the Covtype database from the UCI dataset [4] and a synthetic formal context generated with the IBM synthetic data generator [22].

Where an implementation includes the `NC` postfix there is no compression of bounding sets. Where an implementation includes the `Comp` postfix in its name it employs compression of bounding sets. Note that the compression is only ever applied to the bounding sets on internal pages, never to leaf nodes.

There are two compression techniques used – a generic compression `Comp` and compression relying on FCA knowledge `FCAComp`.

The `RD` and `RD-Comp` use the standard Guttman `picksplit`. The `RD-Comp` is exactly the same RD-Tree GIST implementation that is distributed with the PostgreSQL database server. As such it serves as an effective baseline to compare other implementations with.

The `Asym-NC` uses the asymmetric page split algorithm from Section 3.1, `Shuf-NC` builds on `Asym-NC` by performing a shuffle after the initial allocation in an attempt to subvert the creation of drastically under full pages. Finally `Shuf-Comp` builds on `Shuf-NC` by including compression of bounding sets. These implementations use an allowed cardinality expansion of 1 (see Section 3.1). To demonstrate gains for the compression outlined in Section 4 the `Shuf-FCAComp` takes advantage of the ordinal nature of the scales used to generate the set elements.

The `GuttFCA` implementations use the standard RD-Tree generalized Guttman distribution followed by FCA to provide an asymmetric page split as detailed in Section 3.2. The "R" postfix indicates that the THeap is recalculated after keys from a concept are moved to the target page. The 30p and 50p postfixes indicate a 30% and 50% target for the number of keys to distribute from the source page respectively. All the `GuttFCA` implementations use compression – either generic compression when the `Comp` postfix is used or FCA specific compression when the `FCAComp` postfix is used.

Two major metrics of interest are the tree depth and the number of internal nodes in the GIST. It is highly unlikely that an entire GIST will be resident in RAM. Normally some of the tree can be cached in RAM for a successive search. By minimizing the number of internal nodes and the overall tree depth we can increase the chances that successive searches can find an internal node in the RAM cache.

5.1 Performance on Synthetic Data

The following use synthetic data generated with the IBM synthetic data generator [22]. Parameters include the number of transactions (ntrans), the transaction length (tlen), length of each pattern (patlen), number of patterns (npat) and number of items (nitems). The parameters were as follows: ntrans=1,000,000, nitems=1000, tlen=32, patlen=7, npats=10000.

Index	tree depth	index size (Mb)	leaf node count	internal node count	mean leaf free
rd16					
RD-Comp	3	63.5	8033	95	3950
Shuf-Comp	3	64.9	8233	77	4051
GuttFCA-30p-Comp	3	63.9	8097	80	3982
GuttFCAR-30p-Comp	3	64.1	8118	81	3993
GuttFCA-50p-Comp	3	64.5	8175	77	4021
GuttFCAR-50p-Comp	3	64.5	8181	79	4025
rd32					
RD-Comp	4	67.1	8462	131	3749
Shuf-Comp	4	69.5	8789	103	3910
GuttFCA-30p-Comp	4	68.2	8620	107	3827
GuttFCAR-30p-Comp	4	68.8	8698	104	3864
GuttFCA-50p-Comp	3	69.5	8797	100	3913
GuttFCAR-50p-Comp	3	69.7	8820	99	3924
rd1024					
RD-Comp	5	81.5	9739	697	2804
Shuf-Comp	5	142.5	17577	669	5117
GuttFCA-30p-Comp	5	83.5	9982	710	2928
GuttFCAR-30p-Comp	5	84.8	10169	679	3020
GuttFCA-50p-Comp	5	91.4	11003	695	3384
GuttFCAR-50p-Comp	5	95.2	11599	582	3600

Fig. 6. Overall statistics for various GIST implementations on the IBM data mining synthetic database

The output of the IBM synthetic data generator is a list of *ntrans* transactions. Each transaction contains a number of items. Each item is represented by a unique integer in the range $\{1, ..., nitems\}$. Transactions were imported into an int array field in a PostgreSQL table including only the first n items for $n \in \{16, 32, 1024\}$. Such an arrangement allows an RD-Tree to easily be created on the input data of varying dimensionality.

The static index structure is presented in Fig. 6. Gains can be seen for both the 32 and 64 dimensional RD-Tree `picksplit` customizations. For the 1,024 dimensional data the standard Guttman `picksplit` remains superior. Notice that for the `rd32` the `GuttFCA-30p-Comp` is 1.6% larger on disk with about 2% more leaf nodes though has only 82% the number of internal nodes when compared to `RD-Comp`.

A series of subset queries was posed to the database for one, two and three attributes to test the effectiveness of the index structure. Given that the three indexes contain 16, 32 and 1,024 attributes the single attribute queries were for each of the attributes that the index contains. As nitems=1,000 there are $C_2^{1,000} = 499,500$ combinations of two attribute queries possible. As such, for each of the n-attribute queries and index on the first r attributes, only $r-n$ queries are posed. These are for $i \in \{1, 2, ..., r-n\}$ for the attributes $\{a_i, ..., a_{i+n}\}$. As the input data

Index	in1	lf1	in2	lf2	in3	lf3
rd16						
`RD-Comp`	6147	1851	4515	3362	3772	123
`Shuf-Comp`	2013	597	1041	88	712	31
`GuttFCA-30p-Comp`	2587	745	1197	116	869	43
`GuttFCAR-30p-Comp`	2550	750	1101	117	670	35
`GuttFCA-50p-Comp`	1634	554	789	76	586	22
`GuttFCAR-50p-Comp`	1677	548	766	67	525	20
rd32						
`RD-Comp`	6584	1926	5045	455	4129	168
`Shuf-Comp`	3868	905	1658	188	1044	60
`GuttFCA-30p-Comp`	3887	1051	2000	214	1375	77
`GuttFCAR-30p-Comp`	3503	1002	1846	202	1282	69
`GuttFCA-50p-Comp`	2822	820	1413	158	921	51
`GuttFCAR-50p-Comp`	2543	761	1242	137	819	43
rd1024						
`RD-Comp`	8933	4622	7586	2156	6451	1083
`Shuf-Comp`	8387	2578	4229	889	2432	384
`GuttFCA-30p-Comp`	9507	4580	8409	2206	7425	1156
`GuttFCAR-30p-Comp`	9218	4197	7703	1894	6394	947
`GuttFCA-50p-Comp`	9382	4097	7529	1862	6234	958
`GuttFCAR-50p-Comp`	8619	3457	6209	1383	4666	652

Fig. 7. Average number of internal and leaf keys touched for single, two and three attribute queries on various GIST implementations. Note that i3 is the internal mean and l3 is the leaf mean. Internal counts are exact, leaf counts are expressed as figures rounded to the nearest hundred. ie. a leaf count in the table of n is for a reading of $n \times 100$ leaf keys.

should not be biased toward such queries they should when averaged be fairly representative of any such n-attribute query against the data. Shown in Fig. 7 are the mean number of internal and leaf keys touched while performing these queries.

Note that the FCA `picksplit` both yield superior results for the 16 and 32 attribute indexes. In particular the `GuttFCA-50p-Comp` index touches less than 16% the number of internal keys than `RD-Comp` for 3 attribute queries against the 16 attribute index.

5.2 Performance on UCI Covtype Dataset

This section examines the implementations in the setting of the application of FCA on a large data source. This selected application is particularly difficult due to it having 512 formal attributes as well as each formal object having a relatively large number of attributes.

The UCI covtype database consists of 581,012 tuples (formal objects) with 54 columns of data (many-valued attributes). For this paper two numeric columns were used: the aspect and elevation. The formal attributes, each of which forms a binary relation with the tuples were created for the most frequent 256 values for

Index	tree	index	leaf	internal	mean
	depth	size	node	node	leaf
		(Mb)	count	count	free
		(Mb)	count	count	free
RD-NC	17	66.1	3883	4582	5247
Asym-NC	14	52.2	3581	3100	4799
Shuf-NC	12	47.1	3353	2680	4539
RD-Comp	7	37.7	3846	977	4946
Shuf-Comp	7	33.1	3321	915	4565
GuttFCA-30p-Comp	6	34.3	3652	741	4681
GuttFCA-50p-Comp	6	34.3	3648	739	4677
GuttFCAR-30p-Comp	6	34.3	3652	741	4681
GuttFCAR-50p-Comp	6	34.3	3648	739	4677
Shuf-FCAComp	3	23.4	2912	35	3651
GuttFCA-30p-FCAComp	3	25.3	3207	35	4056
GuttFCA-50p-FCAComp	3	25.3	3207	35	4056
GuttFCAR-30p-FCAComp	3	25.3	3207	35	4056
GuttFCAR-50p-FCAComp	3	25.3	3207	35	4056

Fig. 8. Overall statistics for various GIST implementations on the scaled UCI covtype database mediumscaledcov

both aspect and elevation resulting in 512 binary dimensions. An example formal attribute would be created from a predicate like "elevation $< 3,144$". A smaller table called mediumscaledcov was created with only the first 10,000 tuples.

The static structure of the produced index for various GIST implementations is shown in Fig. 8. As seen in Fig. 8 using an asymmetric page split can reduce the number of internal nodes in the tree by over 30% (Comparing `Asym-NC` with `RD-NC`). Notice that `GuttFCA-50p-Comp` contains 76% and 81% the number of internal nodes when compared to `RD-Comp` and `Shuf-Comp` respectively.

Although the mean leaf fill goes down for the `Shuf-Comp` tree compared with the `RD-Comp`, there are also fewer leaf nodes in all. Considering the statistic of: leaf node count $\times$ mean leaf fill, the `Shuf-Comp` tree has an overall reduction of over 20% compared with `RD-Comp`. As to be expected the custom compression in `Shuf-FCAComp` significantly reduces the number of internal nodes in the tree. By allowing more bounding sets to be stored per internal node the tree itself is reduced in depth. Note that compressing the bounding sets for internal nodes has a significant effect on reducing the tree depth (compare `RD-Comp` and `RD-NC`).

Queries for the extent size of 32 single attributes were executed against each index. The 32 attributes were selected for query x as the $16x^{th}$ formal attribute. Two attribute queries were formed using the 25, 248 and 293rd attributes in combination with every 16th attribute. The results are shown in Fig. 9. It can be seen that using asymmetric page splits and post distribution shuffling lowers the number of internal keys touched. The higher branching factor of the `FCAComp` GIST does require more leaf keys to be touched on average.

Index	single attribute **internal**	single attribute **leaf**	two attribute **internal**	two attribute **leaf**
`RD-NC`	5875	5184	3908	3795
`Asym-NC`	4800	5188	3139	3841
`Shuffle-NC`	4401	5218	2946	3815
`RD-Comp`	3491	5260	2362	3885
`Shuf-Comp`	3107	5249	2083	3873
`GuttFCA-30p-Comp`	3171	5194	2014	3881
`GuttFCA-50p-Comp`	3167	5194	2010	3881
`GuttFCAR-30p-Comp`	3171	5194	2014	3881
`GuttFCAR-50p-Comp`	3167	5194	2010	3881

Fig. 9. Mean number of keys touched for single and double attribute queries

The two attribute queries saw a reduction in the number of internal keys touched with no significant changes in leaf keys touched. Of particular note, the `GuttFCA-50p-Comp` GIST only touched 85% the number of internal keys when compared with `RD-Comp`.

6 Conclusion

This paper explored customizations to the RD-Tree spatial data structure. Testing was performed both in the setting for direct application to improve FCA queries and generic multidimensional data. It has been demonstrated that using an asymmetric distribution during page splitting in a GIST based on the RD-Tree can improve the resulting index structure and thus query resolution time.

The application of FCA as a post process to the standard Guttman distribution at page splitting time can help to improve the clustering in the index structure itself.

The custom GIST presented offers at times a 24% drop in internal node count and in many cases a reduction of the depth of the tree over previous work [19]. These two metrics directly impact the query performance of such a tree index [9]. It should be noted that in many typical applications of FCA many hundred queries are rendered against the index [19] further compounding the above performance advantages.

Further gains can be achieved through the application of compression tailored to take advantage of knowledge of how FCA is being applied to the input data. This can significantly reduce the size of the index structure. In particular the number of internal nodes can be reduced to just 4% of that required by generic compression. This is reflected directly in the depth of the GIST going from 7 to just 3 levels.

This ability to resolve queries in a more timely manner enables bolder applications of FCA. For example, application to data sets the size of a filesystem becomes possible. Due to the indexing structure's ability to more efficiently handle high dimensional data FCA can also be applied with a wider scope. For example, the ability to consider more attributes simultaneously than was previously tractable.

Though the work on applying FCA to obtain superior clustering post page split is still in its infancy this paper has demonstrated superior results for indexes with 16 and 32 dimensions.

References

1. libferris, http://witme.sourceforge.net/libferris.web/. Visited Nov 2005.
2. Mail-sleuth homepage, http://www.mail-sleuth.com/. Visited Jan 2005.
3. Paul M. Aoki. Implementation of extended indexes in POSTGRES. *SIGIR Forum*, 25(1):2–9, 1991.
4. Blake, C., Merz, C. UCI Repository of Machine Learning Databases. [`http://www.ics.uci.edu/~mlearn/MLRepository.html`]. Irvine, CA: University of California, Department of Information and Computer Science, 1998.
5. Richard Cole and Peter Eklund. Browsing semi-structured web texts using formal concept analysis. In *Proceedings 9th International Conference on Conceptual Structures*, number 2120 in LNAI, pages 319–332. Springer Verlag, 2001.
6. Richard Cole and Gerd Stumme. Cem: A conceptual email manager. In *7th International Conference on Conceptual Structures, ICCS'2000*. Springer Verlag, 2000.
7. Sebastien Ferré and Olivier Ridoux. A file system based on concept analysis. In *Computational Logic*, pages 1033–1047, 2000.
8. Sebastien Ferré and Olivier Ridoux. A logical generalization of formal concept analysis. In Guy Mineau and Bernhard Ganter, editors, *International Conference on Conceptual Structures*, August 2000.
9. Michael J. Folk and Bill Zoelick. *File Structures*. Addison-Wesley, Reading, Massachusetts 01867, 1992.
10. Bernhard Ganter and Rudolf Wille. *Formal Concept Analysis — Mathematical Foundations*. Springer–Verlag, Berlin Heidelberg, 1999.
11. David K. Gifford, Pierre Jouvelot, Mark A. Sheldon, and James W. Jr O'Toole. Semantic file systems. In *Proceedings of 13th ACM Symposium on Operating Systems Principles*, ACM SIGOPS, pages 16–25, 1991.
12. Bart Goethals and Mohammed Javeed Zaki. Advances in frequent itemset mining implementations: Report on fimi'03. In Bart Goethals and Mohammed Javeed Zaki, editors, *Proceedings of the ICDM 2003 Workshop on Frequent Itemset Mining Implementations*, volume 90 of *CEUR Workshop Proceedings*, 2003.
13. Antonin Guttman. R-trees: A dynamic index structure for spatial searching. In *Proc. ACM-SIGMOD International Conference on Management of Data*, Boston Mass, 1984.
14. Joseph M. Hellerstein, Jeffrey F. Naughton, and Avi Pfeffer. Generalized search trees for database systems. In Umeshwar Dayal, Peter M. D. Gray, and Shojiro Nishio, editors, *Proc. 21st Int. Conf. Very Large Data Bases, VLDB*, pages 562–573. Morgan Kaufmann, 11–15 1995.
15. Joseph M. Hellerstein and Avi Pfeffer. The RD-Tree: An Index Structure for Sets, Technical Report 1252. University of Wisconsin at Madison, October 1994.
16. Ben Martin. Formal concept analysis and semantic file systems. In Peter W. Eklund, editor, *Concept Lattices, Second International Conference on Formal Concept Analysis, ICFCA 2004, Sydney, Australia, Proceedings*, volume 2961 of *Lecture Notes in Computer Science*, pages 88–95. Springer, 2004.

17. Ben Martin and Peter Eklund. Applying formal concept analysis to semantic file systems leveraging wordnet. In *Australian Document Computing Symposium (ADCS05)*. Sydney University, 2005.
18. Ben Martin and Peter W. Eklund. Asymmetric page split generalized index search trees for formal concept analysis. In Floriana Esposito, Zbigniew W. Ras, Donato Malerba, and Giovanni Semeraro, editors, *ISMIS*, volume 4203 of *Lecture Notes in Computer Science*, pages 218–227. Springer, 2006.
19. Ben Martin and Peter W. Eklund. Spatial indexing for scalability in fca. In Rokia Missaoui and Jürg Schmid, editors, *ICFCA*, volume 3874 of *Lecture Notes in Computer Science*, pages 205–220. Springer, 2006.
20. Yoann Padioleau and Olivier Ridoux. A logic file system. In *USENIX 2003 Annual Technical Conference*, pages 99–112, 2003.
21. Susanne Prediger. Logical scaling in formal concept analysis. In *International Conference on Conceptual Structures*, pages 332–341. Springer, 1997.
22. R. Agrawal, H. Mannila, R Srikant, H. Toivonen and A. Inkeri Verkamo. Fast discovery of association rules. In U. Fayyad et al., editor, *Advances in Knowledge Discovery and Data Mining*, pages 307–328, Menlo Park CA, 1996. AAAI Press.
23. T. Rock and R. Wille. Ein toscana-erkundungssystem zur literatursuche. In G. Stumme and R. Wille, editors, *Begriffliche Wissensverarbeitung, Methoden und Anwendungen*, pages 239–253, Berlin-Heidelberg, 2000. Springer-Verlag.
24. Gerd Stumme, Rafik Taouil, Yves Bastide, Nicolas Pasquier, and Lotfi Lakhal. Computing iceberg concept lattices with titanic. In *J. on Knowledge and Data Engineering (KDE)*, volume 42, pages 189–222, 2002.

The Efficient Computation of Complete and Concise Substring Scales with Suffix Trees

Sébastien Ferré

Irisa/Université de Rennes 1
Campus de Beaulieu, 35042 Rennes cedex, France
ferre@irisa.fr

Abstract. Strings are an important part of most real application multi-valued contexts. Their conceptual treatment requires the definition of *substring scales*, i.e., sets of relevant substrings, so as to form informative concepts. However these scales are either defined by hand, or derived in a context-unaware manner (e.g., all words occuring in string values). We present an efficient algorithm based on suffix trees that produces complete and concise substring scales. Completeness ensures that every possible concept is formed, like when considering the scale of all substrings. Conciseness ensures the number of scale attributes (substrings) is less than the cumulated size of all string values. This algorithm is integrated in Camelis, and illustrated on the set of all ICCS paper titles.

1 Introduction

In information systems, one of the most common datatype is the *string*. For instance, in a bibliographic application, most attributes are string-valued (author names, title, journal or conference name). While these strings usually bring a lot of information, they are hardly exploited in conceptual information systems based on Formal Concept Analysis (FCA) [GW99]. They are most often represented as (1) nominal values, which is right for entry types (e.g., "journal", "inproceedings") but uninteresting for titles, (2) a set of keywords given by hand [CS00], or (3) a set of keywords derived in a context-unaware manner, e.g., all title words [FR01].

An important objective of conceptual information systems is to ensure a tight combination of querying and navigation [GMA93]. In this respect, the manual or context-unaware production of keywords is unsatisfactory because they are fully part of the navigation structure, and so should be automatically derived from the context, like the concept lattice. We consider in this paper the automatic derivation of *substring scales*, whose values are full strings (like titles), and whose attributes are substrings (corresponding to keywords). For instance, in the case of the bibliographic context of all ICCS papers, one would expect to have substrings like "Formal Concept Analysis", "Conceptual Graphs". These substrings play the same role as inequalities and intervals over numeric values (*ordinal* and *interordinal* scales [GW99]), or general terms in taxonomies.

S.O. Kuznetsov and S. Schmidt (Eds.): ICFCA 2007, LNAI 4390, pp. 98–113, 2007.

A substring scale should be *complete* in the sense that every possible concept is derived from the scaled context, like when considering all substrings. A substring scale should also be *concise* enough so as not to overwhelm users during navigation, and be computed *efficiently*.

In Section 2, we present a naive conceptual scaling, and show that it does not satisfy conciseness and efficiency. In Section 3, we introduce a new solution, and show with the help of suffix trees that it has good properties w.r.t. completeness, conciseness and efficiency. Section 4 describes an algorithm for computing a complete substring scale from a set of string values. This algorithm is incremental, and so supports context updates, as required in information systems. It has been integrated into CAMELIS, an implementation of Logical Information Systems (LIS) [FR04], and Section 5 shows its application to a bibliographic context of ICCS paper titles, how many domain keywords are clearly identified, and how they naturally form a taxonomy. This paper ends with a discussion about other datatypes (Section 6), and a conclusion (Section 7).

2 Naive Approach

Suppose we have n objects, each object being decribed by a string over an alphabet Σ. This forms a *string context.*

Definition 1 (string context). *A* string context *is a triple* $D = (\mathcal{O}, \Sigma^*, d)$, *where* $\mathcal{O}$ *is a finite set of objects,* Σ^* *is the domain of strings over a finite alphabet* Σ, *and* d *is a mapping from objects to* Σ*-strings: for every object* $o \in \mathcal{O}$, $d(o) \in \Sigma^*$ *is the description of the object by a string.*

A string context can be seen as a multivalued context with only one attribute, $d(o)$ being the value of this attribute for the object o. All results in this paper also apply to contexts with several attributes, but it is not necessary to consider them explicitly here as each attribute can be treated in isolation.

Example 1. *The following table shows a basic string context that serves as an example in the following.*

o	$d(o)$
1	*abc*
2	*dab*
3	*ac*
4	*dab*

The cover of a substring is the set of objects whose description contains it in a string context. This is equivalent to the definition of extent in logical concept analysis [FR04], where formulas would be strings and substrings.

Definition 2 (cover). *Let* $D = (\mathcal{O}, \Sigma^*, d)$ *be a string context. The* cover *of a string* $s \in \Sigma^*$ *in* D *is defined by (where* $\supseteq$ *denotes the containment relation between strings and substrings)*

$$cover_D(s) = \{o \in \mathcal{O} \mid d(o) \supseteq s\}.$$

For example, in the above string context, the cover of the string "ab" is the set of object $\{1, 2, 4\}$.

We want to apply concept analysis on a string context, in order to group objects in a concept when they share common substrings in their description. There is *a priori* no way to prefer some substrings, so that we define the conceptual scale of all substrings of a string context, which accounts for the subsumption relations that exist between strings and substrings.

Definition 3 (scale of all substrings). *Let $D = (\mathcal{O}, \Sigma^*, d)$ be a string context. The set of all substrings of the string context D is defined as*

$$S(D) = \{s \in \Sigma^* \mid cover_D(s) \neq \emptyset\},$$

and the related conceptual scale is defined by $\mathbb{S}(D) = (d(\mathcal{O}), S(D), \supseteq)$.

Example 2. *The scale of all substrings derived from the string context in Example 1 is given by the following table.*

	abc	*ab*	*bc*	*a*	*b*	*c*		*dab*	*da*	*d*	*ac*
abc	×	×	×	×	×	×	×				
dab		×		×	×		×	×	×	×	
ac				×		×	×				×

A string context and its derived substring scale can be combined in order to form a scaled formal context, from which the concept lattice can ultimately be built.

Definition 4 (scaled context). *Let $D = (\mathcal{O}, \Sigma^*, d)$ be a string context. The scaled context that is derived from D is defined by $K(D) = (\mathcal{O}, S(D), I)$, where*

$$(o, s) \in I \iff d(o) \supseteq s.$$

Example 3. *The scaled context derived from the above string context and substring scale is given by the following table.*

	abc	*ab*	*bc*	*a*	*b*	*c*		*dab*	*da*	*d*	*ac*
1	×	×	×	×	×	×	×				
2		×		×	×		×	×	×	×	
3				×		×	×				×
4		×		×	×		×	×	×	×	

There are now 3 properties we want to consider in the evaluation of this naive approach:

- completeness: *Is every set of objects sharing a common set of substrings a formal concept of $K(D)$?*
- conciseness: *Is the set of all substrings $S(D)$ concise enough so as to be useful for navigation ?*
- efficiency: *Can the formal context $K(D)$ be computed efficiently ?*

Completeness is here trivially ensured because all substrings occuring in the string context are considered. All other strings label only the bottom concept, and can be ignored without practical consequence. Conciseness can be evaluated as the size of the scale $S(D)$. Given a string context made of n strings of length up to k, the number of substrings is in $O(k^2n)$. Efficiency can be evaluated as the cost of computing the scaled context. Each substring must be produced, and checked against already produced substrings. With the help of a lexical tree, this check can be made in $O(k)$, so that the scaled context can be computed in $O(k^3n)$.

To get an idea of these complexities, suppose we have 1000 strings of length up to 100 characters (e.g., a set of paper titles), the number of substrings, and hence the number of attributes in the scale context, would be up to 10^7. This is an awful lot, and in most cases many substrings will be redundant: e.g., the substring "ormal Contex" generally has the same cover as "Formal Context".

3 Maximal Substrings and Suffix Trees

Our objective is to reduce the number of attributes in the scaled context, while retaining all the information, i.e., while deriving the same extents and equivalent intents. This is possible because generally many substrings label the same concept (they have the same extent), i.e., they occur in exactly the same strings. The most concise solution consists in retaining one substring label for each meet-irreducible concept, but this entails a loss of information in intents and an arbitrary choice among substring labels.

3.1 Maximal Substrings

The idea is to retain only the more informative substrings, that is the substrings that cannot be extended (by adding characters at the left or at the right) without reducing their cover. This is equivalent to retaining *maximal* substring labels on each concept. We define a new scale of maximal substrings.

Definition 5 (scale of maximal substrings). *Let $D = (\mathcal{O}, \Sigma^*, d)$ be a string context. The set of maximal substrings of the string context D is defined as*

$$S_{max}(D) = \{s \in S(D) \mid \forall t \in S(D) : t \supset s \Rightarrow cover_D(t) \supsetneq cover_D(s)\},$$

and the related conceptual scale is defined by $\mathbb{S}_{max}(D) = (d(\mathcal{O}), S_{max}(D), \supseteq)$, where $\supseteq$ denotes the containment relation between strings and substrings.

Example 4. *The scale of maximal substrings derived from the string context in Example 1 is given by the following table. Compared to the scale of all substrings given in Example 2, the substring "b" has disappeared because it has the same cover as "ab". The same happens for the substrings "bc", "da", "d" and the empty string "".*

	abc	*ab*	*a*	*c*	*dab*	*ac*
abc	×	×	×	×		
dab		×	×		×	
ac			×	×		×

This scale can be combined with the string context from which it is derived in order to form a scaled context.

Definition 6 (scaled context with maximal substrings)
Let $D = (\mathcal{O}, \Sigma^*, d)$ be a string context. The scaled context that is derived from D is defined by $K_{max}(D) = (\mathcal{O}, S_{max}(D), I)$, where $(o, s) \in I \iff d(o) \supseteq s$.

Example 5. *The scaled context derived from the above string context and substring scale is given by the following table.*

	abc	*ab*	*a*	*c*	*dab*	*ac*
1	×	×	×	×		
2		×	×		×	
3			×	×		×
4		×	×		×	

This scaled context is just a projection of the scale context in Section 2 over the set of substrings $S_{max}(D)$.

3.2 Completeness

An important question is: *Did we loose something by retaining maximal substrings only?* More precisely, we have to show that the concept lattice has the same extents, and equivalent intents. We first prove that every substring has a maximal substring extension with the same cover.

Lemma 1. *Let D be a string context. For all substring $s \in S(D)$, there exists a maximal substring $m \in S_{max}(D)$ such that m is an extension of s, $m \supseteq s$, and has the same cover, $cover_D(m) = cover_D(s)$.*

Proof: We prove by recurence that the lemma is true for every length of s (denoted by $|s|$), starting with the longest (string length is bounded by k), and decreasing it.

1. Base case: $|s| = k$.
   $\nexists t \in S(D) : t \supset s \Longrightarrow s \in S_{max}(D)$.
2. General case: the lemma is assumed true for every substring longer than s.

   – either $\forall t \in S(D) : t \supset s \Rightarrow cover_D(t) \neq cover_D(s)$
   $\Longrightarrow s \in S_{max}(D)$,
   – or $\exists t \in S(D) : t \supset s \wedge cover_D(t) = cover_D(s)$
   $\Longrightarrow |t| > |s|$
   $\Longrightarrow \exists m \in S_{max}(D) : m \supseteq t \wedge cover_D(m) = cover_D(t)$
   $\Longrightarrow \exists m \in S_{max}(D) : m \supseteq s \wedge cover_D(m) = cover_D(s)$. ∎

It follows immediately that every non-maximal substring attribute can be replaced by a maximal substring that contains it, hence no loss of information, and that has the same cover, hence discriminating the same extents.

Theorem 1. *The concept lattices of $K(D)$ and $K_{max}(D)$ are equivalent for conceptual navigation. They have the same extents, and intents are equivalent in the sense that missing substrings in the latter case are redudant, i.e., included in some maximal substring from the same intent.*

For instance, the substring “ormal Contex” is replaced by “Formal Context”, but “al Context” is maximal if it can be extended by either “Formal Context” or “Logical Context” in different strings.

3.3 Conciseness and Efficiency with Suffix Trees

In order to evaluate the improvement of using maximal substrings, we have to bound their number, and compare it with the set of all substrings. We also have to compare the computation complexity for building the scaled context, because a smaller context does not entails necessarily a more efficient computation.

It seems difficult to estimate precise complexities, given the definition of $S_{max}(D)$. However there exists a very interesting data structure for reasoning and computing with sets of strings, namely *suffix trees* [Ukk95, Gus97]. The suffix tree of a string is the compact lexical tree of all suffixes of this string, where *compact* means that branches may be labelled by several characters, and nodes have several children. Figure 1 displays the suffix tree of the string "googol$", where $ is a special final character that is necessary in the computation of the suffix tree. Each leaf represents a suffix, whose position in the string labels the leaf (starting with position 0). Each node represents a repeated substring, whose occuring positions are the leaf labels below this node. For instance, we can read in Figure 1 that the word "go" occurs at positions 0 and 3 in the string "googol$". In addition suffix links point from a node n (representing a substring s) to another node n' (representing the first suffix of s, i.e., s minus its first character). For instance, the suffix link in Figure 1 (dashed arrow) goes from the node "go" to the node "o".

However we are here interested in finding maximal substrings over several strings, and so in building the suffix tree of a set of strings. This is an easy generalisation of suffix trees [Gus97]. It suffices to end each string with a different special character, and to concatenate them. A difference is that each leaf must be labelled by a string in addition to a position so as to determine to which string a suffix belongs to. Figure 2 displays the generalized suffix tree of the set of strings in Example 1. In leaf labels, the first number identifies the string, and the second number gives the suffix position in this string. The sets labelling nodes are the cover of the substrings they represent; and black nodes correspond to maximal substrings. The computation of these 2 informations is explained in Section 4.2.

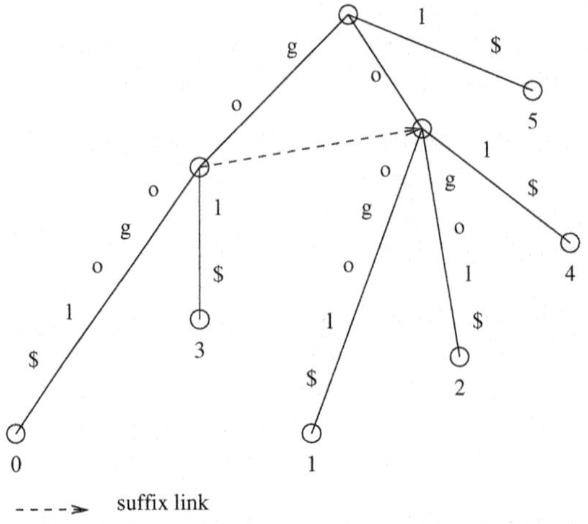

Fig. 1. The suffix tree of the string "googol$"

A known result states that every *maximal repeat* is represented by a node in the suffix tree (Lemma 7.12.1 in [Gus97]). These maximal repeats differ from our maximal substrings in 2 ways:

1. a maximal repeat may occur in a single string at different positions, and in this case cover a single object: these repeats are not maximal because they can be replaced by the full string that contains them (same cover),
2. full strings (from the string context) may not be repeated, but are obviously maximal substrings.

Proposition 1 (maximal substrings in suffix trees). *The set of maximal substrings forms a subset of the suffix tree nodes plus full strings.*

A fundamental result of suffix trees is that the number of nodes (hence the number of maximal repeats) and the computation time are both in $O(\Sigma_{o \in \mathcal{O}} |d(o)|)$, i.e., the cumulated size of all strings. This can be approximated as $O(kn)$, if k is the maximum length of strings, and n the number of strings.

Proposition 2 (number of maximal substrings). *The number of maximal substrings is in* $O(kn)$, *i.e., a* k*-fold improvement compared to the naive scaling.*

In the example of paper titles, this is about 2 orders of magnitude better. The same factor is obtained when computing the concept lattice, or even squared for algorithms that are quadratic in the number of attributes. About the complexity of computing the scaled context with maximal substrings, we need to take into account the selection of maximal substrings among suffix tree nodes, and the computation of covers. We describe an algorithm in next section, and show its complexity is in $O(kn.ln(kn))$, thus adding only an logarithmic factor to the computation of the suffix tree.

4 An Efficient and Incremental Algorithm Based on Suffix Trees

As said in previous section, the maximal substrings of a string context is made of full strings and a subset of the nodes of the suffix tree of all these full strings. The first step is then to compute this suffix tree. We sketch in Section 4.1 the well known Ukkonen's algorithm, which can build such a suffix tree in an incremental manner, and in linear time. In Section 4.2 we refine this algorithm in order to determine which nodes of the suffix tree represent maximal substrings. This requires the computation of substring covers, which is a useful information *per se* in information systems (e.g., for computing answers to queries). The complexity of this refined algorithm is given, as well as practical details on its implementation and integration with the existing logical information system, CAMELIS.

4.1 Ukkonen's Algorithm for Computing Suffix Trees

The rough principle of Ukkonen's algorithm is to have 3 gliding positions on a string s (having length n): the position j of the suffix $s[j,n]$ being added[1], the position $i \geq j$ of the next character to be read, and the position $j \leq p \leq i$ that matches the last node on the path $s[j,i]$ in the suffix tree. Initially i and j are set to 0, and p is the root. While the character $s[i]$ can be read down the tree, position i is incremented and p is updated accordingly. Otherwise a leaf labelled by j is added to the node nd at the end of the path $s[j,i-1]$ (nd is created if this path ends in the middle of a branch), position j is incremented, a suffix link is followed from p in order to reach the path $s[j+1,i-1]$, and if nd has been created, a new suffix link is created from nd to the end of the new path. This is repeated until all suffixes have been added. Because position i is always greater or equal than position j, and at least one of these positions is incremented at each step, the suffix tree can be computed in linear time with respect to the length of s. This impressive result is achieved with the help of additional tricks, which are given in detail in the litterature [Gus97, Ukk95].

4.2 Computing Covers and Maximality

We first give a few definitions and results to help navigating in suffix trees. In these definitions we talk equivalently of nodes and substrings. For instance, we can talk of the cover of a node, or we can talk about the node "ab" in Figure 2.

Lemma 2 (cover of a node). *The* cover *of a suffix tree node is the set of string identifiers that label the leaves below this node.*

Definition 7 (right extensions of a node). *The* right extensions *of a node nd are the children nodes of nd. The substring nd is a proper prefix of every right extension.*

[1] The notation $s[a,b]$ denotes the substring of s from position a to position b.

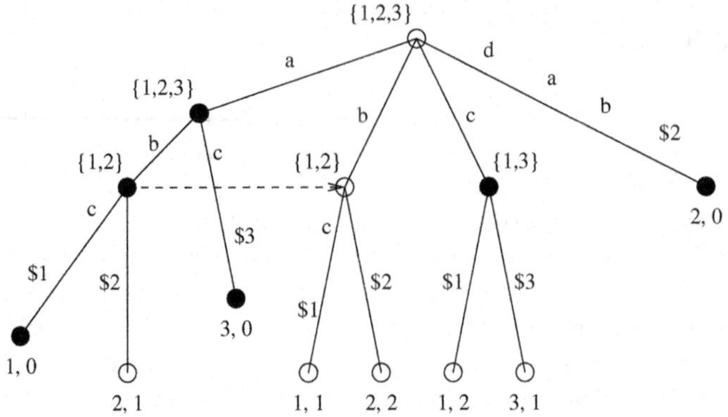

Fig. 2. The generalized suffix tree of the set of strings {"abc", "dab", "ac"}

Definition 8 (left extensions of a node). *The* left extensions *of a node nd are the antecedent nodes of suffix links ending at nd. The substring nd is a proper suffix of every left extension.*

Our objective is to compute the set of maximal substrings, along with their cover. In Section 3.3 it was observed that a maximal substring is either a full string, or a repeated substring such that every other substring that contains it has a smaller cover. Therefore maximal leaves are those labelled with 0 as suffix position (full strings); and maximal nodes are those labelled by a cover strictly greater than any (left or right) extension.

Suppose we have the suffix tree for the $i-1$ first strings of a string context, enriched with covers and maximality, and we want to update the enriched suffix tree with the string i. An important result can be summarized as "once maximal, always maximal".

Lemma 3. *Once a node has been identified as maximal upon the addition of a string, it will remain maximal upon the addition of any other string.*

Proof: We prove the contraposition of this lemma. Suppose a node *nd* is not maximal after inserting the i-th string, and already existed before this insertion. This implies there is an extension nd' of *nd* that has the same cover. Suppose nd' was created during the insertion of string i. This implies that the substring represented by nd' was not a repeated substring, while *nd* was. This means the 2 nodes had different covers, which contradicts the fact they have the same cover after insertion of i. So *nd* and nd' did exist, had the same cover, and nd' was an extension of *nd*. Therefore *nd* was not maximal before the insertion of the string i. ■

Therefore a node becomes maximal when adding string i if it is not yet maximal, covers i, and has no extension covering i. From Lemma 2 the addition of string i in a cover comes from the creation of a leaf labelled by i below this node.

Modification 1. Ukkonen's algorithm is modified so that each time a leaf is added on node *nd*, then the string i is added to the cover of *nd* as well as all its ancestors till the root. There are 2 cases about maximality:

1. if i already belongs to the cover of *nd* (due to the creation of a previous leaf), then *nd* has a right extension that covers i, and so it is not maximal;
2. otherwise i does not belong to any extension of *nd*. First, *nd* is marked as i-maximal, meaning that this node became maximal upon the insertion of string i. Then any ancestor of *nd* that was marked i-maximal is unmarked (this does not contradict Lemma 3 because unmarking occurs at the same stage as marking).

Modification 2. A second modification of Ukkonen's algorithm is necessary. When a suffix link is created from some node nd' to the node *nd*, then a leaf has necessary been added to node nd' (Ukkonen's algorithm), and so i belongs to the cover of nd'. As nd' is a left extension of *nd*, it follows that *nd* must be unmarked but *only if* it has been marked i-maximal.

The complexity of computing the enriched suffix tree is modified compared to Ukkonen's algorithm. This is due to the traversal of the ancestors of a node in the first modification. Given the size of the suffix tree is in $O(kn)$, its height is in $O(ln(kn))$. This traversal is applied for each creation of a leaf, i.e. $O(kn)$ times. Hence the time complexity for computing the enriched suffix tree in $O(kn.ln(kn))$.

4.3 Practical Aspects

The above algorithm, based on Ukkonen's algorithm, has been implemented and integrated in CAMELIS[2], a logical information system [FR04]. CAMELIS makes use of a toolbox of logic components, *logic functors* [FR02], amongst which the functor `String` handles representation and reasoning on strings and substrings. This functor has been extended with suffix tree algorithms so as to compute the maximal substrings in an incremental way. The cumulated complexity of a non-incremental algorithm would not be linear, but quadratic because of repeated computation of maximal substrings upon insertion of new strings. The logic functor also makes it possible to compute the extension of a substring so as to support navigation in CAMELIS, to remove strings from the string context, and to manually hide maximal substrings when judged irrelevant by users (customization). A concrete application example is given in the next section.

5 Example and Application-Specific Improvement

As an illustrative example of our approach, we consider the string context made of the titles of all papers published at ICCS from 1993 to 2005, as they were found on DBLP pages[3]. This string context contains $n = 374$ strings, whose length is bounded by $k = 140$. For string contexts of this size, the worst case number of substring attributes in the scaled context are the following.

[2] `http://www.irisa.fr/lande/ferre/camelis/`

[3] `http://www.informatik.uni-trier.de/~ley/db/conf/iccs/index.html`

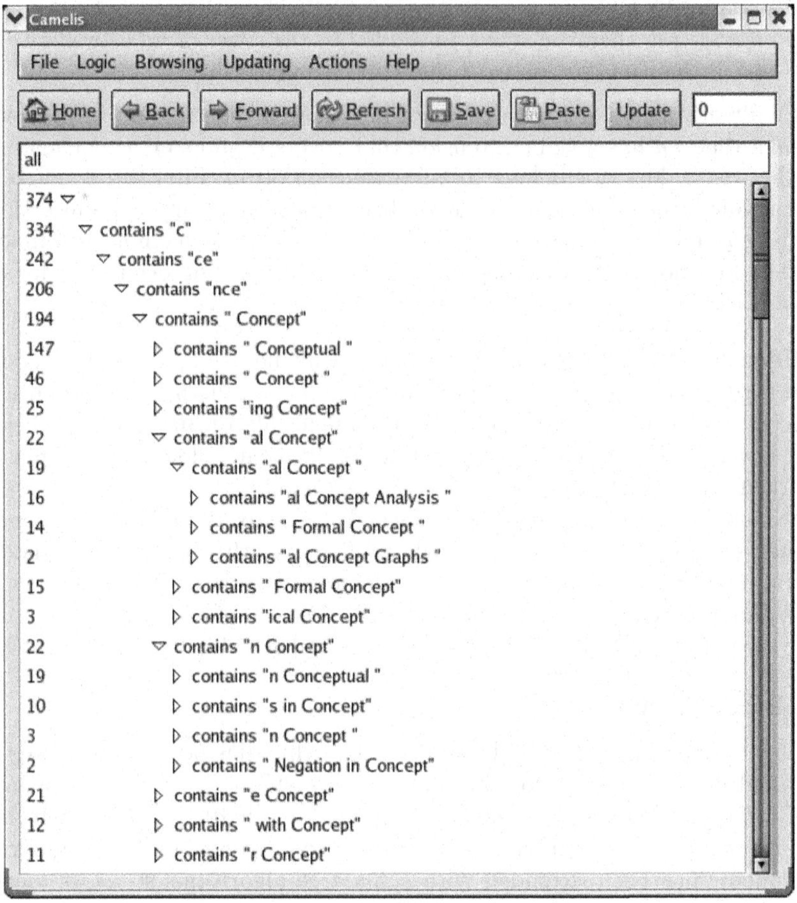

Fig. 3. Navigation tree of maximal substrings in ICCS string context

all substrings (nk^2)	7,330,400
maximal substrings (nk)	52,360

In the case of maximal substrings, the worst case number of substrings is more sharply defined as the cumulated size of all strings, which is 21,412 in the ICCS string context.

We applied our algorithm for computing all maximal substrings of the ICCS string context. The computation time is a few seconds on a standard machine, and the number of maximal substrings is only 3,816. This is to be compared with 569,676 substrings found with the naive approach. This low figure, compared to the worst case, can partly be explained by the homogeneity of the string context, where titles share many common keywords. The size of the scaled context, i.e., the number of crosses, is 44,056; equivalently, objects have on average 117 attributes. This means that each title, whose average length is 57, contains on

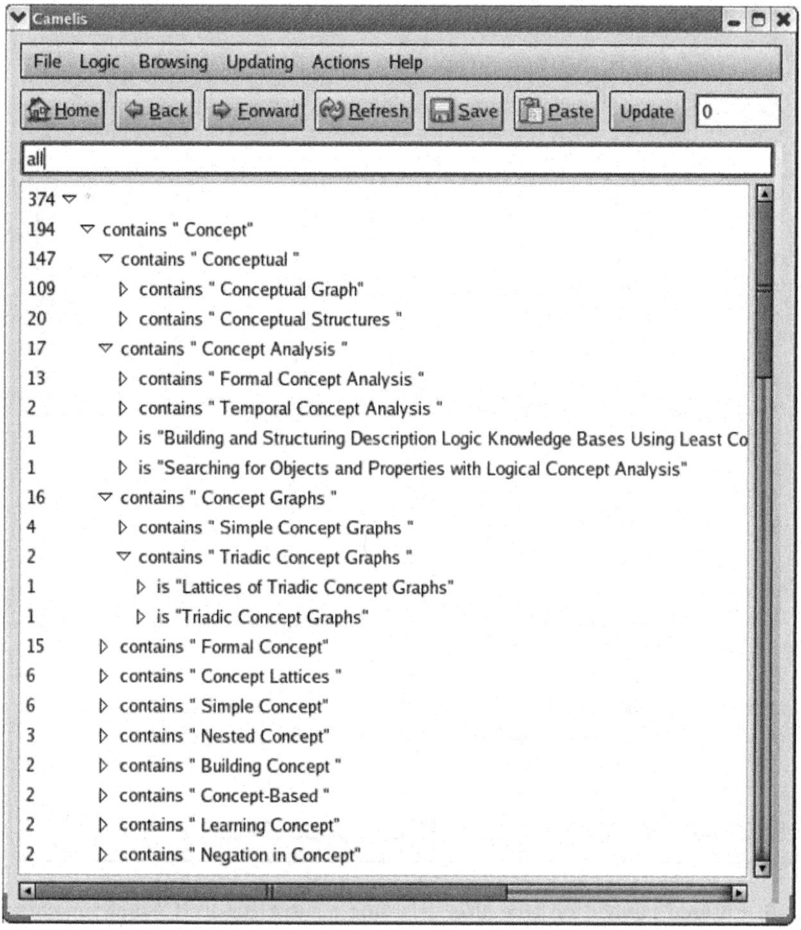

Fig. 4. Filtered navigation tree of maximal substrings in ICCS string context

average 117 maximal substrings. Even if this is small in comparison to $k^2 = 19,600$, this still seems a lot.

Figure 3 shows a navigation tree in CAMELIS made of these maximal substrings, along with their object count. This tree contains several informative substrings, and the tree structure reflects their containment relations[4]: e.g., “Concept”, “Conceptual”, “Concept Analysis”, “Formal Concept”, “Negation in Concept”. But at the same time there are irrelevant substrings, e.g., “c”, “nce”, and redundant substrings, e.g., “al Concept”and “n Concept” w.r.t. “Concept”. The problem here is that the algorithm makes no assumption on the contents

[4] In fact, it is a directed acyclic graph as a substring may be subsumed by several substrings, but it is displayed as a tree in the graphical interface. This implies a substring may occur at different places in the navigation tree.

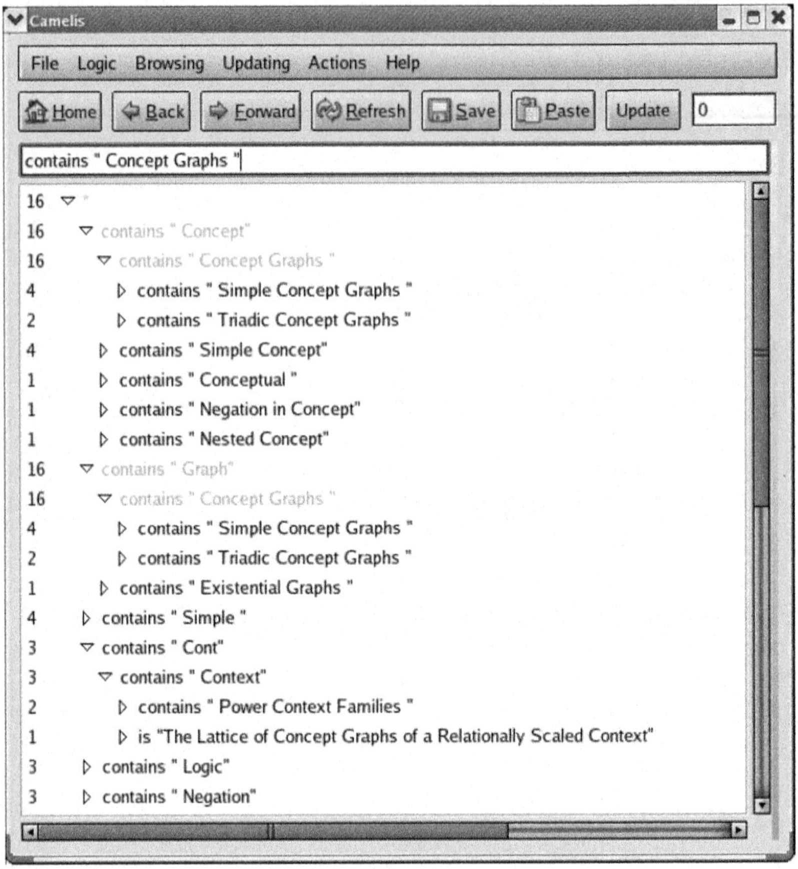

Fig. 5. Filtered navigation tree after selecting a substring in ICCS string context

of strings, and makes no difference between letters and spaces. However, word boundaries are important for the readability and relevance of substrings for users.

We adapted the traversal of the navigation tree so as to allow applications to define a filtering of this tree. The application must determine for each substring which part is relevant. For instance, in the ICCS string context, this part goes from the beginning of the first capitalized word to the end of the last capitalized word, thus neglecting grammatical words at both ends: e.g., the relevant part of "al Concept" is "Concept". Then a substring is filtered out from the navigation tree when its relevant part is equal to the relevant part of its parent node in the tree: e.g., "al Concept" is filtered because it has the same relevant part as " Concept". This filtering entails no change at all in the suffix tree, and all displayed substrings *are* maximal substrings. The consequence is just that some substrings are skipped in the navigation tree, but the containment ordering is kept.

A filtered navigation tree for the ICCS string context is displayed in Figure 4. This time we get a much more readable and informative navigation tree. Note for instance how "Concept" is refined by "Concept Analysis", "Concept Graphs", etc., and how "Concept Analysis" is refined by "Formal Concept Analysis" and "Temporal Concept Analysis" (skipping "al Concept Analysis"). Note also that full strings (prefixed by the keyword `is`) appear as maximal strings, so that full titles can be accessed. After filtering, the number of substrings is only 928 (vs 3,816). If only proper substrings are considered, i.e., if full strings are excluded, this number is a mere 554 substrings. This is the same order of magnitude as the number of objects. The following table summarizes the decreasing of the number of substrings in successives approaches.

all substrings	569,676
maximal substrings	3,816
filtered maximal substrings	928
filtered maximal proper substrings	554

Navigation trees in above figures are dynamic. In conceptual navigation (a.k.a. browsing), they play the same role as attribute lists, and shrink to a subset of relevant substrings each time a substring is selected. Figure 5 displays the shrinked navigation tree after the substring "Concept Graphs" has been selected. Grey-colored substrings are those covering all selected strings: they make up the intent of the current concept. Of course, it is possible to consult the extent of the current concept as a list of titles (or full bibliographical references). Finally, thanks to the logical nature of CAMELIS, it is also possible to use arbitrary substrings in queries, even if they are not maximal.

6 Discussion

Our approach has no pretention w.r.t. Natural Language Processing (NLP). It proves successful on the analysis of a set of paper titles, but it is difficult to say how it would behave on more free text. Our purpose was to provide a solution for analysing strings that is simple and generic (no *a priori* knowledge needed, unlike in NLP), exhaustive (no arbitrary choices), and efficient (for actual use in information systems). However, if linguistic knowledge is available, it can still be used as a preprocessing stage before applying our algorithm: e.g., removal of plurals, replacing words by their root or a dictionnay entry.

The string datatype can be seen as a logic, where formulas are sets of strings and substrings, the deduction relation $\sqsubseteq$ is based on the string containment $\supseteq$, and disjunction $\sqcup$ computes the maximal substrings shared by 2 strings. This kind of logic can be used in the framework of Logical Concept Analysis (LCA) [FR00, FR04], where intents are precisely computed by application of disjunction: $int(O) = \bigsqcup_{o \in O} d(o)$. The set of maximal substrings can thus be computed by applying this disjunction on all subsets of objects in a logical context. A similar approach based on *pattern structures* [GK01] has been applied on graphs and subgraphs for analysing molecules [Kuz99]. These results could in principle be applied on strings, as strings can be represented with graphs. However, the complexity of this approach is totally different because the disjunction operation must be applied in the best case once for each concept. This results in an exponential complexity to be compared with the polynomial complexity of the algorithm presented in Section 4. The equivalent of our approach to graphs (if possible) would be to compute in a polynomial way the set of maximal subgraphs, i.e., all elements of *closed subgraph sets* [GK01], without computing the concepts.

A logic, defined as a language of formulas (representation) partially ordered by a subsumtion relation (reasoning), is definetely valuable for describing objects in a natural way, querying a context in an expressive manner, and organizing navigation features, as demonstrated by our numerous experiments with LIS. However the use of logical disjunction for computing maximal features (e.g., substrings, subgraphs) and concepts is untractable in general. Given the importance of actually computing the concept lattice for many people, this may explain why LCA has not been more widely accepted, while there is an obvious and shared interest in exploiting various and richer datatypes. We hope the results presented in this paper for the string logic will improve its acceptability by the FCA community.

A long-term objective is to design a genuine logical *and* conceptual navigation for all sorts of datatypes. The results presented in this paper are a first step in this direction, and a significant one if we consider that strings cover a large part of many applications. An important way to reduce the problem is the shift from the direct production of the logical concept lattice to the production of maximal features that determine the same lattice. Indeed, producing the full lattice *apriori* is not necessary, as experienced in LIS applications, and it is not manageable in many real applications given its size (computation cost and visualization). It seems sufficient to show neighbour concepts of the current concept when browsing a context. Moreover the availability of maximal features makes it possible to compute the logical concept lattice with regular algorithms.

In LIS applications, logics are formed by the composition of *logic functors* [FR02] corresponding to various datatypes (e.g., strings, integers) and datatype constructors (e.g., sum, tuples, sets). It seems promising to extend logic functors so as to integrate the incremental computation of maximal features. This decomposition allows for highly specialized data structures and algorithms (e.g., suffix trees), and has been applied here for the string datatype.

7 Conclusion

We have defined *scales of maximal substrings* and their computation from multi-valued contexts with string-valued attributes. For each string-valued attribute such a scale is computed from a set of strings with the help of a suffix tree. Scales of maximal substrings are proved complete w.r.t. the formation of concepts. The use of suffix trees enables to bound their size by the cumulated size of strings (kn), and to efficiently compute them ($O(kn.ln(kn))$).

An algorithm is given as an adaptation of Ukkonen's algorithm for computing suffix trees. It has been integrated in logical information systems so as to support logical and conceptual navigation. This is illustrated on the navigation among ICCS papers through the automatic extraction of keywords from titles. We plan to extend these results to other datatypes, like strings with gaps in patterns, XML trees, and graphs, where the challenge is to compute maximal features without having to build the concept lattice, and possibly in a polynomial way like for substrings.

References

[CS00] R. Cole and G. Stumme. CEM - a conceptual email manager. In G. Mineau and B. Ganter, editors, *Int. Conf. Conceptual Structures*, LNCS 1867, pages 438–452. Springer, 2000.

[FR00] S. Ferré and O. Ridoux. A logical generalization of formal concept analysis. In G. Mineau and B. Ganter, editors, *Int. Conf. Conceptual Structures*, LNCS 1867, pages 371–384. Springer, 2000.

[FR01] S. Ferré and O. Ridoux. Searching for objects and properties with logical concept analysis. In H. S. Delugach and G. Stumme, editors, *Int. Conf. Conceptual Structures*, LNCS 2120, pages 187–201. Springer, 2001.

[FR02] S. Ferré and O. Ridoux. A framework for developing embeddable customized logics. In A. Pettorossi, editor, *Int. Work. Logic-based Program Synthesis and Transformation*, LNCS 2372, pages 191–215. Springer, 2002.

[FR04] S. Ferré and O. Ridoux. An introduction to logical information systems. *Information Processing & Management*, 40(3):383–419, 2004.

[GK01] B. Ganter and S. Kuznetsov. Pattern structures and their projections. In H. S. Delugach and G. Stumme, editors, *Int. Conf. Conceptual Structures*, LNCS 2120, pages 129–142. Springer, 2001.

[GMA93] R. Godin, R. Missaoui, and A. April. Experimental comparison of navigation in a Galois lattice with conventional information retrieval methods. *International Journal of Man-Machine Studies*, 38(5):747–767, 1993.

[Gus97] D. Gusfield. *Algorithms on Strings, Trees, and Sequences.* Cambridge University Press, 1997.

[GW99] B. Ganter and R. Wille. *Formal Concept Analysis — Mathematical Foundations.* Springer, 1999.

[Kuz99] S. Kuznetsov. Learning of simple conceptual graphs from positive and negative examples. In J. M. Żytkow and J. Rauch, editors, *Principles of Data Mining and Knowledge Discovery*, LNAI 1704, pages 384–391. Springer, 1999.

[Ukk95] E. Ukkonen. On-line construction of suffix trees. *Algorithmica*, 14(3):249–260, 1995.

A Parameterized Algorithm for Exploring Concept Lattices

Peggy Cellier[1], Sébastien Ferré[1], Olivier Ridoux[1], and Mireille Ducassé[2]

[1] IRISA/University of Rennes 1
[2] IRISA/INSA, Campus universitaire de Beaulieu, 35042 Rennes, France
`firstname.lastname@irisa.fr`
`http://www.irisa.fr/LIS/`

Abstract. Formal Concept Analysis (FCA) is a natural framework for learning from positive and negative examples. Indeed, learning from examples results in sets of frequent concepts whose extent contains only these examples. In terms of association rules, the above learning strategy can be seen as searching the premises of exact rules where the consequence is fixed. In its most classical setting, FCA considers attributes as a non-ordered set. When attributes of the context are ordered, Conceptual Scaling allows the related taxonomy to be taken into account by producing a context completed with all attributes deduced from the taxonomy. The drawback, however, is that concept intents contain redundant information. In this article, we propose a parameterized generalization of a previously proposed algorithm, in order to learn rules in the presence of a taxonomy. The taxonomy is taken into account during the computation so as to remove all redundancies from intents. Simply changing one component, this parameterized algorithm can compute various kinds of concept-based rules. We present instantiations of the parameterized algorithm for learning positive and negative rules.

1 Introduction

Learning from examples is a fruitful approach when it is not possible to a priori design a model. It has been mainly tried for classification purposes [Mit97]. Classes are represented by examples and counter-examples, and a formal model of the classes is learned by a machine.

Formal Concept Analysis (FCA) [GW99] is a natural framework for learning from positive and negative examples [Kuz04]. Indeed, learning from positive examples (respectively negative examples) results in sets of frequent concepts with respect to a minimal support, whose extent contains only positive examples (respectively negative examples). In terms of *association rules* [AIS93, AS94], the above learning strategy can be seen as searching the premises of exact rules where the consequence is fixed. When augmented with statistical indicators like *confidence* and *support* it is possible to extract various kinds of concept-based rules taking into account exceptions [PBTL99, Zak04].

The input of FCA is a formal context that relates objects and attributes. FCA considers attributes as a non-ordered set. When attributes of the context

S.O. Kuznetsov and S. Schmidt (Eds.): ICFCA 2007, LNAI 4390, pp. 114–129, 2007.

are ordered, Conceptual Scaling [GW99] allows the attribute taxonomy to be taken into account by producing a context completed with all attributes deduced from the taxonomy. The drawback is that concept intents contain redundant information. In a previous work [CFRD06], we proposed an algorithm based on Bordat's algorithm [Bor86] to find frequent concepts in a context with taxonomy. In that algorithm, the taxonomy is taken into account during the computation so as to remove all redundancies from intents.

There are several kinds of association rules, and several related issues: e.g. find all association rules with respect to some criteria, compute all association rules with a given conclusion or premise. We propose a generic algorithm to address the above issues. It is a parameterized generalization of our previous algorithm. It learns rules and is able to benefit from the presence of a taxonomy. The advantage of taking a taxonomy into account is to reduce the size of the results. For example, the attributes of contexts about Living Things are intrinsically ordered [1]. For a target such as "suckling", a rule such as "Living Things"$\wedge$ "Animalia" $\wedge$ "Chordata" $\wedge$ "Vertebrata" $\wedge$ "Mammalia" $\rightarrow$ "suckling" is less relevant than the equivalent rule "Mammalia" $\rightarrow$ "suckling" where elements redundant with respect to the taxonomy have been eliminated. The presented algorithm can compute various kinds of concept-based rules by simply changing one component. We present two instantiations which find positive and negative rules. Positive rules predict some given target (e.g. predict a mushroom as poisonous), while negative rules predict its opposite (e.g. edible).

The contributions of this article are twofold. Firstly, it formally defines FCA with taxonomy (FCA-Tax) using the Logical Concept Analysis (LCA) framework [FR04], where the taxonomy is taken into account as a specific logic. Secondly, it specifies a generic algorithm which facilitates the exploration of frequent concepts in a context with taxonomy. Quantitative experiments show that taking a taxonomy into account does not introduce slowdowns. Furthermore, the pruning implemented by our algorithm related to the taxonomy can often improve efficiency.

In the following, Section 2 formally defines FCA with taxonomy (FCA-Tax). Section 3 presents the generalization of the algorithm described in [CFRD06] to filter frequent concepts in a formal context with taxonomy. Section 4 shows how to instantiate the algorithm to learn different kind rules. Section 5 discusses experimental results.

2 A Logical Framework for FCA with Taxonomy

In this section, we formally describe Formal Concept Analysis with Taxonomy (FCA-Tax) using the Logical Concept Analysis (LCA [FR04]) framework. We first present an example of context with taxonomy. Then we briefly introduce LCA, concept-based rules, and we instanciate LCA to FCA-Tax. A taxonomy describes how the attributes of the context are ordered and thus a taxonomy is a kind of logic where the subsumption relation represents this order relation.

[1] http://anthro.palomar.edu/animal/table_humans.htm

Table 1. Context of threatened or endangered bird families in the USA

	Pacific	*Southwest*	*Southeast*	*Region8*	*Hawai*	*Washington*	*Arizona*	*Michigan*	*Mississipi*	*Florida*	*Utah*	*Alaska*	*Oregon*	*California*	*Nevada*	*Endangered*	*Threatened*
Accipitridae										•						•	
Alcedinidae	•															•	
Alcidae						•							•	•			•
Anatidae												•					•
Cathartide														•		•	
Charadriidae								•		•				•		•	•
Corvidae					•											•	
Drepanidinae	•															•	
Emberizidae				•													•
Gruidae		•							•							•	
Icteridae			•													•	
Muscicapidae				•													•
Strigidae						•	•				•		•	•			•
Tyrannidae							•				•			•	•	•	
Vireonidae														•		•	

2.1 Example of Context with a Taxonomy

The context given Table 1 represents the observations of threatened and endangered bird families. Data come from the web site of USFWS[2] (U.S Fish and Wildlife Service). The objects are bird families and each family is described by a set of states and regions where specimens have been observed and by a status: *threatened* or *endangered*. Note that in this context objects are elements of what could be a taxonomy in another context. Figure 1 shows the taxonomy of states and regions of USA.

The context given in Table 1 is the straightforward transcription of the information found on the USFWS site. Note that it is not completed with respect to FCA. For example object *Accipitridae* which has in its description the attribute *Florida* has not the attribute *Southeast*. The information that an object has the attribute *Florida* implies that it has the attribute *Southeast* is implicitly given by the taxonomy. We have observed that this situation is quite common.

2.2 Logical Concept Analysis (LCA)

In LCA the description of an object is a logical formula instead of a set of attributes as in FCA.

Definition 1 (logical context). *A* logical context *is a triple* $(\mathcal{O}, \mathcal{L}, d)$ *where* $\mathcal{O}$ *is a set of objects,* $\mathcal{L}$ *is a logic (e.g. proposition calculus) and* d *is a mapping from* $\mathcal{O}$ *to* $\mathcal{L}$ *that describes each object by a formula.*

[2] http://ecos.fws.gov/tess_public/CriticalHabitat.do?listings=0&nmfs=1

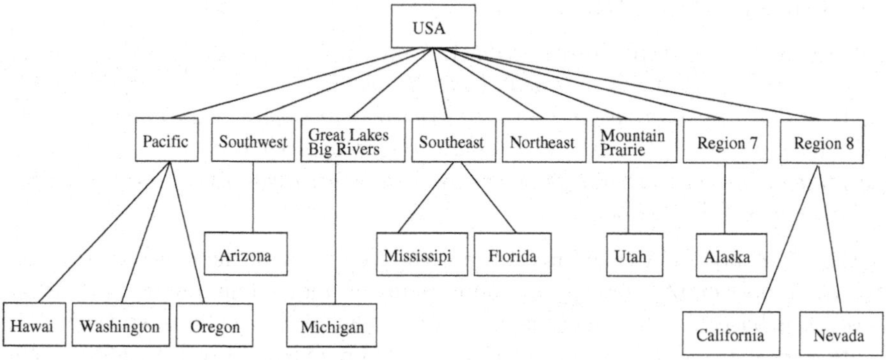

Fig. 1. Taxonomy of USA states and regions

Definition 2 (logic). *A* logic *is a 6-tuple* $\mathcal{L} = (L, \sqsubseteq, \sqcap, \sqcup, \top, \bot)$ *where*

- L *is the language of formulas,*
- $\sqsubseteq$ *is the subsumption relation,*
- $\sqcap$ *and* $\sqcup$ *are respectively conjunction and disjunction,*
- $\top$ *and* $\bot$ *are respectively tautology and contradiction.*

Definition 3 defines the logical versions of *extent* and *intent*. The extent of a logical formula f is the set of objects in $\mathcal{O}$ whose description is subsumed by f. The intent of a set of objects O is the most precise formula that subsumes all descriptions of objects in O. Definition 4 gives the definition of a *logical concept.*

Definition 3 (extent, intent). *Let* $\mathcal{K} = (\mathcal{O}, \mathcal{L}, d)$ *be a logical context. The definition of extent and intent are:*

- $\forall f \in \mathcal{L} \quad ext(f) = \{o \in \mathcal{O} \mid d(o) \sqsubseteq f\}$
- $\forall O \subseteq \mathcal{O} \ int(O) = \bigsqcup_{o \in O} d(o)$

Definition 4 (logical concept). *Let* $\mathcal{K} = (\mathcal{O}, \mathcal{L}, d)$ *be a logical context. A* logical concept *is a pair* $c = (O, f)$ *where* $O \subseteq \mathcal{O}$*, and* $f \in \mathcal{L}$*, such that* $int(O) \equiv f$ *and* $ext(f) = O$*. O is called the extent of the concept c, i.e.* ext_c*, and f is called its intent, i.e.* int_c*.*

The set of all logical concepts is ordered and forms a *lattice*: let c and c' be two concepts, $c \leq c'$ iff $ext_c \subseteq ext_{c'}$. Note also that $c \leq c'$ iff $int_c \sqsubseteq int_{c'}$. c is called a **subconcept** of c'.

The fact that the definition of a logic is left so abstract makes it possible to accommodate non-standard types of logics. For example, attributes can be valued (e.g., integer intervals, string patterns), and each domain of value can be defined as a logic. The subsumption relation allows to order the terms of a taxonomy.

2.3 Concept-Based Rules

Definition 5 (concept-based rules). *Concept-based rules consider a target $W \in \mathcal{L}$ such that W is a logical formula which represents a set of objects, $ext(W)$, that are positive (respectively negative) examples. Concept-based rules have the form $X \rightarrow W$, where X is the intent of a concept. A rule can have exceptions. These exceptions are measured with statistical indicators like support and confidence, defined below.*

Neither FCA nor LCA take into account the frequency of the concepts they define. An almost empty concept is as interesting as a large one. Learning strategy, however, is based on the generalization of frequent patterns. Thus, statistical indicators must be added to concept analysis. There exists several statistical measures like *support*, *confidence*, *lift* and *conviction* [BMUT97]. In the following, support and confidence are used to measure the relevance of the rules, because they are the most widespread. However, the algorithm presented in this article does not depend on this choice.

Definition 6 (support). *The support of a formula X is the number of objects described by that formula. It is defined as:*

$$sup(X) = \|ext(X)\|$$

where $\|Y\|$ denotes the cardinal of a set Y.

The support of a rule is the number of objects described by both X and W. It is defined as:

$$sup(X \rightarrow W) = \|ext(X) \cap ext(W)\|$$

Definition 7 (confidence). *The confidence of a rule $X \rightarrow W$ describes the probability for objects that are described by X to be also described by W. It is defined as:*

$$conf(X \rightarrow W) = \frac{\|ext(X) \cap ext(W)\|}{\|ext(X)\|}$$

The support applies as well to concepts as to rules. In the case of concept it introduces the notion of *frequent concept* as formalized by Definition 8.

Definition 8 (frequent concept). *A concept is called frequent with respect to a min_sup threshold if $sup(int_c)$ is greater than min_sup.*

2.4 Formal Concept Analysis with Taxonomy

FCA-Tax is more general than FCA because the attributes of the context are ordered. It can be formalized in LCA.

Definition 9 (taxonomy). *A taxonomy is a partially ordered set of terms*

$$TAX = < T, \leq >$$

where T is the set of terms and $\leq$ is the partial ordering. Let x and y be attributes in T, $x \leq y$ means that y is more general than x.

Example 1. In Figure 1, *Hawai* $\leq$ *Pacific* $\leq$ *USA*.

Definition 10 (predecessors, successors, Min$_{\text{tax}}$). *Let* $TAX = < T, \leq >$ *be a taxonomy and* X *in* T *be a set of terms.*
The predecessors *of* X *in the taxonomy* TAX *are denoted by*

$$\uparrow_{tax}(X) = \{t \in T \mid \exists x \in X : x \leq t\}$$

The successors *of* X *in the taxonomy* TAX *are denoted by*

$$succ_{tax}(X) = \{t \in T \mid \exists x \in X : x > t \wedge (\nexists t' \in T : x > t' > t)\}$$

The successors represented by $succ_{tax}(X)$ *are immediate ones only.* $succ^{+}_{tax}(X)$ *is the transitive closure of* $succ_{tax}(X)$*, i.e. all successors of* X *in the taxonomy.*

$Min_{tax}(X)$ *is the set of minimal elements of* X *with respect to* TAX.

$$Min_{tax}(X) = \{t \in X \mid \nexists x \in X : x < t\}$$

In fact, $Min_{tax}(X)$ *is* X *minus its elements that are redundant with respect to* TAX.

Example 2. On the context of Table 1, with the taxonomy of Figure 1, examples of predecessors, successors and Min_{tax} are as follows:

- $\uparrow_{tax}(\{\mathit{Mississipi}, \mathit{Arizona}\}) = \{\mathit{Mississipi}, \mathit{Southeast}, \mathit{Arizona}, \mathit{Southwest}, \mathit{USA}\}$

- $succ_{tax}(\{\mathit{USA}\}) = \{\mathit{Pacific}, \mathit{Southeast}, \mathit{GreatLakes/BigRivers}, \mathit{Southwest}, \mathit{Northeast}, \mathit{Mountains/Prairie}, \mathit{Region7}, \mathit{Region8}\}$

- $Min_{tax}(\{\mathit{Florida}, \mathit{Southeast}, \mathit{USA}\}) = \{\mathit{Florida}\}$

Definition 11 ($\mathcal{L}_{\text{tax}}$). *Let* $TAX = < T, \leq >$ *be a taxonomy.* $\mathcal{L}_{tax} = (L_{tax}, \sqsubseteq_{tax}, \sqcap_{tax}, \sqcup_{tax}, \top_{tax}, \bot_{tax})$ *is the logic of FCA-Tax, related to* TAX*:*

- $L_{tax} = 2^T$ *where* T *is the set of terms,*
- $\sqsubseteq_{tax}$ *such that* $X \sqsubseteq Y$ *iff* $\uparrow_{tax}(X) \supseteq \uparrow_{tax}(Y)$,
- $\sqcap_{tax}$ *is* $\cup_{tax}$ *such that* $X \cup_{tax} Y = Min_{tax}(X \cup Y)$,
- $\sqcup_{tax}$ *is* $\cap_{tax}$ *such that* $X \cap_{tax} Y = Min_{tax}(\uparrow_{tax}(X) \cap \uparrow_{tax}(Y))$,
- $\top_{tax} = \{x \in T \mid ext(\{x\}) = \mathcal{O}\}$
- $\bot_{tax} = Min_{tax}(T)$.

$\sqcap$ and $\sqcup$ follow the usage of description logics, where $\sqcap$ (resp. $\sqcup$) corresponds to intersection (resp. union) over sets of objects.

Example 3. Figure 2 shows examples of operations on the context of bird families. The language, L is the powerset of attributes (1). *Florida* implies *Southeast* in the taxonomy, thus all bird families observed in Michigan and Florida can also

$$L = \{\{Hawai\}, \{Oregon\}, \{Pacific\}, \{Southwest\}, ..., \{USA\}, \{Hawai, Southwest\}, ...\} \quad (1)$$

$$\{Michigan, Florida\} \sqsubseteq \{Michigan, Southeast\} \quad (2)$$

$$\{Michigan, Southeast\} \sqcap \{Florida\} = \{Michigan, Florida\} \quad (3)$$

$$\{Mississipi\} \sqcup \{Florida\} = \{Southeast\} \quad (4)$$

$$\top = \{USA\} \quad (5)$$

$$\bot = \{Hawai, Washington, Oregon, Arizona, Michigan, Mississipi, Florida, Utah, Alaska, California, Nevada\} \quad (6)$$

Fig. 2. Examples of operations on the attributes of the taxonomy of Figure 1

be said to have been observed in Michigan and Southeast (2). To be observed in Florida is an information more precise than to be observed in Southeast, thus only the attribute *Florida* is kept in (3). The fact that birds are observed in Michigan or Florida can be generalized by birds which are observed in Southeast (4). The top concept contains only one attribute *USA* (5), and the bottom concept all minimal attributes (6).

All notions defined in LCA apply in FCA-Tax, in particular extent, intent, and concept lattice. Note that the intents of two ordered concepts $c < c'$ differ because one or more attributes are added in $int(c)$, or because an attribute of $int(c')$ is replaced in $int(c)$ by a more specific attribute in the taxonomy, or a combination of both. FCA-Tax differs from FCA in that the intents of concepts in FCA-Tax are without redundancy (see the use of Min_{tax} in the definition of $\sqcup$).

Example 4. $ext(Oregon, California) = \{Alcidae, Strigidae\}$
$int(Alcedinidae, Corvidae) = \{Pacific, Endangered\}$
$sup(Oregon, California \rightarrow Threatened) = 2$
$conf(California \rightarrow Threatened) = 0.5$

Note in the example about the computation of intent that *Corvidae* has not explicitly the property $Pacific$ but the property $Hawai$. But in the taxonomy the property $Hawai$ implies the property $Pacific$.

3 A Parameterized Algorithm for Finding Concept-Based Rules

In a previous article [CFRD06] we described an algorithm for finding frequent concepts in a context with taxonomy. The algorithm is a variant of Bordat's algorithm [Bor86] which takes care of the taxonomy for avoiding redundant intents. In this section, a generalization of this algorithm is described in order to search concept-based rules. In a first step, the relation between frequent concepts and rules is presented. In a second step, the parameterized algorithm with this filter function is described.

3.1 From Frequent Concepts to Rules

Relevant rules are rules that are frequently observed. In the general case, see for example [PBTL99], all frequent rules are searched, without any constraint on the premises and conclusions. However, in the learning case, either the premise or the conclusion is fixed by the learning target. So, one searches for frequent rules where conclusions or premises match the target. For instance, learning sufficient conditions for a target W is to search frequent rules like $X \rightarrow W$. For example, in a context describing mushrooms it can be relevant to search properties implying that a mushroom is poisonous, i.e. $X \rightarrow poisonous$.

As explained in Section 2, the frequency of a rule is evaluated by a statistical measure: the support. A rule cannot be more frequent than its premise or its conclusion. Indeed, let c be a concept then $sup(int_c) \geq sup(int_c \cup W) = sup(int_c \rightarrow W)$. This implies that only the intents of frequent concepts are good candidates to be the premises of the searched rules. Therefore, in order to learn rules, the frequent concepts that are computed by the algorithm presented in [CFRD06] need to be filtered. Only most frequent concepts that form relevant rules with respect to statistical measures are kept. Some statistical measures like the support, are monotonous. For instance, given two concepts c and c', $c < c' \Rightarrow sup(c) < sup(c')$ holds. It means that if the support of a concept c is lower than min_sup, all subconcepts of c have a support lower than min_sup. Thus, it is not relevant to explore subconcepts of c.

A filter function, called **FILTER**, is defined in order to take into account these observations. **FILTER** takes two parameters, two sets of objects: ext_c and $ext(W)$. These two parameters are sufficient to compute all statistical measures like support, confidence, lift and conviction as it is illustrated in Figure 3. Using ext_c and $ext(W)$, we can compute $ext_c \cup ext(W)$, $ext_c \cap ext(W)$, and all complements like $\mathcal{O} \setminus ext_c$, $\mathcal{O} \setminus (ext_c \cup ext(W))$. **FILTER**($ext_c$, $ext(W)$) returns two booleans: **KEEP** and **CONTINUE**. **KEEP** tells whether $int_c \rightarrow W$ is a relevant rule with respect to statistical measures. **CONTINUE** allows monotonous properties to be considered. Indeed, **CONTINUE** tells whether there may be some subconcept c' of c such that $int_{c'} \rightarrow W$ is a relevant rule. Thus **FILTER** gives four possibilities: 1) keep the current concept and explore subconcepts, 2) keep the current concept and do not explore subconcepts, 3) do not keep the current concept and explore subconcepts, 4) do not keep the current concept and do not explore subconcepts. Section 4 illustrates on examples how these four possibilities are used.

3.2 Algorithm Parameterized with Function FILTER

In the first part of this subsection, the data structures used in the algorithm are briefly introduced. In the second part, the algorithm is described. In the last part, the difference with the previous algorithm is given.

The algorithm uses two data structures: $incr_c$ of a concept c and **Exploration**. $incr_c$ contains *increments* of a concept c. Apart from the top concept, each concept s is computed from a concept $p(s)$, called the predecessor of s. Let s be a concept and $p(s)$ be the predecessor of s then there exists a set of attributes, X, such that

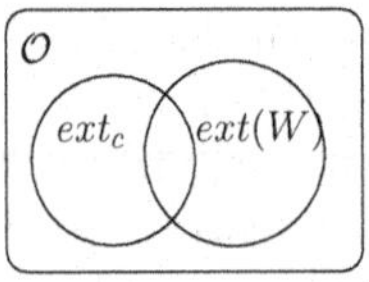

$$sup(int_c \rightarrow W) = \|ext_c \cap ext(W)\|$$

$$conf(int_c \rightarrow W) = \frac{\|ext_c \cap ext(W)\|}{\|ext_c\|}$$

$$lift(int_c \rightarrow W) = \frac{\|ext_c \cap ext(W)\|}{\|ext_c\| * \|ext(W)\|}$$

$$conv(int_c \rightarrow W) = \frac{\|ext_c\| * \|\mathcal{O} \setminus ext(W)\|}{\|ext_c \cap (\mathcal{O} \setminus ext(W))\|}$$

Fig. 3. Computing measures using the extents of a concept c and a target W

$ext_s = ext_{p(s)} \cap ext(X)$. We call X an *increment* of $p(s)$, and we say that X *leads* from $p(s)$ to s. All known increments are kept in a data structure $incr_{p(s)}$ which is a mapping from subconcepts to increments: s maps to X iff X leads to s. Notation $incr_{p(s)}[s \mapsto X]$ means that the mapping is overridden so that s maps to X.

Exploration contains frequent subconcepts that are to be explored. In **Exploration**, each concept s to explore is represented by a triple: $(ext_s \mapsto X, int_{p(s)}, incr_{p(s)})$ where $ext_s = ext(int_{p(s)} \cup X) = ext_{p(s)} \cap ext(X)$ and $\|ext_s\| \geq min\_sup$, which means that X is an increment from $p(s)$ to s.

An invariant for the correction of the algorithm is

$$\forall c \text{ concept} : incr_c \subseteq \{s \mapsto X \mid ext_s = ext_c \cap ext(X) \wedge \|ext_s\| \geq min\_sup\} .$$

Thus, all elements of $incr_c$ are frequent subconcepts of c.

An invariant for completeness is

$$\forall c, s \text{ concepts} : (ext_s \subset ext_c \wedge \|ext_s\| \geq min\_sup \wedge \neg\exists X \subset T : s \mapsto X \in incr_c)$$
$$\Longrightarrow (\exists s' \text{ a concept} : ext_s \subset ext_{s'} \subset ext_c \wedge \exists X \subset T : s' \mapsto X \in incr_c) .$$

Thus, all frequent subconcepts of c that are not in $incr_c$ are subconcepts of a subconcept of c which is in $incr_c$.

Algorithm Explore_concepts allows frequent concepts of a context with taxonomy to be filtered with the generic function **FILTER** previously introduced. Some examples of instantiation of the function **FILTER** are given in section 4. The concept lattice is explored top-down, starting with the top of the lattice, i.e. the concept labelled by all objects (i.e. the $\top$ concept) (line 2). At each iteration of the while loop, an element of **Exploration** with the largest extent is selected (line 4): $(ext_s \mapsto X, int_{p(s)}, incr_{p(s)})$. This element represents a concept s which is tested with the function **FILTER** (line 5). At line 7, the intent of s is computed by supplementing $(int_{p(s)} \cup_{tax} X)$ with successors of X in the taxonomy. The redundant attributes are eliminated thanks to $\cup_{tax}$.

Then the increment of s are computed; $incr_s$ is computed by exploring the increments of $p(s)$ (step 9) and the successors of the attributes of X in the taxonomy (steps 10-15). Note that increments of $p(s)$ can still be increments of s. For instance, if $p(s) > s$, $p(s) > s'$, and $s > s'$, the increment of $p(s)$ which leads to s' is also an increment of s that leads to s'. If these increments are relevant with respect to **FILTER**, they are added to **Exploration** (lines 16-18).

Algorithm 1. Explore_concepts

Require: $\mathcal{K}$, a context with a taxonomy TAX; and *min_sup*, a minimal support
Ensure: **Solution**, the set of all frequent concepts of $\mathcal{K}$ with respect to *min_sup*, and **FILTER**

1: **Solution** := $\emptyset$
2: **Exploration**.add($\mathcal{O} \mapsto \top$, $\emptyset$, $\emptyset$)
3: **while Exploration** $\neq \emptyset$ **do**
4:   **let** $(ext_s \mapsto X, int_{p(s)}, incr_{p(s)}) = max_{ext}($**Exploration**$)$ **in**
5:   (**KEEP**, **CONTINUE**) := **FILTER**(ext_s, $ext(X)$)
6:   **if KEEP** or **CONTINUE then**
7:     $\mathrm{int}_s := (\mathrm{int}_{p(s)} \cup_{tax} X) \cup_{tax} \{y \in succ^{+}_{tax}(X) \mid ext_s \subseteq ext(\{y\})\}$
8:     **if CONTINUE then**
9:       $\mathrm{incr}_s := \{c \mapsto X \mid \exists c' : c' \mapsto X \in incr_{p(s)} \wedge c = ext_s \cap c' \wedge \|c\| \geq min\_sup\}$
10:       **for all** $y \in succ_{tax}(X)$ **do**
11:         **let** $c = ext_s \cap ext(\{y\})$ **in**
12:         **if** $\|c\| \geq min\_sup$ **then**
13:           $incr_s := incr_s[c \mapsto (incr_s(c) \cup \{y\})]$
14:         **end if**
15:       **end for**
16:       **for all** $(ext \mapsto Y)$ in $incr_s$ **do**
17:         **Exploration**.add($ext \mapsto Y$, int_s, incr_s)
18:       **end for**
19:     **end if**
20:     **if KEEP then**
21:       **Solution**.add(ext_s, int_s)
22:     **end if**
23:   **end if**
24: **end while**

Finally if s is relevant with respect to **FILTER**, it is added to **Solution** (line 21).

The difference between the algorithm in [CFRD06] and the algorithm presented here is the addition of the function **FILTER** that allows frequent concepts to be filtered and exploration to be stopped. If function **FILTER** always returns (**KEEP** = true, **CONTINUE**=true), then all frequent concepts are eventually computed.

Example 5. Illustration of the computation of **Exploration** is as follows. The taxonomy is presented in Figure 4; Figure 5 shows the explored lattice. We assume *min_sup*=3. The first 2 steps of computation are explained. Initially, **Exploration** is:

- **Exploration** = $\{(\mathcal{O} \mapsto \top, \emptyset, \emptyset)\}$.

First step: the top of the lattice is explored, i.e. s=c_0. Increments of s are computed from the taxonomy only, as there is no predecessor concept:

- $incr_{c_0} = \{\ \{o_3, o_4, o_7, o_8, o_9, o_{10}\} \mapsto \{\mathrm{b}\}, \{o_1, o_2, o_7, o_8, o_{10}\} \mapsto \{\mathrm{a}\}, \{o_5, o_6, o_7, o_9, o_{10}\} \mapsto \{\mathrm{c}\}\}$

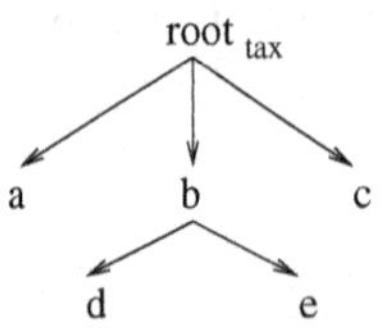

Fig. 4. Taxonomy of the example

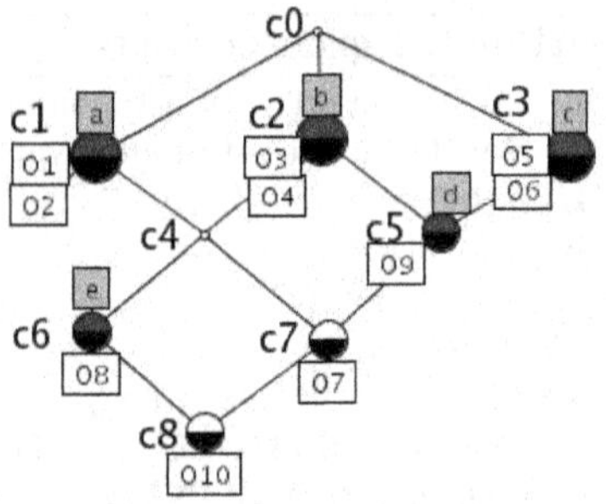

Fig. 5. Concept lattice

- **Exploration** $= \{(\{o_3, o_4, o_7, o_8, o_9, o_{10}\} \mapsto \{\text{b}\}, \emptyset, incr_{c_0}), (\{o_1, o_2, o_7, o_8, o_{10}\} \mapsto \{\text{a}\}, \emptyset, incr_{c_0}), (\{o_5, o_6, o_7, o_9, o_{10}\} \mapsto \{\text{c}\}, \emptyset, incr_{c_0})\}$.

Second step: an element of **Exploration** with the largest possible extent is explored: $s = c_2$, $p(s) = c_0$. In order to compute $incr_{c_2}$, we have to consider the elements of $incr_{c_0}$ and the elements in the taxonomy.

- $incr_{c_2} = \{ \{o_7, o_8, o_{10}\} \mapsto \{\text{a}\}, \{o_7, o_9, o_{10}\} \mapsto \{\text{c, d}\}$~~, {$o_8$, o_{10}} ↦ {e}~~}
- **Exploration** $= \{(\{o_1, o_2, o_7, o_8, o_{10}\} \mapsto \{\text{a}\}, \emptyset, incr_{c_0}), (\{o_5, o_6, o_7, o_9, o_{10}\} \mapsto \{\text{c}\}, \emptyset, incr_{c_0}), (\{o_7, o_9, o_{10}\} \mapsto \{\text{c,d}\}, \{\text{b}\}, incr_{c_2}), (\{o_7, o_8, o_{10}\} \mapsto \{\text{a}\}, \{\text{b}\}, incr_{c_2})\}$.

In the second step, attributes d and e are introduced as successors of attributes b, and attributes a and c are introduced as increments of c_0, the predecessor of c_2.

Increment {e} is eliminated because it leads to an unfrequent concept. Attributes c and d are grouped into a single increment because they lead to the same subconcept. This ensures that computed intents are complete.

3.3 Comparison with Bordat's Algorithm

The algorithm presented in the previous section is based on Bordat's algorithm. Like in Bordat's version, our algorithm starts by exploring the top concept. Then for each concept s explored, the subconcepts of s are computed; this corresponds to the computation of $incr_s$. Our algorithm uses the same data structure to represent **Solution**, i.e. a trie. The differences between our algorithm and Bordat's are: 1) the strategy of exploration, 2) the data structure of **Exploration** and 3) taking into account the taxonomy.

In our algorithm,

1. We first explore concepts in **Exploration** with the largest extent. Thus, it is a top-down exploration of the concept lattice, where each concept is explored only once. Contrary to Bordat, it avoids to test if a concept has already been found when it is added to **Solution**.
2. Whereas in Bordat's algorithm **Exploration** is represented by a queue that contains subconcepts to explore, in our version the data structure of

Exploration elements is a triple: $(ext_s \mapsto X, int_{p(s)}, incr_{p(s)})$ where $incr_{p(s)}$ avoids to test all attributes of the context when computing increments of s. Indeed, if an increment X is not relevant for $p(s)$ it cannot be relevant for s, because $ext_s \subset ext_{p(s)}$ and thus $(ext_s \cap ext(X)) \subset (ext_{p(s)} \cap ext(X))$. Thus, the relevant increments of the predecessor of s are stored in $incr_{p(s)}$ and are potential increments of s. In **Exploration**, a concept cannot be represented several times. When a triple representing a concept c is added to **Exploration**, if c is already represented in **Exploration**, the new triple erases the previous one.

3. The attributes can be structured in a taxonomy. This taxonomy is taken into account during the computation of the increments and the intents. Using $\cup_{tax}$ instead of the plain set-theoretic union operation allows computed intents to be without redundancy. Another benefit is that it makes computation more efficient. For instance, when the taxonomy is deep, a lot of attributes can be pruned without testing.

4 Instantiations of the Parameterized Algorithm

In the previous section, the presented algorithm is parameterized with a filter function in order to permit the computation of concept-based rules in FCA-Tax. In this section, we show two instantiations of the filter function to learning. The first one allows *sufficient conditions* to be computed. Sufficient conditions are premises of rules where the conclusion (the target), W, represents the positive examples. In other words, the searched rules are of the form $X \rightarrow W$.

The second one allows *incompatible conditions* to be computed. Incompatible conditions are premises of rules where the conclusion (the target), W, is the negation of the intent of positive examples. In other words, the searched rules are of the form $X \rightarrow \neg W$.

4.1 Computing Sufficient Conditions

In the generation of sufficient conditions, the learning objective, W, is the target of the rules to be learned. For computing sufficient conditions, one must determine what conjunctions of attributes, X, implies W, i.e. the intents X of concepts such that the rule $X \rightarrow W$ has a support and a confidence greater than the thresholds min_sup and min_conf. Sufficient conditions can be computed with the algorithm described in Section 3 that filters all frequent concepts of a context, by instantiating the filter function with:

$$\begin{aligned}\textbf{FILTER}_{sc}(ext_s, ext(W)) = (\textbf{KEEP} &= sup(int_s \rightarrow W) \geq min\_sup \wedge \\ &\qquad conf(int_s \rightarrow W) \geq min\_conf, \\ \textbf{CONTINUE} &= sup(int_s \rightarrow W) \geq min\_sup)\end{aligned}$$

It implies three possible behaviours of the algorithm.

1. The concept s is relevant, i.e. the rule $int_s \rightarrow W$ has a support and a confidence greater than the thresholds. s is a solution and has to be kept, and its subconcepts are potential solutions and have to be explored. Thus $\mathbf{FILTER}_{sc}(ext_s)$ returns $(true, true)$.
2. The concept s has a support greater than min_sup but a confidence lower than min_conf. s is not a solution but the confidence is not monotonous and thus subconcepts of s can be solutions. $\mathbf{FILTER}_{sc}(ext_s)$ returns $(false, true)$.
3. The concept s has a support lower than min_sup. s is not a solution. As the support is monotonous, i.e. the support of a subconcept of s is lower than the support of s, subconcepts of s cannot be solutions. $\mathbf{FILTER}_{sc}(ext_s)$ returns $(false, false)$.

$\mathbf{FILTER}_{sc}(ext_s, ext(W))$ can be expressed in terms of extent by simply applying the definitions of support and confidence seen in Section 2. The filter function can therefore be defined by :

$$\begin{aligned}\mathbf{FILTER}_{sc}(ext_s, ext(W)) = (\mathbf{KEEP} &= \|ext_s \cap ext(W)\| \geq min\_sup \wedge \\ &\frac{\|ext_s \cap ext(W)\|}{\|ext_s\|} \geq min\_conf, \\ \mathbf{CONTINUE} &= \|ext_s \cap ext(W)\| \geq min\_sup)\end{aligned}$$

Note that it is easy to use other statistical indicators like *lift* or *conviction*, by adding conditions in the evaluation of **KEEP**.

4.2 Computing Incompatible Conditions

In the case of incompatible conditions, the learning objective, W, is the negation of the target of the rules to be learned. For computing incompatible conditions, one must determine what conjunctions of attributes, X, implie $\neg W$. In other words, one looks for the intents X of concepts such that the rule $X \rightarrow \neg W$ has a support and a confidence greater than the thresholds min_sup and min_conf. Incompatible conditions can be computed with the algorithm described in Section 3 by instantiating the filter function with:

$$\begin{aligned}\mathbf{FILTER}_{ic}(ext_s, ext(W)) = (\mathbf{KEEP} &= sup(int_s \rightarrow \neg W) \geq min\_sup \wedge \\ &conf(int_s \rightarrow \neg W) \geq min\_conf, \\ \mathbf{CONTINUE} &= sup(int_s \rightarrow \neg W) \geq min\_sup)\end{aligned}$$

As for sufficient conditions, this can be expressed in terms of extents, using the definitions of support and confidence. The filter function to compute incompatible conditions can thus be defined by :

$$\begin{aligned}\mathbf{FILTER}_{ic}(ext_s, ext(W)) = (\mathbf{KEEP} &= \|ext_s \cap (\mathcal{O} \setminus ext(W))\| \geq min\_sup \wedge \\ &\frac{\|ext_s \cap (\mathcal{O} \setminus ext(W))\|}{\|ext_s\|} \geq min\_conf, \\ \mathbf{CONTINUE} &= \|ext_s \cap (\mathcal{O} \setminus ext(W))\| \geq min\_sup)\end{aligned}$$

5 Experiments

Experimental settings. The algorithm is implemented in CAML (a functional programming language of the ML family) inside the Logic File System-LISFS [PR03]. LISFS implements the notion of Logical Information Systems (LIS) as a native Linux file system. Logical Information Systems (LIS) are based on LCA. In LISFS, attributes can be ordered to create a taxonomy (logical ordering). The data structures which are used allow taxonomies to be easily manipulated. For more details see [PR03]. We ran experiments on an Intel(R) Pentium(R) M processor 2.00GHz with Fedora Core release 4, 1GB of main memory.

We tested the parameterized algorithm with two contexts: "Mushroom" and "Java". The Mushroom benchmark[3] contains 8 416 objetcs which are mushrooms. The mushrooms are described by properties such as whether the mushroom is edible or poisonous, the ring number, or the veil color. The context has 127 properties. The Java context[4] taken from [SR06], contains 5 526 objects which are the methods of java.awt. Each method is described by its input and output types, visibility modifiers, exceptions, keywords extracted from its identifiers, and keywords from its comments. The context has 1 624 properties, and yields about 135 000 concepts.

Quantitative Experiments. Many efficient algorithms computing association rules exist[5]. None of them allow a taxonomy to be taken into account. We do not pretend that our algorithm is faster. The objectives of this section are rather to show 1) that our algorithm is reasonably efficient, 2) that given an algorithm which searches for association rules, the additional mechanisms described in Sections 3 and 4 do not cause any significant slowdown in the presence of a taxonomy, and 3) that they can even improve efficiency in certain cases.

We evaluate our algorithm on the mushroom benchmark, computing frequent concepts. It enables us to compare our algorithm's performance with Pasquier *et al.*'s CLOSE algorithm, using the figures published in Pasquier's PhD thesis [Pas00]. For the same task, execution times are of the same order for both algorithms. For instance, with $min\_sup = 5\%$ CLOSE computation time is about 210s and our algorithm computation time is 290.8s; with $min\_sup = 20\%$ CLOSE computation time is about 50s and our algorithm computation time is only 26.2s. Hence, even without a taxonomy our algorithm does not introduce any slowdown.

On the Java context, we measured the computation time of the frequent concepts, in order to show the impact of the taxonomy on the computation. The taxonomy is derived, for the largest part, from the class inheritance graph, and, for a smaller part, from a taxonomy of visibility modifiers that is predefined in Java. The result is given in Table 2. The third column of that table contains the execution times in seconds of algorithm when the taxonomy is taken into account. The fourth column corresponds to the execution times when the context is a priori

[3] Available at ftp://ftp.ics.uci.edu/pub/machine-learning-databases/mushroom/

[4] Available at http://lfs.irisa.fr/demo-area/awt-source/

[5] Some of them are available on the FIMI web site http://fimi.cs.helsinki.fi/src

Table 2. Number of concepts and execution times (in seconds) for different values of *min_sup*

min_sup (%)	Number of concepts	Execution time (with taxonomy) (s)	Execution time (without taxonomy) (s)
20	35	4.5	12.2
15	48	4.8	13.2
10	54	5.2	13.6
7.5	86	5.8	14.8
5	189	7.9	16.3
2.5	2300	55.0	54.1

completed with the elements of the taxonomy. When $min\_sup = 5\%$, 189 frequent concepts are computed in 8s with taxonomy, whereas they are computed in 16s without taxonomy. When $min\_sup = 2.5\%$, 2300 frequent concepts are computed in 55s with taxonomy, whereas they are computed in 54s without taxonomy. It is explained by the fact that the Java context has few very frequent concepts. Thus pruning using the taxonomy when min_sup is greater than 5% occurs very early in the traversal of the concept lattice. For $min\_sup = 2.5\%$, pruning are less impact, computation time is therefore equivalent with or without taxonomy.

Note that from a qualitative point of view, with this context, using the taxonomy allows the number of irrelevant attributes in the intents to be reduced by 39% for $min\_sup = 5\%$.

6 Conclusion

In this article we have proposed an algorithm parameterized by a filter function to explore frequent concepts in a context with taxonomy in order to learn association rules. We have described how to instantiate the filter function in order to find premises of rules where the target are positive examples (sufficient conditions) and negative examples (incompatible conditions). Quantitative experiments have shown that, in practice, taking a taxonomy into account does not negatively impact the performance and can even make the computation more efficient.

The advantage of the presented method is to avoid the redundancies that a taxonomy may introduce in the intents of the frequent concepts.

References

[AIS93] Rakesh Agrawal, Tomasz Imielinski, and Arun Swami. Mining associations between sets of items in massive databases. In *Proc. of the ACM-SIGMOD 1993 Int. Conf. on Management of Data*, pages 207–216, May 1993.

[AS94] Rakesh Agrawal and Ramakrishnan Srikant. Fast algorithms for mining association rules. In *Proc. 20th Int. Conf. Very Large Data Bases, VLDB*, pages 487–499, 12–15 1994.

[BMUT97] Sergey Brin, Rajeev Motwani, Jeffrey D. Ullman, and Shalom Tsur. Dynamic itemset counting and implication rules for market basket data. In Joan Peckham, editor, *Proc. ACM SIGMOD Int. Conf. on Management of Data*, pages 255–264. ACM Press, 05 1997.

[Bor86] J. Bordat. Calcul pratique du treillis de Galois d'une correspondance. *Mathématiques, Informatiques et Sciences Humaines*, 24(94):31–47, 1986.

[CFRD06] Peggy Cellier, Sébastien Ferré, Olivier Ridoux, and Mireille Ducassé. An algorithm to find frequent concepts of a formal context with taxonomy. In *Concept Lattices and Their Applications*, 2006.

[FR04] Sébastien Ferré and Olivier Ridoux. An introduction to logical information systems. *Information Processing & Management*, 40(3):383–419, 2004.

[GW99] Bernhard Ganter and Rudolf Wille. *Formal Concept Analysis: Mathematical Foundations.* Springer-Verlag, 1999.

[Kuz04] Sergei O. Kuznetsov. Machine learning and formal concept analysis. In *International Conference Formal Concept Analysis*, pages 287–312, 2004.

[Mit97] Tom Mitchell. *Machine Learning.* McGraw-Hill, 1997.

[Pas00] Nicolas Pasquier. *Data Mining : Algorithmes d'extraction et de réduction des régles d'association dans les bases de données.* Computer science, Université Blaise Pascal - Clermont-Ferrand II, January 2000.

[PBTL99] Nicolas Pasquier, Yves Bastide, Rafik Taouil, and Lotfi Lakhal. Discovering frequent closed itemsets for association rules. In *Proc. of the 7th Int. Conf. on Database Theory*, pages 398–416. Springer-Verlag, 1999.

[PR03] Yoann Padioleau and Olivier Ridoux. A logic file system. In *Proc. USENIX Annual Technical Conference*, 2003.

[SR06] Benjamin Sigonneau and Olivier Ridoux. Indexation multiple et automatisée de composants logiciels. *Technique et Science Informatiques*, 2006.

[Zak04] Mohammed J. Zaki. Mining non-redundant association rules. *Data Mining Knowl. Discov.*, 9(3):223–248, 2004.

About the Lossless Reduction of the Minimal Generator Family of a Context

T. Hamrouni[1,2,3], P. Valtchev[2,3], S. Ben Yahia[1], and E. Mephu Nguifo[4]

[1] URPAH, Faculté des Sciences de Tunis, Tunis, Tunisie
{tarek.hamrouni, sadok.benyahia}@fst.rnu.tn
[2] LATECE, Université du Québec à Montréal, Montréal, Québec, Canada
petko.valtchev@uqam.ca
[3] DIRO, Université de Montréal, Montréal, Québec, Canada
{hamrount, valtchev}@iro.umontreal.ca
[4] CRIL-CNRS, IUT de Lens, Lens, France
mephu@cril.univ-artois.fr

Abstract. Minimal generators (MGs), *aka* minimal keys, play an important role in many theoretical and practical problem settings involving closure systems that originate in graph theory, relational database design, data mining, etc. As minima of the equivalence classes associated to closures, MGs underlie many compressed representations: For instance, they form premises in *canonical* implication/association rules – with closures as conclusions – that losslessly represent the entire rule family of a closure system. However, MGs often show an intra-class combinatorial redundancy that makes an exhaustive storage and use impractical. In this respect, the *succinct system of minimal generators* (SSMG) recently introduced by Dong *et al.* is a first step towards a lossless reduction of this redundancy. However, as shown elsewhere, some of the claims about SSMG, *e.g.*, its invariant size and lossless nature, do not hold. As a remedy, we propose here a new succinct family which restores the losslessness by adding few further elements to the SSMG core, while theoretically grounding the whole. Computing means for the new family are presented together with the empirical evidences about its relative size *w.r.t.* the entire MG family and similar structures from the literature.

1 Introduction and Motivations

Minimal generators (MGs) [1], *aka* minimal keys, play an important role in many theoretical and practical problem settings involving closure systems that originate in graph theory, database design and data mining, to cite but a few. Standing at the "antipodes" of the closures within their respective equivalence classes in the Boolean lattice [2] – MGs are the smallest elements of a class while the closures are the largest – they help delimit the classes and hence ease their detection/traversal.

From a computational viewpoint, the MG set is often the intermediate step in the construction of structures that are either larger or lay higher in the Boolean lattice: the *frequent* itemset family [3], *frequent* closed itemset (CI) family [4,5], an iceberg lattice [6], etc. Underlying their computational importance is a structural property of the

S.O. Kuznetsov and S. Schmidt (Eds.): ICFCA 2007, LNAI 4390, pp. 130–150, 2007.

MG set, *i.e.*, its *order ideal* shape [5]. Indeed, the MGs are the first elements of their respective equivalence classes to be reached by a breadth-first climb in the Boolean lattice. This fact is essentially exploited both by level-wise [4,7] and depth-first [8] itemset mining algorithms in achieving better performances.

Beside their impact on computation and efficiency, the use of MGs brings gains on the semantic level, especially in decision support environments (*e.g.*, medical diagnosis). As they are usually strictly smaller than the closures (unless themselves closed), MGs offer minimal combinations of tests/exams/answers necessary to identify a class of situations and hence reduce the economic cost of the decision process. For example, they were shown to be highly instrumental in applications involving rule induction and classification [8].

On the structural side, various concise representations of the frequent itemset family have been defined in terms of MGs [9,10,11]. More interestingly, MGs underly a variety of compact subsets of the implication/association rule families of a context [1,12,13], which are hence called by some *generic bases*. Traditionally, a generic basis is considered as an irreducible nucleus of the underlying rule family, although some redundancy clearly persists. In fact, given two MGs g_1 and g_2 of the same equivalence class, there is a one-to-one correspondence between rules of the basis involving g_1 and those involving g_2.

In this respect, a study of intra-class redundancies in MGs was initiated by Dong *et al.*, who recently proposed a way to derive MGs from other ones in the same equivalence class [14]. The overall reduction principle may be grossly summarized as follows: an arbitrary total order is defined on the itemset family and the unique minimal members of the respective equivalence classes are kept. This results in a split of the global MG family into *succinct* and *redundant* parts. Thus, the succinct system of minimal generators (SSMG) was introduced as a concise representation from which the entire MG family can be retrieved without any information loss.

However, contrary to the authors' claim, the SSMG as defined in [14] proved to be loss-prone, *i.e.*, in some cases *a priori* redundant MGs are impossible to derive. Furthermore, the different SSMGs of a context (emerging through different orders) do not necessarily share the same size, again contradicting what was stated in [14]. As an attempted improvement on both issues, a new construct was hence proposed by Hamrouni *et al.* in [15]. Unfortunately, the new family lost the order ideal structure what greatly complicates its extraction.

In this paper, we propose a third system that overcomes the worst limitations of the previous ones. We present its definition and show that it preserves the precious order ideal property together with further structural properties that underly a lossless reduction mechanism. The presentation is organized as follows: the next section defines the basic constructs to be used throughout the remainder of the text. Section 3 is a detailed study of the SSMG as defined by Dong *et al.*, whereas Section 4 sketches that of Hamrouni *et al.* Section 5 expands on our own definition as well as its structural properties. An algorithm extracting the family is sketched in Section 6, while the empirical evidences about the utility of the approach is provided in Section 7.

2 Basic Concepts

In this section, basic constructs used in the remainder of the paper are presented.

Definition 1. (EXTRACTION CONTEXT) *An extraction context is a triplet* $\mathcal{K} = (\mathcal{O}, \mathcal{I}, \mathcal{R})$*, where* $\mathcal{O}$ *represents a finite set of objects,* $\mathcal{I}$ *is a finite set of items and* $\mathcal{R}$ *is a binary (incidence) relation (i.e.,* $\mathcal{R} \subseteq \mathcal{O} \times \mathcal{I}$*). Each couple* $(o, i) \in \mathcal{R}$ *indicates that the object* $o \in \mathcal{O}$ *contains the item* $i \in \mathcal{I}$*.*

Example 1. Consider the extraction context in Table 1 where $\mathcal{O} = \{1, 2, 3, 4\}$ and $\mathcal{I} = \{a, b, c, d, e, f, g\}$. The couple $(2, d) \in \mathcal{R}$ since it is crossed in the matrix.

Table 1. An extraction context $\mathcal{K}$

	a	b	c	d	e	f	g
1			×	×	×	×	×
2	×	×	×	×	×		
3	×	×	×			×	×
4	×	×	×	×	×		×

For arbitrary sets $X \subseteq \mathcal{I}$ and $Y \subseteq \mathcal{O}$, the following derivation operators are defined [16]:

- $X' = \{o \in \mathcal{O} \mid \forall\, i \in X, (o, i) \in \mathcal{R}\}$,
- $Y' = \{i \in \mathcal{I} \mid \forall\, o \in Y, (o, i) \in \mathcal{R}\}$.

Let $''$ be the resulting closure operator on $(2^{\mathcal{I}}, \subseteq)$. Once applied, this operator induces an equivalence relation on $2^{\mathcal{I}}$ portioning it into distinct subsets called *equivalence classes* [3]. In the remainder, the latter will be denoted γ*-equivalence classes*. Elements of each γ-equivalence class share the same closure. The largest one is called a *closed itemset* (CI) [4] while smallest ones are called *minimal generators* (MGs) [1]. The notions of *closed itemset* and of *minimal generator* are defined as follows:

Definition 2. (CLOSED ITEMSET) *An itemset* $I \subseteq \mathcal{I}$ *is said to be closed* ***iff*** $I'' = I$.

Example 2. Given the extraction context depicted by Table 1, the itemset *cdeg* [1] is a closed itemset since it is the maximal set of items common to the set of objects $\{1, 4\}$. The itemset *cdg* is not a closed itemset since all objects containing *cdg* also contain the item *e*.

Definition 3. (MINIMAL GENERATOR) *An itemset* $g \subseteq \mathcal{I}$ *is said to be a minimal generator (*MG*) of a* CI f ***iff*** $g'' = f$ *and* $\nexists\, g_1 \subset g$ *s.t.* $g_1'' = f$.

The set of the MGs associated to a CI f (resp. an extraction context $\mathcal{K}$) will further be denoted MG_f (resp. $\mathcal{MG}_{\mathcal{K}}$).

[1] We use a separator-free form for the sets, *e.g.*, the set *cdeg* stands for $\{c, d, e, g\}$.

Example 3. Consider the CI *cdeg* described by the previous example. *cdeg* has *dg* as a MG. Indeed, $(dg)'' = cdeg$ and the closure of every subset of *dg* is different from *cdeg*. Indeed, $(\emptyset)'' = c$, $(d)'' = cde$ and $(g)'' = cg$. *cdeg* has also another MG which is *eg*. Hence, $\text{MG}_{cdeg} = \{dg, eg\}$. *cdeg* is then the largest element of its γ-equivalence class, whereas *dg* and *eg* are the smallest ones. All these itemsets share the set of objects $\{1, 4\}$.

Since in practice, we are mainly interested on itemsets that occur at least in a given number of objects, we introduce the notion of support.

Definition 4. (SUPPORT) *The support of an itemset $I \subseteq \mathcal{I}$, denoted by Supp(I), is equal to the number of objects in $\mathcal{K}$ that contain I. I is said to be frequent in $\mathcal{K}$ if Supp(I) is greater than or equal to a minimum support threshold, denoted minsupp.*

Example 4. Consider the itemset *cf* of the extraction context depicted by Table 1. Both objects 1 and 3 contain *cf*. Hence, Supp(*cf*) = 2.

For *minsupp* = 1, Table 2 shows, for each frequent CI, its MGs and its support value.

Table 2. The frequent CIs extracted from $\mathcal{K}$ and for each one, the corresponding MGs and support

	Frequent CI	MGs	Support
1	*c*	$\emptyset$	4
2	*abc*	*a*, *b*	3
3	*cde*	*d*, *e*	3
4	*cg*	*g*	3
5	*cfg*	*f*	2
6	*abcde*	*ad*, *ae*, *bd*, *be*	2
7	*abcg*	*ag*, *bg*	2
8	*abcfg*	*af*, *bf*	1
9	*cdeg*	*dg*, *eg*	2
10	*cdefg*	*df*, *ef*	1
11	*abcdeg*	*adg*, *aeg*, *bdg*, *beg*	1

To the best of our knowledge, only two approaches were proposed dealing with the succinct system of minimal generators. We will scrutinize the main characteristics of both approaches. Then, we will concentrate on our new approach towards a lossless reduction of the minimal generator family while preserving the order ideal property.

3 The Original Definition

We present the original definition of the succinct system of minimal generators (SSMG) introduced in [14] [2]. We then clarify the aspects of the definition that remained unclear and show its flaws.

[2] Please notice that we mainly refer to the SSMG_MINER algorithm proposed by the authors. In fact, the concrete examples of its function provided by the authors are the only source of precise information about several aspects of the target structure.

3.1 Original Definition

In [14], Dong *et al.* showed that the minimal generator (MG) set may contain redundant information as some MGs associated to a CI can be derived from other ones by a subset substitution process. They hence tried to remove the redundancy within the MG set and to achieve a succinct representation of MGs. Thus, Dong *et al.* introduced the succinct system of minimal generators (SSMG) as a concise representation of the MG set. The main idea was then to remove the redundant information by choosing one (*e.g.* the smallest *w.r.t.* a given total order) MG of a CI and to elect it as its *representative* MG, and discarding those containing at least a non-representative MG [14]. The purpose is to only maintain, in each γ-equivalence class, minimal generators that cannot be derived from other ones belonging to the same γ-equivalence class. The authors hence proposed to set up a relation between itemsets. This relation is defined as follows:

Definition 5. *Let f be a closed itemsets. Let X and Y be two itemsets. X and Y are called f-equivalent, denoted $X \approx_f Y$, if:*
(i) X and Y are two MG*s of a* CI *f_1 s.t. $f_1 \subset f$.*
(ii) X can be obtained from Y by replacing a subset Z_1 of X $(Z_1 \subset X)$ by a subset Z_2 of Y $(Z_2 \subset Y)$ s.t. $Z_1 \approx_f Z_2$.

Example 5. To illustrate this definition, consider the extraction context in Table 1. The relation between itemsets given in the case (i) is verified by a and b *w.r.t.* the CI $abcde$. Indeed, both are MGs of abc which is included in $abcde$. Hence, $a \approx_{abcde} b$. That of the case (ii) is satisfied by ad and bd also *w.r.t.* the CI $abcde$. Indeed, by replacing a by b in ad, we obtain bd. This replacement is correct since $a \approx_{abcde} b$.

Surprisingly enough, $\approx_f$ is not an equivalence one since the transitivity property is not fulfilled, as this will be shown in Subsection 3.3. Dong *et al.* aimed at using this relation to split the MGs associated to a given CI into different equivalence classes. To avoid confusion with the γ-equivalence classes induced by the closure operator $''$, the latter will be denoted *σ-equivalence classes*. The achievement of the goal of deriving a minimal non-redundant subset of MGs is achieved by only maintaining one MG for each σ-equivalence class. The choice of the representative member of a σ-equivalence class is of paramount importance. Dong *et al.* proposed to freely choose a representative MG for the smallest CIs. For the other CIs, the authors proposed to choose one of the canonical MGs, *i.e.*, those that do not contain any non-representative MG of a smaller CI. Even though the authors do not give a precise way to choose the representative MG, the examples of the paper indicate that shorter sets are considered as smaller and hence favored [14]. In other words, the cardinality of MGs constitutes the first criterion when choosing a representative MG among a set of canonical ones belonging to the same σ-equivalence class.

3.2 Unveiling Imprecise Aspects

Several aspects remain unclear in the presentation of the SSMG in [14]. For instance, the selection of the representative itemset for each closure seems to be defined procedurally rather than analytically: A climb in the Boolean lattice is used to guide the choice which is some way randomly performed for the minimal closed sets. On upper levels of the closed itemset lattice, the representative is chosen among the sets that are canonical *w.r.t.* already fixed part of the representative MG family (thus enforcing the order ideal structure of the target SSMG). After cross-checking with the algorithmic description, it becomes clear that a global order on items is used which makes all the choices on the lowest level deterministic. Moreover, the choice among canonicals on upper levels is fixed by a preference for smaller-size sets. The following definition of a total order on itemsets summarizes this:

Definition 6. (TOTAL ORDER RELATION) *Let $\preceq$ be a total order relation among item literals, i.e., $\forall\ i_1,\ i_2 \in \mathcal{I}$, we have either $i_1 \preceq i_2$ or $i_2 \preceq i_1$. This relation is extended to also cope with itemsets of different sizes by first considering their cardinality. This is done as follows: Let X and Y be two itemsets and let $|X|$ and $|Y|$ be their respective cardinalities. We then have:*

- *If $|X| < |Y|$, then $X \prec Y$.*
- *If $|X| = |Y|$, then X and Y are compared using the lexicographic order.*

Example 6. If we consider the alphabetic order on items as the basis for the total order relation $\preceq$ on itemsets [3], then:

- $|d| < |be| \Longrightarrow d \prec be$.
- $abd \preceq abe \Longleftrightarrow abd \cup \{c\} \preceq abe \cup \{c\}$ (*i.e.*, $abcd \preceq abce$).

Please notice that the cardinality factor preserves the spirit of MGs as the smallest itemset in an γ-equivalence class is necessarily a MG. Three categories of MGs emerge [15]:

Definition 7. (MINIMAL GENERATORS CATEGORIES) *The set* MG_f, *of the* MGs *associated to a* CI f, *can be portioned into three distinct subsets as follows:*

1. $\text{MG}\mathtt{rep}_f = \{g \in \text{MG}_f \mid \nexists\ g_1 \in \text{MG}_f \text{ s.t. } g_1 \prec g\}$, *hence it contains the smallest* MG, *given a total order relation $\preceq$, which constitutes the* ***representative*** MG *of f.*
2. $\text{MG}\mathtt{can}_f = \{g \in \text{MG}_f \mid g \notin \text{MG}\mathtt{rep}_f,\ \forall\ g_1 \subset g,\ g_1 \in \text{MG}\mathtt{rep}_{f_1} \text{ with } f_1 = g_1''\}$, *hence it contains* ***canonical*** MGs *of f. A canonical* MG *is not the smallest one in* MG_f *and, hence, is not the representative* MG *of f. Nevertheless, all its subsets are the representative* MGs *of their respective closures.*
3. $\text{MG}\mathtt{red}_f = \{g \in \text{MG}_f \mid \exists\ g_1 \subset g,\ g_1 \notin \text{MG}\mathtt{rep}_{f_1} \text{ with } f_1 = g_1''\}$, *hence it contains the* ***redundant*** MGs *of f.*

It was proven in [15] that the subsets of a representative MG are also representative ones. Indeed, the admission of the contrary, *i.e.*, the existence of a subset which is not a representative, would lead to a contradiction with the minimal status of a representative MG. A MG is said to be ***succinct*** if it is either a *representative* or a *canonical* one [14]. Hence, the set $\text{MG}\mathtt{suc}_f$, composed by the *succinct* MGs associated to the CI f, is

[3] In the remainder of the paper, we will only mention the criterion used to order items (alphabetic order, ascending support order, etc). The latter is then extended to be the total order relation on itemsets.

equal to the union of MGrep_f and MGcan_f: $\text{MGsuc}_f = \text{MGrep}_f \cup \text{MGcan}_f = \{g \in \text{MG}_f \mid \forall\, g_1 \subset g, g_1 \in \text{MGrep}_{f_1} \text{ with } f_1 = g_1''\}$. The set $\mathcal{MG}\text{suc}_\mathcal{K}$ of the succinct MGs that can be extracted from a context $\mathcal{K}$ is then an order ideal (or down-set) of $(2^\mathcal{I}, \subseteq)$ [15] [4].

Example 7. Consider the extraction context $\mathcal{K}$ depicted by Table 1 and the alphabetic order be the total order relation $\preceq$. This relation is used to sort the MGs associated to the CIs of Table 2. Note that for **11** CIs, there are **23** MGs, from which only **13** are *succinct* ones (**11** are *representative* MGs and only **2** are *canonical* ones and which are b and e). The MG ad is a *representative* one since it is the smallest MG *w.r.t.* $\preceq$, among those of $abcde$. Indeed, $ad \preceq ae$, $ad \preceq bd$ and $ad \preceq be$. The MG e is not the *representative* of its CI cde since $d \preceq e$. Nevertheless, its unique subset (*i.e.*, $\emptyset$) is the *representative* MG of its CI c. Hence, e is a *canonical* MG. Finally, the MG bdg is a *redundant* one since at least one of its subsets is not a *representative* MG (bg, for example).

The definition of a succinct system of minimal generators (SSMG) according to Dong *et al.* is as follows:

Definition 8. *A succinct system of minimal generators (*SSMG*) w.r.t. a total order relation $\preceq$, consists of, for each closed itemset, the representative minimal generator and a possibly void set of canonical minimal generators.*

Noteworthily, for a given extraction context, there may be several SSMGs depending on the choice of the order relation $\preceq$ (*e.g.*, the extraction context shown in Table 3 (left)). It is also important to point out that the SSMG is clearly a generalization of the clone items framework which focusses on items playing symmetric roles within CIs [18]. Indeed, instead of simple items, the SSMG considers subsets of items.

3.3 Problems in the Original Definition

In [14], the authors made the following claims:

Claim 1: A SSMG is a *lossless representation* of the MG set, *i.e.*, if g is a redundant MG, then g can be inferred from the SSMG without loss of information.

Claim 2: The cardinality of a SSMG is invariant *w.r.t.* the considered total order relation $\preceq$.

To infer the redundant MGs of each γ-equivalence class, Dong *et al.* proposed to replace the subsets (one or more) of its *succinct* MGs by *non-representative* MGs having, respectively, the same closures as those of the replaced subsets [14]. For example, the redundant MG bdg, extracted from the context in Table 1, can be inferred from the *succinct* MG adg by replacing its subset ad by bd (since both MGs ad and bd have the same closure (*cf.* Table 2)).

To be satisfied, both claims closely rely on how maintaining representative members of the different σ-equivalence classes, from where the remainder can be derived using

[4] This interesting property allowed us to propose an efficient algorithm to extract the SSMG according to the definition of Dong *et al.* (see [17] for more details).

the relation $\approx_f$. However, localizing such members by pruning redundant elements, containing non-representative subsets, can lead to:

• An σ-equivalence class without representative member: It is the case of the σ-equivalence class $\mathcal{S}_1 = \{ecf, edf, acb, abd, abf, cbf, bdf\}$ associated to the closed itemset *eacbdf* (*cf.* Table 3 (left) for the ascending support order). Indeed, each element of this σ-equivalence class contains at least a non-representative MG. Such σ-equivalence class will not be taken into account and all its elements will then not be derived, which presents a loss of information, in the contrary to the statement of **Claim 1**.

• An σ-equivalence class with more than one candidate for being the representative member: It is the case of the σ-equivalence class $\mathcal{S}_2 = \{bdf, bda, bfa, bfc, bac, dfe, fce\}$ associated to the closed itemset *bdface* (*cf.* Table 3 (left) for the descending support order). Indeed, *bdf*, *bda* and *bfa* have all their subsets as representative MGs. Even one can choose the smallest candidate and set it as the representative member, the definition given by Dong *et al.* lacks the important part allowing to delete the remaining candidates.

It is also worth noting that the elements of $\mathcal{S}_1$ and $\mathcal{S}_2$ are exactly the same while being sorted according to two different order relations. In $\mathcal{S}_1$, there is no representative member while in $\mathcal{S}_2$, there are three possible. This fact shows that the cardinality of the SSMG closely depends on the selected total order relation, in the contrary to the statement of **Claim 2**.

In addition, as mentioned in Subsection 3.1, the application of the relation $\approx_f$ does not induce an equivalence relation on the minimal generator set of a CI. Indeed, if we consider the extraction context sketched by Table 1 and the CI *abcde*, we have $ad \approx_{abcde} bd \approx_{abcde} be$ but $ad \not\approx_{abcde} be$ (*cf.* Table 2). Hence, this relation is not an equivalence one since the transitivity property is not fulfilled.

4 The Improvement of Hamrouni *et al.*

In this section, we describe the improved SSMG defined in [15].

4.1 Redefinition of the Succinct System of Minimal Generators

In their attempt to fix the main flaws of the original SSMG, the authors proposed a relation, denoted $\models$, allowing to divide the set of MGs associated to a given closed itemset (CI) f into σ-equivalence classes via a substitution process. The latter uses a substitution operator denoted *Subst* allowing to replace a subset of an itemset X, say Z_1, by another itemset Z_2 belonging to the same γ-equivalence class as Z_1 (*i.e.*, $Z_1'' = Z_2''$). This operator works as follows: *Subst*$(X, Z_1, Z_2) = (X \backslash Z_1) \cup Z_2$. It was shown in [15] that X and *Subst*(X, Z_1, Z_2) have the same closure.

For each γ-equivalence class $\mathcal{C}$ (or equivalently, for each CI f), the substitution operator induces an equivalence relation on the set MG_f of the MGs of f. The concept of minimal generators redundancy within each σ-equivalence class is then defined as follows.

Definition 9. (MINIMAL GENERATORS REDUNDANCY) *Let g and g_1 be two* MG*s belonging to the same γ-equivalence class.*

- *g is said to be a* ***direct redundant*** *(resp. derivable) with respect to (resp. from) g_1, denoted $g_1 \vdash g$, if Subst(g_1, g_2, g_3) = g with $g_2 \subset g_1$ and $g_3 \in \mathcal{MG}_{\mathcal{K}}$ s.t. $g_3'' = g_2''$.*
- *g is said to be a* ***transitive redundant*** *with respect to g_1, denoted $g_1 \vDash g$, if it exists a sequence of n MGs ($n \geq 2$), gen_1, gen_2, ..., gen_n, such that $gen_i \vdash gen_{i+1}$ ($i \in$ [1..(n-1)]) with $gen_1 = g_1$ and $gen_n = g$.*

Property 1
- The relation $\vdash$ is reflexive, symmetric but not necessarily transitive.
- The relation $\vDash$ is reflexive, symmetric and transitive.

According to Definition 9, if $g \in \text{MG}_f$, then the σ-equivalence class of g, denoted by $[g]$, is the subset of MG_f consisting of all elements that are transitive redundant *w.r.t.* g. In other words, we have: $[g] = \{g_1 \in \text{MG}_f \mid g \vDash g_1\}$. To uniquely define a *succinct* MG in each σ-equivalence class, the authors adopted the same total order relation described in Definition 6. Its smallest MG will then be considered as the ***succinct*** MG. While the other MGs will be tagged as ***redundant*** MGs.

Example 8. Consider the extraction context $\mathcal{K}$ depicted by Table 1. The total order relation $\preceq$ is set to the alphabetic order. The MG adg is a *succinct* one, since it is the smallest MG, *w.r.t.* $\preceq$, among those of $abcdeg$. Indeed, when extracting the first σ-equivalence class associated to $abcdeg$, the whole MG set associated to $abcdeg$ is considered. We then have: $adg \preceq aeg$, $adg \preceq bdg$ and $adg \preceq beg$. The MG aeg is a redundant one since *Subst*(adg, ad, ae) = $aeg \in \text{MG}_{abcdeg}$ ($adg \vdash aeg$ and, hence, $adg \vDash aeg$). It is the same for the MGs bdg and beg since $adg \vDash bdg$ and $adg \vDash beg$.

The succinct system of minimal generators (SSMG) is then redefined as follows:

Definition 10. (SUCCINCT SYSTEM OF MINIMAL GENERATORS) *A succinct system of minimal generators* (SSMG) *is the set of all succinct* MGs *of the* CIs.

Since the definition of the relation $\vDash$ is independent from any adopted total order relation, the different SSMGs that may be drawn from a given extraction context share the same size. More interestingly, the SSMG becomes a lossless representation of the MG set. Hence, the derivation of the whole set of redundant MGs is ensured to be complete. The impact of the newly defined SSMG on generic bases of association rules is discussed in [19].

4.2 Problems in the Redefinition

Unfortunately, the approach given in [15] still presents some drawbacks. For example, the interesting order ideal property is not preserved. Indeed, if we consider the extraction context given in Table 3 (left) and the ascending support order as a total order relation, the MG ecf will be characterized as a *succinct* MG since it is the smallest one in its σ-equivalence classes. However, the subset cf of ecf is not the smallest one in its γ-equivalence class (or equivalently, is not a representative MG *w.r.t.* the original definition given by Dong *et al.*). Hence, additional tests have to be performed to guess whether a MG is a *succinct* one or not. Thus, the compaction of the SSMG is conditioned by an exhaustive calculation of substitutions between MGs.

Table 3. (**left**) An extraction context $\mathcal{K}'$. (**right**) The CIs extracted from $\mathcal{K}$ and for each one, the corresponding MGs for different total order relations. The *succinct* MGs, according to the original definition of Dong *et al.*, are indicated with bold letters. Those, *w.r.t.* the definition of Hamrouni *et al.*, are underlined. Intersecting elements are thus in bold letters and underlined.

	a	*b*	*c*	*d*	*e*	*f*
1	×	×				
2	×		×			×
3	×			×		
4		×	×	×	×	
5	×	×	×	×	×	×
6			×	×		
7		×				×
8				×		×
9		×			×	×

	alphabetic order		*ascending support* order		*descending support* order	
	CI	MGs	CI	MGs	CI	MGs
1	∅	**∅**	∅	**∅**	∅	**∅**
2	*a*	***a***	*a*	***a***	*a*	***a***
3	*b*	***b***	*b*	***b***	*b*	***b***
4	*c*	***c***	*c*	***c***	*c*	***c***
5	*d*	***d***	*d*	***d***	*d*	***d***
6	*be*	***e***	*eb*	***e***	*be*	***e***
7	*f*	***f***	*f*	***f***	*f*	***f***
8	*ab*	***ab***	*ab*	***ab***	*ba*	***ba***
9	*acf*	***ac***, ***af***, ***cf***	*acf*	***ac***, ***af***, ***cf***	*fac*	***fa***, ***fc***, ***ac***
10	*ad*	***ad***	*ad*	***ad***	*da*	***da***
11	*abcdef*	***ae***, ***abc***, *abd*, *abf*, ***acd***, *adf*, *bcf*, *bdf*, *cdf*, *cef*, *def*	*eacbdf*	***ea***, *ecf*, *edf*, *acb*, ***acd***, *abd*, *abf*, *adf*, *cbf*, *cdf*, *bdf*	*bdface*	***ae***, ***bdf***, ***bda***, ***bfa***, *bfc*, *bac*, ***dfa***, *dfc*, *dfe*, *dac*, *fce*
12	*bcde*	***bc***, ***bd***, ***ce***, ***de***	*ecbd*	***ec***, ***ed***, ***cb***, ***bd***	*bdce*	***bd***, ***bc***, ***de***, ***ce***
13	*bf*	***bf***	*bf*	***bf***	*bf*	***bf***
14	*cd*	***cd***	*cd*	***cd***	*dc*	***dc***
15	*df*	***df***	*df*	***df***	*df*	***df***
16	*bef*	***ef***	*ebf*	***ef***	*bfe*	***fe***

5 A New Succinct Minimal Generator Family

The idea behind our own definition is to "repair" the flaws in the proposal of Dong *et al.* We start with a summary of the relative merits of both SSMG systems which motivates the developments presented in the remainder of the paper.

5.1 Analysis

First, both definitions rely on the same substitution mechanism which amounts to a special case of the well-known Armstrong axiom of *pseudo-transitivity* [20]:

$$\frac{X \rightarrow Y; WY \rightarrow Z}{WX \rightarrow Z}.$$

In the substitution used by Dong *et al.* and later formalized by Hamrouni *et al.*, the following constraints on the above general rule apply: (i) WY, X and Y are MGs, (ii) X and Y belong to the same γ-equivalence class (hence $Y \rightarrow X$), and (iii) Z is obviously the closure of WY.

Next, Dong *et al.* looked for a classical way to reduce a set, *i.e.*, by breaking it into equivalence classes that are further shrunk to a unique, representative, element. Thus, they based their system definition on a substitution-based $\approx_f$ relation which, although $\approx_f$ wrongfully assumed to be an equivalence, did not result in a major flaw in the construction of the SSMG since a correct equivalence could be easily derived (as done by Hamrouni *et al.*). Independently, in the original definition of their SSMG, the authors put, implicitly, the requirement for it being an order ideal (through the definition of a canonical element). We have shown above that this ideal can be assimilated to a total order on itemsets, itself induced by an order on items. The exact composition of the ideal, however, depends strongly on the chosen total order: different orders could result in different sets becoming representative and canonical.

What Dong *et al.* seem to have miscalculated is the interplay between the ideal and the partition of the Boolean lattice into substitution-based σ-equivalence classes. Indeed, they hastily concluded that whatever the order, there will always be at least one canonical element per σ-equivalence class, which is of course wrong. Moreover, their claim of invariance for the SSMG size upon the choice of the total order seems to come from either (i) belief that there will be a unique canonical element in each σ-equivalence class, also wrong, or (ii) a discrepancy between the analytical description and the SSMG_MINER algorithm which clearly keeps all the canonical elements that are found.

To keep the basic idea of finer σ-equivalence classes of MGs, Hamrouni *et al.* proposed a straightforward remedy of some unfortunate repercussions. Their own definition relies on an explicit identification (and exhaustive substitution-based computation!) of the same σ-equivalence classes as in Dong *et al.* The novelty in their work is the choice of the unique class member to keep in the SSMG, which has been disconnected from any order. This brought a constant size to the system, but also made it lose its order ideal structure, and hence, the necessity of more extensive computation effort in the construction of the system, in particular, for testing reducibility between MGs.

A first observation is that three different phenomena interact in the above definitions: the Armstrong axiom of pseudo-transitivity, or substitution, the minimal generator status of a set within its γ-equivalence class which induces an order ideal of its own, and an additional ordering on the Boolean lattice of itemsets which induces a different, yet somehow connected to the previous one, order ideal, made of the representative itemsets. The latter ideal is completed with its "shell" of canonical elements which constitute its outside frontier, or *negative border* (actually a subset of it), and the result is another, a bit larger ideal combining both sorts of MGs. Our initial analysis is that it is impossible to always have all three constructs "aligned", *i.e.*, that all substitution-based classes of MGs intersect the second order ideal to a unique element. In practice, it may happen that several such elements belong to the same substitution class (which is not a problem as one could always pick a single one at random), just as there could be none in some classes.

It is worth pointing out that the picture gets more regular on the higher granularity level, *i.e.*, within a γ-equivalence class. Indeed, whatever the used item order for its generation, the ideal of representative/canonical MGs has at least one element in each γ-equivalence class. This fact, used unproven in the referred publications, admits an immediate proof based on the same induction employed in the completeness proof for our expansion procedure (Subsection 5.2). Moreover, one can easily show that given a γ-equivalence class, the *representative*, *i.e.*, minimal set *w.r.t.* any linear extension of the

$\subseteq$-induced order, what is the case here [5], is necessarily a MG. As canonical elements from the border are also required to be MGs, one might (too) easily conclude that the entire border is in the MG family. Unsurprisingly, this does not hold in the general case: there will be non-MG elements whose every subset is representative. The existence of these elements seems to have been missed by Dong *et al.*, although they play the same role in the substitution mechanism as the canonical MGs. In fact, this is the main reason for their expansion mechanism to be incomplete, *i.e.*, to fail in the recovery of some of the redundant MGs from the SSMG.

One may now question the interplay between the substitution and the total order, *i.e.*, in what sense the representative/canonical sets are *irreducible* for substitution? After all, the substitution is a reversible operator, so that any MG within a σ-equivalence class could have been chosen as its distinguished element to be kept (as Hamrouni *et al.* actually do). Although [14] says little on that point, our analysis shows that the representative/canonical order ideal structure is crucial. Indeed, it works like a magnetic nucleus for substitution in the sense that when properly performed, *i.e.*, in the *right* direction, it transforms an arbitrary itemset into a member of the representative ideal or of its (complete) border. Here the right direction is the substitution of a subset Z_1 in the argument X by the representative Z_2 in the γ-equivalence class of Z_1. We prove below that this inevitably "attracts" the result within the aforementioned set where such substitutions can no more be performed.

Our proposal is about completing the succinct representation with all those non-MGs from the border of the ideal order of representative MGs, as in many cases they are the only point from which some of the redundant MGs can be reached by substitution. The details of our approach come in the following paragraphs.

5.2 The DSFS Family

Here is an illustration of the above arguments: Assume the ascending support order on items in Table 3 (left) and consider the MGs of *eacbdf*. As pointed out above, it is impossible to derive *ecf* from the resulting SSMG. Indeed, its subset *cf* is a non-representative MG (*ac* is the representative in its γ-equivalence class). Hence, it remains outside the system, whereas neither *ea* nor *acd* have a derivation chain that ends at *ecf*. If we look the case other way round, the only sensible substitution from *cf* backwards is *cf*/*ac*. This produces *eac*, a curious set whose every subset is a representative (hence it belongs to the border of the corresponding ideal) without the set itself being even a MG. Clearly, adding *eac* to the SSMG would restore its completeness. This requires a larger definition of the canonicity which we provide below.

Definition 11. *Let $\mathcal{MG}\mathtt{rep}_\mathcal{K}$ be the set of the representative* MGs *that can be extracted from a context $\mathcal{K}$. The negative border of $\mathcal{MG}\mathtt{rep}_\mathcal{K}$ is:* Border($\mathcal{MG}\mathtt{rep}_\mathcal{K}$) = $\{X \subseteq \mathcal{I} \mid \forall Y \subset X,\ Y \in \mathcal{MG}\mathtt{rep}_\mathcal{K} \wedge X \notin \mathcal{MG}\mathtt{rep}_\mathcal{K}\}$.

Since canonical itemsets form the negative border of the representative ideal, the old canonical MGs of Dong *et al.* are obviously included in it ($\mathcal{MG}\mathtt{can}_\mathcal{K} \subseteq$ *Border*

[5] We leave aside the question of whether or not the choice of the representative MG in [14] amounts to fix a single linear extension of the set inclusion order and adopt, for the time being, the interpretation of Hamrouni *et al.*

($\mathcal{MG}\texttt{rep}_{\mathcal{K}}$)), together with the canonical non-MG elements. Moreover, the frequency constraint further splits it into four subsets.

In order to formalize the irreducibility status of the above sets, we rely on a constrained substitution operator. Actually, we distinguish two complementary "directions" for the substitution depending on the status of the involved sets. Thus, a positive (resp. negative) substitution for an itemset X amounts to replace a subset Z_1 of X by a set Z_2 of the same closure as Z_1 which is larger (resp. smaller) *w.r.t.* the itemset order. We focus on the relations between the set X and the set Y induced by one of the aforementioned substitutions.

Definition 12. *Let* $X, Y \subseteq \mathcal{I}$, $Z_1 \subset X$ *and* $Z_2 \subseteq \mathcal{I}$ *s.t.* $Z_1 \neq Z_2$, $Z_1'' = Z_2''$, *and Subst(*X, Z_1, Z_2*)* = Y*. The positive/negative substitution relations are defined as follows:*

- $X \vdash_d^+ Y$ *iff* $Z_1 \preceq Z_2$,
- $X \vdash_d^- Y$ *iff* $Z_2 \preceq Z_1$.

It is noteworthy that each substitution is either positive or negative, *i.e.*, there is no neutral substitution. Moreover, positive substitutions produce results that are larger *w.r.t.* $\preceq$ than the initial sets and hence have bigger ranks in the order ($X \vdash_d^+ Y$ implies $X \preceq Y$), while the negative ones have the opposite effect. In particular, if the replaced set is a representative, then the substitution is necessarily positive, while, conversely, if a representative replaces another set, then it is negative.

Consider now the irreducible elements for the negative substitution, *i.e.*, elements for which such substitution could not be applied. We call them *directed substitution-free* sets (denoted DSFSs).

Definition 13. *Let* $\mathcal{DSFS}_{\mathcal{K}}$ *be the collection of the directed substitution-free sets (DSFSs) that can be extracted from a context* $\mathcal{K}$. $\mathcal{DSFS}_{\mathcal{K}} = \{I \subseteq \mathcal{I} \mid \forall\, I_1 \subset I, \forall\, I_2 \subseteq \mathcal{I}, (I_1'' = I_2'' \Longrightarrow I_1 \preceq I_2)\}$.

Example 9. Consider the context in Table 3 with the ascending support order on items. The itemset *eac* is a DSFS, as mentioned above, whereas the family comprises *ea* and *acd*, but not *ecf*.

Clearly, the set of DSFSs equals the union of representative MGs and their negative border ($\mathcal{DSFS}_{\mathcal{K}} = \mathcal{MG}\texttt{rep}_{\mathcal{K}} \cup$ *Border*($\mathcal{MG}\texttt{rep}_{\mathcal{K}}$)). The next proposition is therefore immediate.

Proposition 1. *The set* $\mathcal{DSFS}_{\mathcal{K}}$ *is an order ideal of* $(2^{\mathcal{I}}, \subseteq)$.

Given its structure, the DSFS family can be easily constructed by a level-wise algorithm that, additionally, enumerates itemsets in the order $\preceq$. Thus, all the DSFSs at a particular level are easily recognizable since, all their subsets (in particular the maximal ones) belong to the already discovered part of the family. An additional effort is necessary to identify the representative itemsets among all the family members. To that end, the order properties are exploited: In fact, a representative is the first itemset to be examined within its γ-equivalence class. Hence, to establish that a DSFS is a representative, it is enough to check that its closure has not been produced by a previously extracted DSFS.

So far, we have established that any total order on itemsets generates a core ideal in the Boolean lattice that works as an irreducible nucleus for swapping subsets with equivalent ones. On the reverse side of the question, there is the expansion process: It starts with the DSFSs and retrieves the entire MG family. Unsurprisingly, the positive substitution is used to that end. Moreover, as for each negative substitution there is a reverse positive one, and *vice versa*, every itemset from the Boolean lattice is necessarily reachable by at least one chain of positive substitutions starting from a DSFS. In particular, redundant MGs are reachable in this way.

Following to the above arguments, we claim that every redundant MG can be derived from a DSFS of the same closure, using positive substitutions. More specifically, starting from the DSFS X, and operating successive substitutions of a representative subset Z_1 by a non-representative set Z_2 from the same γ-equivalence class will necessarily result in the generation of the entire MG family. Hence we can assert that the above retrieval mechanism, in its most general form is a complete means for calculating the MGs.

Proposition 2. *The expansion process is complete.*

Proof. (Sketch) Formally, it can be shown by induction on the rank of a MG, *w.r.t.* the total order relation on itemsets $\preceq$, that each redundant MG contracts by successive negative directed substitutions to a DSFS. The key fact of the proof says that the result of a negative directed substitution has a smaller rank *w.r.t.* $\preceq$ than the original set.

Another concern with retrieval is the correctness of the mechanism, *i.e.*, the warranty that only MGs will be retrieved. To that end, we employ a straightforward support test: An itemset is a MG whenever its support is strictly lower than the support of all its strict subsets. For efficiency reasons, the test is limited to maximal subsets only. Obviously, such a test requires a level-wise traversal of the Boolean lattice, which is a classical approach of frequent itemset mining. Consequently, we may assert that the expansion is correct as well.

Proposition 3. *The expansion process is correct.*

Proof. By construction, all derived elements from a DSFS are explicitly checked for being MGs.

Theorem 1 states the adequacy of our global approach.

Theorem 1. *The set $\mathcal{DSFS}_{\mathcal{K}}$ of the directed substitution-free sets (*DSFS*s) is a lossless representation of the minimal generator set.*

To sum up, the DSFS family is the complete structure necessary to ensure that every MG can be reached by a substitution-based expansion process that is well directed and hence cycle-free. The DSFSs can be efficiently mined due to the order ideal form of the family as it does not even require the discovery of all MGs. Despite the significant progress with respect to the previous two studies, there are issues with our framework that are yet to be clarified.

First, while both SSMGs only rely on MGs, our construct involves further sets from (yet laying not too far in) the Boolean lattice. The impact of these elements on the size of the representation needs to be examined. Some clues on how many non-MG DSFSs could appear are provided in Section 7.

Another issue, somewhat related to the previous one, concerns the expansion mechanism. An important feature thereof would be to limit all substitutions to MG subsets. In other terms, it would be much simpler and more efficient to always replace a representative MG Z_1 by a non-representative one Z_2 (and not an arbitrary set from the same γ-equivalence class). Although we do not have a formal proof that such restriction preserves the completeness of the expansion, we conjecture that it is the case.

Finally, minimality of the DSFS family is an issue as well. Whereas it is definitely minimal for the entire Boolean lattice, it could be in some cases that a strict subset of the DSFSs suffices to generate all the MGs. For instance, if there are more than one DSFS in the same σ-equivalence class, then clearly only the smallest of them *w.r.t.* the total order is indispensable. Moreover, some non-MG DSFSs may be of no use for expansion towards all MGs, so it could be useful to remove them from the effective representation. Provided a method for eliminating unnecessary DSFSs is designed, the trade-off between reduction rate and cost should also be looked at.

6 The DSFS_MINER Algorithm

In this section, we sketch the key ideas related to an algorithm allowing the extraction of the DSFSs. This algorithm, called DSFS_MINER, uses a breadth-first (or level-wise) browsing of the search space. It hence treats minimal generator (MG) candidates by ascending size. For a given size, the associated candidates are sorted *w.r.t.* the total order relation $\preceq$. This is naturally obtained as soon as items are ordered *w.r.t.* $\preceq$. Indeed, the procedure we use to generate candidates of size (i+1), using those of size i, respects the total order relation since it combines each time two itemsets X and Y, *s.t.* $X \prec Y$, sharing their first (i-1) items. The latter items will be augmented by the remaining item in X and, then, by the remaining one in Y. Hence, the total order relation will always be respected.

The pseudo-code of the DSFS_MINER algorithm is given by Algorithm 1. $\mathcal{FDSFS}_{\mathcal{K}}$ denotes the frequent directed substitution-free sets that can be extracted from the extraction context $\mathcal{K}$. While the set of frequent closed itemsets that can be extracted from $\mathcal{K}$ is denoted $\mathcal{FCI}_{\mathcal{K}}$. Also in the pseudo-code, the set of candidates, to be tested in the ith iteration whether they are frequent representative MGs or not, will be denoted $\mathcal{FMG}\mathtt{rep}_i$. Each candidate c is characterized by the following fields: (1) *Supp*: its support, (2) *Clos*: its closure, (3) *Immediate-subsets*: its list of immediate subsets. Since by definition, the representative MG is the smallest one in its γ-equivalence class, *w.r.t.* $\preceq$, c is a representative if it is the first one to produce the associated closure of its γ-equivalence class (*cf.* Algorithm 1, lines 14-15).

Algorithm 1. DSFS_MINER

Data: - An extraction context $\mathcal{K}$ where items are sorted *w.r.t.* the total order relation $\preceq$ and the threshold of support *minsupp*.

Results: - The set $\mathcal{FDSFS}_{\mathcal{K}}$.

Begin
  $\mathcal{FDSFS}_{\mathcal{K}} = \{\emptyset\}$;
  $\mathcal{FCI}_{\mathcal{K}} = \{\emptyset''\}$;
  $\mathcal{FMG}\mathtt{rep}_1 = \mathcal{I} \setminus \{\emptyset''\}$;
  For *(i=1 ; $\mathcal{FMG}\mathtt{rep}_i \neq \emptyset$; i++)* **do**
    closures in $\mathcal{FMG}\mathtt{rep}_i = \emptyset$;
    supports in $\mathcal{FMG}\mathtt{rep}_i = 0$;
    $\mathcal{FMG}\mathtt{rep}_i$ = GEN-CLOSURE($\mathcal{FMG}\mathtt{rep}_i$); /*The procedure GEN-CLOSURE produces closures as done in [4]*/
    Forall *($c \in \mathcal{FMG}\mathtt{rep}_i$)* **do**
      If *(c.Supp < minsupp)* **then**
        $\mathcal{FMG}\mathtt{rep}_i = \mathcal{FMG}\mathtt{rep}_i \setminus \{c\}$;
      Else
        $\mathcal{FDSFS}_{\mathcal{K}} = \mathcal{FDSFS}_{\mathcal{K}} \cup \{c\}$;
        If *(c.Clos $\notin \mathcal{FCI}_{\mathcal{K}}$)* **then**
          $\mathcal{FCI}_{\mathcal{K}} = \mathcal{FCI}_{\mathcal{K}} \cup \{c.Clos\}$;
        Else
          $\mathcal{FMG}\mathtt{rep}_i = \mathcal{FMG}\mathtt{rep}_i \setminus \{c\}$;
    $\mathcal{FMG}\mathtt{rep}_{i+1}$ = GEN-REPRESENTATIVE($\mathcal{FMG}\mathtt{rep}_i$);
  Return $\mathcal{FDSFS}_{\mathcal{K}}$;
End

To generate candidates of size $(i+1)$ starting from frequent representative MGs of size i, DSFS_MINER uses the GEN-REPRESENTATIVE procedure whose pseudo-code is given by Algorithm 2. The latter is illustrated by Example 10.

Example 10. Consider the extraction context given by Table 3 (left). Let *minsupp* be equal to 1 and the total order relation be the ascending support one. We will mainly focus on how our new definition is able to take in consideration an itemset such as *eac* thanks to the tests used in Algorithm 2. Indeed, when generating the set of representative MG candidates of size 3, we have the 2-frequent representative MGs ea and ec that have their first item in common. Hence, by composing them we obtain the candidate *eac* (*cf.* lines 2-6). After that, *eac* will be tested to check whether all its subsets are frequent representative MGs. It is the case. Hence, the value of the variable `Is-deleted` remains equal to 0. While that of `Is-covered` will change and become equal to 1 since *eac* is included in the closure of its subset *ea*, equal to *eacbdf* (*cf.* lines 7-16). After this test, we have the information that *eac* has all its subsets as frequent representative MGs but is not a MG since it has the same closure than one of its subsets. Hence, it is a frequent non-MG DSFS. Thus, *eac* belongs to the frequent part of the negative border. It will then be retained as an element of the representation (*cf.* lines 17-19).

Algorithm 2. GEN-REPRESENTATIVE

Data: - The set $\mathcal{FMG}\mathtt{rep}_i$.
Results: - The set $\mathcal{FMG}\mathtt{rep}_{i+1}$.
Begin
/*The combinatorial phase of APRIORI-GEN [7] *w.r.t.* the total order relation $\preceq$*/
insert into $\mathcal{FMG}\mathtt{rep}_{i+1}$
select $p[1], p[2], ..., p[i\text{-}1], p[i], q[i]$
from $\mathcal{FMG}\mathtt{rep}_i\ p$, $\mathcal{FMG}\mathtt{rep}_i\ q$
where $p[1] = q[1], p[2] = q[2], ..., p[i\text{-}1] = q[i\text{-}1], p[i] \prec q[i]$;

Forall *($c \in \mathcal{FMG}\mathtt{rep}_{i+1}$)* **do**
  Is-deleted = 0; /*This variable checks whether c is deleted because one of its immediate subsets is not a frequent representative MG of its γ-equivalence class.*/
  Is-covered = 0; /*This variable checks whether c is covered by the closure of one of its immediate subsets.*/
  Forall *($c_1 \in$ c.Immediate-subsets)* **do**
    If *($c_1 \notin \mathcal{FMG}\mathtt{rep}_i$)* **then**
      $\mathcal{FMG}\mathtt{rep}_{i+1} = \mathcal{FMG}\mathtt{rep}_{i+1} \setminus \{c\}$;
      Is-deleted = 1;
    Else
      If *($c \subseteq c_1.Clos$)* **then**
        Is-covered = 1;

  If *(Is-deleted = 0 and Is-covered = 1)* **then**
    $\mathcal{FDSFS}_\mathcal{K} = \mathcal{FDSFS}_\mathcal{K} \cup \{c\}$;
    $\mathcal{FMG}\mathtt{rep}_{i+1} = \mathcal{FMG}\mathtt{rep}_{i+1} \setminus \{c\}$;

Return $\mathcal{FMG}\mathtt{rep}_{i+1}$;
End

7 Experimental Results

In these experiments, we compare the cardinality of $\mathcal{FDSFS}_\mathcal{K}$ to that of the frequent succinct MGs as defined by Dong *et al.* (denoted $\mathcal{FMG}\mathtt{suc}_\mathcal{K}$) and to that of the whole set of frequent MGs (denoted $\mathcal{FMG}_\mathcal{K}$). For the sake of clarity, we give the cardinality of $\mathcal{FDSFS}_\mathcal{K}$ as a sum of those of its components, *i.e.*, the frequent representative MGs (denoted $\mathcal{FMG}\mathtt{rep}_\mathcal{K}$) [6], the frequent canonical MGs (denoted $\mathcal{FMG}\mathtt{can}_\mathcal{K}$) and the frequent canonical non-MG elements. The latter set represents the difference between $\mathcal{FDSFS}_\mathcal{K}$ and $\mathcal{FMG}\mathtt{suc}_\mathcal{K}$. It will hence be denoted $\mathcal{DIFF}_\mathcal{K}$. Experiments were conducted on several benchmark datasets [7] and for different total order relations. We will give some representative results obtained from the PUMSB, MUSHROOM, CHESS and T40I10D100K datasets. The first three datasets are commonly considered as dense

[6] Since, a frequent representative MG is unique in its γ-equivalence class, the number of frequent closed itemsets (CIs) is equal to that of frequent representative MGs and, hence, it will not be given.

[7] Downloadable from *http://fimi.cs.helsinki.fi/data*

Table 4. The size of the different sets for benchmark datasets

PUMSB						
minsupp (%)	$\|\mathcal{FMG}_{\mathcal{K}}\|$	$\|\mathcal{FMG}\mathtt{rep}_{\mathcal{K}}\|$	$\|\mathcal{FMG}\mathtt{can}_{\mathcal{K}}\|$	$\|\mathcal{DIFF}_{\mathcal{K}}\|$	$\|\mathcal{FDSFS}_{\mathcal{K}}\|$	$\frac{\|\mathcal{FMG}_{\mathcal{K}}\|}{\|\mathcal{FDSFS}_{\mathcal{K}}\|}$
90	2, 032	1, 467	3	*0*	1, 470	**1.38**
85	13, 795	8, 514	5	*7*	8, 526	**1.62**
80	67, 860	33, 308	5	*8*	33, 321	**2.04**
75	248, 406	101, 083	5	*8*	101, 096	**2.48**
70	658, 565	241, 259	6	*11*	241, 276	**2.70**

MUSHROOM						
minsupp (%)	$\|\mathcal{FMG}_{\mathcal{K}}\|$	$\|\mathcal{FMG}\mathtt{rep}_{\mathcal{K}}\|$	$\|\mathcal{FMG}\mathtt{can}_{\mathcal{K}}\|$	$\|\mathcal{DIFF}_{\mathcal{K}}\|$	$\|\mathcal{FDSFS}_{\mathcal{K}}\|$	$\frac{\|\mathcal{FMG}_{\mathcal{K}}\|}{\|\mathcal{FDSFS}_{\mathcal{K}}\|}$
10	7, 631	4, 897	471	*545*	5, 913	**1.29**
5	21, 160	12, 854	1, 207	*1, 251*	15, 312	**1.38**
3	37, 973	22, 230	1, 943	*1, 911*	26, 084	**1.46**
2	57, 728	31, 767	2, 644	*2, 479*	36, 890	**1.56**
1	103, 517	51, 672	3, 818	*3, 576*	59, 066	**1.75**

CHESS						
minsupp (%)	$\|\mathcal{FMG}_{\mathcal{K}}\|$	$\|\mathcal{FMG}\mathtt{rep}_{\mathcal{K}}\|$	$\|\mathcal{FMG}\mathtt{can}_{\mathcal{K}}\|$	$\|\mathcal{DIFF}_{\mathcal{K}}\|$	$\|\mathcal{FDSFS}_{\mathcal{K}}\|$	$\frac{\|\mathcal{FMG}_{\mathcal{K}}\|}{\|\mathcal{FDSFS}_{\mathcal{K}}\|}$
90	504	504	0	*2*	506	**0.99**
80	5, 114	5, 114	0	*6*	5, 120	**0.99**
70	23, 992	23, 992	0	*16*	24, 008	**0.99**
60	98, 804	98, 778	1	*28*	98, 807	**0.99**
50	372, 604	369, 451	2	*63*	369, 516	**1.01**

T40I10D100K						
minsupp (%)	$\|\mathcal{FMG}_{\mathcal{K}}\|$	$\|\mathcal{FMG}\mathtt{rep}_{\mathcal{K}}\|$	$\|\mathcal{FMG}\mathtt{can}_{\mathcal{K}}\|$	$\|\mathcal{DIFF}_{\mathcal{K}}\|$	$\|\mathcal{FDSFS}_{\mathcal{K}}\|$	$\frac{\|\mathcal{FMG}_{\mathcal{K}}\|}{\|\mathcal{FDSFS}_{\mathcal{K}}\|}$
5	317	317	0	*0*	317	**1**
2.50	1, 222	1, 222	0	*0*	1, 222	**1**
2	2, 294	2, 294	0	*0*	2, 294	**1**
1.50	6, 540	6, 540	0	*0*	6, 540	**1**
1	65, 237	65, 237	0	*0*	65, 237	**1**

ones while the fourth dataset is considered as a sparse one. The ascending support order is chosen as an example of a total order relation $\preceq$. The results on these datasets are summarized in Table 4.

For the PUMSB and MUSHROOM datasets, we notice an important lossless reduction reaching a peak of $\mathbf{2.70}$ and $\mathbf{1.75}$ times, respectively. For the PUMSB dataset, the number of canonical elements is too small. Hence, we can assume that there is an average of $\mathbf{1}$ frequent DSFS per γ-equivalence class. For the MUSHROOM dataset, this number is larger than that of the first dataset. Nevertheless, *w.r.t.* the number of frequent CIs, that of canonical elements is still very low which makes it possible to get, in average, only $\mathbf{1.18}$ frequent DSFS per γ-equivalence class. Noteworthily, the numbers of canonical elements which are, respectively, MGs and non-MGs, are nearly equal.

The case of the CHESS dataset is also very interesting. At a glance, for *minsupp* values higher than $\mathbf{60}\%$, statistics are far from indicating that CHESS is a dense dataset.

Indeed, each γ-equivalence class only contains a unique frequent MG, *i.e.*, the representative one. Nevertheless, the latter is in general different from its closure which leads to the existence of some canonical elements in $\mathcal{FDSFS}_{\mathcal{K}}$. These elements are necessarily non-MGs (*i.e.*, belong to $\mathcal{DIFF}_{\mathcal{K}}$). Hence, for these values of *minsupp*, the size of $\mathcal{FDSFS}_{\mathcal{K}}$ is slightly greater than that of $\mathcal{FMG}_{\mathcal{K}}$. For *minsupp* = **50%**, the size of $\mathcal{FDSFS}_{\mathcal{K}}$ becomes lower than that of $\mathcal{FMG}_{\mathcal{K}}$ since, in this case, there are frequent redundant MGs whose number (equal to ($|\mathcal{FMG}_{\mathcal{K}}|$ - $|\mathcal{FMG}\texttt{rep}_{\mathcal{K}}|$ - $|\mathcal{FMG}\texttt{can}_{\mathcal{K}}|$), *i.e.*, **3531**) is by far greater than the cardinality of $\mathcal{DIFF}_{\mathcal{K}}$ (equal to **63**).

For the sparse dataset T40I10D100K, each itemset is equal to its closure. Hence, the set $\mathcal{FDSFS}_{\mathcal{K}}$ is simply equal to the set $\mathcal{FMG}\texttt{rep}_{\mathcal{K}}$.

8 Conclusion

The rapidly increasing reliance on MGs in association rule mining motivated a push towards deeper understanding of their structural properties, computational behavior, connections to other constructs, etc. Reducing the combinatorial variations within the family is a central issue for its results in smaller-size storage and eases further manipulations. We introduced here a generation operator for MGs and a family of irreducible elements, called DSFSs, for the operator which jointly constitute a concise yet lossless representation of the entire MG family. Our work follows an original idea from the literature that was developed to a theoretically sound construct and provided with both deeper structural results and computational means. First empirical evidences for the benefits of the new family have been obtained as well.

At its current stage, our study focuses on the efficient generation of the DSFS family. Thus, besides our own method, other algorithms from the literature working with MGs could be adapted for this task, both breadth-first search ones, *e.g.*, TITANIC [5], and depth-first ones, *e.g.*, the right-to-left search algorithm GR-GROWTH [8]. Of course, it will be interesting to compare performances of these algorithms on different datasets. The next step in this direction is the design of efficient expansion methods, *i.e.*, ones yielding to the entire MG family from the DSFSs.

We also plan to investigate the connection between the DSFSs and other set families studied in FCA or data mining, *e.g.*, pseudo-closed sets [21,22], non-derivable itemsets [23], etc. The implications of our definition for constructs similar to MGs from fields like Boolean functions and hyper-graphs, will also be examined.

Acknowledgements

We are grateful to the anonymous reviewers for their useful remarks and suggestions. The first author would like to thank *Le fonds québécois de la recherche sur la nature et les technologies* (FQRNT) for its financial support. The work was also partially supported by the second author's NSERC Discovery grant and by the French-Tunisian project CMCU 05G1412.

References

1. Bastide, Y., Pasquier, N., Taouil, R., Stumme, G., Lakhal, L.: Mining minimal non-redundant association rules using frequent closed itemsets. In: Proceedings of the 1st International Conference on Computational Logic (DOOD 2000), Springer-Verlag, LNAI, volume 1861, London, UK. (2000) 972–986
2. Nehme, K., Valtchev, P., Hacene, M.R., Godin, R.: On computing the minimal generator family for concept lattices and Icebergs. In Ganter, B., Godin, R., eds.: Proceedings of the 3rd International Conference on Formal Concept Analysis (ICFCA 2005), Springer-Verlag, LNCS, volume 3403, Lens, France. (2005) 192–207
3. Bastide, Y., Taouil, R., Pasquier, N., Stumme, G., Lakhal, L.: Mining frequent patterns with counting inference. In: Proceeding of the 6th ACM-SIGKDD International Conference on Knowledge Discovery and Data Mining, volume 2(2), Boston, Massachusetts, USA. (2000) 66–75
4. Pasquier, N., Bastide, Y., Taouil, R., Lakhal, L.: Efficient mining of association rules using closed itemset lattices. J. of Information Systems, volume 24 (1) (1999) 25–46
5. Stumme, G., Taouil, R., Bastide, Y., Pasquier, N., Lakhal, L.: Computing iceberg concept lattices with TITANIC. J. on Knowledge and Data Engineering (KDE) **2** (2002) 189–222
6. Hamrouni, T., Ben Yahia, S., Slimani, Y.: PRINCE: An algorithm for generating rule bases without closure computations. In Tjoa, A.M., Trujillo, J., eds.: Proceedings of the 7th International Conference on Data Warehousing and Knowledge Discovery (DAWAK 2005), Springer-Verlag, LNCS, volume 3589, Copenhagen, Denmark. (2005) 346–355
7. Agrawal, R., Srikant, R.: Fast algorithms for mining association rules. In Bocca, J.B., Jarke, M., Zaniolo, C., eds.: Proceedings of the 20th International Conference on Very Large Databases, Santiago, Chile. (1994) 478–499
8. Li, J., Li, H., Wong, L., Pei, J., Dong, G.: Minimum description length principle: Generators are preferable to closed patterns. In: Proceedings of the 21st National Conference on Artificial Intelligence (AAAI 2006), Boston, Massachusetts, USA. (2006) 409–414
9. De Raedt, L., Ramon, J.: Condensed representations for inductive logic programming. In Dubois, D., Welty, C., eds.: Proceedings of the Ninth International Conference on Principles of Knowledge Representation and Reasoning (KR 2004), Whistler, Canada. (2004) 438–446
10. Li, H., Li, J., Wong, L., Feng, M., Tan, Y.: Relative risk and odds ratio: a data mining perspective. In: Proceedings of the 24th ACM SIGMOD-SIGACT-SIGART symposium on Principles Of Database Systems (PODS 2005), Baltimore, Maryland, USA. (2005) 368–377
11. Calders, T., Rigotti, C., Boulicaut, J.F.: A survey on condensed representations for frequent sets. In: Constraint Based Mining, Springer-Verlag, LNAI, volume 3848. (2006) 64–80
12. Zaki, M.J.: Mining non-redundant association rules. J. of Data Mining and Knowledge Discovery (DMKD), volume 9. (2004) 223–248
13. Li, J.: On optimal rule discovery. In IEEE J. on Transactions on Knowledge and Data Engineering (TKDE) **18 (4)** (2006) 460–471
14. Dong, G., Jiang, C., Pei, J., Li, J., Wong, L.: Mining succinct systems of minimal generators of formal concepts. In: Proceedings of the 10th International Conference on Database Systems for Advanced Applications (DASFAA 2005), Springer-Verlag, LNCS, volume 3453, Beijing, China. (2005) 175–187
15. Hamrouni, T., Ben Yahia, S., Mephu Nguifo, E.: Succinct system of minimal generators: A thorough study, limitations and new definitions. In: Proceedings of the 4th International Conference on Concept Lattices and their Applications (CLA 2006), Hammamet, Tunisia. (2006) 139–153
16. Ganter, B., Wille, R.: Formal Concept Analysis. Springer-Verlag (1999)

17. Hamrouni, T., Ben Yahia, S., Mephu Nguifo, E.: Redundancy-free generic bases of association rules. In: Proceedings of the 8th French Conference on Machine Learning (CAp 2006), Presses Universitaires de Grenoble, Trégastel, France. (2006) 363–378
18. Gély, A., Medina, R., Nourine, L., Renaud, Y.: Uncovering and reducing hidden combinatorics in Guigues-Duquenne bases. In Ganter, B., Godin, R., eds.: Proceedings of the 3rd International Conference on Formal Concept Analysis (ICFCA 2005), Springer-Verlag, LNCS, volume 3403, Lens, France. (2005) 235–248
19. Hamrouni, T., Ben Yahia, S., Mephu Nguifo, E.: Generic association rule bases: Are they so succinct? In: Proceedings of the 4th International Conference on Concept Lattices and their Applications (CLA 2006), Hammamet, Tunisia. (2006) 155–169
20. Armstrong, W.W.: Dependency structures of database relationships. In: Proceedings of IFIP Congress, Geneva, Switzerland. (1974) 580–583
21. Guigues, J.L., Duquenne, V.: Familles minimales d'implications informatives résultant d'un tableau de données binaires. Mathématiques et Sciences Humaines. **24** (1986) 5–18
22. Valtchev, P., Duquenne, V.: Towards scalable divide-and-conquer methods for computing concepts and implications. Accepted in the Jounal of Discrete Applied Mathematics (2006)
23. Calders, T., Goethals, B.: Mining all non-derivable frequent itemsets. In Elomaa, T., Mannila, H., Toivonen, H., eds.: Proceedings of the 6th European Conference on Principles and Practice of Knowledge Discovery in Databases (PKDD 2002), Springer-Verlag, LNCS, volume 2431, Helsinki, Finland. (2002) 74–85

Some Notes on Pseudo-closed Sets*

Sebastian Rudolph

Institute AIFB
University of Karlsruhe (TH)
Germany
rudolph@aifb.uni-karlsruhe.de

Abstract. Pseudo-intents (also called pseudo-closed sets) of formal contexts have gained interest in recent years, since this notion is helpful for finding minimal representations of implicational theories. In particular, there are some open problems regarding complexity. In our paper, we compile some results about pseudo-intents which contribute to the understanding of this notion and help in designing optimized algorithms. We provide a characterization of pseudo-intents based on the notion of a formal context's incrementors. The latter are essentially non-closed sets which – when added to a closure system – do not enforce the presence of other new attribute sets. In particular, the provided definition is non recursive. Moreover we show that this notion coincides with the notion of a quasi-closed set that is not closed, which enables to reuse existing results and to formulate an algorithm that checks for pseudo-closedness. Later on, we provide an approach for further optimizing those algorithms based on a result which correlates the set of pseudo-intents of a formal context with the pseudo-intents of this context's reduced version.

1 Introduction

Pseudo-intents are of significant interest in formal concept analysis. One central result ([5]) states, that the implication set $\{P \to P^{II} \mid P \text{ pseudo-intent of } \mathbb{K}\}$ (called *stem base*) constitutes a so-called *implicational base*, i.e., a minimal set of implications generating the implicational theory of the formal context $\mathbb{K}$. In this regard it is also important to note that for an arbitrary implication, checking whether it is semantically entailed by a set of implications can be decided in linear time ([2, 8]). Thus, pseudo-intents become relevant for problems related to small (yet quick to query) representation of implicative knowledge.

The complexity of determining for a given context $\mathbb{K} = (G, M, I)$ and attribute set $A \subseteq M$, whether A is a pseudo-intent (or: pseudo-closed) with respect to $\mathbb{K}$ is still an open problem (see [9]). The prevailing assumption seems to be that the problem's complexity is rather high (at least beyond polynomial time). Partial results ([7, 6]) show that it is in coNP.

In our paper, we compile some results about pseudo-intents and provide optimized algorithms for checking for pseudo-closedness.

In detail, we will proceed as follows: In Section 2, we recall the fundamental definitions and propositions of FCA needed to specify and deal with the topic. Section 3

* Supported by the Deutsche Forschungsgemeinschaft (DFG) under the ReaSem project.

S.O. Kuznetsov and S. Schmidt (Eds.): ICFCA 2007, LNAI 4390, pp. 151–165, 2007.

provides and verifies an algorithm which allows to convert an arbitrary set of implications into a stem base. Section 4 introduces the notion of *incrementor* and shows how it can be used to provide a non-recursive characterization of pseudo-intents. In the end, this notion shows to have a direct correspondence to that of a *quasi intent* introduced in [3]. Resulting from these preceding considerations, Section 5 presents an algorithm which checks for pseudo-closedness. Section 6 shows how pseudo-closedness can be checked even by examining the reduced version of the considered context and provides a corresponding algorithm. Finally, Section 7 concludes and outlines possible directions for further research.

2 Preliminaries

In this section, we will introduce the notions from formal concept analysis necessary for our work.

First of all, note that we use the notation "$\subset$" to indicate the *strict* subset, i.e. $A \subset B$ means $A \subseteq B$ and $A \neq B$.

Deviating from the usual line of presentation, we will introduce implications and pseudo-closed sets just on the basis of closure operators. This allows to talk about those notions independently from concrete formal concepts and facilitates the presentation of some results in the sequel. However, note that this is not a proper generalization, since every closure operator can be represented by the $(.)^{II}$-operator of an appropriately chosen formal context (e.g. the context $(\{A \mid \varphi(A) = A\}, M, \ni)$). Thus, the cited definitions and results – although defined on basis of a formal context – carry over to our way of introducing those notions.

The following considerations are based on an arbitrary set M. We will first define the fundamental notion of a closure operator M. Roughly spoken, applying such an operator to a set can be understood as a minimal extension of that set in order to fulfill certain properties.

Definition 1. *Let M be an arbitrary set. A function $\varphi : \mathcal{P}(M) \to \mathcal{P}(M)$ (where $\mathcal{P}(M)$ denotes the powerset of M) will be called*

- EXTENSIVE, *if $A \subseteq \varphi(A)$ for all $A \subseteq M$,*
- MONOTONE, *if from $A \subseteq B$ follows $\varphi(A) \subseteq \varphi(B)$ for all $A, B \subseteq M$, and*
- IDEMPOTENT, *if $\varphi(\varphi(A)) = \varphi(A)$ for all $A \subseteq M$.*

If φ is extensive, monotone, and idempotent, we will call it a CLOSURE OPERATOR. *In this case, we will additionally call*

- *$\varphi(A)$ the* CLOSURE *of A,*
- *A* CLOSED *(or φ-*CLOSED*), if $A = \varphi(A)$.*

The family of all closed sets is also called CLOSURE SYSTEM. Furthermore, any closure system constitutes a *lattice* with set inclusion as the respective order relation.

In the sequel, we show, in which way closure operators are closely related to implications.

Definition 2. *Let M be an arbitrary set. An* IMPLICATION *on M is a pair (A, B) with $A, B \subseteq M$. To support intuition, we write $A \to B$ instead of (A, B).*[1]
A set $C \subseteq M$ RESPECTS *an implication $A \to B$ if*

$$A \subseteq C \quad \textit{implies} \quad B \subseteq C.$$

Furthermore, for $C \subseteq M$ and a set $\mathfrak{I}$ of implications on M, let $C^{\mathfrak{I}}$ denote the smallest set with

- *$C \subseteq C^{\mathfrak{I}}$ and*
- *$C^{\mathfrak{I}}$ respects $\mathfrak{i}$ for every implication $\mathfrak{i} \in \mathfrak{I}$.*[2]

It is well known, that the operation $(.)^{\mathfrak{I}}$ is a closure operator on M. So, according to Definition 1, if $C = C^{\mathfrak{I}}$, we call C ($\mathfrak{I}$-)CLOSED.

Definition 3. *We say $\mathfrak{I}$* ENTAILS *$A \to B$ (written: $\mathfrak{I} \models A \to B$), if every $C \subseteq M$ that respects all implications of $\mathfrak{I}$ also respects $A \to B$.*

An implication set $\mathfrak{I}$ will be called NON-REDUNDANT, *if for any $\mathfrak{i} \in \mathfrak{I}$, we have that $\mathfrak{I} \setminus \{\mathfrak{i}\}$ does not entail $\mathfrak{i}$.*

An implication set $\mathfrak{I}$ will be called an IMPLICATION BASE *for a closure operator φ if*

- *it is* NON-REDUNDANT, *i.e. for any $\mathfrak{i} \in \mathfrak{I}$, we have that $\mathfrak{I} \setminus \{\mathfrak{i}\}$ does not entail $\mathfrak{i}$,*
- *it is* SOUND, *i.e., any implication on M entailed by $\mathfrak{I}$ is respected by all φ-closed sets, and*
- *it is* COMPLETE, *i.e., any implication on M respected by all φ-closed sets is entailed by $\mathfrak{I}$.*

Well-known facts concerning the entailment of implications are

- $\mathfrak{I} \models A \to B$ exactly if $B \subseteq A^{\mathfrak{I}}$ and
- $\mathfrak{I}$ is non-redundant iff $B \not\subseteq A^{\mathfrak{I} \setminus \{A \to B\}}$ for all $A \to B \in \mathfrak{I}$.

Below, we will now define the central notion of this paper. Opposed to the usual way of presentation, we will define the notion of pseudo-closedness independently from a particular formal context, just referring to a given closure operator. Besides the more general definition this will facilitate our considerations in section 3.[3]

Definition 4. *For a given closure operator φ, a set $P \subseteq M$ will be called* PSEUDO-CLOSED *if $\varphi(P) \neq P$ and $\varphi(Q) \subseteq P$ holds for every pseudo-closed $Q \subset P$.*

Note that this definition is recursive. Since the set M is always assumed to be finite in the sequel, it is nevertheless correct. However, directly using this definition to check whether an attribute set is a pseudo-intent requires a recursion as well and is therefore computationally costly. This led to the complexity questions mentioned in the introduction.

Regarding pseudo-closed sets, we give corollaries of the Propositions 24 and 25 from [4].

[1] To facilitate reading we will occasionally omit the parentheses, i.e., we will write $a, b \to c$ instead of $\{a, b\} \to \{c\}$.

[2] Note, that this is well-defined, since the mentioned properties are closed wrt. intersection.

[3] Trivially, this coincides with the notion of pseudo-intent of a formal context if we set $\varphi = (.)^{II}$.

Proposition 1. *If P and Q are closed or pseudo-closed sets with $P \not\subseteq Q$ and $Q \not\subseteq P$, then $P \cap Q$ is a closed set.*

The first Proposition directly yields the fact that the set of all closed and pseudo-closed sets (of a closure operator φ) constitute a closure system themselves (for another closure operator ψ).

Proposition 2. *Every (wrt. a closure operator φ) sound and complete set of implications contains an implication $A \to B$ with $A \subseteq P$ and $\varphi(A) = \varphi(P)$ for every pseudo-closed set P.*

Moreover, for every closure operator, the family of its pseudo-closed sets can be used to define a canonical implication base called *stem base* ([5]):

Theorem 1. *Let φ be a closure operator. Then the set*

$$\mathfrak{SB} := \{P \to \varphi(P) \mid P \text{ pseudo-closed for } \varphi\}$$

is an implication base of φ.

In the remainder of this section, we will very briefly recall well-known basic facts from FCA for later reference.

Proposition 3. *Properties of the derivation operator $(.)^I$.*

- *$(.)^{II}$ is a closure operator on G as well as on M, i.e., it is extensive (**ext**^{II}), monotone (**mon**^{II}) and idempotent (**idp**^{II}).*
- *for all $A \neq \emptyset$, $A^I = \bigcap_{a \in A} a^I$.* **decomp**

We use $\mathcal{I}(\mathbb{K})$ to denote the family of all concept intents of $\mathbb{K}$.

The concept intents of a formal concept are exactly those attribute sets closed wrt. $(.)^{II}$, i.e., $\mathcal{I}(\mathbb{K}) = \{A \mid A = A^{II} \subseteq M\}$. In other words, the set $\mathcal{I}(\mathbb{K})$ coincides with the closure system generated by $(.)^{II}$ on M. Consequently, the family $\mathcal{I}(\mathbb{K})$ of all concept intents of a formal context is closed wrt. intersection (**clos**$\cap$).

We proceed by giving a Proposition which is the dual of Proposition 30 from [4].

Proposition 4. *If $G \subseteq H$ then every intent of $(G, M, I \cap (G \times M))$ is an intent of (H, M, I).*

In words, the preceding proposition just states that adding an object with arbitrary intent to a context preserves all previous intents.

3 Generating Stem Bases from Implication Sets

In this section, we present an algorithm which is a slight modification of the one presented in [1] and provide a self-contained proof for its correctness.

Given an arbitrary finite set $\mathfrak{I} = \{\mathfrak{i}_1, \ldots, \mathfrak{i}_n\}$ of implications on an attribute set M, the algorithm from Fig. 1 will convert this set into a stem base $\mathfrak{SB}$ with $\mathfrak{SB} \models A \to B$ exactly if $\mathfrak{I} \models A \to B$.

```
function:  stembase(𝔍)
 1. Set 𝔖𝔅 := ∅.

 2. For every A → B ∈ 𝔍
    substitute A → B by A → (A ∪ B)^𝔍.

 3. As long as 𝔍 ≠ ∅,

    (a) select an A → B from 𝔍,
    (b) delete A → B from 𝔍,
    (c) calculate A^(𝔍∪𝔖𝔅),
    (d) if A^(𝔍∪𝔖𝔅) ≠ B then
        add A^(𝔍∪𝔖𝔅) → B to 𝔖𝔅.

 4. Output 𝔖𝔅 and terminate.
```

Fig. 1. Algorithm `stembase(𝔍)` for calculating the stem base of the implicational theory generated by $\mathfrak{J}$

Theorem 2. *The algorithm* `stembase` *computes a stembase for the closure operator* $(.)^{\mathfrak{J}}$.

Proof: We have to show two properties: For any set $\mathfrak{J}$ of implications on M, we have

- $(.)^{\mathfrak{J}} = (.)^{\texttt{stembase}(\mathfrak{J})}$ and
- `stembase`$(\mathfrak{J})$ is a stembase.

The first property will be proved by iteratively showing that every single action carried out by the algorithm does not change the closure operator $(.)^{\mathfrak{J}\cup\mathfrak{S}\mathfrak{B}}$. By "concatenating" those arguments and with the observation that $\mathfrak{S}\mathfrak{B} = \emptyset$ in the beginning and $\mathfrak{J} = \emptyset$ in the end, we can conclude that this first property indeed holds.

So, first, we consider the actions carried out in line 2. Let $\mathfrak{H} = \mathfrak{J} \setminus \{A \to B\} \cup \{A \to (A \cup B)^{\mathfrak{J}}\}$ for an arbitrary $A \to B \in \mathfrak{J}$. Now consider an arbitrary $C \subseteq M$. We have to show, that C respects all implications from $\mathfrak{J}$ exactly if it respects all implications from $\mathfrak{H}$.
"$\Leftarrow$": This is trivial, since $B \subseteq (A \cup B)^{\mathfrak{J}}$.
"$\Rightarrow$": Assume C respects all implications of $\mathfrak{J}$. Now, the only way for C to not respect all implications of $\mathfrak{H}$ would obviously be $A \subseteq C$ and $B^{\mathfrak{J}} \not\subseteq C$. On the other hand, since C respects $A \to B$, we know that $B \subseteq C$. Furthermore, $B^{\mathfrak{J}}$ is by definition the smallest set (wrt. set inclusion) containing B and respecting all implications of $\mathfrak{J}$. Hence, we have $B^{\mathfrak{J}} \subseteq C$, leading to a contradiction.

Now, we consider the actions of point 3. Let $\mathfrak{J}$ and $\mathfrak{S}\mathfrak{B}$ be the sets before carrying out an a-b-c-d block and $\mathfrak{J}^*$ and $\mathfrak{S}\mathfrak{B}^*$ the respective values afterwards. Again, considering an arbitrary $C \subseteq M$, we have to show, that C respects all implications from $\mathfrak{J} \cup \mathfrak{S}\mathfrak{B}$ exactly if it respects all implications from $\mathfrak{J}^* \cup \mathfrak{S}\mathfrak{B}^*$.
"$\Rightarrow$": This is obvious, since clearly for every implication $A \to B$ from $\mathfrak{J}^* \cup \mathfrak{S}\mathfrak{B}^*$ we have an implication $D \to B$ from $\mathfrak{J} \cup \mathfrak{S}\mathfrak{B}$ with $D \subseteq A$.

"$\Leftarrow$": Suppose C respects all implications from $\mathfrak{I}^* \cup \mathfrak{SB}^*$. Assuming that is does not respect all implications of $\mathfrak{I} \cup \mathfrak{SB}$ would imply $A \subseteq C$ and $B \not\subseteq C$. Yet, knowing that C respects $A^{\mathfrak{I}\cup\mathfrak{SB}} \to B$ (being also trivially true for $A^{\mathfrak{I}\cup\mathfrak{SB}} = B$), we have to conclude that $A^{\mathfrak{I}\cup\mathfrak{SB}} \not\subseteq C$. But, again by definition, $A^{\mathfrak{I}\cup\mathfrak{SB}}$ is the smallest set containing A and respecting all implications from $\mathfrak{I} \cup \mathfrak{SB}$, enforcing $A^{\mathfrak{I}\cup\mathfrak{SB}} \subseteq C$ and therefore yielding a contradiction.

Let $\mathfrak{SB} = \texttt{stembase}(\mathfrak{I})$. We prove the second property by showing that for all $A \to B \in \mathfrak{SB}$, the set A is pseudo-closed wrt. $(.)^{\mathfrak{SB}}$. Note that from the construction of the algorithm and the previous proof (including the fact that $(.)^{\mathfrak{I}\cup\mathfrak{SB}}$ remains constant) follows that, for all $A \to B \in \mathfrak{SB}$,

$$A = A^{\mathfrak{SB}\setminus\{A\to B\}}. \quad (*)$$

Now we assume A were not pseudo-closed for an $A \to B \in \mathfrak{SB}$. Obviously, it is not closed either. So there must exist a pseudo-closed set $P \subset A$ with $P^{\mathfrak{SB}} \not\subseteq A$. Now, consider $Q := P^{\mathfrak{SB}\setminus\{A\to B\}}$. By monotonicity, we then have $Q \subseteq A$. So the only possibility to make $P^{\mathfrak{SB}} \not\subseteq A$ true is that Q does not respect $A \to B$. Yet, this would imply $Q = A$ and consequently $P^{\mathfrak{SB}} = B$. Now due to Proposition 2, we know, that $\mathfrak{SB}$ has to contain an implication $C \to D$ with $C \subseteq P$ and $C^{\mathfrak{SB}} = P^{\mathfrak{SB}} = B$. Moreover, due to the construction we know that $D = C^{\mathfrak{SB}} = B$. Since $(A \to B) \neq (C \to B)$, we have that $C \to B \in \mathfrak{SB} \setminus \{A \to B\}$. Yet, from this and $C \subseteq A$ follows $B \subseteq A^{\mathfrak{SB}\setminus\{A\to B\}}$, contradicting the equation (*). □

Calculating the $\mathfrak{I}$-closure (without preprocessing) can be done in time $O(|\mathfrak{I}|)$ due to [8]. Hence, the presented algorithm runs in $O(|\mathfrak{I}|^2)$ i.e. quadratic time (this complexity bound for the task accomplished by the algorithm had already been shown in [10]). Mark that this algorithm naturally also determines all pseudo-intents (being just the premises of the implications of $\mathfrak{SB}$).

4 Characterizing Pseudo-intents

Now, we will introduce notions that are essential for our aim to characterize pseudo-intents non-recursively.

Definition 5. *Let $\mathbb{K} = (G, M, I)$ be a formal context and let $P \subseteq M$. We define $\mathbb{K}[P] := (H, M, I_P)$ (say: $\mathbb{K}$ AUGMENTED BY P) as follows:*

- *$H := G \cup \{g_P\}$ (where we presume $g_P \notin G$) and*
- *$I_P := I \cup (\{g_P\} \times P)$*

The following results are immediate consequences of this definition:

Lemma 1. *Let $\mathbb{K} = (G, M, I)$ be a formal context and let $P \subseteq M$. Then*

- *for all $g \in G$, we have $g^{I_P} = g^I$,* **cons1**
- *for all $A \subseteq G$, we have $A^{I_P} = A^I$,* **cons2**
- *for all $A \subseteq M$, we have $A^I = A^{I_P} \setminus \{g_P\}$,* **cons3**
- *for all $A \subseteq M$ with $g_P \notin A^{I_P}$, we have $A^{I_P I_P} = A^{II}$, and* **cons4**
- *for all $A \subseteq M$ with $g_P \in A^{I_P}$, we have $A^{I_P I_P} = P \cap A^{II}$.* **cons5**

Proof

- **cons1** This is trivial, since $\{m \mid gI_A m\} = \{m \mid gIm\}$.
- **cons2** Due to **decomp**, we have $A^{I_A} = \bigcap_{g\in A} g^{I_A}$. Due to **cons1**, this equals $\bigcap_{g\in A} g^I = A^I$.
- **cons3** Consider an arbitrary $g \in G = H \setminus \{g_P\}$. The statement $g \in A^{I_P}$ is equivalent to $gI_P m$ for all $m \in A$. Since - due to the definition - I_P and I coincide on all objects but g_P, this is equivalent to gIm for all $m \in A$, which in turn is the same as $g \in A^I$.
- **cons4** From **cons3**, we conclude $A^{I_P I_P} = (A^{I_P} \setminus \{g_P\})^{I_P} = A^{II_P}$ and by **cons2** follows $A^{II_P} = A^{II}$.
- **cons5** From $A^{I_P} = \{g_P\} \cup (A^{I_P} \setminus \{g_P\})$, we can conclude $A^{I_P I_P} = \bigcap_{g\in A^{I_P}} g^{I_P} = g_P^{I_P} \cap \bigcap_{g\in A^{I_P}\setminus\{g_P\}} g^{I_P} = P \cap (A^{I_P} \setminus \{g_P\})^{I_P}$. Due to **cons3**, this equals $P \cap A^{II_P}$ and due to **cons2** this is just $P \cap A^{II}$.

Proposition 5. *Properties of augmentations.*

$A \in \mathcal{I}(\mathbb{K}[A])$, i.e., A is an intent of $\mathbb{K}[A]$, **cont[]**
$\mathcal{I}(\mathbb{K}) \subseteq \mathcal{I}(\mathbb{K}[A])$, i.e., every formal concept intent of $\mathbb{K}$ is also a concept intent of $\mathbb{K}[A]$, **mon[]**
If $A \in \mathcal{I}(\mathbb{K})$ then $\mathcal{I}(\mathbb{K}[A]) = \mathcal{I}(\mathbb{K})$, i.e., if an object is added to the context, the intent of which is already an intent of $\mathbb{K}$, the overall set of intents remains unchanged. **intid[]**

Proof

cont[]: Obviously, $({g_A}^{I_A I_A}, A)$ is a formal concept of $\mathbb{K}[A]$.
mon[]: This property follows directly from Proposition 4.
intid[]: Assume the contrary, i.e., there were a $B \in \mathcal{I}(\mathbb{K}[A]) \setminus \mathcal{I}(\mathbb{K})$. Due to **decomp**, we know $B = B^{I_A I_A}$. Obviously, g_A has to be in B^{I_A}, since otherwise $B^{I_A I_A} = B^{II}$ by **cons4**, contradicting $B \notin \mathcal{I}(\mathbb{K})$. Thus, due to **cons5**, $B = B^{I_A I_A} = A \cap B^{II}$. Yet, knowing that A is an intent of K and due to **clos**$\cap$, the intersection of the two closed sets A and B^{II} has again to be closed, we have found a contradiction to the assumption.

To facilitate the intuition about context augmentations, consider Fig. 2, which shows some augmentations of a small context and the impact of this on the set of concept intents.

For our further line of argumentation, the motivating intuitive idea is (also conveyed by the name) that a pseudo-intent is "almost an intent". Since we know that augmenting a context by an intent does not change the corresponding intent set, we could expect that adding a pseudo-intent would result in just a very slight change. Considering the slightest change possible we define the notion of an *incrementor*.

Definition 6. *We say that P is an* INCREMENTOR *of $\mathbb{K}$, if*

- *P is not a concept intent of $\mathbb{K}$ and*
- *for every concept intent $A \subseteq M$ of $\mathbb{K}[P]$ we have $B = P$ or B is a concept intent of $\mathbb{K}$.*

$\mathbb{K}$

$\mathbb{K}$	m_1	m_2	m_3
g_1	×	×	
g_2		×	×

$\mathcal{I}(\mathbb{K})$

$\{m_2\}$,
$\{m_1, m_2\}$,
$\{m_2, m_3\}$,
$\{m_1, m_2, m_3\}$,

$\mathbb{K}[\{m_1, m_3\}]$

	m_1	m_2	m_3
g_1	×	×	
g_2		×	×
$g_{\{m_1,m_3\}}$	×		×

$\mathcal{I}(\mathbb{K}[\{m_1, m_3\}])$

$\underline{\emptyset}, \underline{\{m_1\}}, \{m_2\}, \underline{\{m_3\}}$,
$\{m_1, m_2\}, \underline{\{m_1, m_3\}}, \{m_2, m_3\}$,
$\{m_1, m_2, m_3\}$

$\mathbb{K}[\{m_2\}]$

	m_1	m_2	m_3
g_1	×	×	
g_2		×	×
$g_{\{m_2\}}$		×	

$\mathcal{I}(\mathbb{K}[\{m_2\}])$

$\{m_2\}$,
$\{m_1, m_2\}$,
$\{m_2, m_3\}$,
$\{m_1, m_2, m_3\}$

$\mathbb{K}[\emptyset]$

	m_1	m_2	m_3
g_1	×	×	
g_2		×	×
$g_\emptyset$			

$\mathcal{I}(\mathbb{K}[\emptyset])$

$\underline{\emptyset}, \{m_2\}$,
$\{m_1, m_2\}$,
$\{m_2, m_3\}$,
$\{m_1, m_2, m_3\}$

Fig. 2. Examples for context augmentations and their consequences for the set of concept intents. Intents added by the augmentation are underlined.

Looking back at Fig. 2, we see, that in this case, the empty set would be an incrementor of $\mathbb{K}$. Moreover, it takes little consideration to verify that it is also a pseudointent of $\mathbb{K}$. The following theorem partly justifies our intuition by ensuring that every pseudointent is indeed an incrementor.

Theorem 3. *Let $\mathbb{K}$ be a formal context and P be a pseudo-intent of $\mathbb{K}$. Then P is an incrementor of $\mathbb{K}$.*

Proof: Consider the context $\mathbb{K}[P]$. Let (A, B) be a formal concept of $\mathbb{K}[P]$. We know, that $B = A^{I_P} = \bigcap\{a^{I_P} \mid a \in A\}$ (due to **decomp**).

Obviously, if $g_P \notin A$, we have that (A, B) is a concept of $\mathbb{K}$ as well, since $a^{I_P} = a^I$ for all $a \neq g_P$.

If $g_P \in A$, we have that $B = g_P^{I_P} \cap \bigcap\{a^{I_P} \mid a \in A \setminus \{g_P\}\}$ which yields $B = P \cap \bigcap\{a^I \mid a \in A \setminus \{g_P\}\}$.

Supposing pseudoclosedness of P, from Proposition 1 follows that B is an intent of $\mathbb{K}$, provided there exists some $a \in A \setminus \{g_P\}$ with $P \not\subseteq a^I$. In the other case, we would have $B = P$. Thus (P^{I_P}, P) is the only additional formal concept of $\mathbb{K}[P]$ compared to $\mathbb{K}$. This shows that P is an incrementor. □

Now it remains to investigate, whether this necessary condition for being a pseudo-intent is also sufficient. Unfortunately, this is not the case as Fig. 3 illustrates: in this

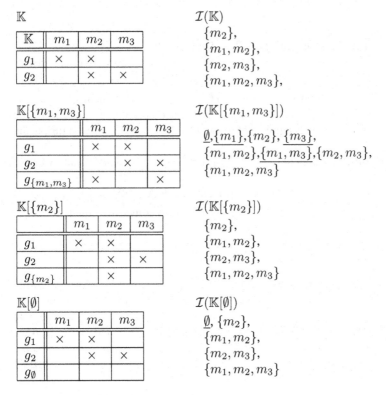

$\mathbb{K}$

	m_1	m_2	m_3	m_4
g		×	×	×

$\mathcal{I}(\mathbb{K})$

$\{m_2, m_3, m_4\}$,
$\{m_1, m_2, m_3, m_4\}$,

$\mathbb{K}[\{m_2, m_3\}]$

	m_1	m_2	m_3	m_4
g		×	×	×
$g_{\{m_2,m_3\}}$		×	×	

$\mathcal{I}(\mathbb{K}[\{m_2, m_3\}])$

$\underline{\{m_2, m_3\}}$,
$\{m_2, m_3, m_4\}$,
$\{m_1, m_2, m_3, m_4\}$,

$\mathbb{K}[\{m_2\}]$

	m_1	m_2	m_3	m_4
g		×	×	×
$g_{\{m_2\}}$		×		

$\mathcal{I}(\mathbb{K}[\{m_2\}])$

$\underline{\{m_2\}}$,
$\{m_2, m_3, m_4\}$,
$\{m_1, m_2, m_3, m_4\}$,

Fig. 3. Counterexample for the coincidence of pseudo-intents and incrementors

example, $\{m_2, m_3\}$ is an incrementor of $\mathbb{K}$ but not a pseudo-intent since it contains the pseudo-intent $\emptyset$ but not its closure $\{m_2, m_3, m_4\}$.

Yet, examining this counter-example a bit further, we see that the set being an in-rementor but not a pseudo-intent contains a set being again an incrementor (namely $\{m_2\}$) – with no intent "in between". This justifies to strengthen the condition accordingly. Yet, prior to proving that this leads to the desired characterization, we show a lemma that will facilitate the subsequent proof.

Lemma 2. *Let $\mathbb{K}$ be a formal context and A be an incrementor of $\mathbb{K}$. Then for any pseudo-intent Q of $\mathbb{K}$ with $Q \subset A$ and $Q^{II} \not\subseteq A$ we even have $A \subseteq Q^{II}$.*

Proof: Assume the contrary, i.e. $A \not\subseteq Q^{II}$. Then, considering $B := A \cap Q^{II}$, we see that $B \subset A$.

By **cont[]** and **mon[]**, respectively, we know $A, Q^{II} \in \mathcal{I}(\mathbb{K}[A])$ and hence by **clos**$\cap$ also $B \in \mathcal{I}(\mathbb{K}[A])$.

On the other hand, B can not be an intent of $\mathbb{K}$, since $Q \subset B$ (following from $Q \subset A$ and $Q \subset Q^{II}$ – the latter by **ext**^{II}) but $Q^{II} \not\subseteq B$ (this is because $Q \subset B$ implies $Q^{II} \subseteq B^{II}$ by **mon**^{II} and B being an intent of $\mathbb{K}$ would mean $B = B^{II}$)

So B must be an intent of $\mathbb{K}[A]$ that is neither A itself nor an intent of $\mathbb{K}$. Yet, this contradicts the assumption of A being an incrementor. □

Now we will provide and prove the announced non-recursive characterization for pseudo-intents.

Theorem 4. *Let $\mathbb{K} = (G, M, I)$ be a formal context and let $P \subseteq M$. P is a pseudo-intent of $\mathbb{K}$ if and only if*

P is an incrementor of $\mathbb{K}$ and **inc**
for every incrementor $Q \subset P$, there is an intent R with $Q \subset R \subset P$. **min**

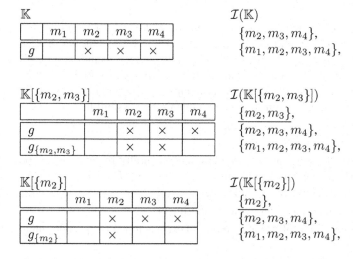

Proof: "⇒"
That every pseudo-intent is an incrementor has already been shown by Theorem 3.

We will prove the second condition **min** indirectly. Thus, we assume we have a pseudo-intent P violating **min**, i.e., there is a $Q \subset P$ being an incrementor and for all R with $Q \subset R \subset P$, the set R is not an intent. Note that, from Theorem 3, we know that P is an incrementor as well. Q cannot be an intent (as this would contradict the definition of incrementor), thus we consider the two remaining possibilities:

- Suppose Q is a pseudo-intent.
  This would (due to the definition of pseudo-intent) naturally require Q^{II} to be contained in P. Altogether this would mean: $Q \subset Q^{II} \subset P$ contradicting our assumption. Hence, R cannot be a pseudo-intent.
- Now, suppose Q is neither an intent nor a pseudo-intent.
  Then – due to the definition of pseudo-intent – there has to exist a pseudo-intent $S \subset Q$ with $S^{II} \not\subseteq Q$.
  From Lemma 2 then additionally follows $Q \subseteq S^{II}$.
  Since the definition of pseudo-intent requires P to contain S^{II}, we have the setting: $Q \subset S^{II} \subset P$. Yet, again, this obviously contradicts our assumption.
  Thus, it is impossible that Q is neither closed nor pseudo-closed wrt. $\mathbb{K}$.

Concluding, Q can be neither an intent nor a pseudo-intent nor none of both. Hence, the assumption of its existence must be false.
"⇐"
Assume the contrary, i.e., both conditions **inc** and **min** be fulfilled and yet P not be a pseudo-intent. Obviously P is not an intent either (otherwise, it would not be an incrementor by definition).

Therefore, P must be neither closed nor pseudo-closed. Then, by the definition of pseudo-closedness, there must be a pseudo-closed set $Q \subset P$ with $Q^{II} \not\subseteq P$.

From Lemma 2 follows that $P \subset Q^{II}$. Then, we have the setting $Q \subset P \subset Q^{II}$.

Furthermore, note that $Q \subseteq P \subseteq Q^{II}$ entails $Q^{II} \subseteq P^{II} \subseteq (Q^{II})^{II}$ via $\mathbf{mon}^{II}$ which together with $\mathbf{idp}^{II}$ yields $Q^{II} = P^{II}$. But then, the very same argument yields $S^{II} = Q^{II}$ for every S with $Q \subset S \subset P$ (and therefore $S \neq S^{II}$). Clearly, this contradicts the initial assumption **min**. □

After having established those results, it takes little consideration to see (referring to [3] and [7]) that the incrementors of a formal context are just those quasi-intents which are not intents. This allows to reuse the corresponding results. In particular, the following corollary to the Proposition 2 from [7] can be used to check whether a given set is an incrementor in polynomial time.

Theorem 5. *P is an incrementor of $\mathbb{K}$ if and only if*

- *P is not an intent of $\mathbb{K}$ and*
- *for all $g \in G$, we have $P \subseteq g^I$ or $g^I \cap P$ is an intent of $\mathbb{K}$.*

```
function:  incrementor(A,𝕂)
  -- Calculate A^II.  If A^II = A,
     output "NO" and terminate.
  -- For all g ∈ G,
       - Calculate Ã := g^I ∩ A.
         If Ã = A then continue with next g.
       - Calculate Ã^II.
       - If Ã ≠ Ã^II then
         output "NO" and terminate.
  -- Output "YES" and terminate.
```

Fig. 4. Algorithm incrementor $(A, \mathbb{K})$ for checking whether A is an incrementor of $\mathbb{K}$

```
function:  scan(A,𝕂,check(.))
  -- For all a ∈ A,
     add A\{a} to (previously empty) list ℒ
  -- Starting from the ℒ's first element
     for every B from ℒ
       - If B^II ≠ A^II,
         continue with next list element.
       - If check(B),
         output "YES" and terminate.
       - Otherwise, for every b ∈ B, append B\{b} to ℒ
         if not already contained.
  -- if ℒ processed,
     output "NO" and terminate.
```

Fig. 5. Algorithm scan for determining whether for a given $A \subseteq M$ there is a $B \subset A$ with $B^{II} = A^{II}$ and check (B)

5 An Algorithm for Checking Pseudo-closedness

Applying the results cited and presented in the preceding sections, we will now provide an algorithm for checking pseudo-closedness and analyze its complexity.[4] We start by giving an algorithm computing whether for a given formal context $\mathbb{K} = (G, M, I)$, a given attribute set $A \subseteq M$ is an incrementor of $\mathbb{K}$. This algorithm is shown in Fig. 4.

[4] We expect the reader to be familiar with the basic notions from complexity theory.

```
function:  pseudoIntent(A,𝕂)
  -- Check whether incrementor(A,𝕂).
     If not so, output "NO" and terminate.

  -- If scan(A,𝕂,incrementor(.,𝕂)),
     output "NO" and terminate,
     otherwise, output "YES" and terminate.
```

Fig. 6. Algorithm `pseudoIntent(A,𝕂)` for checking whether A is a pseudo-intent of $\mathbb{K}$

It is well-known, that the time complexity for computing the closure A^{II} of a given attribute set A is in $O(|G| \cdot |M|)$ while comparing two sets or computing g^I for a given object is less costly. Thus, regarding the time costs, the `incrementor` function consists essentially of the $|G|+1$-fold calculation of the closure, hence its time complexity is in $O(|G|^2 \cdot |M|)$.

Next, we provide an algorithm which for a given attribute set A, "scans" whether there exists a set $B \subset A$ with $B^{II} = A^{II}$ fulfilling an arbitrary computable criterion (denoted by the function `check`). This algorithm is shown in Fig. 5.

In general the time complexity of this algorithm is bounded by $2^{|M|}$ times the complexity of `check`.

Finally, we employ the `incrementor` and the `scan` functions to formulate the algorithm which actually checks for pseudo-closedness. This algorithm is displayed in Fig. 6.

Resulting from the earlier complexity considerations, we find that its the time complexity is in $O(2^{|M|}$.

6 Optimization: Operating on the Reduced Context

We will now discuss in which way this algorithm can be optimized. One of the straightforward issues to think about would be whether the problem of identifying pseudo-intents of a formal context $\mathbb{K}$ can be solved by checking for pseudo-closedness in the reduced version of $\mathbb{K}$. This should be possible, since – roughly spoken – a reduced context contains the same implicative information as the original one.

Theorem 6. *Let $\mathbb{K} = (G, M, I)$ be a formal context and $\mathbb{K}^* = (H, N, J)$ (with $H \subseteq G$ and $N \subseteq M$ as well as $J = I \cap (H \times N)$) the corresponding reduced context. Let furthermore $m^* = m^{II} \cap N$ for any $m \in M$. The fact that $\mathbb{K}^*$ is a reduced version of $\mathbb{K}$ then yields $m^{*II} = m^{II}$.*

A set $P \subseteq M$ is a pseudo-intent of $\mathbb{K}$ exactly if one of the following is true:

- *there is a pseudo-intent P^* of $\mathbb{K}^*$ such that $P = P^* \cup \{m \in M \setminus N \mid m^* \subset P^*\}$,*
- *$P = \{m\} \cup \emptyset^{II}$ for an $m^* \neq \emptyset$, or*
- *$P = m^* \cup \{\overline{m} \in M \setminus N \mid \overline{m}^* \subset m^*\}$ for an $m \in M \setminus N$ if there is no pseudo-intent Q^* of $\mathbb{K}^*$ with $Q^{*JJ} = m^*$.*

Proof: First note that the set

$$\mathfrak{I} := \mathfrak{SB}^* \cup \{\{m\} \to m^* \mid m \in M \setminus N\} \cup \{m^* \to \{m\} \mid m \in M \setminus N\}$$

(where $\mathfrak{S}\mathfrak{B}^*$ is the stembase of $\mathbb{K}^*$) is sound and complete for the closure operator $(.)^{II}$. So we will just show, that applying the `stembase`-algorithm from Section 3 just yields an implication set where the premises are exactly the sets presented above.

First we consider the result of point 2 of the algorithm:

- Every $P^* \to P^{*JJ} \in \mathfrak{S}\mathfrak{B}^*$ will be transformed to

$$P^* \to P^{*JJ} \cup \{m \mid m^* \subseteq P^{*JJ}\}.$$

- Every $\{m\} \to m^*$ will be transformed to

$$\{m\} \to m^* \cup \{\overline{m} \mid \overline{m}^* \subseteq m^*\}.$$

- Every $m^* \to \{m\}$ will be transformed to

$$m^* \to m^* \cup \{\overline{m} \mid \overline{m}^* \subseteq m^*\}.$$

Now consider point 3:

- Every $P^* \to P^{*II}$ will be transformed to

$$P^* \cup \{m \in M \setminus N \mid m^* \subset P^*\} \to P^{*II}.$$

```
function:  pseudoIntentRed(A,𝕂)
  -- Check whether incrementor(A,𝕂).
     If not so, output ''NO'' and terminate.

  -- Calculate reduced context 𝕂* = (H,N,J)


  -- Calculate ∅^II and m* for all m ∈ M\N.

  -- If A\∅^II = {a} ⊆ M\N then
     output ''YES'' and terminate.

  -- Calculate P* := A∩N.

  -- Check whether A = P* ∪ {m ∈ M\N | m* ⊂ P*}.
     If so, check whether pseudoIntent(P*,𝕂*).
     If so, output ''YES'' and terminate.

  -- Check whether A = m* ∪ {m̄ ∈ M\N | m̄* ⊂ m*}
     for an m ∈ M\N.
     If so, check whether
     scan(A,𝕂*,incrementor(.,𝕂*)).
     If not so, output ''YES'' and terminate.

  -- Output "NO" and terminate.
```

Fig. 7. Algorithm `pseudoIntentRed` for checking whether a set A is a pseudo-intent of a formal context $\mathbb{K}$

- Every $\{m\} \to m^{*II}$ will be
  - deleted if $m^* = \emptyset$ or
  - otherwise, transformed to $\{m\} \cup \emptyset^{II} \to m^{*II}$.
- Every $m^* \to \{m\}^{II}$ will be
  - deleted, if there is a pseudo-intent Q^* of $\mathbb{K}^*$ with $Q^{*JJ} = m^*$ or
  - otherwise, transformed to

$$m^* \cup \{\overline{m} \in M \setminus N \mid \overline{m}^* \subset m^*\} \to \{m\}^{II}.$$

□

These observations allow an optimization of the pseudo-closedness checking algorithm from Fig. 6 in Section 5.

Considering the complexity, we can state the following. Due to [4], a formal context can be reduced in $O((|G| + |M|) \cdot |G| \cdot |M|)$ time (and is hence rather cheap). Therefore, this optimization could be potentially beneficiary in cases where the context is not already reduced, since the upper bound for the time complexity is decreased to $O(2^{|N|})$.

7 Conclusions and Further Work

In our paper, we presented several results regarding pseudo-intents. We showed how an arbitrary implication set can be turned into a stem base (the premises of which are per definitionem just the pseudo-intents). Furthermore, based on a characterization of pseudo-intents via incrementors and using known results about quasi-intents, we provided an algorithm which allows to decide for a given formal context $\mathbb{K}$ and an attribute set P whether P is a pseudo-intent of $\mathbb{K}$ Moreover, we showed how this algorithm can be further optimized by calculating with the reduced version of the considered context.

Although the complexity questions mentioned in the beginning remain unsolved, we hope that the structural insights presented in this paper might contribute to their solution. Of course this would be a main goal of further research.

On the other hand, comprehensive experiments would be the next step to investigate how the algorithms proposed here perform in practical cases (albeit not having substantial evidence for this, our conjecture is that the average complexity would be much better than suggested by our worst-case analyses).

Finally, if this should be the case, the provided algorithms could be used for developing new data-mining and exploration methods.

References

[1] Alan Day. The lattice theory of functional dependencies and normal decompositions. *International Journal of Algebra and Computation*, 2(4):409–431, 1992.

[2] William F. Dowling and Jean H. Gallier. Linear-time algorithms for testing the satisfiability of propositional Horn formulae. *J. Log. Program.*, 1(3):267–284, 1984.

[3] Bernhard Ganter. Two basic algorithms in concept analysis. Technical Report 831, FB4, TH Darmstadt, 1984.

[4] Bernhard Ganter and Rudolf Wille. *Formal Concept Analysis: Mathematical Foundations*. Springer-Verlag New York, Inc., Secaucus, NJ, USA, 1997. Translator-C. Franzke.

[5] J.-L. Guigues and Vincent Duquenne. Familles minimales d'implications informatives resultant d'un tableau de données binaires. *Math. Sci Humaines*, 95:5–18, 1986.

[6] S. O. Kuznetsov. On the intractability of computing the Duquenne-Guigues base. *Journal of Universal Computer Science*, 10(8):927–933, 2004.

[7] Sergei O. Kuznetsov and Sergei A. Obiedkov. Counting pseudo-intents and #P-completeness. In Rokia Missaoui and Jürg Schmid, editors, *ICFCA*, volume 3874 of *Lecture Notes in Computer Science*, pages 306–308. Springer, 2006.

[8] David Maier. *The Theory of Relational Databases.* Computer Science Press, 1983.

[9] Uta Priss. Some open problems in formal concept analysis. `http://www.upriss.org.uk/fca/problems06.pdf`, FEB 2006.

[10] Marcel Wild. Implicational bases for finite closure systems. In Wilfried Lex, editor, *Arbeitstagung Begriffsanalyse und Künstliche Intelligenz*, pages 147–169. Springer, 1991.

Performances of Galois Sub-hierarchy-building Algorithms

Gabriela Arévalo[1,3], Anne Berry[2], Marianne Huchard[1], Guillaume Perrot[1], and Alain Sigayret[2]

[1] LIRMM - CNRS UMR 5506 - Université de Montpellier II - Montpellier (France)
{huchard, arevalo, perrot}@lirmm.fr
[2] LIMOS - CNRS UMR 6158 - Univ. Blaise Pascal - Clermont-Ferrand II (France)
{berry, sigayret}@isima.fr
[3] LIFIA - Facultad de Informática (UNLP) - La Plata (Argentina)
garevalo@sol.info.unlp.edu.ar

Abstract. The Galois Sub-hierarchy (GSH) is a polynomial-size representation of a concept lattice which has been applied to several fields, such as software engineering and linguistics.

In this paper, we analyze the performances, in terms of computation time, of three GSH-building algorithms with very different algorithmic strategies: Ares, Ceres and Pluton. We use Java and C++ as implementation languages and Galicia as our development platform.

Our results show that implementations in C++ are significantly faster, and that in most cases Pluton is the best algorithm.

Keywords: Galois Sub-hierarchy, AOC-Poset, Performance Analysis.

1 Introduction

Formal concept analysis (FCA) has been used in a broad spectrum of research fields, such as knowledge representation, data mining, machine learning, software engineering and databases. The main drawback of concept lattices is that the number of concepts may be of much larger size than the relation (or even exponential in the size of the relation). It is therefore feasible, when this problem is encountered, to use a polynomial-size representation of the lattice while preserving the most relevant information.

One of the approaches, which has proved useful in practice, is to restrict the lattice to the concepts which introduce a new object or a new property. This idea is the basis for two very similar structures called the *Galois Sub-hierarchy (GSH)* and the *Attribute Object Concept poset (AOC-poset)*. The Galois Sub-hierarchy has been introduced in the software engineering field by Godin et al. [GM93] for class hierarchy reconstruction and successfully applied in later research work [GMM+98], [AYLCB96], [HDL00], [DHL+02]. The AOC-poset has been used in applications of FCA to non-monotonic reasoning and domain theory [Hit04] and to produce classifications from linguistic data [OP02], [Pet01].

S.O. Kuznetsov and S. Schmidt (Eds.): ICFCA 2007, LNAI 4390, pp. 166–180, 2007.

These structures are interesting not only as a feasible alternative to oversized concept lattices, but also as a conceptual improvement, as human perception of a problem is enhanced by an easy visualization of a restricted number of elements.

As the size of the input may still be large, naturally it is important to have efficient Galois Sub-hierarchy-building algorithms to work with. There are several efficient Galois Sub-hierarchy-building algorithms, with very different algorithmic strategies, and with theoretical worst-time complexity analyses which are difficult to compare. Kuznetsov et al. [KO02] propose a rather extensive implementative comparative analysis of lattice-building algorithms, but to our knowledge the only existing work on comparing algorithms related to GSH-building algorithms is proposed by Godin et al. [GC99], comparing ARES and ISGOOD, which is restrictive, as it builds only the attribute elements of the Galois Sub-hierarchy.

In this paper we address the issue of comparing the execution times of the three main Galois Sub-hierarchy-building algorithms: ARES, CERES and PLUTON, in order to determine which algorithm can be recommended to a user and in which case. This choice is meaningful because these three algorithms are used as tools with a strong user-based interaction, where the response time is a very important factor. The performance factors we tested are the density of the relation and the number of objects and attributes.

The paper is structured as follows: Section 2 introduces the main terminology of Galois Sub-hierarchy. Section 3 explains how the three Galois Sub-hierarchy-building algorithms work. Section 4 details the experimental approach which we used. Section 5 presents our evaluation of the results. We conclude in Section 6.

2 Notations and Definitions

In this section, we introduce the main terminology necessary to understanding how the Galois Sub-hierarchy algorithms work. We do not explain in detail the basics of FCA features but focus more on Galois Sub-hierarchy definitions. We refer the reader to Ganter et al. [GW99] for a complete introduction to partial orders and lattices.

In FCA, a formal context is a triple $\mathbb{K} = (G, M, I)$ where G and M are sets (objects and attributes respectively) and I is a binary relation, i.e., $I \subseteq G \times M$. Figure 1(left) shows context $\mathbb{K} = (\{1, 2, 3, 4, 5, 6\}, \{a, b, c, d, e, f, g, h\}, I)$.

For a set $A \subseteq G$ of objects, we define $A' := \{m \in M | gIm \text{ for all } g \in A\}$ (the set of attributes common to the objects in A). Correspondingly, for a set $B \subseteq M$, we define $B' := \{g \in G | gIm \text{ for all } m \in B\}$ (the set of objects which have all attributes in B). Then, a formal concept of the context (G, M, I) is a pair (A, B) with $A \subseteq G$, $B \subseteq M$, $A' = B$ and $B' = A$. A is called the *extent* and B the *intent* of the concept (A, B). $\mathfrak{B}(G, M, I)$ denotes the set of all concepts of the context (G, M, I). Figure 1 (right) shows the concept lattice corresponding to our example.

The concepts $C^O = \{\gamma o = (o'', o') | o \in G\}$ are called the *object concepts* of o, and the concepts $C^A = \{\mu a = (a', a'') | a \in A\}$ are called the *attribute concepts*.

	a	b	c	d	e	f	g	h
1		×	×	×	×			
2	×	×	×				×	×
3	×	×				×	×	×
4				×	×			
5			×	×				
6	×							×

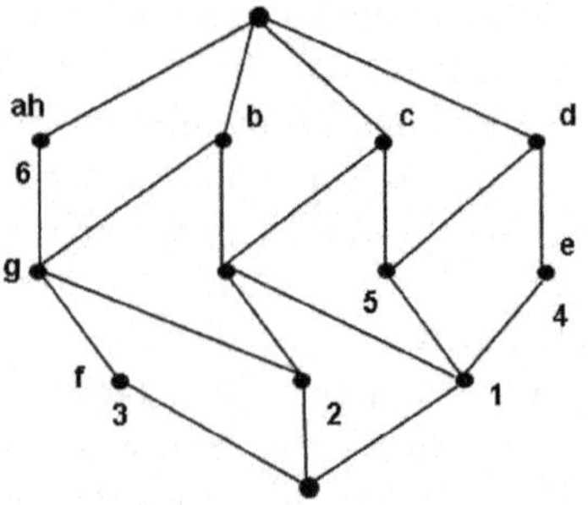

Fig. 1. Left: Binary relation of a context $\mathbb{K}$ - Right: Concept lattice $\underline{\mathfrak{B}}(G, M, I)$

The object concept which corresponds to object o, γo, is the smallest concept with o in its extent, and dually, the attribute concept which corresponds to attribute a, μa, is the greatest concept with a in its intent. The *Galois Sub-hierarchy* is the sub-order of the lattice made out of the set $C^O \cup C^A$ and the restriction of the lattice order to that set [HDL00]. Figure 1 (right) shows the lattice corresponding to context $\mathbb{K}$. Figure 2 (left) shows the Galois Sub-hierarchy of the context introduced in Figure 1. As we see, in the Hasse diagram of a Galois Sub-hierarchy, empty concepts are omitted. Thus, in the paper we use the notation (4,e) instead of (14,de), and use the terms *simplified intent* (for (4)) and *simplified extent* (for (e)), as well as *simplified concept* (for $(4, e)$) as shown in Figure 2.

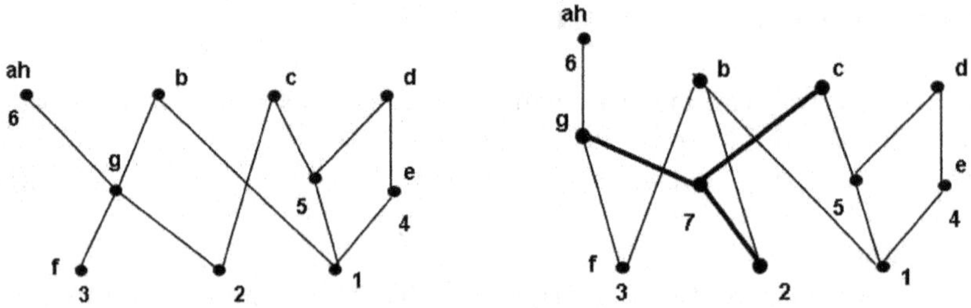

Fig. 2. Left: Galois Sub-hierarchy GSH(I). (b) denotes the concept (123,b) or its simplified form (∅,b) - Right: Galois Sub-hierarchy after addition of object 7 (Refer to Section 3 in ARES algorithm)

3 The Algorithms

This section briefly explains the basic features of the Galois Sub-hierarchy-building algorithms which we analyze in this paper. The reader is referred to the cited papers for further details on the corresponding algorithms. We will use the example from Figure 1 to illustrate our explanations.

PLUTON

PLUTON is composed of three algorithms: TOMTHUMB, TOLINEXT, TOGSH. TOMTHUMB produces an ordered list of the simplified extents and intents in Berry et al. [BHM+05]. This ordered list maps to a linear extension of the Galois Sub-hierarchy. Algorithm TOLINEXT then searches the ordered list to merge pairs consisting of a simplified extent and a simplified intent pertaining to the same concept, in order to reconstruct the elements of the Galois Sub-hierarchy. Algorithm TOGSH is then used to compute the edges of the Hasse diagram (transitive reduction) of the Galois Sub-hierarchy.

Algorithm TomThumb uses a sub-algorithm which computes either an ordered partition of objects into simplified extents, or dually a partition of attributes into simplified intents. The order on the simplified closed sets (simplified extents or intents) maps to a linear extension of the Galois Sub-hierarchy.

Algorithm TOMTHUMB uses partition refinement to construct ordered partitions into simplified intents and extents, by using a list of attributes to partition a list of objects (or vice-versa). If for example the list of attributes (a, b, c, d, e, f, g, h) is used to refine the class of objects (123456), the first step using attribute a will split class (123456) into the two classes $(236, 145)$, as $a' = \{2, 3, 6\}$. The process is then repeatedly applied by the next attribute to each of the current classes. Berry et al. [BHM+05] give full details as well as a detailed example.

Algorithm TOMTHUMB proceeds in three steps:

- Computation of an ordered partition $\mathcal{L}_e$ of the simplified extents, using any ordering of the properties as input.
  For example, using the ordering (a, b, c, d, e, f, g, h) of the attributes, the output is $\mathcal{L}_e = \{(2),(3),(6),(1),(5),(4)\}$
- Computation of an ordered partition $\mathcal{L}_i$ of the simplified intents, using $\mathcal{L}_e$ as input. In our example, using $\mathcal{L}_e = \{(2),(3),(6),(1),(5),(4)\}$ as input, the output is $\mathcal{L}_i = \{(\text{d}),(\text{e}),(\text{c}),(\text{b}),(\text{ah}),(\text{g}),(\text{f})\}$
- The two partitions are then merged to produce a list of simplified closed sets which can be mapped to a linear extension of the Galois Sub-hierarchy, *e.g.* LIST = {(2),(3),(f),(g),(6),(ah),(1),(b),(5),(c),(4),(e),(d)}, which represents the linear extension {(2),(3,f),(g),(6,ah),(1),(b),(5),(c),(4,e),(d)} of the Galois Sub-hierarchy.

Algorithm ToLinext assembles simplified (non-empty) extent E_r and simplified intent I_r pertaining to a same concept, that is such that for complete extent and intent, $E_r = (I_r)'$. Only pairs formed by an extent directly followed by an intent need be considered. For a simplified extent $E_r = \text{LIST}[i]$, we check that: $E = (\text{LIST}[i])' = I = (\text{LIST}[i + 1])''$. In our example, the result is:
$L = \{(2),(3,\text{f}),(\text{g}),(6,\text{ah}),(1),(\text{b}),(5),(\text{c}),(4,\text{e}),(\text{d})\}$, but to apply the algorithm TOGSH we consider a form of L where simplified empty sets are added:
$L = \{(2,\emptyset),(3,\text{f}),(\emptyset,\text{g}),(6,\text{ah}),(1,\emptyset),(\emptyset,\text{b}),(5,\emptyset),(\emptyset,\text{c}),(4,\text{e}),(\emptyset,\text{d})\}$.

Algorithm ToGSH builds the Hasse diagram of the Galois Sub-hierarchy by computing the edges of the graph. The ordering into an linear extension L is used to reduce the number of comparisons, as by definition of a linear extension, an edge can only go from a concept (for example (2,∅) to a concept which appears to its right in the list (for example (∅,c))). Once an edge is detected, sub-concepts of the origin are marked in order to avoid already visited concepts linked by transitive edges.

Theoretical complexity. In Berry et al. [BHM+05], Tom Thumb's time complexity is analyzed as in $\mathcal{O}(|J|)$. A brute force implementation of ToLinExt has a complexity in $\mathcal{O}((|O|+|A|)^3)$. Fura et al. [FLPP05] evaluates the complexity of ToGSH as $\mathcal{O}((|O|+|A|)^2 \times max(|O|,|A|)^2)$. It is worth noting that in the Galicia implementation of Pluton, whole extents and intents are computed in a simple pass of the Galois Sub-hierarchy.

CERES

Ceres mixes the computation of the concepts and that of the Hasse diagram. Concepts are computed respecting an order which maps to a linear extension of the Galois Sub-hierarchy. First, the columns of I are sorted by decreasing number of crosses to generate the concepts of C^A by decreasing extent size. In the example shown in Figure 1, columns could be ordered as follows: a, b, c, d, h, e, g, f. The strategy is then to compute C^A by groups of concepts which have the same extent and adding concepts of $C^O \setminus C^A$ to the GSH under construction, when their intent is covered by the intents of the C^A concepts previously computed. Extents and simplified intents of C^A concepts, as well as closed sets of C^O concepts, are computed using I. Extent inclusion is used to compute edges during a top-down traversal of the current Hasse diagram. Simplified extents and intents of C^O concepts are computed by propagation after edge construction. A simplified execution of the algorithm on our example could be:

- Column size = 3: concepts (6,ah),(b),(c),(d) are generated and included in the Hasse diagram (no edges at this step).
- Concept (5) can be added because attributes c and d have already been found. It is linked to (c) and (d).
- Column size = 2: concepts (4,e) and (g) are generated and linked respectively to (d), and (6,ah), (b).
- Concepts (2) and (1) are added and linked respectively to (g),(c) and (b),(5), (4,e).
- Column size = 1: concept (3,f) is generated and linked to concept (g).

Theoretical complexity. The time complexity of Ceres is in $\mathcal{O}(|O| \times (|O| + |A|)^2)$ [Leb00].

ARES

ARES constructs the Galois Sub-hierarchy in an incremental fashion. At each step, it considers the Galois Sub-hierarchy $GSH(I)$ associated with (G, M, I) as well as a new formal object o given with its attribute set $A_o = o'$. The result of the algorithm is the Galois Sub-hierarchy $GSH(I')$ for (G, M', I'), $A' = A \cup \{o\}$, $I' = I \cup \{(o, x) | x \in A_o\}$. The initial GSH is traversed using a linear extension, ensuring that a concept is explored after all its superconcepts. Let us denote by C the current (explored) concept and by RI_o the attribute set which at the end is the simplified intent of the concept introducing o (o is in its simplified extent). Discarding cases such that the intersection between C's intent and o' is empty, four main cases may occur:

- Case 1: C's intent is exactly o'. o is added to C's extent. The Hasse diagram remains unchanged.
- Case 2: C's intent is strictly included in o'. C is or will be a superconcept of o' (the algorithm stores this information). o is added to C's extent. The attributes of C are removed from RI_o.
- Case 3: C's intent includes o'. C is a sub-concept of γo. C either inherits all o's attributes, or some of o's attributes are in C's simplified intent. A new concept C_o with the intent o' is created if needed. C_o is introduced in the Hasse diagram between C and the C's superconcepts which also satisfied Case 2. Intents and extents, as well as their simplified forms, are updated.
- Case 4: C's intent and o' cannot be compared by set inclusion. In this situation, a new concept can be needed to factorize the common attributes not inherited by C. This new concept is introduced in the Hasse diagram between C and the C's superconcepts which also satisfied Case 2. RI_o, intents and extents, as well as their simplified forms, are updated.

If during the exploration, the algorithm did not find an initial concept whose intent is o', it is necessary to create a new concept $C_o = o'$. C_o's extent is o, C_o's simplified intent is RI_o (which is o' minus the attributes found during the GSH exploration). C_o is linked to initial or newly created concepts when their intent is included in o'. In every modification of the Hasse diagram, the algorithm removes transitivity edges as necessary. Meanwhile, when simplified intents are modified, the algorithm checks if for a concept both extent and intent are empty, and if so, the concept is removed.

Let us examine the addition of new object 7 with $A_7 = 7' = \{a, c, g, h\}$. The Galois Sub-hierarchy is traversed by successively analyzing: concept (6,ah) (Case 2), concept (b) (no intersection between intents), (c) (Case 2), (d) (no intersection between intents), (g) (Case 4), a new super-concept of concept (g) and o' is created to factorize the common attribute g, it is attached as a sub-concept of (6,ah) and (c) and as a super-concept of (g) which becomes extent-empty and intent-empty and will be removed at the end), concept (2) (Case 3), concept (3,f) (Case 4, but the common attributes are inherited), etc. The Galois Sub-hierarchy integrating object 7 is shown in Figure 2.

Theoretical complexity. The time complexity of ARES is in $\mathcal{O}(|O| \times |(w \times a + m))$, where w is the width of the Galois Sub-hierarchy (i.e. the maximum number of pairwise non-comparable elements), a is the maximum size for an intent and m the number of edges of the Hasse diagram [DDHL94].

4 Experimental Setup

In this section, we give the parameters we use in our experimentations, and explain our approach.

Parameters used. We have done the experiments using Galicia [Gal]. Galicia is a Java-based platform dedicated to constructing lattices. It offers to FCA researchers advanced tools for performance studies and an open environment to new lattices-related techniques. Galicia is implemented in Java because it ensures a high portability of the entire system. Thus Java was our first choice in the implementation of algorithms ARES, CERES and PLUTON. Because of the first results in performances, we considered C++ as a second choice, as it is known as a language with a good processing speed.

In the rest of the paper, we name as `Ares`, `Ceres` and `Pluton` the algorithms implemented in Java, and `Ares++`, `Ceres++` and `Pluton++` the algorithms implemented in C++. We must remark that all the algorithms were implemented by the same programmer [Per05], so that differences in the implementation style should not be a factor.

Tests: Random Generation of Binary Relations. To perform our experiments we generate a test suite using randomly generated binary relations. Similarly to Kuznetsov et al. [KO02], the binary relations were randomly generated using the following parameters: the number of objects, the numbers of attributes, and the binary relation density defined as follows:

$$\frac{|J|}{|O| \times |A|} \times 100$$

In fact, we use the complexity of the binary relation, defined as follows:

$$\sqrt{\left(\frac{|J|}{|O|}\right)^2 + \left(\frac{|J|}{|A|}\right)^2}$$

which is equal to the density multiplied by $\sqrt{|O|^2 + |A|^2}$ divided by 100.

Test Suite. We generate a test suite considering the variability of the density, the number of objects and the number of attributes, as follows:

- Square binary relations (the same number of objects and attributes) with variable density. In this case, the numbers of objects and attributes are 500 and the density varies from 2 to 82.

- Variable number of objects with fixed number of attributes and density. In this case, the number of attributes is 500, the density is 50 and the number of objects varies from 1000 to 4800 incremented by 200.
- Variable number of attributes with fixed number of objects and density. In this case, the number of objects is 500, the density is 50 and the number of attributes varies from 1000 to 4800 incremented by 200.

Evolutive Approach of Experiments. We developed the experiments in three phases:

- First Phase: We compared the results between Ares, Ceres and Pluton only implemented in Java. In this specific phase, we used HashSet and ArrayList of the Java language API as our main data structures.
- Second Phase: We compared the results between the implementations in Java (Ares, Ceres and Pluton) and those in C++ (Ares++, Ceres++ and Pluton++). In this specific phase, we used `set<>` and `vector<>` of the C++ language API as our main data structures.
- Third Phase: We compared the results between Ares, Ceres, Pluton, Ares++, Ceres++, Pluton++ and their dual versions, where we transposed the matrices representing the binary relations, meaning that the rows of the non-dual versions are made into columns in the dual versions, and viceversa.

5 Evaluation and Results[1]

5.1 First Phase

In this first phase, we will only consider the Java implementations of the algorithms. Our results show that each algorithm is interesting in its own right for certain input parameters:

- Pluton is the best if there is no difference between the number of objects and the number of attributes (shown in Figure 3).
- Ares is the best if we vary the number of objects (shown in Figure 4).
- Ceres is the best if we vary the number of attributes (shown in Figure 5).

In this first phase, when considering square matrices, we see a common intersection point (ca. (180,120)) where all the algorithms converge, and afterwards we observe major differences in terms of performance. The common point represents a density of around 30% of the binary relation, corresponding to the complexity of 180. This means that up to 30% of the binary relation, Ceres and Ares (with a small interval of difference) are the best algorithms. But with a larger density, both algorithms increase their time, whereas Pluton keeps its monotone shape. From this first phase, with a square matrix, we can conclude that Pluton is not influenced by the density, and that with a low density, Ares and Ceres are the most suitable.

[1] The interested reader can find a colored version of the figures in the original version of the paper in http://www.lirmm.fr/~huchard/Huchard/pub.frametop.html

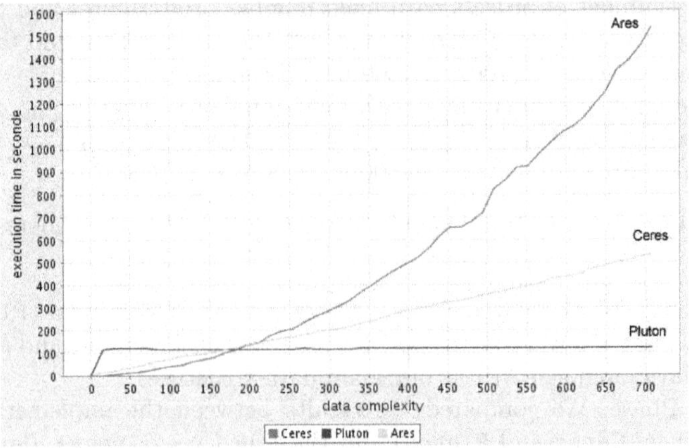

Fig. 3. Test results with a square matrix with variable density

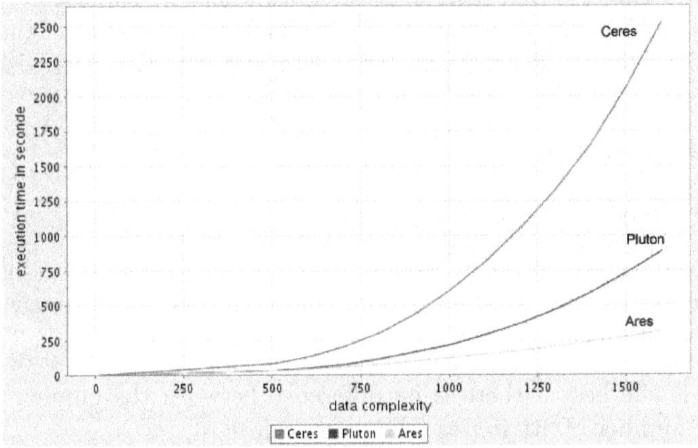

Fig. 4. Test results with a variable number of objects

When considering matrices with a variable number of attributes, we see that the differences in performance between the three algorithms are minimal when we vary the number of attributes between 1000 and 2000 (meaning 500 and 1000 attributes per object) with a density of 50% of the binary relation. After this point, we see that `Ceres` is almost a monotone function, while `Ares` and `Pluton` increase their times. From this we can infer that `Ceres` is the most suitable algorithm.

When considering matrices with a variable number of objects, we see no differences in performance between `Ares` and `Pluton` with a complexity from 0 to 680 with a density of 50% and 50 attributes. However, `Ceres` increases its time as a power function (with an exponent larger than 1) from the minimal complexity.

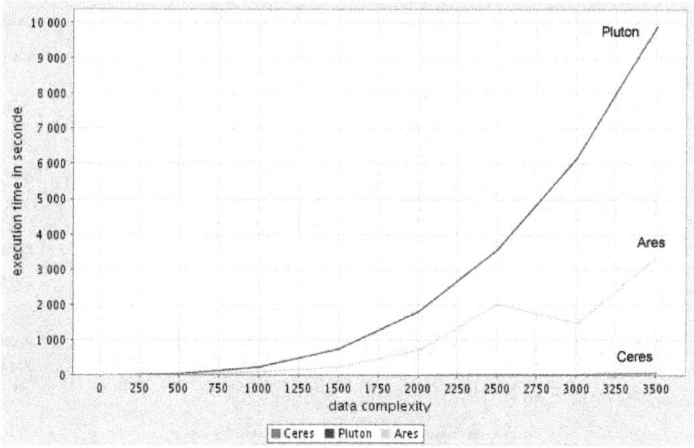

Fig. 5. Test results with a variable number of attributes

In the last two cases, we do not see critical points where the algorithms change their performances.

5.2 Second Phase

In this second phase, we consider Java and C++ implementations of the algorithms: `Ares`, `Ceres`, `Pluton`, `Ares++`, `Ceres++` and `Pluton++` by completing the results of the first phase with C++ implementations. We see in Figures 6, 7 and 8 that:

- `Pluton++` is the best algorithm when considering square matrices (shown in Figure 6), and when considering a variable number of objects (shown in Figure 7).
- `Pluton++` and `Ceres` are the best ones when considering a variable number of attributes (shown in Figure 8).

Let us discuss the main issues we discover in this phase. Within the context of square matrices (shown in Figure 6), for low densities (up to a complexity of 150, which means 22% of density of binary relation), all the algorithms - except `Pluton` and `Pluton++` - present a monotone increasing function, while `Pluton` and `Pluton++` present almost constant functions. From the complexity of 150, we see three branches of algorithms: `Ares++` and `Ceres`, `Ares` and `Ceres++`, `Pluton` and `Pluton++`. In the first group, the performances of the algorithms increase as power functions (with an exponent larger than 1). In the second group, the performances are monotone functions; and in the third group, `Pluton` and `Pluton++` remain as almost constant functions. Let us now compare the versions of the algorithms implemented in Java and in C++. If we consider large densities, we see that there is less difference between `Ares` and `Pluton` than between `Ares++` and `Pluton++`, but the situation is reversed in the case of `Ceres` and

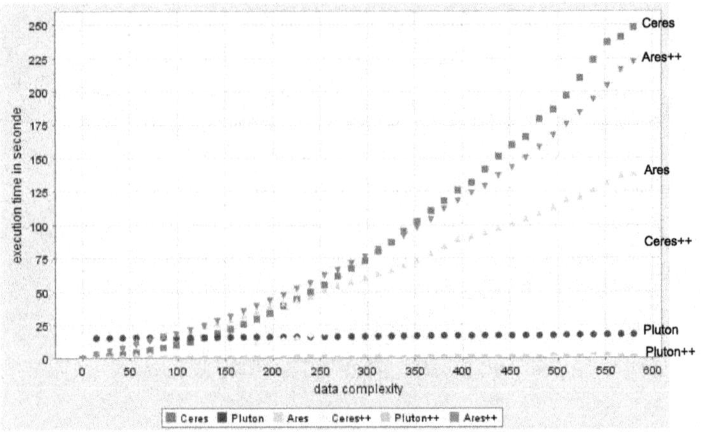

Fig. 6. Test results with a square matrix

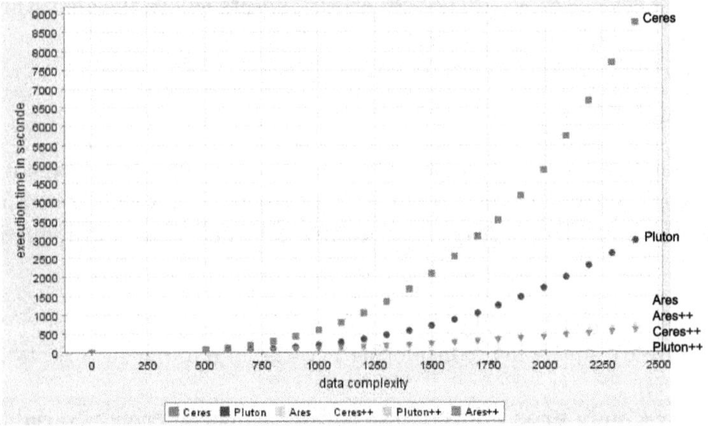

Fig. 7. Test results with a variable number of objects

`Pluton`. There is less difference between `Ceres++` and `Pluton++` than between `Ceres` and `Pluton`.

Generally, we see an improvement of the C++-based algorithms compared to their versions in Java, except in the case of `Ares++`. The major improvement in `Pluton++` illustrates how the API in C++ influences the performances.

When considering a variable number of objects, we see the results in Figure 7. We observe that the implementations in C++ significantly improves the results, as the slowest algorithm in C++ (`Ares++`) has a better performance than the fastest one in Java (`Ares`). Compared to the results of the square matrix, `Ares++` improves its performance. `Ceres++` and `Ares++` seem to have equivalent performances and `Pluton++` remains - so far - the fastest. As a last issue, we see that there is little difference between the implementations in C++ compared to the implementations in Java, although the difference between the algorithms is very significant.

Regarding a variable number of attributes, Figure 8 shows the results. We observe that there are pairs of algorithms (`Ares++` and `Ares` , `Ceres` and `Ceres++`) that have the same performance with the same variations (with some improvements in the Java version). We also see that the difference between `Pluton` and `Pluton++` is significant. `Pluton` is the slowest while `Pluton++` is the fastest. In addition to this, `Ceres++` seems influenced by the increase in the number of attributes while `Ceres` is not. We should remark that, around the complexity of 2400 (meaning 50 objects, 4800 attributes and a density of 50%), there is an important difference in the performances of `Ares` and `Ares++` regarding complexities smaller than 2400.

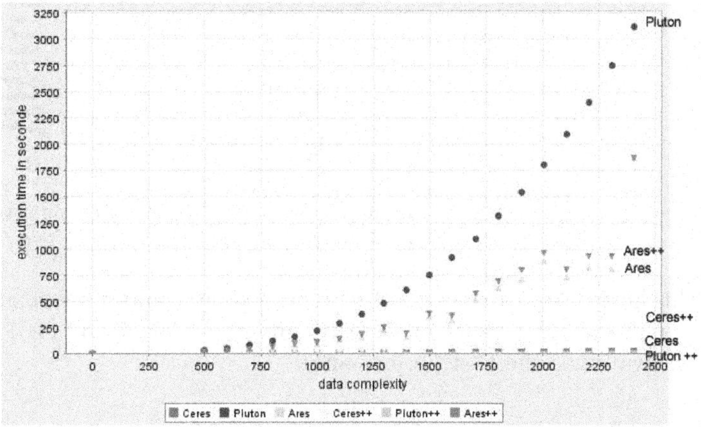

Fig. 8. Test results with a variable number of attributes

As a summary of this phase, we confirm that the API in C++ mostly has a meaningful positive influence on the performances of the algorithms.

5.3 Third Phase

In this phase we decided to test the performances of the algorithms considering the dual versions of the binary relations. When we talk about the dual versions -as said previously-, it does not imply changes in the algorithms, but in the way we deal with the data. In the dual versions, we transpose the matrices representing the binary relations. To perform the experiments, we vary the number of attributes -as we have done in the previous phases. Afterwards, we obtain the dual versions, meaning that we obtain matrices with a variable number of objects. Then we run the algorithms for non-dual binary relations, and for dual binary relations. Figure 9 shows the corresponding results. In order to detail the local results in the lower part of Figure 9, we provide a zoomed version in Figure 10. It is worth remarking that in this case, we do not analyze the square matrices, because the matrix transposition (to analyze the dual version) has the same characteristics in our experiments. Despite the fact that Figure 10 is crowded, we see that it represents the superposition of Figure 7 and Figure 8.

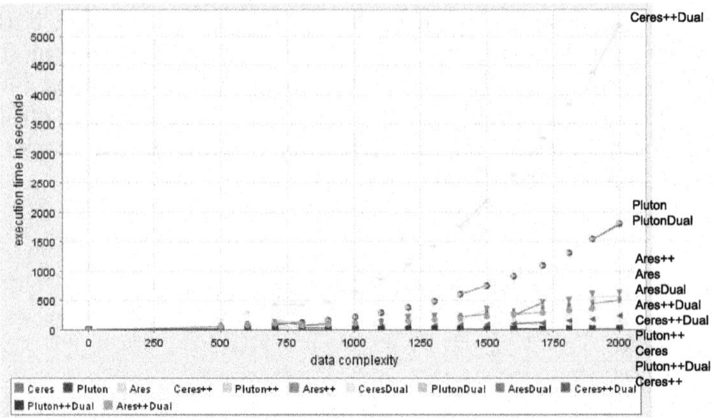

Fig. 9. Test results with a variable number of attributes and their dual versions

From this phase, we observe that `Pluton` has the same performance if we vary the number of attributes or objects, and the same holds for `Pluton++`. `Pluton++` always has the best performance.

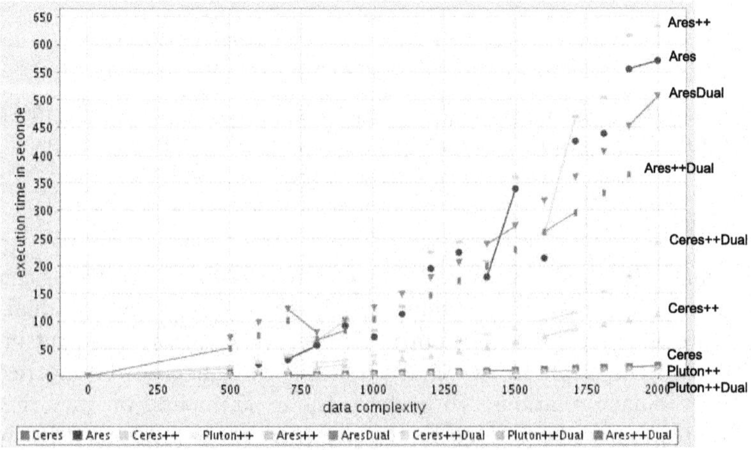

Fig. 10. Zoom of Figure 9

However, `Ceres` changes its performance considerably if we vary the number of objects or the number of attributes. It does not have a good performance when the objects vary but it is faster than the C++ version in the case of attributes. However, `Ceres++` is slower when the number of objects varies than when the number of attributes varies. `Ares` and `Ares++` have the same performance. `Ares++` is better when the number of objects varies.

6 Conclusions and Future Work

This paper compares three different Galois Sub-hierarchy-building algorithms (Ares, Ceres and Pluton) implemented in Java and C++. We see that in most cases, `Pluton++` is the most efficient and stable algorithm. We also see that the API in C++ affects the results of the computation time of these algorithms. It is worth mentioning that for low densities, all algorithms are useful, and the significant differences in performance occur when the binary relations have large densities. Clearly this analysis can guide the user in his/her choice.

In the future, we plan to extend the analysis to similar algorithms, such as the incremental algorithm ISGOOD [GMM95] and the global algorithm proposed in the work of Mineau et al. [MG95], which build only C^A concepts. Besides this, we propose a profiling of all these algorithms, in order to see which are the critical parts that influence the concepts calculation and performance, taking into account the fact that each algorithm follows a different lattice-building algorithm. As a last issue, we plan to implement the algorithms on another platform (such as Smalltalk/VisualWorks) to see if the C++ implementations are still the best in terms of performance.

Acknowledgements. Gabriela Arévalo gratefully acknowledges the financial support of the Swiss National Foundation for the Project: "Advanced Object-Oriented Reverse Engineering using Formal Concept Analysis" SNF Project No. PBBE2-111194.

References

[AYLCB96] S. Amer-Yahia, L. Lakhal, R. Cicchetti, and J.-P. Bordat. iO2 — An Algorithmic Method for Building Inheritance Graphs in Object Database Design. In *Proc. of ER'96*, volume 1157 of *LNCS*, pages 422–437. Springer-Verlag, 1996.

[BHM+05] A. Berry, M. Huchard, R. M. McConnell, A. Sigayret, and J. P. Spinrad. Efficiently computing a linear extension of the sub-hierarchy of a concept lattice. In *Proc. of ICFCA'05*, volume 3403 of *LNCS*, pages 208–222. Springer-Verlag, 2005.

[DDHL94] H. Dicky, C. Dony, M. Huchard, and T. Libourel. ARES, un algorithme d'ajout avec restructuration dans les hiérarchies de classes. *Proc. of LMO'94, L'Objet*, pages 125–136, 1994.

[DHL+02] M. Dao, M. Huchard, T. Libourel, C. Roume, and H. Leblanc. A New Approach to Factorization: Introducing Metrics. In *Proc. of Metrics '02*, pages 227–236. IEEE Computer Society, 2002.

[FLPP05] L. Fura, G. Laplace, A. Le Provost, and G. Perrot. Algorithme de construction d'une sous-hiérarchie de Galois. Technical report, Université de Montpellier II, 2005.

[Gal] GaLicia: Galois lattice interactive constructor. Université de Montréal. http://www.iro.umontreal.ca/~galicia.

[GC99] R. Godin and T.-T. Chau. Comparaison d'algorithmes de construction de hiérarchies de classes. *L'Objet*, 5(3/4), 1999.

[GM93] R. Godin and H. Mili. Building and Maintaining Analysis-Level Class Hierarchies using Galois Lattices. In *Proc. of OOPSLA '93*, volume 28, pages 394–410. ACM Press, October 1993.

[GMM95] R. Godin, G. Mineau, and R. Missaoui. Incremental structuring of knowledge bases. In G. Ellis, R. A. Levinson, A. Fall, and V. Dahl, editors, *Proc. of KRUSE'95*, pages 179–193, University of California at Santa Cruz, 1995. Department of Computer Science.

[GMM+98] R. Godin, H. Mili, G. W. Mineau, R. Missaoui, A. Arfi, and T.-T. Chau. Design of Class Hierarchies based on Concept (Galois) Lattices. *Theory and Practice of Object Systems*, 4(2):117–134, 1998.

[GW99] B. Ganter and R. Wille. *Formal Concept Analysis: Mathematical Foundations*. Springer Verlag, 1999.

[HDL00] M. Huchard, H. Dicky, and H. Leblanc. Galois Lattice as a Framework to specify Algorithms Building Class Hierarchies. *Theoretical Informatics and Applications*, 34:521–548, 2000.

[Hit04] P. Hitzler. Default reasoning over domains and concept hierarchies. In *Proc. of KI 2004*, volume 3238 of *LNCS*, pages 351–365. Springer Verlag, 2004.

[KO02] Sergei O. Kuznetsov and Sergei A. Obiedkov. Comparing performance of algorithms for generating concept lattices. *Journal of Experimental & Theoretical Artificial Intelligence*, 14(2-3):189–216, 2002.

[Leb00] H. Leblanc. *Sous-hiérarchies de Galois: un modèle pour la construction et l'évolution des hiérarchies d'objets*. PhD thesis, Univ. de Montpellier II, 2000.

[MG95] G. W. Mineau and R. Godin. Automatic structuring of knowledge bases by conceptual clustering. *IEEE Trans. Knowl. Data Eng.*, 7(5):824–828, 1995.

[OP02] R. Osswald and W. Petersen. Induction of classifications from linguistic data. In *Proc. of ECAI'02 Workshop*, pages 75–84. Université de Lyon I, July 2002.

[Per05] G. Perrot. Implémentation d'algorithmes de construction de sous-hiérarchies de Galois et étude des performances, 2005.

[Pet01] W. Petersen. A set-theoretical approach for the induction of inheritance hierarchies. In *Electronic Notes in Theoretical in Computer Science*, volume 51. Elsevier, July 2001.

Galois Connections Between Semimodules and Applications in Data Mining

Francisco J. Valverde-Albacete and Carmen Peláez-Moreno*

Dpto. de Teoría de la Señal y de las Comunicaciones
Universidad Carlos III de Madrid
Avda. de la Universidad, 30. Leganés 28911. Spain
{fva,carmen}@tsc.uc3m.es

Abstract. In [1] a generalisation of Formal Concept Analysis was introduced with data mining applications in mind, $\mathcal{K}$-Formal Concept Analysis, where incidences take values in certain kinds of semirings, instead of the standard Boolean carrier set. A fundamental result was missing there, namely the second half of the equivalent of the main theorem of Formal Concept Analysis. In this continuation we introduce the structural lattice of such generalised contexts, providing a limited equivalent to the main theorem of $\mathcal{K}$-Formal Concept Analysis which allows to interpret the standard version as a privileged case in yet another direction. We motivate our results by providing instances of their use to analyse the confusion matrices of multiple-input multiple-output classifiers.

1 Motivation: The Exploration of Confusion Matrices with $\mathcal{K}$-Formal Concept Analysis

In pattern recognition tasks, when a classifier is provided training data in the form of feature vectors tagged with an *input pattern set* and produces for each vector a tag within an *output pattern set*, the performance of the classifier can be gleaned from the collection of pairs (g_i, m_j) of one input tag, g_i, for the input data and one output tag, m_j , produced by the classifier. These results are aggregated into a *confusion matrix*, T, whose element T_{ij} gives a "measure" of the joint event $(G = g_i, M = m_j)$, "providing an input pattern g_i to the classifier who then produces an output pattern m_j".

In the pattern recognition community we often encounter methods that use confusion matrices to analyse classification results. However, most of the times the analysis is manual and limited to the (human-based) pondering of a confusion matrix-representation like the one depicted in figure 1, where the warmer, brighter (resp. cooler, darker) colour hues are designed to be related to high occurrence (resp. to low occurrence) of events. Often, this type of analysis is used

* This work has been partially supported by two grants for "Estancias de Tecnólogos Españoles en el International Computer Science Institute, año 2006" of the Spanish Ministry of Industry and a Spanish Government-Comisión Interministerial de Ciencia y Tecnología project TEC2005-04264/TCM.

S.O. Kuznetsov and S. Schmidt (Eds.): ICFCA 2007, LNAI 4390, pp. 181–196, 2007.

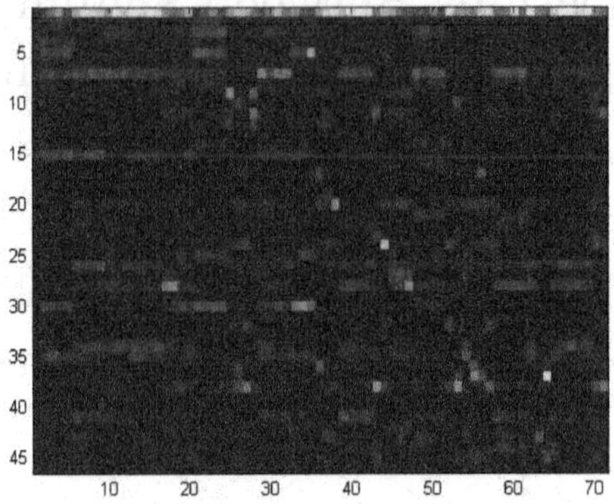

Fig. 1. Confusion matrix of the desired transformation of English phoneme labels of speech frames versus their true Mandarin phoneme labels

to bootstrap existing classifiers in order to obtain even better classification figures or simply to understand the underlying principles of the methods employed in designing the classification. In particular, in speech recognition, the designer of a system is challenged to find in this type of representation meaningful or systematic confusions to determine to what degree the behaviour of an automatic system differs from human performance.

$\mathcal{K}$-Formal Concept Analysis was introduced in [1] as a generalisation of standard Formal Concept Analysis in the sense that incidences $R \in \mathcal{K}^{n\times p}$ represented as matrices may take values in an idempotent, reflexive semifield $\mathcal{K}$ and we take $R(i,j) = \lambda$ to mean "object g_i has attribute m_j in degree λ." Adequate analogues of basic objects in Formal Concept Analysis become therefore available.

Two serious obstacles may prevent widespread adoption of $\mathcal{K}$-Formal Concept Analysis as a data exploration technique complementary to the standard theory: on the one hand, the $\mathcal{K}$-Formal Concept Analysis analogue of the main theorem of Formal Concept Analysis is incomplete and this may worry the user willing to be on a sound mathematical ground; on the other hand, [1] did not provide an algorithm for constructing the lattice of a $\mathcal{K}$-valued formal context, which prevents its use as a data-intensive exploration procedure.

In this paper, we try to explore further whether $\mathcal{K}$-Formal Concept Analysis is a proper generalisation of standard Formal Concept Analysis for finite contexts and to pave the way for the completion of the main theorem. In order to do so we introduce the *structural lattice* of a $\mathcal{K}$-Formal Context and try to relate it to the Concept Lattice of a Formal Context.

In section 2 we first review the theory of idempotent semirings and their semimodules with a view to providing the necessary objects for our discussion. In section 3.1 we present a summary of the theory of $\mathcal{K}$-Formal Concept Analysis presented in ([1], §. 3) and add a new theoretical construct, the *structural lattice* of a semimodule over an idempotent, reflexive semiring. We demonstrate in section 4 the use of this new tool to analyse confusion matrices of multiple input-multiple output classifiers, which turn out to be amenable to $\mathcal{K}$-Formal Concept Analysis modelling, and finish with a summary of contributions and an outlook.

2 Mathematical Preliminaries: Semimodules over Idempotent, Reflexive Semifields as Vector Spaces

2.1 Idempotent Semirings

A *semiring* $\mathcal{K} = \langle K, \oplus, \otimes, \epsilon, e \rangle$ is an algebraic structure whose additive structure, $\langle K, \oplus, \epsilon \rangle$, is a commutative monoid and the multiplicative one, $\langle K, \otimes, e \rangle$, a monoid whose multiplication distributes over addition from right and left and whose neutral element is absorbing for $\otimes$, $\epsilon \otimes x = \epsilon, \forall x \in K$ [2] . On any semiring $\mathcal{K}$ left and right multiplications can be defined:

$$\begin{aligned} L_a : K &\to K & \qquad R_a : K &\to K \\ b &\mapsto L_a(b) = ab & \qquad b &\mapsto R_a(b) = ba \end{aligned} \tag{1}$$

A *commutative semiring* is a semiring whose multiplicative structure is commutative, and a *semifield* one whose multiplicative structure over $K \backslash \{\epsilon\}$ is a group. Thus, compared to a ring, a semiring which is not a ring lacks additive inverses.

An *idempotent semiring* $\mathcal{K}$ is a semiring whose addition is idempotent: $\forall a \in K, a \oplus a = a$. All idempotent commutative monoids $(K, \oplus, \epsilon)$ are endowed with a *natural order* $\forall a, b \in K, a \leq b \iff a \oplus b = b$, which turns them into join-semilattices with least upper bound defined as $a \vee b = a \oplus b$. Moreover, for the additive structure of and idempotent semiring $\mathcal{K}$ the neutral element is the infimum for this natural order, $\epsilon_{\mathcal{K}} = \bot$.

An idempotent semiring $\mathcal{K}$ is *complete*, if it is complete as a naturally ordered set and left (L_a) and right (R_a) multiplications are lower semicontinuous, that is, they commute with joins over any subset of $\mathcal{K}$. Therefore, complete idempotent semirings, as join-semilattices with infimum are automatically complete lattices [3] with join ($\vee$, max or sup) and meet ($\wedge$, min or inf) connected by the equivalences: $\forall a, b \in K, a \leq b \iff a \vee b = b \iff a \wedge b = a$.

Example 1. *1. The Boolean semiring* $\mathcal{B} = \langle \mathbb{B}, \vee, \wedge, 0, 1 \rangle$, *with* $\mathbb{B} = \{0, 1\}$, *is complete, idempotent and commutative.*

2. *The completed Maxplus semiring* $\overline{R}_{max,+} = \langle \mathbb{R} \cup \{\pm\infty\}, \max, +, -\infty, 0 \rangle$, *is a complete, idempotent semifield when defining* $-\infty + \infty = -\infty$, *so that* $\epsilon_{\mathcal{K}} \otimes \top_{\mathcal{K}} = \epsilon_{\mathcal{K}}$ *for* $\mathcal{K} \equiv \overline{R}_{max,+}$
3. *The completed Minplus semiring* $\overline{R}_{min,+} = \langle \mathbb{R} \cup \{\pm\infty\}, \min, +, \infty, 0 \rangle$ *is a complete, idempotent semifield with a similar completion to that of ex. 2 with* $\infty + (-\infty) = \infty$, *that is* $\epsilon_{\mathcal{K}} \otimes \top_{\mathcal{K}} = \epsilon_{\mathcal{K}}$ *for* $\mathcal{K} \equiv \overline{R}_{min,+}$.

2.2 Idempotent Semimodules: Basic Definitions

A semimodule over a semiring is defined in a similar way to a module over a ring [4,5,6][1]: a *left $\mathcal{K}$-semimodule*, $\mathcal{X} = \langle X, \oplus, \epsilon_X \rangle$, is an additive commutative monoid endowed with a map $(\lambda, x) \mapsto \lambda \cdot x$ such that for all $\lambda, \mu \in K$, $x, z \in X$, and following the convention of dropping the symbol for the scalar action and multiplication for the semiring we have:

$$(\lambda\mu)x = \lambda(\mu x) \qquad \epsilon_K x = \epsilon_X \qquad (2)$$
$$\lambda(x \oplus z) = \lambda x \oplus \lambda z \qquad e_K x = x$$

The definition of a *right $\mathcal{K}$-semimodule*, $\mathcal{Y}$, follows the same pattern with the help of a *right action*, $(\lambda, y) \mapsto y\lambda$ and similar axioms to those of (2.)

A *$(\mathcal{K}, \mathcal{S})$-semimodule* is a set M endowed with left $\mathcal{K}$-semimodule and a right $\mathcal{S}$-semimodule structures, and a *$(\mathcal{K}, \mathcal{S})$-bisemimodule* a $(\mathcal{K}, \mathcal{S})$-semimodule such that the left and right multiplications commute. For a left $\mathcal{K}$-semimodule, $\mathcal{X}$, the left and right multiplications are defined as:

$$L_\lambda^{\mathcal{K}} : X \to X \qquad R_x^{\mathcal{X}} : K \to X \qquad (3)$$
$$x \mapsto L_\lambda^{\mathcal{K}}(x) = \lambda x \qquad \lambda \mapsto R_x^{\mathcal{X}}(\lambda) = \lambda x$$

And similarly, for a right K-semimodule. If $\mathcal{X}$, $\mathcal{Z}$ are left semimodules a *morphism of left semimodules or left linear map* $F : \mathcal{X} \to \mathcal{Z}$ is a map that preserves finite sums and commutes with the action: $F(\lambda v \oplus \mu w) = \lambda F(v) \oplus \mu F(w)$, and similarly, *mutatis mutandis* for right linear maps of right semimodules.

The elements of a semimodule may be conceived as *vectors*[2]. Given a semiring $\mathcal{K}$ and a left $\mathcal{K}$-semimodule $\mathcal{X}$, for each finite, non-void set $W \subseteq X$, there exists an homomorphism $\alpha : K^W \to X,\ f \mapsto \bigoplus_{w \in W} f(w)w$. Moreover, α induces a congruence of semimodules $\equiv_\alpha$ on K^W, by $f \equiv_\alpha g \iff \alpha(f) = \alpha(g)$. Then W is a *set of generators* or a *generating family* precisely when α is surjective, in which case any element $x \in X$ can be written as $x = \bigoplus_{w \in W} \lambda_w w$, and we will write $\mathcal{X} = \langle W \rangle_{\mathcal{K}}$, that is, $\mathcal{X}$ *is the span of* W. A semimodule is *finitely generated* if it has a finite set of generators.

For individual vectors, we say that $x \in W$ is *dependent (in W)* if $x = \bigoplus_{w \in W \setminus \{x\}} \lambda_w w$ otherwise, we say that it is *free (in W)*. The set W is *linearly independent* if and only if $\equiv_\alpha$ is the trivial congruence, that is, when $\bigoplus_{w \in W} f(w)w = \bigoplus_{w \in W} h(w)w \iff f = h$, otherwise, W is *linearly dependent*. Let $\ker \alpha = \{ f \in K^W \mid \alpha(f) = 0 \}$; then W is *weakly linearly independent* if and only if $\ker \alpha = \{0\}$, otherwise it is *weakly linearly dependent*.

A *basis for $\mathcal{X}$ (over $\mathcal{K}$)* is a linearly-independent set of generators, and a semimodule generated by a basis is *free*. By definition, in a free semimodule $\mathcal{X}$ with with basis $\{ x_i \}_{i \in I}$ each element $x \in X$ can be uniquely written as $x = \bigoplus_{i \in I} \alpha_i x_i$, with $[a_i]_{i \in I}$ the *co-ordinates of x with respect to the basis*. A

[1] We are following essentially the notation of [4].

[2] Most of the material in this section is from [5], §17, and [7,8,9].

weakly linearly-independent set of generators for $\mathcal{X}$ is a *weak basis for* $\mathcal{X}$ *(over* $\mathcal{K}$*)*. The cardinality of a (weak) basis is the *(weak) rank* of the semimodule.

In this framework, notions in usual vector spaces have to be imported with care. For instance, the *image* of a linear map $F : \mathcal{X} \to \mathcal{Y}$ is simply the semimodule $\mathrm{Im}F = \{ F(x) \mid x \in X \}$, but it is in general **not free**.

Given a free semimodule $\mathcal{X}$ with basis $\{ x_i \}_{i \in I}$, for each family $\{ y_i \}_{i \in I}$ of elements of an arbitrary semimodule $\mathcal{Y}$ there is a unique morphism of semimodules $F : \mathcal{X} \to \mathcal{Y}$ such that $F(x_i) = y_i, \forall i \in I$, namely $F\left(\bigoplus_{i \in I} \lambda_i x_i\right) = \bigoplus_{i \in I} \lambda_i y_i$ and all the linear maps $Lin(\mathcal{X}, \mathcal{Y})$ are obtained in this way ([7], prop. §73; [5], prop. §17.12). That is, linear maps from free semimodules are characterised by the images of the elements of a basis.

On the other hand, a semiring $\mathcal{K}$ has the *linear extension property* if for all free, finitely generated $\mathcal{K}$-semimodules $\mathcal{X}, \mathcal{Y}$, for all finitely generated subsemimodules $\mathcal{Z} \subset X$ and for all $F \in Lin(\mathcal{Z}, \mathcal{Y})$, there exists $H \in Lin(\mathcal{X}, \mathcal{Y})$ such that $\forall x \in X, H(x) = F(x)$. The importance of this property derives from the fact that when the linear extension property holds, each linear map between finitely generated subsemimodules of free semimodules is represented by a matrix. In particular, when it holds for free, finitely generated (left) semimodules, $\mathcal{X}$ and $\mathcal{Y}$ with bases $\{ x_i \}_{i \in I}$ and $\{ y_j \}_{j \in J}$, each linear map is characterised by the $n \times p$-matrix $R = (F(x_i)_j)$, which sends vector $x = \{x_i\}_{i=1}^n$ to the vector $F(x) \simeq ((xR)_1, \ldots, (xR)_p)$.

2.3 Semimodules over Idempotent Semirings

In this section all semimodules will be defined over an idempotent semifield. Recall that examples of these are $\mathcal{B}$, the Boolean semifield and the completed maxplus and minplus semifields.

Idempotency and Natural Order in Semimodules. A left, right $\mathcal{K}$-semimodule $\mathcal{X}$ over an idempotent semiring $\mathcal{K}$ inherits the idempotent law, $v \oplus v = v, \forall v \in X$, which induces a *natural order* on the semimodule by $v \leq w \iff v \oplus w = w, \forall v, w \in X$ whereby it becomes a $\vee$-semilattice, with $\epsilon_{\mathcal{X}}$ the minimum. In the following we systematically equate idempotent $\mathcal{K}$-semimodules and semimodules over an idempotent semiring $\mathcal{K}$. When $\mathcal{K}$ is a complete idempotent semiring, a left $\mathcal{K}$-semimodule, $\mathcal{X}$ is *complete (in its natural order)* if it is complete as a naturally ordered set and its left and right multiplications are (lower semi)continuous. Trivially, it is also a complete lattice, with join and meet operations given by: $v \leq w \iff v \vee w = w \iff v \wedge w = v$. This extends naturally to right- and bisemimodules.

Example 2. *1. Each semiring, $\mathcal{K}$, is a left (right) semimodule over itself, with the semiring product as left (right) action. Therefore, it is a ($\mathcal{K}$, $\mathcal{K}$)-bisemimodule over itself, because both actions commute by associativity. Such is the case for the Boolean ($\mathcal{B}$, $\mathcal{B}$)-bisemimodule, the Maxplus and the Minplus bisemimodules. These are all complete and idempotent.*

2. *For $n, m \in \mathbb{N}$, the set of matrices $K^{n\times p}$ is a $(K^{n\times n}, K^{p\times p})$-bisemimodule with matrix multiplication-like left and right actions and component-wise addition, the set of column vectors $K^{p\times 1}$ is a $(K^{p\times p}, K)$-bisemimodule and the set of row vectors $K^{1\times n}$ a $(K, K^{n\times n})$-bisemimodule with similarly defined operations. If $\mathcal{K}$ is idempotent (resp. complete), then all are idempotent (resp. complete) with the component-wise partial order their natural order.*

As in the semiring case, because of the natural order structure, the actions of idempotent semimodules admit residuation: given a complete, idempotent left $\mathcal{K}$-semimodule, $\mathcal{X}$, we define for all $x, z \in X$, $\lambda \in K$ the residuals:

$$
\begin{aligned}
\left(L_\lambda^{\mathcal{K}}\right)^{\#} &: X \to X \qquad & \left(L_\lambda^{\mathcal{K}}\right)^{\#}(z) &= \bigvee\{\, x \in X \mid \lambda x \le z \,\} = \lambda \backslash z \\
\left(R_x^{\mathcal{X}}\right)^{\#} &: X \to K \qquad & \left(R_x^{\mathcal{X}}\right)^{\#}(z) &= \bigvee\{\, \lambda \in K \mid \lambda x \le z \,\} = z / x
\end{aligned}
\tag{4}
$$

and likewise for a right semimodule, $\mathcal{Y}$.

There is a remarkable operation that changes the character of a semimodule while at the same time reversing its order by means of residuation:

Definition 3. *Let $\mathcal{K}$ be a complete, idempotent semiring, and $\mathcal{Y}$ be a complete right $\mathcal{K}$-semimodule, its* opposite semimodule *is the complete* left *$\mathcal{K}$-semimodule $\mathcal{Y}^{op} = \langle Y, \overset{op}{\oplus}, \overset{op}{\rightarrow} \rangle$ with the same underlying set Y, addition defined by $(x, y) \mapsto x \overset{op}{\oplus} y = x \wedge y$ where the infimum is for the natural order of $\mathcal{Y}$, and left action:*

$$K \times Y \to Y \quad (\lambda, y) \mapsto \lambda \overset{op}{\rightarrow} y = y / \lambda$$

Consequently, the order of the opposite is the dual *of the original order.*

For the opposite semimodule the residual definitions are:

$$
\begin{aligned}
\lambda \overset{op}{\backslash} x &= \left(L_\lambda^{\mathcal{Y}^{op}}\right)^{\#}(x) = \bigwedge\{\, y \in Y \mid x \le y/\lambda \,\} = x \cdot \lambda \\
x \overset{op}{/} y &= \left(R_y^{\mathcal{Y}^{op}}\right)^{\#}(x) = \bigvee\{\, \lambda \in K \mid x \le y/\lambda \,\} = x \backslash y
\end{aligned}
\tag{5}
$$

Note that we can define *mutatis mutandis* the opposite semimodule of a left $\mathcal{K}$-semimodule, $\mathcal{X}$, with right action $x \overset{op}{\leftarrow} \lambda = \lambda \backslash x$. Also, noticing that the first residual in eq. 5 is in fact an involution we may conclude that the operation of finding the opposite of a complete (left, right) $\mathcal{K}$-semimodule is an involution: $(\mathcal{Y}^{op})^{op} = \mathcal{Y}$.

Constructing Galois Connections in Idempotent Semimodules. The following construction is due to Cohen et al. [4]. Let $\mathcal{K}$ be a complete idempotent semiring; for a bracket $\langle \cdot \mid \cdot \rangle : X \times Y \to Z$ between left and right $\mathcal{K}$-semimodules, $\mathcal{X}$ and $\mathcal{Y}$ respectively, onto a $\mathcal{K}$-bisemimodule $\mathcal{Z}$ and an arbitrary element $\varphi \in Z$, which we call the *pivot*, define the maps:

$$
\begin{aligned}
\cdot_\varphi^- &: X \to Y \qquad & x_\varphi^- &= L_x^{\#}(\varphi) = \bigvee\{\, y \in Y \mid \langle x \mid y \rangle \le \varphi \,\} \\
{}_\varphi^-\cdot &: Y \to X \qquad & {}_\varphi^- y &= R_y^{\#}(\varphi) = \bigvee\{\, x \in X \mid \langle x \mid y \rangle \le \varphi \,\}
\end{aligned}
\tag{6}
$$

We have $\langle x \mid y \rangle \leq \varphi \iff y \leq x^{-}_{\varphi} \iff x \leq {}_{\varphi}^{-}y$, whence the pair is a Galois connection between $\mathcal{Y}$ and $\mathcal{X}$, $(\cdot^{-}_{\varphi}, {}_{\varphi}^{-}\cdot) : \mathcal{X} \multimap \mathcal{Y}$. This construction is affected crucially by the choice of a suitable pivot φ: if we consider the bracket to reflect a *degree of relatedness* between the elements of each pair, only those pairs $(x, y) \in X \times Y$ are considered by the connection whose degree amounts *at most* to φ . Therefore we can think of the pivot as a *maximum degree of existence* allowed for the pairs.

Recall $\mathcal{X}$ and $\mathcal{Y}$ are both complete lattices as well as free vector spaces. Note that the closure lattices $\overline{X} = {}_{\varphi}^{-}(\mathcal{Y})$ and $\underline{Y} = (\mathcal{X})^{-}_{\varphi}$ do not agree with their ambient vector spaces in their joins, but only in their meets. To improve on this, the notion of a *left (resp. right) reflexive*, $(\mathcal{K}, \varphi)$, semiring is introduced in [4] as a complete idempotent semiring such that $(\langle \cdot \mid \cdot \rangle : K \times K \to K, \varphi)$ with $\langle \lambda \mid \mu \rangle = \lambda\mu$ induces a *perfect Galois connection*[3] under construction (6) for all $\lambda \in K$, ${}^{-}(\lambda^{-}) = \lambda$ (resp. $({}^{-}\lambda)^{-} = \lambda$.)[4] The interest in reflexive semirings stems from the fact that in such semirings $\overline{X}$ and $\overline{Y}$ are actually subsemimodules (that is their suprema coincide with those) of the corresponding spaces ([4], prop. 28).

Note that φ need not be unique: if $(\mathcal{K}, \varphi)$ is right (or left) reflexive, for any $\lambda \in K$ invertible, $(\mathcal{K}, \varphi\lambda)$ is left reflexive (and $(\mathcal{K}, \lambda\varphi)$ is right reflexive.) Finally, Cohen et al. [4] prove that idempotent semifields are left and right reflexive, and suggest that for the Boolean semiring we must choose $\varphi = 0_{\mathcal{B}}$, the bottom in the order. For other semifields any invertible element may be chosen, e.g. $\varphi = e_{\mathcal{K}}$.

Idempotent Semimodules as Vector Spaces. When $\mathcal{K}$ is an idempotent semiring if a $\mathcal{K}$-semimodule has a (weak) basis, it is unique up to a permutation and re-scaling of the axes, that is a *scaling endomorphism* ([9], Th. §3.1), $x'_i = \lambda_i x_i$, and every finitely generated $\mathcal{K}$-semimodule has a weak basis ([9], Coroll. §3.6). In particular, let $\mathcal{K}$ be an idempotent semifield, then the free idempotent semimodule with n generators is isomorphic to $\mathcal{K}^n$. Essentially, such free idempotent semimodules are generated by the bases $E_n \triangleq \{e_i\}_{i=1}^{n}$, $e_i = (\delta_{i1}, \delta_{i2}, \ldots, \delta_{in})$, where δ_{ij} is the Kronecker symbol over $\mathcal{K}$, $\delta_{ii} = e_{\mathcal{K}}, \delta_{ij} = \epsilon_{\mathcal{K}}, i \neq j$.

Importantly, the linear property holds in every idempotent semiring which is a distributive lattice for the natural order ([7], Th. §83). This is the case for the semifields $\mathbb{B}$ (the Boolean semiring), $\overline{R}_{\max,+}$ and $\overline{R}_{\min,+}$. Therefore, in such semimodules, modulo a choice of bases for $\mathcal{X}$ and $\mathcal{Y}$, we may identify $\mathcal{X} \cong \mathcal{K}^{1\times n}$ and $\mathcal{Y} \cong \mathcal{K}^{1\times p}$, and linear maps to matrix transformations $\mathrm{Lin}(\mathcal{X}, \mathcal{Y}) \cong \mathcal{K}^{n\times p}$, $R : \mathcal{K}^{1\times n} \to \mathcal{K}^{1\times p}, x \mapsto xR$. When passing from left to right semimodules this should read $\mathcal{K}^{p\times 1} \to \mathcal{K}^{n\times 1}, y \mapsto Ry$.

Idempotent semimodules have additional properties which make them easier to work with as spaces: when $\mathcal{X}$ is a vector space over an idempotent semiring $\mathcal{K}$, for a set of vectors, $W \subseteq X$, the set of finite sums $W^{+} \triangleq \{\bigoplus_i w_i \mid w_i \in W\}$, is a $\vee$-subsemilattice of $\langle W \rangle_{K}$. Therefore, the $\vee$-irreducibles of W, generate the span of W, $\langle \mathcal{J}(W) \rangle_{\mathcal{K}} = \langle W \rangle_{\mathcal{K}}$. This makes the $\vee$-irreducibles an interesting set to obtain a basis.

[3] That is, a pair of mutually inverse isomorphisms.

[4] When the pivot is the multiplicative unit $\varphi = e$ we drop it.

The Projective Space and the Structural Semilattice. Let $\mathcal{X}$ be a left $\mathcal{K}$-semimodule over an idempotent semiring $\mathcal{K}$. The relation $x \preccurlyeq y \Leftrightarrow \exists \lambda \in K, x \leq \lambda \otimes y$ defines a quasi-order $\langle \mathcal{X}, \preccurlyeq \rangle$. Since any basis $W_{\mathcal{X}}$ is unique up to a re-scaling map, the Hasse diagram of $(W_{\mathcal{X}}, \preccurlyeq)$ is independent of the choice of basis.

Now define the equivalence relation ([7], p. 41), $x \simeq y \overset{\Delta}{\Leftrightarrow} x \preccurlyeq y$ and $y \preccurlyeq x$. This relation appears already in ([10], p. 2018) and was later considered under the name of *siblinghood relation* [11] where two vectors v and w are *siblings* if $w = \lambda \otimes v$ for some $\lambda \in K$. This is a congruence of $\vee$-semilattices, therefore [7], the *projective space* is the quotient set $\mathbb{P}(\mathcal{X}) \triangleq \{[x]_{\simeq} \mid x \in X\}$ (where $[x]_{\simeq}$ the equivalence class of $x \in X$, is also called the *ray* of x or the *sibling class* of x), which is also a $\vee$-semilattice, $\langle \mathbb{P}(\mathcal{X}), \preccurlyeq \rangle$ with the induced order.

For any subset $W \subseteq X$, let a *section of the quotient set* $W/_{\simeq}$, $\sigma : 2^X \to X, W \mapsto \sigma(W)$ be a set obtained by choosing a single representative from each sibling class. Note that a section has the order directly induced by $\preccurlyeq_{\mathcal{X}}$ [11] . It is now clear that a section of the quotient set of the join irreducibles of a set of vectors is a (weak) basis of their span $\sigma\left[\mathcal{J}(W)\right] = \langle W \rangle_{\mathcal{K}}$.

Next, consider the siblinghood relation above and a basis $W_{\mathcal{X}}$:

Definition 4 (Wagneur [10]). *Let $\mathcal{X}$ be a left $\mathcal{K}$-semimodule over and idempotent semifield $\mathcal{K}$ with a basis $W_{\mathcal{X}}$. The* structural ($\vee$-)semilattice *of $\mathcal{X}$, $S(\mathcal{X})$ is the quotient set of ${W_{\mathcal{X}}}^+$ through the siblinghood relation $S(\mathcal{X}) \triangleq {W_{\mathcal{X}}}^+/_{\simeq}$.*

The following theorem states that the quotient set ${W_{\mathcal{X}}}^+/_{\simeq}$ is an intrinsic invariant of $\mathcal{X}$.

Theorem 1 ([10], Th. 2). *For any basis $W_{\mathcal{X}}$ of a left $\mathcal{K}$-semimodule over and idempotent semifield $\mathcal{K}$, the quotient map $\pi : {W_{\mathcal{X}}}^+ \to {W_{\mathcal{X}}}^+/_{\simeq}, w \mapsto [w]_{\simeq}$ is an epimorphism of $\vee$-semilattices and ${W_{\mathcal{X}}}^+/_{\simeq}$ is independent of the particular choice of basis $W_{\mathcal{X}}$.*

Since π is an epimorphism of $\vee$-semilattices and $\simeq$ a $\vee$-congruence, the quotient set of the basis through the siblinghood relation $W_{\mathcal{X}}/_{\simeq} = \{[w]_{\simeq} \mid w \in W_{\mathcal{X}}\}$ is the set of $\vee$-irreducibles of the quotient set, $\mathcal{J}\left({W_{\mathcal{X}}}^+/_{\simeq}\right) = W_{\mathcal{X}}/_{\simeq}$ [10].

3 The Structural Lattice of a $\mathcal{K}$-Concept Lattice

3.1 $\mathcal{K}$-Formal Concept Analysis, a Reminder

The following has been adapted from [1] to emphasise the fact that the theory does not cover the case of unbounded cardinalities[5].

Definition 5 ($\mathcal{K}$-valued formal context). *For $n, p \in \mathbb{N}$, given two sets of objects $G = \{g_i\}_{i=1}^n$, and attributes $M = \{m_j\}_{j=1}^p$, an idempotent semiring, $\mathcal{K}$, and a $\mathcal{K}$-valued matrix, $R \in \mathcal{K}^{n \times p}$, where $R(i,j) = \lambda$ reads as "object g_i has attribute m_j in degree λ" and dually "attribute m_j is manifested in object g_i to degree λ", the triple $(G, M, R)_{\mathcal{K}}$ is called a $\mathcal{K}$-valued formal context.*

[5] This section follows in the tracks of §1.1 of [12].

Clearly single objects are isomorphic to elements of the space $\mathcal{K}^{1\times p}$, that is *rows of* R or *object descriptions*, vectors of as many values as attributes. And dually, single attributes are isomorphic to elements of the space $\mathcal{K}^{n\times 1}$, *columns of* R or *attribute descriptions*. We model ($\mathcal{K}$-valued) sets of objects as row vectors in a left $\mathcal{K}$-semimodule, $x \in \mathcal{X} \cong \mathcal{K}^{1\times n}$, and sets of attributes as column vectors in a right $\mathcal{K}$-semimodule, $y \in \mathcal{Y} \cong \mathcal{K}^{p\times 1}$ as generalisations of characteristic functions in the power sets $\mathbf{2}^G, \mathbf{2}^M$, respectively.

The proof of the following proposition is crucial for future argumentation, hence we reproduce it in full:

Proposition 2. *Let* $(\mathcal{K}, \varphi)$ *be a reflexive, idempotent semiring. For a* $\mathcal{K}$*-valued formal context* $(G, M, R)_{\mathcal{K}}$*, with* $n, p \in \mathbb{N}$*, there is at least one Galois connection between the lattices of (*$\mathcal{K}$*-valued) sets of objects* $\mathcal{K}^{1\times n}$ *and attributes* $\mathcal{K}^{p\times 1}$.

Proof. Recall that $\mathcal{X} = \mathcal{K}^{1\times n}$ is a left semimodule and $\mathcal{Y} = \mathcal{K}^{p\times 1}$ a right semimodule, whence $\mathcal{X}^{op}$ and $\mathcal{Y}^{op}$ are right and left semimodules, respectively, whose multiplications are $R \overset{op}{\leftarrow} x = x^t \backslash R$ and $y \overset{op}{\rightarrow} R = R / y^t$. We build a new bracket over the opposite semiring $\mathcal{K}^{op}$ as given by $\langle y \mid x \rangle_R = y \overset{op}{\rightarrow} R \overset{op}{\leftarrow} x = x^t \backslash R / y^t$. Therefore, by the construction of section 2.3 the following maps form a Galois connection $(\cdot_{\bar{\varphi}}, {}_{\bar{\varphi}}\cdot) : \mathcal{Y}^{op} \multimap \mathcal{X}^{op}$:

$$y_{\bar{\varphi}} = \bigwedge\{x \in X \mid \langle y \mid x\rangle_R \geq \varphi\} = \left(y \overset{op}{\rightarrow} R\right) \overset{op}{\backslash} \varphi \tag{7}$$

$$_{\bar{\varphi}}x = \bigwedge\{y \in Y \mid \langle y \mid x\rangle_R \geq \varphi\} = \varphi \overset{op}{/} \left(R \overset{op}{\leftarrow} x\right)$$

In fact, in an idempotent semifield we are guaranteed enough φ to build as many connections as necessary: choose any invertible $\lambda \in K$, so that $\varphi = \lambda \otimes e_{\mathcal{K}}$. □

Definition 6 (φ-polars). *Given a reflexive, idempotent semiring* $(\mathcal{K}, \varphi)$ *and a* $\mathcal{K}$*-valued formal context* $(G, M, R)_{\mathcal{K}}$ *satisfying the conditions of proposition 2, we call* φ-polars *the dually adjoint maps of the corresponding Galois connection of equation (7.)*

However, in this dualised construction the pivot describes a *minimum degree of existence* required for pairs $(x, y) \in X \times Y$ to be considered for operation.

Definition 7 (Formal φ-concepts and φ-Concept Lattices). *Given a reflexive, idempotent semiring* $(\mathcal{K}, \varphi)$*, a* $\mathcal{K}$*-valued formal context* $(G, M, R)_{\mathcal{K}}$ *with* $n, p \in \mathbb{N}$*, and* $\mathcal{K}$*-valued vector spaces of rows* $\mathcal{X} \cong \mathcal{K}^{1\times n}$ *and columns* $\mathcal{Y} \cong \mathcal{K}^{p\times 1}$

1. *A (formal)* φ*-concept of the formal context* $(G, M, R)_{\mathcal{K}}$ *is a pair* $(a, b) \in \mathcal{X} \times \mathcal{Y}$ *such that* ${}_{\bar{\varphi}}a = b$ *and* $b_{\bar{\varphi}} = a$. *We call* a *the* extent *and* b *the* intent *of the concept* (a, b)*, and* φ *its* (minimum) degree of existence.
2. *If* (a_1, b_1) (a_2, b_2) *are* φ*-concepts of a context, they are ordered by the relation* $(a_1, b_1) \leq (a_2, b_2) \iff a_1 \leq a_2 \iff b_1 \overset{op}{\leq} b_2$*, called the* hierarchical order. *The set of all concepts ordered in this way is called the* φ-concept lattice, $\underline{\mathfrak{B}}^{\varphi}(G, M, R)_{\mathcal{K}}$, of the $\mathcal{K}$-valued context $(G, M, R)_{\mathcal{K}}$.

The nomenclature introduced in definition 7 is supported by the following:

Theorem 3 (Fundamental theorem of $\mathcal{K}$-valued Formal Concept Analysis, finite version, 1st half). *Given a reflexive, idempotent semiring $(\mathcal{K}, \varphi)$, the φ-concept lattice $\mathfrak{B}^{\varphi}(G, M, R)_{\mathcal{K}}$ of a $\mathcal{K}$-valued formal context $(G, M, R)_{\mathcal{K}}$ with $n, p \in \mathbb{N}$, is a (finite, complete) lattice in which infimum and supremum are given by:*

$$\bigwedge_{t \in T} (a_t, b_t) = \left(\overset{op}{\bigoplus_{t \in T}} a_t, \ {}_{\varphi}^{-}\!\left[\overset{op}{\bigoplus_{t \in T}} a_t \right] \right) \quad \bigvee_{t \in T} (a_t, b_t) = \left(\left[\overset{op}{\bigoplus_{t \in T}} b_t \right]_{\varphi}^{-}, \overset{op}{\bigoplus_{t \in T}} b_t \right) \tag{8}$$

In [1] the question was posed whether this theorem could be completed in the direction of §1.1 of [12] . This will be looked into next.

3.2 The Structural Lattice of a $\mathcal{K}$-Concept Lattice

This section contains this paper's theoretical contributions to the characterisation of the semimodules over an idempotent, reflexive semifield $\mathcal{K}$ that allow to define the anti-isomorphic lattices of theorem 3.

From definition 4 and theorem 1 the notion of structural semilattice emerges as important to characterise semimodules over an idempotent semifield. We may wonder whether more interesting characterisations may be possible when the set of generators comes from a homomorphism of spaces of finite dimension.

For that purpose, recall that in the Galois connection of equation (7), $(\cdot_{\varphi}^{-}, {}_{\varphi}^{-}\cdot) : \mathcal{Y}^{op} \multimap \mathcal{X}^{op}$, the dually isomorphic closure lattices are:

$$\overline{\mathcal{Y}^{op}} = \left\{ \varphi \overset{op}{/} \left(R \overset{op}{\leftarrow} x \right) \mid x \in X \right\} \qquad \overline{\mathcal{X}^{op}} = \left\{ \left(y \overset{op}{\rightarrow} R \right) \overset{op}{\backslash} \varphi \mid y \in Y \right\} \tag{9}$$

where $\mathcal{X}$ is the free space of object sets and $\mathcal{Y}$ is the free space of attribute sets. Now let the singleton sets of objects (row vectors), $\boldsymbol{g}_i = [\epsilon \ \cdots e_i \cdots \ \epsilon]$, and attributes (column vectors), $\boldsymbol{m}_j = [\epsilon \ \cdots e_j \cdots \ \epsilon]^T$, which are bases of their respective spaces, be mapped through the polars $W_{\overline{\mathcal{Y}^{op}}} \triangleq {}_{\varphi}^{-}(\{g_i\}_{i=1}^{n})$ and $W_{\overline{\mathcal{X}^{op}}} \triangleq (\{m_j\}_{j=1}^{p})_{\varphi}^{-}$, to obtain generator sets for the closure lattices: $\langle W_{\overline{\mathcal{Y}^{op}}} \rangle_{\mathcal{K}^{op}} = \overline{\mathcal{Y}^{op}}$ $\langle W_{\overline{\mathcal{X}^{op}}} \rangle_{\mathcal{K}^{op}} = \overline{\mathcal{X}^{op}}$.

But note that the generation process is directed by the algebra of the *opposite semiring* $\mathcal{K}^{op}$, that is, the generation process is carried out using the addition in the opposite semimodules, $\overset{op}{\oplus} = \wedge$. As we know that the $\overset{op}{\vee}$-irreducibles are included in each set of generators we may test the latter to find the former:

$$\mathcal{J}\left(\overline{\mathcal{Y}^{op}}\right) \subseteq \langle W_{\overline{\mathcal{Y}^{op}}} \rangle_{\mathcal{K}^{op}} = \overline{\mathcal{Y}^{op}} \qquad \mathcal{J}\left(\overline{\mathcal{X}^{op}}\right) \subseteq \langle W_{\overline{\mathcal{X}^{op}}} \rangle_{\mathcal{K}^{op}} \tag{10}$$

Next recall that the Galois connection of equation (7) is $\overset{op}{\vee}$-inverting, in other words, $\wedge$-inverting, therefore, the images of the $\overset{op}{\vee}$-irreducibles are $\overset{op}{\wedge}$-irreducibles:

$$\mathcal{M}\left(\overline{\mathcal{X}^{op}}\right) = \left(\mathcal{J}\left(\overline{\mathcal{Y}^{op}}\right)\right)_{\varphi}^{-} \qquad \mathcal{M}\left(\overline{\mathcal{Y}^{op}}\right) = {}_{\varphi}^{-}\left(\mathcal{J}\left(\overline{\mathcal{X}^{op}}\right)\right) \tag{11}$$

Alternatively we may think of the product of each pair of join- and meet-irreducible sets as being comprised of $\vee$-irreducible concepts $\wedge$-irreducible concepts with definitions resembling those of the standard theory:

$$\tilde{\gamma}^{\varphi}(\boldsymbol{g}_i) = \left(\left(\bar{\varphi}(\boldsymbol{g}_i) \right)^{-}_{\varphi}, \bar{\varphi}(\boldsymbol{g}_i) \right) \qquad \tilde{\mu}^{\varphi}(\boldsymbol{m}_j) = \left((\boldsymbol{m}_j)^{-}_{\varphi}, \bar{\varphi}\left((\boldsymbol{m}_j)^{-}_{\varphi} \right) \right) \tag{12}$$

Definition 8 (Structural Lattice). *The structural lattice $\underline{\mathfrak{B}}(G, M, I^{\varphi}_{R})$ of a $\mathcal{K}$-Concept Lattice $\underline{\mathfrak{B}}^{\varphi}(G, M, R)$ is the Concept Lattice of the context, (G, M, I^{φ}_{R}) , where*

$$I^{\varphi}_{R}(i, j) = \tilde{\gamma}^{\varphi}(\boldsymbol{g}_i) \leq \tilde{\mu}^{\varphi}(\boldsymbol{m}_j) \tag{13}$$

Note that we have not used the siblinghood relation to define the structural lattice. The coalescing of different join- or meet-irreducibles in the same ray to obtain a basis is in this case counterproductive because in any section of a partitioned $\mathcal{M}\left(\overline{\mathcal{X}^{op}}\right)$, $\mathcal{J}\left(\overline{\mathcal{X}^{op}}\right)$ or their images some join- or meet-irreducibles may be missing, for instance if structural lattice has the appearance of the N_5 lattice. We believe this is one more instance of the differences between idempotent semimodules and traditional vector spaces.

4 Example: The Analysis of Cross-Lingual Classifier Adaptation Systems

In this example we analyse a particular problem: the cross-lingual adaptation of an automatic speech recogniser trained to recognise English phonemes into a system capable of recognising Mandarin phonemes. Our aim in this task is to analyse several ways of mapping the English outputs of such classifiers into Mandarin phonemes by observing whether the mapping has an intuitive, meaningful structure. We will compare two ways to accomplish this:

- An *original system* trained with English-speech data with a particular classifier-building technique.
- An *enhanced system* which uses the previous system as a start point and improved afterwards by using some Mandarin-speech data to learn to map the English outputs into Mandarin phonemes.

In both cases, the input to our algorithm will be the confusion matrix (more properly called the *translation matrix*, T, in this context) between English phonemes (outputs, $n = 46$) and Mandarin phonemes (inputs, $p = 71$) by observing the English labels that both networks assign to the Mandarin speech frames and confronting it with the true Mandarin labels.

4.1 Lattice Construction

This section describes the algorithm employed to obtain the structural lattice of relation T for a range of degrees of existence φ as defined in previous sections.

For that purpose, we first transform the event counts of the confusion matrix into a maximum-likelihood estimate of the probability of the true Mandarin label g_i given that the output of the classifier is the English label m_j , $P(G = g_i \mid M = m_j)$. We then take logarithms to transform probabilities in $[0,1]$ into log-likelihoods, that is $\overline{\mathbb{R}}_{\max,+}$ costs to obtain $R \in \overline{\mathbb{R}}_{\max,+}^{n\times p}$ The following algorithm then obtains the structural lattice of the cost matrix, R:

Step 1. Compute the closures of the n unitary row vectors of dimension $1 \times n$, $\boldsymbol{g}_i = [\epsilon \cdots e_i \cdots \epsilon]$ and p unitary column vectors of dimension $p\times 1$, $\boldsymbol{m}_j = [\epsilon \cdots e_j \cdots \epsilon]^T$ that stand for the characteristic functions of singleton sets of objects and attributes, respectively. The φ-polars of definition 6 allow us to obtain, the $\vee$- and $\wedge$-irreducible concepts of the structural lattice using equation (12).

Step 2. Build the standard context associated to those concepts and the structural lattice by comparing the previous concepts, $\underline{\mathfrak{B}}(G, M, I_R^{\varphi})$ where I_R^{φ} is the incidence with $\{0,1\}$ entries in formula (13.)

Step 3. Once the standard context adequate for the structural lattice is obtained for each particular φ we used ConExp [13] to obtain the standard lattices.

Because the Galois connection that obtains the formal concepts depends on the pivot, φ, typically the above algorithm must be carried out a number of times, one for each choice of φ that is deemed interesting.

4.2 Lattice Exploration

The Influence of the Enhancement stage: Choosing φ. Our aim now is to explore the behaviour of the $\mathcal{K}$-Concept Lattice for a particular $\mathcal{K}$-formal context with varying φ . For this purpose, we have found the standard context of the structural lattice with the algorithm above for each φ and worked out the number of concepts resulting for the standard Formal Contexts of the original and enhanced systems. Figure 2 shows this evolution where we have chosen to sample the curve more frequently as we approach the right end (i.e. $\varphi = 0$) by using the tangent function of a uniform sampling. A logarithmic scale has been used in the vertical axis to improve the comparison of the two curves given the notorious differences in the number of concepts of the two examples we are evaluating here.

For a perfect translation between two phonemic systems of identical cardinality, the best system would show a diagonal matrix in the $\mathcal{K}$-Formal Context, equivalent to a diamond lattice of as many $\vee$- and $\wedge$-irreducibles as phonemes. We expect to find systems that do a worse translation further and further from this structure and with increasing concept counts. Indeed, the most significant observation we can gather from the plot above is the reduction in the number of concepts achieved by the enhanced system. We can infer, therefore, that the *enhancing stage* improves the translation in such direction.

We notice that the overall shapes of the curves are very similar. For smaller φ the number of concepts remains constant for each matrix being evaluated.

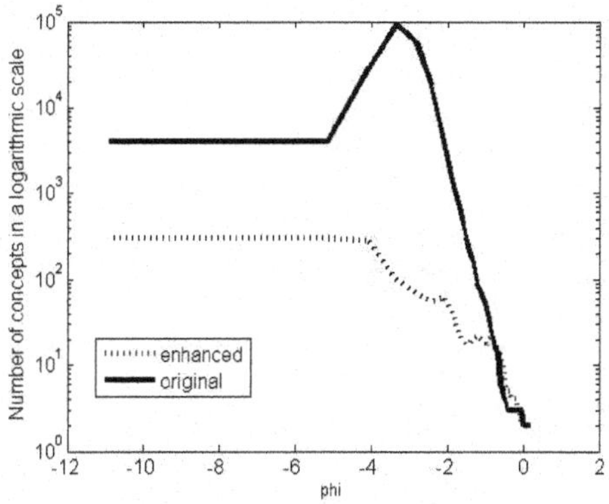

Fig. 2. Number of concepts vs. φ

For $\varphi \geq \max_{i,j} R_{ij}$, the incidence matrix is everywhere null $I_R = 0$ leading to a two-concept lattice for both curves. In between these ranges we see how the enhanced system shows less and less concepts while the original system's number of concepts reaches a really high peak (around 10^5) and then quickly diminishes.

Reading Structural Lattices. We now try to understand what kind of information can be gleaned from structural lattices. We begin by observing the most salient properties of the systems, that is, those lattices obtained with the higher values of the pivot. Afterwards we try to bring more detail into the picture by decreasing the value of the pivot so as to vary the number of concepts from right to left as suggested by figure 2. We thus obtain a sequence of structural lattices starting from the least complex (and the least number of concepts) and gradually increasing the complexity as new concepts appear.

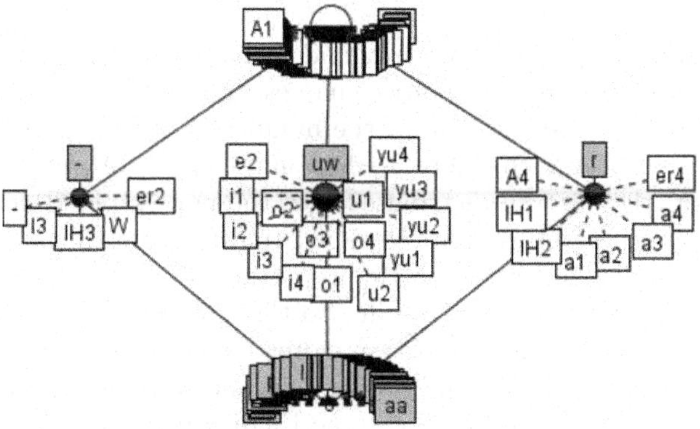

Fig. 3. Structural lattice of the enhanced system with a $\varphi = -0.40$ and 5 concepts

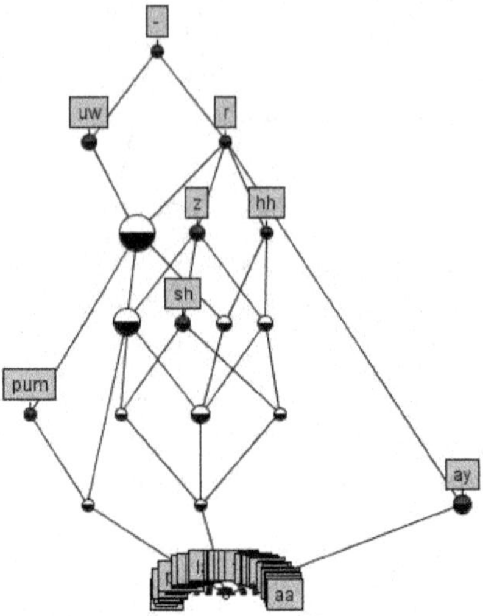

Fig. 4. Structural lattice of the enhanced system with a $\varphi = -0.99$ and 18 concepts

The first thing we can notice in both the sequence from the original and the enhanced system is the appearance of the *silence* attribute concept (tagged '-' in the figures). This is a well-known peculiarity of systems such as those we explore in this example and it is therefore a good sign that our analysis is progressing correctly.

Figure 3 shows a more advanced stage of analysis for a pivot $\varphi = -0.40$. In it some groups of Mandarin objects are assigned just to three English attribute concepts. Both *silence* and r are attributed to several different Mandarin phonemes which can be interpreted as an error of the system. However, uw is always attributed to Mandarin vowel sounds which, at this level of detail, seems to be a good choice.

As we continue our analysis we can find lattices such as those in figures 4 and 5. We have omitted here the object names that clutter the picture but the radius of the nodes is proportional to the number of objects pertaining just to them. Here, in contrast with figure 3 where most of the objects were assigned the top concept, they are all distributed into the non-extremal nodes. We can also observe that these objects are grouped (most of the times) meaningfully from the point of view of their acoustic properties.

Although we show no picture here due to the large number of concepts, it is interesting to consider the lattice corresponding to the leftmost constant portion of figure 2. Despite the difficulty of drawing any conclusion from such a big lattice the most salient characteristic is that some English phonemes remain attached to the bottom concept which could be interpreted as the systems being unable to assign those English phonemes to any of the Mandarin, either due to limitations of the system or to some intrinsic characteristics of these phoneme sets.

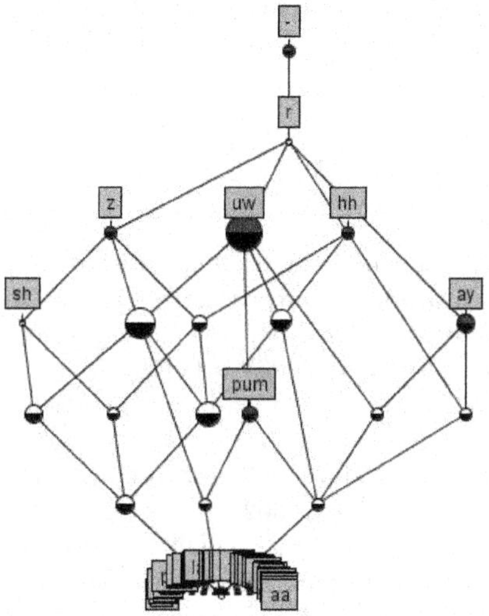

Fig. 5. Structural lattice of the enhanced system with a $\varphi = -1.09$ and 20 concepts

5 Conclusion

We have presented an attempt at the solution of two problems of $\mathcal{K}$-Formal Concept Analysis of different ilk: first, the lack of an analogue for the second half of the basic theorem of Formal Concept Analysis, and second, the lack of a building procedure for the $\mathcal{K}$-Concept Lattice.

For the first purpose we have introduced the concept of the structural lattice of the $\mathcal{K}$-Concept Lattice based in the similar structural semilattice, as featured in idempotent algebra. Thus we expect the structural lattice to provide the "scaffolding" for the bigger $\mathcal{K}$-concept lattice. As to the relation of the structural lattice to the lattice described in the second half of the Basic Theorem of Formal Concept Analysis, we recall that (for $n, p \in \mathbb{N}$) Formal Concept Analysis may be taken to be the particular case of $\mathcal{K}$-Formal Concept Analysis when the idempotent reflexive semiring is the Boolean semiring, and the pivot is $\bot_{\mathbb{B}^{op}} = \mathbf{1}$, $\underline{\mathfrak{B}}(G, M, I) = \underline{\mathfrak{B}}^{\mathbf{1}}(G, M, I)_{\mathbb{B}}$. It is easy to see that in that case the definition of the structural lattice and the lattice of the second part of Theorem 3 of [12] coincide.

Secondly, we have provided an algorithm to build the structural lattice by reducing its calculations to those of a standard Concept Lattice, and have used such construction to analyse the behaviour of the confusion matrices of multiple input-multiple output classifiers. We have also discussed the role of the pivot φ introduced in [1] to modulate the Galois connection between the spaces of (multi-valued) sets of objects and attributes in such a setting and we have tried to argue how the performance of the classifier relates to its $\mathcal{K}$-Concept Lattice conforming to a particular, expected shape.

Acknowledgements. We would like to thank J. Frankel for providing the classifier systems and to N. Morgan for comments on the intuitions afforded by the analysis technique.

References

1. Valverde-Albacete, F.J., Peláez-Moreno, C.: Towards a generalisatioin of formal concept analysis for data mining purposes. In Ganter, B., Kwuida, L., eds.: Proceedings of the International Conference on Formal Concept Analysis, ICFCA06. Volume 3874 of LNAI., Springer (2006) 161–176
2. Baccelli, F., Cohen, G., Olsder, G., Quadrat, J.: Synchronization and Linearity. Wiley (1992)
3. Davey, B., Priestley, H.: Introduction to lattices and order. 2nd edn. Cambridge University Press, Cambridge, UK (2002)
4. Cohen, G., Gaubert, S., Quadrat, J.P.: Duality and separation theorems in idempotent semimodules. Linear Algebra and Its Applications **379** (2004) 395–422
5. Golan, J.S.: Semirings and Their Applications. Kluwer Academic (1999)
6. Golan, J.S.: Power Algebras over Semirings. With Applications in Mathematics and Computer Science. Volume 488 of Mathematics and its applications. Kluwer Academic, Dordrecht, Boston, London (1999)
7. Gaubert, S.: Two lectures on max-plus algebra. Support de cours de la 26–iéme École de Printemps d'Informatique Théorique (1998) http://amadeus.inria.fr/gaubert/papers.html
8. Gaubert, S., the Maxplus Group: Methods and applications of $(max, +)$ linear algebra. Technical Report 3088, INRIA – (1997)
9. Wagneur, E.: Moduloïds and pseudomodules 1. dimension theory. Discrete Mathematics **98** (1991) 57–73
10. Wagneur, E.: The geometry of finite dimensional pseudomodules. In: Proceedings of the 34th Conference on Decision & Control, New Orleans, LA (1995)
11. Cunninghame-Green, R., Butković, P.: Bases in max-algebra. Linear Algebra and its Applications **389** (2004) 107–120
12. Ganter, B., Wille, R.: Formal Concept Analysis: Mathematical Foundations. Springer, Berlin, Heidelberg (1999)
13. Yevtushenko, S.A.: System of data analysis "Concept Explorer". In: Proceedings of the 7th national conference on Artificial Intelligence KII-2000, Russia, ACM (2000) 127–134 (In Russian) http://sourceforge.net/projects/conexp

On Multi-adjoint Concept Lattices: Definition and Representation Theorem

J. Medina, M. Ojeda-Aciego, and J. Ruiz-Calviño

Dept. Matemática Aplicada. Universidad de Málaga*
{jmedina,aciego,jorgerucal}@ctima.uma.es

Abstract. Several fuzzifications of formal concept analysis have been proposed to deal with uncertainty or incomplete information. In this paper, we focus on the new paradigm of multi-adjoint concept lattices which embeds different fuzzy extensions of concept lattices, our main result being the representation theorem of this paradigm. As a consequence of this theorem, the representation theorems of the other paradigms can be proved more directly. Moreover, the multi-adjoint paradigm enriches the language providing greater flexibility to the user.

Keywords: concept lattices, multi-adjoint lattices, Galois connection, implication triples.

1 Introduction

The study of reasoning methods under uncertainty, imprecise data or incomplete information has shown to be an important topic in the recent years. Most of the current research areas are receiving this message and it is frequent to see *fuzzified* versions of several well-known standard structures. In this paper, we focus on the area of formal concept analysis and, specifically, on the generalization of the classical definition of concept lattice to the fuzzy case.

A number of different approaches have been proposed to generalize the classical concept lattices given by Ganter and Wille [10] allowing some uncertainty in data, a recent survey and comparison of approaches to fuzzy concept lattices is presented in [6].

One of these approaches was proposed by Burusco and Fuentes-González [7] where fuzzy concept lattices were first presented, and later further developed by Pollandt [23] and Bělohlávek [2] who use complete residuated lattices as structures for the truth degrees. For the latter approach, the main theorem was proved in two ways: firstly, by reduction to the crisp version of main theorem (this was proved independently in [23] and, more generally, via representation of fuzzy Galois connections in [4]); secondly, working directly in a fuzzy setting in [3].

Bělohlávek, in [5], later extended this to the case when a fuzzy partial order is considered on a fuzzy concept lattice instead of on an ordinary partial order.

* Partially supported by Spanish research projects TIC2003-09001-C02-01 and TIN2006-15455-C03-01.

S.O. Kuznetsov and S. Schmidt (Eds.): ICFCA 2007, LNAI 4390, pp. 197–209, 2007.

Georgescu and Popescu extended this framework to non-commutative logic and similarity in a series of papers [11, 12, 13, 14]; in a different direction, it was also extended in an asymmetric way, although only for the case of classical equality ($L = \{0,1\}$), by Krajči, which introduced the so-called generalized concept lattices in [17, 18].

In the context of general logical frameworks, a recent approach so-called multi-adjoint has been recently introduced and is receiving considerable attention [16, 21]. The multi-adjoint framework was originated as a generalization of several non-classical logic programming frameworks, its semantic structure is the multi-adjoint lattice, in which a lattice is considered together with several conjunctors and implications making up adjoint pairs [20].

In [22], with the idea of providing a general framework in which the different approaches stated above could be conveniently accommodated, the authors considered a general non-commutative environment; this naturally leads to the consideration of adjoint triples, also called pre-implication triples [1] or bi-residuated structures [19] as the main building blocks of our multi-adjoint concept lattices.

The aim of the paper is to construct so-called multi-adjoint concept lattices in order to generalise different fuzzy extensions of concept lattices. The main result is a representation theorem which characterises those complete lattices which are isomorphic to multi-adjoint concept lattices. The notion of a multi-adjoint concept lattice is demonstrated by a detailed example.

The plan of this paper is the following: in Section 2 we recall the basics about Galois connection and the notion of multi-adjoint concept lattice is introduced, in Section 3 contains the proof of the representation theorem; in Section 4 an example of the multi-adjoint framework is presented; the paper ends with some conclusions and prospects for future work.

2 Multi-adjoint Concept Lattice

A basic notion in formal concept analysis is that of *Galois connection*, we start this section recalling a result which proves that each Galois connection has an associated complete lattice, called *Galois lattice* or *concept lattice*.

Definition 1. *Let* $(P_1, \leq_1)$ *and* $(P_2, \leq_2)$ *be posets, and let* ${}^{\downarrow}\colon P_1 \to P_2$ *and* ${}^{\uparrow}\colon P_2 \to P_1$ *be mappings, the pair* $({}^{\uparrow}, {}^{\downarrow})$ *forms a* Galois connection *between* P_1 *and* P_2 *if:*

1. ${}^{\uparrow}$ *and* ${}^{\downarrow}$ *are order-reversing.*
2. $x \leq_1 x^{\downarrow\uparrow}$ *for all* $x \in P_1$.
3. $y \leq_2 y^{\uparrow\downarrow}$ *for all* $y \in P_2$.

If P_1 and P_2 are complete lattices then the following theorem can be established, see [9], which will be used in order to prove that our construction of multi-adjoint concept lattices actually leads to a complete lattice.

Theorem 1. *Let $(L_1, \preceq_1)$, $(L_2, \preceq_2)$ be complete lattices, $(^\uparrow, ^\downarrow)$ a Galois connection between L_1, L_2 and $\mathcal{C} = \{\langle x, y\rangle \mid x^\uparrow = y, x = y^\downarrow; x \in L_1, y \in L_2\}$ then $\mathcal{C}$ is a complete lattice, where*

$$\bigwedge_{i\in I} \langle x_i, y_i\rangle = \langle \bigwedge_{i\in I} x_i, (\bigvee_{i\in I} y_i)^{\downarrow\uparrow}\rangle \quad \text{and} \quad \bigvee_{i\in I} \langle x_i, y_i\rangle = \langle (\bigvee_{i\in I} x_i)^{\uparrow\downarrow}, \bigwedge_{i\in I} y_i\rangle$$

Firstly, a generalization of multi-adjoint lattices will be introduced in order to admit different sorts, in which we allow non-commutative conjunctors as in [1, 11, 19]. To begin with, the adjoint pairs are generalized into adjoint triples, the basic blocks of multi-adjoint concept lattices, as follows:

Definition 2. *Let $(P_1, \leq_1)$, $(P_2, \leq_2)$, $(P_3, \leq_3)$ be posets and $\&\colon P_1 \times P_2 \longrightarrow P_3$, $\swarrow\colon P_3 \times P_2 \longrightarrow P_1$, $\nwarrow\colon P_3 \times P_1 \longrightarrow P_2$ be mappings, then $(\&, \swarrow, \nwarrow)$, is a* adjoint triple *with respect to P_1, P_2, P_3 if:*

- *$\&$ is order-preserving in both arguments.*
- *$\swarrow$ and $\nwarrow$ are order-preserving in the first argument and order-reversing in the second.*
- *$x \leq_1 z \swarrow y$ iff $x \,\&\, y \leq_3 z$ iff $y \leq_2 z \nwarrow x$, where $x \in P_1$, $y \in P_2$ and $z \in P_3$.*

This last property is known as *adjoint property* and can be seen as a generalisation of the *modus ponens* rule in a non-commutative multi-valued setting, see [15] for the case of adjoint pairs. Notice that no boundary condition is required, in difference to the usual definition of multi-adjoint lattice [21] or implication triples [1].

In order to introduce a Galois connection which generalizes that given in the classical case, the usual motivation underlying the multi-adjoint framework [16, 21] is applied to that of adjoint triples, and leads to the following definition of multi-adjoint frame.

Definition 3. *A* multi-adjoint frame *$\mathcal{L}$ is a tuple*

$$(L_1, L_2, P, \preceq_1, \preceq_2, \leq, \&_1, \swarrow^1, \nwarrow_1, \dots, \&_n, \swarrow^n, \nwarrow_n)$$

where $(L_1, \preceq_1)$ and $(L_2, \preceq_2)$ are complete lattices, $(P, \leq)$ is a poset and, for all $i = 1, \dots, n$ the tuple $(\&_i, \swarrow^i, \nwarrow_i)$ is an adjoint triple with respect to L_1, L_2, P.

A multi-adjoint frame as above will be denoted as $(L_1, L_2, P, \&_1, \dots, \&_n)$, for short. It is convenient to note that, in principle, L_1, L_2 and P could be simply posets, the reason to consider complete lattices is that multi-adjoint frames will be used as the underlying lattice on which the operations will be made; hence, general joins and meets are required.

A *context* for a given frame will mean a tuple (A, B, R, σ) defined as below where, following the usual terminology, A is to be considered as a set of attributes and B as a set of objects.

Definition 4. *A* context *for a given frame $(L_1, L_2, P, \&_1, \ldots, \&_n)$ is a tuple (A, B, R, σ) such that A and B are non-empty sets, R is a P-fuzzy relation $R\colon A \times B \longrightarrow P$ and σ is a mapping which associates any object in B (or attribute in A) with some particular adjoint triple in the frame, that is, $\sigma\colon B \to \{1, \ldots, n\}$ (or $\sigma\colon A \to \{1, \ldots, n\}$).*

The fact that in a multi-adjoint context each object (or attribute) has an associated implication is interesting in that subgroups with different degrees of preference can be established in a convenient way. From now on, we will consider in the context the association $\sigma\colon B \to \{1, \ldots, n\}$.

Now, given a frame and a context for that frame, the following mappings ${}^{\uparrow_\sigma}\colon L_2^B \longrightarrow L_1^A$ and ${}^{\downarrow^\sigma}\colon L_1^A \longrightarrow L_2^B$ can be defined:

$$g^{\uparrow_\sigma}(a) = \inf\{R(a,b) \swarrow^{\sigma(b)} g(b) \mid b \in B\}$$
$$f^{\downarrow^\sigma}(b) = \inf\{R(a,b) \nwarrow_{\sigma(b)} f(a) \mid a \in A\}$$

Notice that these mappings generalise those given in [5,18] and, as proved below, generate a Galois connection.

Proposition 1. *Given a multi-adjoint frame $(L_1, L_2, P, \&_1, \ldots, \&_n)$ and a context (A, B, R, σ), the pair $({}^{\uparrow_\sigma}, {}^{\downarrow^\sigma})$ is a Galois connection between L_1^A and L_2^B.*

Proof. From now on, to improve readability, we will write $({}^\uparrow, {}^\downarrow)$ instead of $({}^{\uparrow_\sigma}, {}^{\downarrow^\sigma})$ and $\swarrow^b$, $\nwarrow_b$ instead of $\swarrow^{\sigma(b)}$, $\nwarrow_{\sigma(b)}$.

By definition, we have to prove that:

1. ${}^\uparrow$ and ${}^\downarrow$ are order-reversing.
   This is trivial since the implications are order-reversing in the second argument.
2. $g \le g^{\uparrow\downarrow}$ for all $g \in L_2^B$.
   Given $a \in A$ and $b \in B$ the following chain of inequalities holds because of the definition of $g^\uparrow(a)$ as an infimum and the adjoint property:

$$\begin{aligned} g^\uparrow(a) \preceq_1 R(a,b) \swarrow^b g(b) &\iff g^\uparrow(a) \&_b\, g(b) \le R(a,b) \\ &\iff g(b) \preceq_1 R(a,b) \nwarrow_b g^\uparrow(a) \end{aligned}$$

   As the inequality above holds for all $a \in A$, by using properties of the infimum, it can be obtained that

$$g(b) \preceq_2 \inf\{R(a,b) \nwarrow_b g^\uparrow(a) \mid a \in A\} = g^{\uparrow\downarrow}(b)$$

3. $f \le f^{\downarrow\uparrow}$ for all $f \in L_1^A$.
   The proof is similar. □

Now, a *concept* is a pair $\langle g, f\rangle$ satisfying that $g \in L_2^B$, $f \in L_1^A$ and that $g^\uparrow = f$ and $f^\downarrow = g$; with $({}^\uparrow, {}^\downarrow)$ being the Galois connection defined above.

Definition 5. *The* multi-adjoint concept lattice *associated to a multi-adjoint frame $(L_1, L_2, P, \&_1, \ldots, \&_n)$ and a context (A, B, R, σ) is the set of concepts:*

$$\mathcal{M} = \{\langle g, f\rangle \mid g \in L_2^B, f \in L_1^A \text{ and } g^\uparrow = f, f^\downarrow = g\}$$

with the ordering $\langle g_1, f_1\rangle \preceq \langle g_2, f_2\rangle$ if and only if $g_1 \preceq_2 g_2$ (equivalently $f_2 \preceq_1 f_1$).

Note that, by Theorem 1, the poset $(\mathcal{M}, \preceq)$ defined above is a complete lattice, since the arrows $(^{\uparrow}, ^{\downarrow})$ form a Galois connection between the complete lattices L_1^A and L_2^B.[1]

3 The Representation Theorem

An extension of the representation (or fundamental) theorem on the classical concept lattice [10] for the multi-adjoint framework is presented below. The proof follows the lines of those given for previous extensions of the classical concept lattices, but the presentation has been simplified. To begin with, we need to introduce some definitions and preliminary results.

Definition 6. *Given a set A, a poset P with bottom element $\bot$, and elements $a \in A$, $x \in P$, the characteristic mapping $@^x_a \colon A \to P$ is defined as:*

$$@^x_a(a') = \begin{cases} x, & \text{if } a' = a \\ \bot, & \text{otherwise} \end{cases}$$

The following lemma gives a technical property which will be needed later.

Lemma 1. *In the concept lattice $(\mathcal{M}, \preceq)$, given $a \in A$, $b \in B$, $x \in L_1$ and $y \in L_2$, the following equalities hold:*

$$@^{x\downarrow}_a(b') = R(a,b') \nwarrow_{b'} x \qquad \text{for all} \quad b' \in B$$
$$@^{y\uparrow}_b(a') = R(a',b) \swarrow^{b} y \qquad \text{for all} \quad a' \in A$$

Proof. By definition of $@^{x\downarrow}_a$:

$$@^{x\downarrow}_a(b') = \inf\{R(a',b') \nwarrow_{b'} @^x_a(a') \mid a' \in A\} = R(a,b') \nwarrow_{b'} x$$

where the last inequality follows because $R(a',b) \nwarrow_b \bot_1 = \top_2$ (this fact is a consequence of the adjoint property, since $\bot_1 \preceq_1 R(a',b) \swarrow^b \top_2$).

The other equality follows similarly. □

The following definitions introduce properties which will be used in the statement of Proposition 2.

Definition 7. *Given a complete lattice L, a subset $K \subseteq L$ is infimum-dense (resp. supremum-dense) if and only if for all $x \in L$ there exists $K' \subseteq K$ such that $x = \inf(K')$ (resp. $x = \sup(K')$).*

Definition 8. *Let $(\mathcal{M}, \preceq)$ be a multi-adjoint concept lattice, $(V, \sqsubseteq)$ a complete lattice and $\alpha \colon A \times L_1 \to V$, $\beta \colon B \times L_2 \to V$ two maps. We say that β* is (V, R)-related with α *if we have that:*

1a) $\alpha[A \times L_1]$ is infimum-dense;

[1] In the rest of the paper we will assume a fixed multi-adjoint frame and context.

1b) $\beta[B \times L_2]$ *is supremum-dense; and*
2) for each $a \in A$, $b \in B$, $x \in L_1$ *and* $y \in L_2$*:*

$$\beta(b, y) \sqsubseteq \alpha(a, x) \quad \text{if and only if} \quad x \,\&_b\, y \leq R(a, b)$$

Proposition 2. *Given a multi-adjoint concept lattice* $(\mathcal{M}, \preceq)$*, a complete lattice* $(V, \sqsubseteq)$ *and two maps* $f \in L_1^A$, $g \in L_2^B$*, if there exist two mappings* $\beta\colon B \times L_2 \to V$, $\alpha\colon A \times L_1 \to V$*, where* β *is* (V, R)*-related with* α *we have that:*

1. β *is order-preserving in the second argument.*
2. α *is order-reversing in the second argument.*
3. $g^{\uparrow}(a) = \sup\{x \in L_1 \mid v_g \sqsubseteq \alpha(a, x)\}$*, where* $v_g = \sup\{\beta(b, g(b)) \mid b \in B\}$.
4. $f^{\downarrow}(b) = \sup\{y \in L_2 \mid \beta(b, y) \sqsubseteq v_f\}$*, where* $v_f = \inf\{\alpha(a, f(a)) \mid a \in A\}$.
5. *If* $g_v(b) = \sup\{y \in L_2 \mid \beta(b, y) \sqsubseteq v\}$*, then* $\sup\{\beta(b, g_v(b)) \mid b \in B\} = v$.
6. *If* $f_v(a) = \sup\{x \in L_1 \mid v \sqsubseteq \alpha(a, x)\}$*, then* $\sup\{\alpha(a, f_v(a)) \mid a \in A\} = v$.

Proof. We give the proofs for items 1, 3 and 5, since the others are similar.
1. Let $y_1 \preceq_2 y_2 \in L_2$, as $\beta(b, y_2) \in V$ and $\alpha[A \times L_1]$ is infimum-dense there exists a set of indices Λ and $K = \{(a_j, x_j) \mid j \in \Lambda\} \subseteq A \times L_1$ such that $\beta(b, y_2) = \inf\{\alpha(a_j, x_j) \mid j \in \Lambda\}$, so $\beta(b, y_2) \sqsubseteq \alpha(a_j, x_j)$ for all $j \in \Lambda$. Now, by Def. 8 property 2, it follows that $x_j \,\&_b\, y_2 \leq R(a_j, b)$ for all j and, as $y_1 \preceq_2 y_2$,

$$x_j \,\&_b\, y_1 \leq x_j \,\&_b\, y_2 \leq R(a_j, b) \text{ for all } j$$

Therefore, $\beta(b, y_1) \sqsubseteq \alpha(a_j, x_j)$ for all j and, as $\beta(b, y_2)$ is the infimum, $\beta(b, y_1) \sqsubseteq \beta(b, y_2)$, so β is order-preserving in the second argument.
3. Given $x \in L_1$, by the adjoint property the inequality $x \preceq_1 R(a, b) \swarrow^b g(b)$ is equivalent to $x \,\&_b\, g(b) \leq R(a, b)$ which is also equivalent, by Def. 8 property 2, to $\beta(b, g(b)) \sqsubseteq \alpha(a, x)$ for all $b \in B$, therefore by properties of the supremum

$$v_g = \sup\{\beta(b, g(b)) \mid b \in B\} \sqsubseteq \alpha(a, x)$$

Thus, we obtain the equality of the sets:

$$\{x \in L_1 \mid x \preceq_1 R(a, b) \swarrow^b g(b) \text{ for all } b \in B\} = \{x \in L_1 \mid v_g \sqsubseteq \alpha(a, x)\}$$

Therefore:

$$\begin{aligned} g^{\uparrow}(a) &= \inf\{R(a, b) \swarrow^b g(b) \mid b \in B\} \\ &\overset{(*)}{=} \sup\{x \in L_1 \mid x \preceq_1 R(a, b) \swarrow^b g(b) \text{ for all } b \in B\} \\ &= \sup\{x \in L_1 \mid v_g \sqsubseteq \alpha(a, x)\} \end{aligned}$$

where $(*)$ is given from the adjoint property.
5. Firstly we will show that, for any $v \in V$, $\sup\{\beta(b, g_v(b)) \mid b \in B\} \sqsubseteq v$, and let us write $Y_b = \{y \in L_2 \mid \beta(b, y) \sqsubseteq v\}$ for any $b \in B$, so that $g_v(b) = \sup Y_b$.

Given $v \in V$, as $\alpha[A \times L_1]$ is infimum-dense, there is a set of indices Λ and $K = \{(a_j, x_j) \mid j \in \Lambda\} \subseteq A \times L_1$ such that $v = \inf\{\alpha(a_j, x_j) \mid j \in \Lambda\}$.

If $Y_b = \varnothing$, then $g_v(b) = \bot_2$ and we have the next chain of equivalences:

$$g_v(b) \preceq_2 R(a_j,b) \nwarrow_b x_j \text{ iff } x_j \mathbin{\&_b} g_v(b) \leq R(a_j,b) \text{ iff } \beta(b, g_v(b)) \sqsubseteq \alpha(a_j, x_j) \quad (1)$$

Otherwise, if Y_b is non-empty, then, by Def. 8 property 2, we have for all $j \in \Lambda$ and $y \in Y_b$:

$$\beta(b,y) \sqsubseteq v \sqsubseteq \alpha(a_j,x_j) \text{ iff } x_j \mathbin{\&_b} y \leq R(a_j,b) \text{ iff } y \preceq_2 R(a_j,b) \nwarrow_b x_j$$

by computing the supremum on y, we get to $g_v(b) = \sup Y_b \preceq_2 R(a_j,b) \nwarrow_b x_j$, and then the rest of equivalences in (1) apply.

Recalling that $v = \inf\{\alpha(a_j,x_j) \mid j \in \Lambda\}$ we obtain than $\beta(b, g_v(b)) \sqsubseteq v$ for all $b \in B$. Finally, taking supremum on the left hand side, we get

$$\sup\{\beta(b, g_v(b)) \mid b \in B\} \sqsubseteq v$$

For the other inequality, as $\beta[B \times L_2]$ is supremum-dense we have that $v = \sup\{\beta(b_j, y_j) \mid (b_j, y_j) \in A \times L_2, j \in \Lambda'\}$. Then, for any $j \in \Lambda'$ we have that $y_j \in Y_{b_j}$ and, moreover, $y_j \preceq_2 \sup Y_{b_j} = g_v(b_j)$. Since β is order-preserving in the second argument, by item 1, we obtain:

$$\beta(b_j, y_j) \sqsubseteq \beta(b_j, g_v(b_j)) \sqsubseteq \sup\{\beta(b_j, g_v(b_j)) \mid j \in \Lambda\} \sqsubseteq \sup\{\beta(b, g_v(b)) \mid b \in B\}$$

As v is the supremum on j of $\beta(b_j, y_j)$, we get $v \sqsubseteq \sup\{\beta(b, g_v(b)) \mid b \in B\}$. □

We can now state and prove the representation theorem for multi-adjoint concept lattices.

Theorem 2 (Representation theorem). *Given a complete lattice $(V, \sqsubseteq)$ and a multi-adjoint concept lattice $(\mathcal{M}, \preceq)$, we have that V is isomorphic to $\mathcal{M}$ if and only if there exist mappings $\alpha\colon A \times L_1 \to V$, $\beta\colon B \times L_2 \to V$ such that β is (V,R)-related to α.*

Proof. Given an isomorphism $\varphi\colon \mathcal{M} \to V$, the mappings $\alpha\colon A \times L_1 \to V$ and $\beta\colon B \times L_2 \to V$ can be naturally defined, for every $a \in A$, $b \in B$, $x \in L_1$ and $y \in L_2$, as follows:

$$\alpha(a,x) = \varphi(\langle @_a^{x\downarrow}, @_a^{x\downarrow\uparrow}\rangle) \qquad \beta(b,y) = \varphi(\langle @_b^{y\uparrow\downarrow}, @_b^{y\uparrow}\rangle)$$

Let us prove that β is (V,R)-related to α:

Firstly, let us show that $\alpha[A \times L_1]$ is infimum-dense. By definition, we have to prove that given $v \in V$ there exists $K \subseteq A \times L_1$ such that $v = \inf(\alpha[K])$.

If $\varphi^{-1}(v) = \langle g, f\rangle \in \mathcal{M}$, we define $K = \{(a, f(a)) \mid a \in A\} \subseteq A \times L_1$. Since φ is an isomorphism, it is sufficient to prove that

$$\langle g, f\rangle = \inf\{\langle @_a^{f(a)\downarrow}, @_a^{f(a)\downarrow\uparrow}\rangle \mid a \in A\}$$

Let us prove, for instance, that $g(b) = \inf\{@_a^{f(a)\downarrow}(b) \mid a \in A\}$. By Lemma 1, we have that $@_a^{f(a)\downarrow}(b) = R(a,b) \nwarrow_b f(a)$, thus

$$\inf\{@_a^{f(a)\downarrow}(b) \mid a \in A\} = \inf\{R(a,b) \nwarrow_b f(a) \mid a \in A\} = f^{\downarrow}(b) = g(b)$$

Similarly, we can prove that $\beta[B \times L_2]$ is supremum-dense.

It only remains to prove that given $a \in A$, $b \in B$, $x \in L_1$ and $y \in L_2$, we have that $\beta(b,y) \sqsubseteq \alpha(a,x)$ iff $x \,\&_b\, y \leq R(a,b)$.

For the direct implication, as φ is an order-isomorphism, we have that $\beta(b,y) \sqsubseteq \alpha(a,x)$ is equivalent to $\langle @_b^{y\uparrow\downarrow}, @_b^{y\uparrow} \rangle \leq \langle @_a^{x\downarrow}, @_a^{x\downarrow\uparrow} \rangle$ and, in particular, to $@_b^{y\uparrow\downarrow} \leq @_a^{x\downarrow}$. From the properties of Galois connection, Lemma 1, and the adjoint property we obtain the following chain:

$$y = @_b^y(b) \preceq_2 @_b^{y\uparrow\downarrow}(b) \preceq_2 @_a^{x\downarrow}(b) = R(a,b) \nwarrow_b x \quad \text{iff} \quad x \,\&_b\, y \leq R(a,b)$$

For the other implication, it is sufficient to prove that $@_a^x \leq @_b^{y\uparrow}$ as this is equivalent to $@_b^{y\uparrow\downarrow} \leq @_a^{x\downarrow}$ which finally implies $\beta(b,y) \sqsubseteq \alpha(a,x)$, from the definition of α and β, and φ order-preserving.

But this is clear because, if $a' \in A$ with $a' \neq a$, then $@_a^x(a') \preceq_1 @_b^{y\uparrow}(a')$ holds because $@_a^x(a') = \bot_1$. If $a' = a$, as $x \,\&_b\, y \leq R(a,b)$ applying the adjoint property and Lemma 1 we obtain that:

$$@_a^x(a) = x \preceq_1 R(a,b) \swarrow^b y = @_b^{y\uparrow}(a)$$

Now, conversely, assume we have mappings $\alpha\colon A \times L_1 \to V$, $\beta\colon B \times L_2 \to V$ where β is (V,R)-related to α, and let us construct an isomorphism $\varphi\colon \mathcal{M} \to V$. We define the mapping φ for every $\langle g,f \rangle \in \mathcal{M}$ as follows:

$$\varphi(\langle g,f \rangle) = \sup\{\beta(b,g(b)) \mid b \in B\}$$

To prove that it is a lattice isomorphism we will construct the inverse mapping $\psi\colon V \to \mathcal{M}$ of φ.

The mapping ψ is defined for each $v \in V$ as $\psi(v) = \langle g_v, f_v \rangle$, where, for each $b \in B$ and $a \in A$, $g_v(b)$ and $f_v(a)$ are defined as in Proposition 2. This proposition shows that ψ is well-defined as well, that is, $\langle g_v, f_v \rangle$ is a concept. The argument is as follows:

$$g_v{}^{\uparrow}(a) = \sup\{x \in L_1 \mid v_{g_v} \sqsubseteq \alpha(a,x)\} = \sup\{x \in L_1 \mid v \sqsubseteq \alpha(a,x)\} = f_v$$

where the first equality is obtained from item 3 and, from item 5 we have the other equality because $v_{g_v} = \sup\{\beta(b, g_v(b)) \mid b \in B\} = v$. The equality $f_v^{\downarrow} = g_v$ is proved analogously.

To prove the equality $\psi(\varphi(\langle g,f \rangle)) = \langle g,f \rangle$, it is sufficient to prove that $f = f_{v_\varphi}$, where $v_\varphi = \varphi(\langle g,f \rangle)$, but this follows from Proposition 2 (item 3) since $v_g = \sup\{\beta(b,g(b)) \mid b \in B\} = \varphi(\langle g,f \rangle) = v_\varphi$ and

$$g^{\uparrow}(a) = \sup\{x \in L_1 \mid v_g \sqsubseteq \alpha(a,x)\}$$

The other composition gives the identity as well, that is, $v = \varphi(\psi(v)) = \varphi(\langle g_v, f_v \rangle) = \sup\{\beta(b, g_v(b)) \mid b \in B\}$ for all $v \in V$, as a mapping of item 5 of Proposition 2.

To finish the proof it is sufficient to prove that φ it is order-preserving and order-reflecting, see [9]. Given $\langle g_1, f_1 \rangle$, $\langle g_2, f_2 \rangle$ in $\mathcal{M}$ with $\langle g_1, f_1 \rangle \leq \langle g_2, f_2 \rangle$, we have that $g_1 \leq g_2$ and therefore $\beta(b, g_1(b)) \sqsubseteq \beta(b, g_2(b))$ for all $b \in B$, since β is order-preserving in the second argument. Thus, by definition of φ, we obtain that:

$$\varphi(\langle g_1, f_1\rangle) \sqsubseteq \varphi(\langle g_2, f_2\rangle)$$

To prove that it is order-reflecting it is sufficient to check that the inverse mapping is order-preserving, but this is straightforward. □

Regarding an improvement of a previous representation theorem: let us notice that, in Proposition 2 it is proved directly that the function α is order-reversing and β is order-preserving in their second argument, hence these hypotheses, which are explicitly required for the representation theorem of [17], can be dropped.

Let us finish this section with a further proposition which relates the behaviour of the mappings α and β.

Proposition 3. *Given a multi-adjoint concept lattice $(\mathcal{M}, \preceq)$, a concept $\langle g, f\rangle \in \mathcal{M}$ and two mappings $\beta\colon B \times L_2 \to \mathcal{M}$, $\alpha\colon A \times L_1 \to \mathcal{M}$, where β is $(\mathcal{M}, R)$-related to α, we have that:*

$$\sup\{\beta(b, g(b)) \mid b \in B\} = \inf\{\alpha(a, f(a)) \mid a \in A\}$$

Proof. Given $a \in A$, we have that

$$f(a) = g^{\uparrow}(a) = \inf\{R(a, b) \swarrow^b g(b) \mid b \in B\}$$

then $f(a) \preceq_1 R(a, b) \swarrow^b g(b)$ for all $b \in B$ and applying the adjoint property and Property 2 we have that $\beta(b, g(b)) \sqsubseteq \alpha(a, f(a))$ for all $b \in B$. Therefore if we apply the supremum and infimum properties we obtain the inequality:

$$\sup\{\beta(b, g(b)) \mid b \in B\} \sqsubseteq \inf\{\alpha(a, f(a)) \mid a \in A\}$$

Let $v_\beta = \sup\{\beta(b, g(b)) \mid b \in B\} \in V$ be, as $\alpha[A \times L_1]$ is infimum-dense there exists a set of indices Λ and $K = \{(a_j, x_j) \mid j \in \Lambda\} \subseteq A \times L_1$ such that $v_\beta = \inf\{\alpha(a_j, x_j) \mid j \in \Lambda\}$ and, for all $j \in \Lambda$ and $b \in B$, we have that $\beta(b, g(b)) \sqsubseteq \alpha(a_j, x_j)$ which leads us, from Property 2, to $x_j \preceq_1 R(a_j, b) \swarrow^b g(b)$ and, using that $f = g^{\uparrow}$, to $x_j \preceq_1 f(a_j)$ for all $j \in \Lambda$. Hence we have the following chain which provides the required equality:

$$\begin{aligned} v_\beta = \sup\{\beta(b, g(b)) \mid b \in B\} &\sqsubseteq \inf\{\alpha(a, f(a)) \mid a \in A\} \\ &\sqsubseteq \inf\{\alpha(a_j, f(a_j)) \mid j \in \Lambda\} \\ &\overset{(*)}{\sqsubseteq} \inf\{\alpha(a_j, x_j) \mid j \in \Lambda\} \\ &= v_\beta \end{aligned}$$

where $(*)$ holds because $x_j \preceq_1 f(a_j)$ for all $j \in \Lambda$ and α is order-reversing in the second argument. □

4 A Detailed Example

Now, we apply the language capabilities of the multi-adjoint concept lattices in an example introduced by Umbreit (later Pollandt, after her marriage) [24] and used in [8]. Furthermore, in the multi-adjoint concept lattice framework the user can express in a better way his necessities.

Example 1. Let $([0,1],[0,1],[0,1],\leq,\leq,\leq,\&_G,\&_L)$ be the multi-adjoint frame where $\&_G$ and $\&_L$ are the commutative Gödel and Łukasiewicz conjunctors respectively, so the residuated implications are defined as:

$$b \nwarrow_L a = b \swarrow^L a = \min\{1, 1+b-a\}$$

$$b \nwarrow_G a = b \swarrow^G a = \begin{cases} 1, & \text{if } b \geq a; \\ b, & \text{otherwise.} \end{cases}$$

The different contexts considered later have the same set of objects and attributes:

$$A = \{\text{warm}, \text{cold}, \text{poor in rain}, \text{calm wind}\}$$

$$B = \{\text{Mon}, \text{Tue}, \text{Wed}, \text{Thu}, \text{Fri}, \text{Sat}, \text{Sun}\}$$

and the same relation $R\colon A \times B \to P$, defined in Table 1.

Table 1. Data for Example 1

R	Mon	Tue	Wed	Thu	Fri	Sat	Sun
warm	0.5	1	0.5	0.5	0	0	0
cold	0.5	0	0.5	0.5	1	1	1
poor in rain	1	1	0	1	0	0.5	1
calm wind	1	1	0	0	0	0	1

Now, if we consider the contexts (A, B, R, σ_1), (A, B, R, σ_2), where $\sigma_1(b) = \&_G$ and $\sigma_2(b) = \&_L$ for every $b \in B$ we can check that we obtain the same result as [8]. We can see this in the concrete example of the problem of *walking time*, that is defined in [24] as a day of the week not much warm or cold and with no rain, so the fuzzy notion can be expressed by the fuzzy subset $f\colon A \to [0,1]$ defined as:

$$f(\text{warm}) = 0.5,\ f(\text{cold}) = 0.5,\ f(\text{poor in rain}) = 1,\ f(\text{calm wind}) = 0.5$$

and represented as: $f = \{\text{warm}/0.5, \text{cold}/0.5, \text{poor in rain}/1, \text{calm wind}/0.5\}$. Let us compute a multi-adjoint concept which represents the situation given by f.

With the first context we have that

$$\begin{aligned} f^{\downarrow}(\text{Mon}) &= \inf\{R(a,\text{Mon}) \nwarrow_G f(a) \colon a \in A\} \\ &= \inf\{0.5 \nwarrow_G 0.5, 0.5 \nwarrow_G 0.5, 1 \nwarrow_G 1, 1 \nwarrow_G 0.5\} \\ &= 1 \end{aligned}$$

If we make the same computation for the other days the value 0 is obtained.

In a similar way $f^{\downarrow\uparrow}$ is calculated:

$$\begin{aligned}
&f^{\downarrow\uparrow}(\text{warm}) = \\
&\quad = \inf\{R(\text{warm}, b) \swarrow^G f^{\downarrow}(b) \colon b \in B\} \\
&\quad = \inf\{0.5 \swarrow^G 1, 1 \swarrow^G 0, 0.5 \swarrow^G 0, 0.5 \swarrow^G 0, 0 \swarrow^G 0, 0 \swarrow^G 0, 0 \swarrow^G 0\} \\
&\quad = 0.5
\end{aligned}$$

If the same is done for the other attributes we have that $f^{\downarrow\uparrow}(\text{cold}) = 0.5$ and that $f^{\downarrow\uparrow}(\text{poor in rain}) = f^{\downarrow\uparrow}(\text{calm wind}) = 1$. So, the best days for walking time (with definition given above) is Monday while the others are bad days.

If we use the second context we obtain the concept formed by the two components below:

$$\begin{aligned}
f^{\downarrow} &= \{\text{Mon}/1, \text{Tue}/0.5, \text{Wed}/0, \text{Thu}/0.5, \text{Fri}/0, \text{Sat}/0.5, \text{Sun}/0.5\} \\
f^{\downarrow\uparrow} &= \{\text{warm}/0.5, \text{cold}/0.5, \text{poor in rain}/1, \text{calm wind}/0.5\}
\end{aligned}$$

In this case the best day is also Monday, but Tuesday, Thursday, Saturday and Sunday are good ones, while Wednesday and Friday are bad ones. Hence, as stated above, the concepts obtained $\langle f^{\downarrow}, f^{\downarrow\uparrow}\rangle$ are the same as in [8].

However, we can consider a multi-adjoint context where we can adapt the definition of *walking time* in order to consider some restriction in the objects (or attributes). If we think of a modified problem of *walking time* in which the preferences are modified to add "better at weekends", we can consider the context (A, B, R, σ_3), where $\sigma_3(b) = \&_G$ for every $b \in B_1$ and $\sigma_3(b) = \&_L$ for every $b \in B_2$, where $B_1 = \{\text{Mon}, \text{Tue}, \text{Wed}, \text{Thu}, \text{Fri}\}$ and $B_2 = \{\text{Sat}, \text{Sun}\}$, this way we make the computation differently on weekends by using the following definition by cases:

$$\begin{aligned}
f^{\downarrow}(b_1) &= \inf\{R(a, b_1) \nwarrow_G f(a) \colon a \in A\} \quad \text{for } b_1 \in B_1 \\
f^{\downarrow}(b_2) &= \inf\{R(a, b_2) \nwarrow_L f(a) \colon a \in A\} \quad \text{for } b_2 \in B_2
\end{aligned}$$

and obtain $f^{\downarrow} = \{\text{Mon}/1, \text{Tue}/0, \text{Wed}/0, \text{Thu}/0, \text{Fri}/0, \text{Sat}/0.5, \text{Sun}/0.5\}$. Similarly, the computation of $f^{\downarrow\uparrow}$ takes into account the relationship between objects and implications:

$$\begin{aligned}
&f^{\downarrow\uparrow}(\text{warm}) = \\
&= \inf(\{R(\text{warm}, b_1) \swarrow^G f^{\downarrow}(b_1) \colon b_1 \in B_1\} \cup \{R(\text{warm}, b_2) \swarrow^L f^{\downarrow}(b_2) \colon b_2 \in B_2\} \\
&= 0.5
\end{aligned}$$

If we make the same computation for the other attributes we obtain:

$$f^{\downarrow\uparrow} = \{\text{warm}/0.5, \text{cold}/0.5, \text{poor in rain}/1, \text{calm wind}/0.5\}$$

Now, though the user prefers weekends, Monday is still the best day, but now, Saturday and Sunday are better days than the others. Remind that fuzzy notions related to the attributes can be given, for example *weather in weekends* can be studied, represented by the fuzzy set:

$$g = \{\mathrm{Mon}/0, \mathrm{Tue}/0, \mathrm{Wed}/0, \mathrm{Thu}/0, \mathrm{Fri}/0, \mathrm{Sat}/1, \mathrm{Sun}/1\}$$

and fixed the attention in the attributes 'warm' and 'poor in rain', considering different implications, that is, the context could be (A, B, R, τ) where τ is defined as:

$$\tau(\text{warm}) = \tau(\text{poor in rain}) = \&_L \; ; \; \tau(\text{cold}) = \tau(\text{calm wind}) = \&_G$$

5 Conclusions and Future Work

Multi-adjoint concept lattices have been introduced as a generalization of different existing approaches to fuzzified and/or generalized versions of the classical concept lattice. One of the interesting features is that in a multi-adjoint context each object (or attribute) has an associated implication and, thus, subgroups with different degrees of preference can be easily established.

The representation theorem for multi-adjoint concept lattices has been shown by taking advantage of the relationship between Galois connections and concept lattices given in [9]. This fact, in particular, shows that the "concepts" defined in [17] form a complete lattice without having to rely on the particular definitions of the Galois connections.

The multi-adjoint concept lattice embeds the generalized concept lattice [18] and, as a consequence, other different fuzzy extensions of the classical concept lattice [10], such as the fuzzy concepts of [7] and of [5] for the case of $\{0, 1\}$-equality and crisp ordering.

Continuing with the comparison of the multi-adjoint frame with other fuzzy approaches, one future work would be to study the relationship between the concepts given in [11]. Another point to take into account is the consideration of fuzzy ordering in order to completely embed the fuzzy concept lattice of [5].

Acknowledgements

We are grateful to the anonymous referees who helped us to improve the paper.

References

1. A. Abdel-Hamid and N. Morsi. Associatively tied implicacions. *Fuzzy Sets and Systems*, 136(3):291–311, 2003.
2. R. Bělohlávek. Fuzzy concepts and conceptual structures: induced similarities. In *Joint Conference on Information Sciences*, pages 179–182, 1998.
3. R. Bělohlávek. Lattices of fixed points of fuzzy galois connections. *Mathematical Logic Quartely*, 47(1):111–116, 2001.
4. R. Bělohlávek. Reduction and a simple proof of characterization of fuzzy concept lattices. *Fundamenta Informaticae*, 46(4):277–285, 2001.

5. R. Bělohlávek. Concept lattices and order in fuzzy logic. *Annals of Pure and Applied Logic*, 128:277–298, 2004.
6. R. Bělohlávek and V. Vychodil. What is a fuzzy concept lattice? In *3rd Intl Conf on Concept Lattices and their Applications*, pages 34–45, 2005.
7. A. Burusco and R. Fuentes-González. The study of *L*-fuzzy concept lattice. *Mathware & Soft Computing*, 3:209–218, 1994.
8. A. Burusco and R. Fuentes-González. Concept lattices defined from implication operators. *Fuzzy Sets and Systems*, 114:431–436, 2000.
9. B. Davey and H. Priestley. *Introduction to Lattices and Order.* Cambridge University Press, second edition, 2002.
10. B. Ganter and R. Wille. *Formal Concept Analysis: Mathematical Foundation.* Springer Verlag, 1999.
11. G. Georgescu and A. Popescu. Concept lattices and similarity in non-commutative fuzzy logic. *Fundamenta Informaticae*, 55(1):23–54, 2002.
12. G. Georgescu and A. Popescu. Non-commutative fuzzy galois connections. *Soft Comput.*, 7(7):458–467, 2003.
13. G. Georgescu and A. Popescu. Non-dual fuzzy connections. *Arch. Math. Log.*, 43(8):1009–1039, 2004.
14. G. Georgescu and A. Popescu. Similarity convergence in residuated structures. *Logic Journal of the IGPL*, 13(4):389–413, 2005.
15. P. Hájek. *Metamathematics of Fuzzy Logic.* Trends in Logic. Studia Logica Library. Kluwer Academic Publishers, 1998.
16. P. Julián, G. Moreno, and J. Penabad. On fuzzy unfolding: A multi-adjoint approach. *Fuzzy Sets and Systems*, 154:16–33, 2005.
17. S. Krajči. The basic theorem on generalized concept lattice. In V. Snášel and R. Bělohlávek, editors, *ERCIM workshop on soft computing*, pages 25–33, 2004.
18. S. Krajči. A generalized concept lattice. *Logic Journal of IGPL*, 13(5):543–550, 2005.
19. J. Medina, M. Ojeda-Aciego, A. Valverde, and P. Vojtáš. Towards biresiduated multi-adjoint logic programming. *Lect. Notes in Artificial Intelligence*, 3040:608–617, 2004.
20. J. Medina, M. Ojeda-Aciego, and P. Vojtáš. Multi-adjoint logic programming with continuous semantics. In *Logic Programming and Non-Monotonic Reasoning, LPNMR'01*, pages 351–364. Lect. Notes in Artificial Intelligence 2173, 2001.
21. J. Medina, M. Ojeda-Aciego, and P. Vojtáš. Similarity-based unification: a multi-adjoint approach. *Fuzzy Sets and Systems*, 146(1):43–62, 2004.
22. J. Medina and J. Ruiz-Calviño. Towards multi-adjoint concept lattices. In *Information Processing and Management of Uncertainty for Knowledge-Based Systems, IPMU'06*, pages 2566–2571, 2006.
23. S. Pollandt. *Fuzzy Begriffe.* Springer, Berlin, 1997.
24. S. Umbreit. *Formale Begriffsanalyse mit unscharfen Begriffen.* PhD thesis, Halle, Saale, 1995.

Base Points, Non-unit Implications, and Convex Geometries

Bernhard Ganter and Heiko Reppe

Institute for Algebra
Dresden University of Technology
D-01062 Dresden
Bernhard.Ganter@tu-dresden.de, Heiko.Reppe@mailbox.tu-dresden.de

Abstract. We study the "non-unit implications" of a formal context and investigate the closure system induced by these implications. It turns out that this closure system is the largest closure system on the same base set containing the given one as a complete sublattice. This was studied by other authors with special emphasis on semidistributivity and convex geometries. We present some of their results in FCA language.

The complete lattice refinements of a closure system form an interval within the lattice of all closure systems. We describe the reduced context for this interval.

For better compatibility with the literature, we dualize and consider implications between objects, not attributes.

1 Complete Lattice Refinements of Closure Systems

We study closure systems on a set G. For simplicity, and w.l.o.g., we assume each such closure system to be given as the system $\text{Ext}(\mathbb{K})$ of extents of some formal context $\mathbb{K} := (G, M, I)$. The corresponding closure operator will be written as $X \mapsto X^{II}$, or simply as $X \mapsto X''$.

A *refinement* of a closure system is another closure system on the same base set, containing the first as a subset. Such refinements can be obtained as the systems of extents of appositions $\mathbb{K} \mid \mathbb{L}$, where $\mathbb{L}$ is an arbitrary formal context with object set G, cf. [GW99, Definition 30]. The lattice $(\text{Ext}(\mathbb{K}), \subseteq)$ is always a $\bigwedge$-subsemilattice of $(\text{Ext}(\mathbb{K} \mid \mathbb{L}), \subseteq)$, but usually not a sublattice, because joins may not be preserved. In the exceptional case that $(\text{Ext}(\mathbb{K}), \subseteq)$ is a (complete) sublattice of $(\text{Ext}(\mathbb{K} \mid \mathbb{L}), \subseteq)$, we speak of a *(complete) lattice refinement.*

In the finite case, the two notions (of lattice refinement and complete lattice refinement) conincide iff $\emptyset$ is closed. Lattice refinements of finite closure systems have been studied by Adaricheva and Nation [AN03], [N04].

Lemma 1. *Let* $\mathbb{K} := (G, M, I)$ *and* $\mathbb{L} := (G, N, J)$. *Then* $\text{Ext}(\mathbb{K} \mid \mathbb{L})$ *is a complete lattice refinement of* $\text{Ext}(\mathbb{K})$ *if and only if for each* $n \in N$ *the set*

$$\Theta(n) := \{g \in G \mid g^{II} \subseteq n^J\}$$

is an extent of $\mathbb{K}$.

S.O. Kuznetsov and S. Schmidt (Eds.): ICFCA 2007, LNAI 4390, pp. 210–220, 2007.

Proof. Clearly $\Theta(n)^{II}$ is an extent of $\mathbb{K}$. More precisely, it is the supremum of all $g^{II} \subseteq n^J$ in $\mathrm{Ext}(\mathbb{K})$. In $\mathrm{Ext}(\mathbb{K} \mid \mathbb{L})$, n^J is an upper bound of $\{g \in G \mid g^{II} \subseteq n^J\}$, and in order to preserve joins it is necessary that $\Theta(n)^{II} \subseteq n^J$. Then we have $\Theta(n)^{II} = \Theta(n)$.

Now consider an arbitrary family $\mathcal{F} \subseteq \mathrm{Ext}(\mathbb{K})$ and let $U := \bigcup \mathcal{F}$. Then U^{II} is the join of $\mathcal{F}$ in $\mathrm{Ext}(\mathbb{K})$, and the join in $\mathrm{Ext}(\mathbb{K} \mid \mathbb{L})$ is different iff there exists some attribute $n \in N$ such that n^J contains U but not U^{II}. Since U is a union of extents, we have that

$$g \in U \Rightarrow g^{II} \subseteq U,$$

and thus $U \subseteq \Theta(n) \subseteq n^J$. If this is an extent of $\mathbb{K}$, it is an upper bound of $\mathcal{F}$ and thus contains U^{II}, a contradiction. □

It is quite evident (and follows from Theorem 1) that the intersection of complete lattice refinements of $\mathrm{Ext}(\mathbb{K})$ again yields a complete lattice refinement. The lattice refinements of $\mathrm{Ext}(\mathbb{K})$, ordered by inclusion, therefore form a complete lattice. (However, lattice refinements of $\mathrm{Ext}(\mathbb{K})$ need not be lattice refinements of each other, see Example 1). We shall discuss some properties of this lattice in Section 2.

Let us call a closure system on G *elementary*, if each singleton set $\{g\}$ is closed ($g \in G$). If $|G| > 1$, then a closure system on G can only be elementary if the empty set $\emptyset$ is closed. The following proposition states that a closure system has a proper complete lattice refinement, i.e. $\mathrm{Ext}(\mathbb{K}) \subsetneq \mathrm{Ext}(\mathbb{K} \mid \mathbb{L})$, iff $\mathrm{Ext}(\mathbb{K})$ is not elementary.

Proposition 1. *Each closure system containing $\emptyset$ has an elementary complete lattice refinement. No elementary closure system admits a proper complete lattice refinement.*

Proof. The condition of Lemma 1 is automatically fulfilled when the empty set is closed and $|n^J| = 1$. Therefore, $\mathrm{Ext}(\mathbb{K} \mid (G, G, =))$ always is an elementary complete lattice refinement if $\emptyset$ is closed.

If the closure system $\mathrm{Ext}(\mathbb{K})$ is elementary then $g = g^{II}$ and thus $\{g \in G \mid g^{II} \subseteq n^J\} = n^J$. The condition of Lemma 1 requires in this case that all attribute extents of $\mathbb{L}$ are already extents of $\mathbb{K}$. □

Remark 1. Proposition 1 does not hold for lattice refinements in general (without the prefix "complete"). As an example, consider the closure system consisting of the set of natural numbers $\mathbb{N}$ and of all its finite subsets. It is elementary, and the closure system of all subsets of $\mathbb{N}$ is a proper lattice refinement.

Example 1. Example 1 shows, on the example of $\underline{N}_5$, an elementary complete lattice refinement as given by Proposition 1 (middle, $\underline{\mathfrak{B}}(\widetilde{\mathbb{K}})$). The diagram on the right, $\underline{\mathfrak{B}}(\widehat{\mathbb{K}})$ shows another elementary complete lattice refinement. Note that $\mathrm{Ext}(\widehat{\mathbb{K}})$ refines $\mathrm{Ext}(\widetilde{\mathbb{K}})$, but is no lattice refinement. In fact, according to Proposition 1, $\mathrm{Ext}(\widetilde{\mathbb{K}})$ cannot have a proper complete lattice refinement.

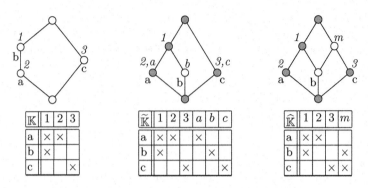

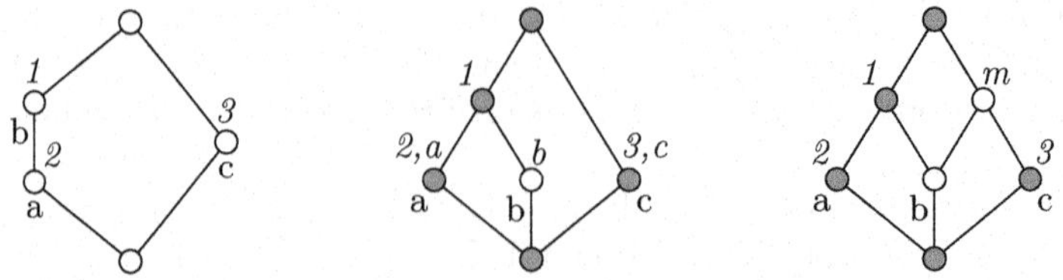

$\mathbb{K}$	1	2	3
a	×	×	
b	×		
c			×

$\widetilde{\mathbb{K}}$	1	2	3	a	b	c
a	×	×		×		
b	×				×	
c			×			×

$\widehat{\mathbb{K}}$	1	2	3	m
a	×	×		
b	×			×
c			×	×

Fig. 1. The lattice $\underline{N}_5 = \underline{\mathfrak{B}}(\mathbb{K})$, and two lattice refinements $\underline{\mathfrak{B}}(\widetilde{\mathbb{K}})$ and $\underline{\mathfrak{B}}(\widehat{\mathbb{K}})$. The latter two closure systems are both elementary and contain $\underline{N}_5$ as sublattice. The marked nodes in the latter diagrams show the embeddings of $\underline{N}_5$ as closure system on the set $\{a, b, c\}$.

A more appealing characterization of complete lattice refinements can be formulated in terms of (object) implications. Let $\mathbb{K} := (G, M, I)$ be a formal context and let

$$\text{oImp}(\mathbb{K}) := \{A \to B \mid A \subseteq G, B \subseteq A^{II}\}.$$

A set $A \subseteq G$ is called *1-closed* if

$$A = \bigcup_{a \in A} a^{II},$$

i.e., if $a \in A$ aways implies $a^{II} \subseteq A$. All closed sets of course are 1-closed, but there may be others. For example, if $P \subseteq G$ is pseudo-closed and $|P| > 1$, then P also is 1-closed. Let

$$\mathcal{N} := \{A \to B \mid A \text{ 1-closed, but not closed}, B \subseteq A^{II}\}.$$

The implications in $\mathcal{N}$ are closely related to the *non-unit implications* considered by Gély and Nourine [GN06]. If G is finite, it is clear that every implication of $\mathbb{K}$ follows from $\mathcal{N}$ together with all *unit implications* $g \to g^{II}$, $g \in G$. This remains true if we only use implications from the stem base of $\mathbb{K}$ (for object implications).

Note that the set $\emptyset$ is 1-closed. It does not respect the implication $\emptyset \to \emptyset^{II}$ iff $\emptyset^{II} \neq \emptyset$. In this case $\emptyset \to \emptyset^{II} \in \mathcal{N}$.

Theorem 1. $\text{Ext}(\mathbb{K} \mid \mathbb{L})$ *is a complete lattice refinement of* $\text{Ext}(\mathbb{K})$ *iff* $\text{Ext}(\mathbb{L})$ *respects all implications in* $\mathcal{N}$.

Proof. The condition for a set $X \subseteq G$ to respect $\mathcal{N}$ is that if $A \subseteq X$ is 1-closed, then $A^{II} \subseteq X$. Every set X has a largest 1-closed subset, namely $\{g \in X \mid g^{II} \subseteq X\}$. The closure of this set is also 1-closed and therefore must be the same, if X respects $\mathcal{N}$. Thus n^J satisfies the condition of Lemma 1 iff $\Theta(n) = \Theta(n)^{II}$. □

Using implications, it is now easy to characterize two important lattice refinements of $\text{Ext}(\mathbb{K})$, provided that $\emptyset$ is closed:

- The coarsest (i.e., smallest) elementary complete lattice refinement of Ext(𝕂) is
$$\mathrm{Ext}(\mathbb{K} \mid (G, G, =)).$$
Its object implications are generated by all object implications of 𝕂 that have a premise with more than one element.
- The finest complete lattice refinement of Ext(𝕂) has its object implications generated by $\mathcal{N}$. This refinement is elementary as well.

A characterization of the meet-irreducibles in the large refinement has been described by Gély and Nourine [GN06].

Example 2. Another example of the maximal complete lattice refinement is given in Figure 2. Starting from a small lattice, we obtain its finest complete lattice refinement by computing its stem base, removing the implications with one-element premise (and thereby obtaining a basis for $\mathcal{N}$). The sets respecting these implications form the finest complete lattice refinement. The shaded points indicate the original lattice as a complete sublattice.

𝕂						
a		×	×		×	×
b			×			
c	×	×		×	×	×
d	×	×				×
e	×	×		×		

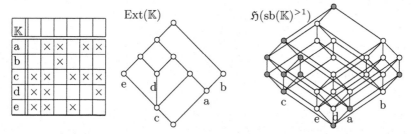

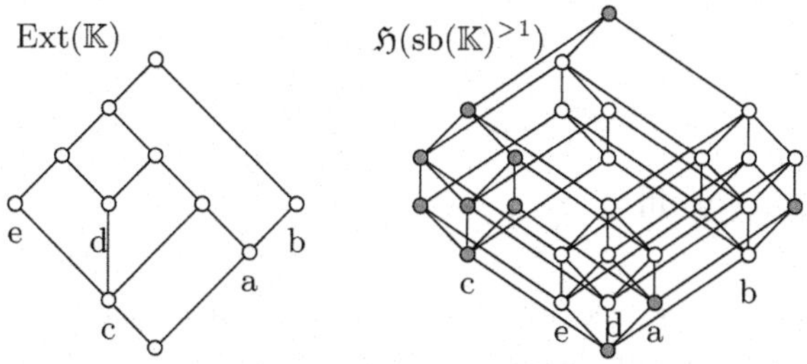

Fig. 2. A lattice (Ext(𝕂), ⊆) and its finest lattice refinement. The stem base is $\mathrm{sb}(\mathbb{K}) = \{\{b\} \to \{a\}, \{d\} \to \{c\}, \{e\} \to \{c\}, \{a,c,e\} \to \{d\}, \{a,b,c\} \to \{d,e\}\}$. The derived closure system is defined by the non-unit implications in the stem base.

2 The Lattice of Complete Lattice Refinements

It follows from Theorem 1 that the lattice of all complete lattice refinements of Ext(𝕂) is an interval in the lattice of all closure systems on the base set G.

It is not difficult to give a formal context for this lattice. Such a context can have as objects all sets satisfying the conditions of Lemma 1, i.e. all $\mathcal{N}$-closed sets. As attributes we can take all object-implications that hold in 𝕂; the incidence relation will be $\models$. The extents of this context then are precisely the complete lattice refinements of Ext(𝕂), its intents are the corresponding implicational theories.

However, this context is quite large already for small 𝕂, and it is desirable to reduce its size without changing the lattice structure. A first easy step is to restrict the objects to all $\mathcal{N}$-closed sets, which are not closed (in Ext(𝕂)), and to restrict the attribute set to implications of the form $A \to b$, where A is $\mathcal{N}$-closed and $b \in A'' \setminus A$.

More challenging is to determine the irreducible objects and attributes. In order to achieve this, we must study the object- and attribute order and the arrow relations.

Lemma 2. *Suppose* $\text{Ext}(\mathbb{K})$ *is a closure system containing* $\emptyset$ *and let* X, Y *be sets which are* $\mathcal{N}$*-closed, but not in* $\text{Ext}(\mathbb{K})$, *and let* $A \to b$, $C \to d$ *be object implications where* A, C *are* $\mathcal{N}$*-closed,* $b \in A'' \setminus A$ *and* $d \in C'' \setminus C$. *Then*

1. $X^{\models} \subseteq Y^{\models} \iff Y \subseteq X$ *and* $(Y'' \setminus Y) \cap X = \emptyset$.
2. $(A \to b)^{\models} \subseteq (C \to d)^{\models} \iff A \subseteq C$ *and* $d \in \mathcal{N}(C \cup \{b\})$.

Proof. 1. We have

$$\begin{aligned} X^{\models} \subseteq Y^{\models} &\iff X \models A \to b \text{ always implies } Y \models A \to b \\ &\iff A \subseteq Y, b \notin Y \text{ always implies } A \subseteq X, b \notin X. \end{aligned}$$

To see that $Y \subseteq X$ is necessary, choose $A := Y$. But with $Y \subseteq X$, the condition $A \subseteq X$ is immediate from $A \subseteq Y$, so that $b \notin Y$ must imply $b \notin X$. This is the condition in 1).

2. $(A \to b)^{\models} \subseteq (C \to d)^{\models}$ holds iff for all X

$$X \models A \to b \text{ implies } X \models C \to d,$$

which is true iff

$$C \subseteq X, d \notin X \text{ implies } A \subseteq X, b \notin X.$$

Letting $X := C$, we obtain that $A \subseteq C$ is necessary. But for $A \subseteq C$ the condition reduces to the following:

$$C \subseteq X, d \notin X \text{ implies } b \notin X,$$

or, equivalently,

$$C \subseteq X, b \in X \text{ implies } d \in X,$$

which is equivalent to $d \in \mathcal{N}(C \cup \{b\})$. □

Lemma 3. *Let* $\text{Ext}(\mathbb{K})$ *be a closure system containing* $\emptyset$. *If* X, A *are* $\mathcal{N}$*-closed sets and* $b \in A'' \setminus A$ *then:*

1. $X \swarrow (A \to b)$ *if and only if* $A \subseteq X$ *and* $(Y'' \setminus Y) \cap X \neq \emptyset$ *for all* $\mathcal{N}$*-closed* Y *with* $A \subseteq Y \subsetneq X$,
2. $X \nearrow (A \to b)$ *if and only if* $A = X$ *and* $X \cup \{b\}$ *is* $\mathcal{N}$*-closed.*

Proof. 1. From the definition of $\swarrow$ we get $X \swarrow (A \to b) \iff$

$$A \subseteq X, b \notin X, \text{ and if } Y \subsetneq X,\ (Y'' \setminus Y) \cap X = \emptyset, \text{ then } A \not\subseteq Y \text{ or } b \in Y.$$

For $A = X$ this condition is always fulfilled. The possibility that it is fulfilled for some $A \subset X$ requires attention: A violation in this case could only occur for a set Y with $A \subseteq Y \subsetneq X$. For such a Y the conclusion of the condition is never

true. Thus to avoid a violation of the condition, the premise must also be false. That is what is stated in the lemma.
2. Again we unfold the definition of $\nearrow$ in this case:

$$X \nearrow (A \to b) \iff A \subseteq X,\ b \notin X,\ \text{and } (C \not\subseteq X \text{ or } d \in X)$$
$$\text{for all } C \to d \neq A \to b \text{ with } A \subseteq C \text{ and } d \in \mathcal{N}(C \cup \{b\}).$$

If $A = X$, then the only possible choice for a violation of the condition is $C = X$, $d \in \mathcal{N}(X \cup \{b\})$. To avoid such a violation, we must require that $\mathcal{N}(X \cup \{b\})$ contains no elements outside X, except b.

If $A \subsetneq X$, we can choose $C := X$, $d := b$ and obtain a violation since then

$$C \subseteq X, d \notin X, A \subsetneq C, d \in \mathcal{N}(C \cup \{b\}).$$ □

The following definition is adapted from Edelmann and Jamison [EJ85]:

Definition 1. *A closure system* Ext($\mathbb{K}$) *on a set G is called* convex geometry *iff for all $A \in$* Ext($\mathbb{K}$) *and for all $g, h \in G \setminus A$ with $g'' \neq h''$ we have:*

$$g \in (A \cup \{h\})'' \Rightarrow h \notin (A \cup \{g\})''.$$

This condition is called anti-exchange property. ◊

Corollary 1. *The lattice of all complete lattice refinements of* Ext($\mathbb{K}$) *is a convex geometry.*

This follows from [GW99, Theorem 44], and the fact that for each attribute $A \to b$ there is at most one object X such that $X \nearrow A \to b$, namely $X = A$. It is not surprising, since the lattice is an interval in the lattice of all closure operators, and the latter is known to be a convex geometry. More about convex geometries will be said in Theorem 3 and in Section 4.

Example 3. The formal context in Figure 3 demonstrates the results of Lemma 3. Apparently all of the objects and most of the attributes are irreducible. This is elaborated in Theorem 2. The concept lattice of this context has 2215 elements.

Theorem 2. *For a closure system* Ext($\mathbb{K}$) *containing $\emptyset$ the reduced context for the lattice of complete lattice refinements is given as follows:*

1. *The irreducible objects are the subsets of G which are $\mathcal{N}$-closed but not in* Ext($\mathbb{K}$).
2. *The irreducible attributes are the implications of the form $A \to b$, where A is $\mathcal{N}$-closed, but not in* Ext($\mathbb{K}$), *$b \in A'' \setminus A$, and $A \cup \{b\}$ is $\mathcal{N}$-closed.*

The incidence relation is $\models$.

This is immediate from Lemma 3, due to the well known fact that an object or attribute is irreducible iff there is an arrow "pointing to it" [GW99, Prop. 13].

	$\{b\}\to\{a\}$	$\{d\}\to\{c\}$	$\{e\}\to\{c\}$	$\{a,d\}\to\{c\}$	$\{a,e\}\to\{c\}$	$\{a,e\}\to\{d\}$	$\{b,c\}\to\{a\}$	$\{b,c\}\to\{d\}$	$\{b,c\}\to\{e\}$	$\{b,d\}\to\{a\}$	$\{b,d\}\to\{c\}$	$\{b,d\}\to\{e\}$	$\{b,e\}\to\{a\}$	$\{b,e\}\to\{c\}$	$\{b,e\}\to\{d\}$	$\{d,e\}\to\{c\}$	$\{a,b,d\}\to\{c\}$	$\{a,b,d\}\to\{e\}$	$\{a,b,e\}\to\{c\}$	$\{a,b,e\}\to\{d\}$	$\{a,d,e\}\to\{c\}$	$\{b,c,d\}\to\{a\}$	$\{b,c,d\}\to\{e\}$	$\{b,c,e\}\to\{a\}$	$\{b,c,e\}\to\{d\}$	$\{b,d,e\}\to\{a\}$	$\{b,d,e\}\to\{c\}$	$\{a,b,d,e\}\to\{c\}$	$\{b,c,d,e\}\to\{a\}$
$\{b\}$	⤢	×	×	×	×	×	×	×	×	×	×	×	×	×	×	×	×	×	×	×	×	×	×	×	×	×	×	×	×
$\{d\}$	×	⤢	×	×	×	×	×	×	×	×	×	×	×	×	×	×	×	×	×	×	×	×	×	×	×	×	×	×	×
$\{e\}$	×	×	⤢	×	×	×	×	×	×	×	×	×	×	×	×	×	×	×	×	×	×	×	×	×	×	×	×	×	×
$\{a,d\}$	×		×	⤢	×	×	×	×	×	×	×	×	×	×	×	×	×	×	×	×	×	×	×	×	×	×	×	×	×
$\{a,e\}$	×	×		×	↙	⤢	×	×	×	×	×	×	×	×	×	×	×	×	×	×	×	×	×	×	×	×	×	×	×
$\{b,c\}$		×	×	×	×	×	↙	⤢	⤢	×	×	×	×	×	×	×	×	×	×	×	×	×	×	×	×	×	×	×	×
$\{b,d\}$			×	×	×	×	×	×	×	↙	⤢	⤢	×	×	×	×	×	×	×	×	×	×	×	×	×	×	×	×	×
$\{b,e\}$		×		×	×	×	×	×	×	×	×	×	⤢	⤢	⤢	×	×	×	×	×	×	×	×	×	×	×	×	×	×
$\{d,e\}$	×			×	×	×	×	×	×	×	×	×	×	×	×	⤢	×	×	×	×	×	×	×	×	×	×	×	×	×
$\{a,b,d\}$	×		×		×	×	×	×	×	×	×	×	×	×	×	×	↙	⤢	×	×	×	×	×	×	×	×	×	×	×
$\{a,b,e\}$	×	×		×			×	×	×	×	×	×	×	↙	↙	×	×	×	⤢	⤢	×	×	×	×	×	×	×	×	×
$\{a,d,e\}$	×				↙	×	×	×	×	×	×	×	×	×	×		×	×	×	×	⤢	×	×	×	×	×	×	×	×
$\{b,c,d\}$		×	×	×	×	×	↙	×	↙	↙	×	↙	×	×	×	×	×	×	×	×	×	↙	⤢	×	×	×	×	×	×
$\{b,c,e\}$		×	×	×	×	×	↙	↙	×	×	×	×	↙	×	↙	×	×	×	×	×	×	×	×	↙	⤢	×	×	×	×
$\{b,d,e\}$				×	×	×	×	×	×	↙	↙	×	↙	↙	×		×	×	×	×	×	×	×	×	×	⤢	↙	×	×
$\{a,b,d,e\}$	×					×	×	×	×	×	↙	×	×	↙	×		↙	×	↙	×		×	×	×	×	×	↙	⤢	×
$\{b,c,d,e\}$		×	×	×	×	×	↙	×	×	↙	×	×	↙	×	×	×	×	×	×	×	×	↙	×	↙	×	×	×	×	⤢

Fig. 3. The formal context for the complete lattice refinements of the lattice Ext($\mathbb{K}$) from Figure 2, with the arrow relations. The object set contains all 1-closed sets, which are not elements of Ext($\mathbb{K}$) and is therefore reduced. The attribute set contains all implications of $\mathcal{N}$ with a singleton as conclusion and may be further reduced.

3 Special Elements of Closed Sets

Definition 2. *Let* $\mathbb{K} := (G, M, I)$ *be a formal context and let* $A \in \text{Ext}(\mathbb{K})$ *be a closed set. An object* $g \in A$ *is called*

- *an* extremal point *of* A *iff*

$$g \notin (A \setminus \{h \in G \mid g'' = h''\})'',$$

- *a* base point *of* A *iff*

$$g \notin (A \setminus \{h \in G \mid g'' \subseteq h''\})'', \text{ and}$$

- *a* maximal base point *of* A *iff* g *is a base point of* A *and there is no base point* h *of* A *with* $g'' \subsetneq h''$. ◇

It is immediate from the definition that extremal points are special base points, and it is not difficult to prove that every base point g defines a join-irreducible object-extent g'' of the lattice $(\text{Ext}(\mathbb{K}), \subseteq)$.

Definition 3. *An element* x *of a complete lattice* L *is called* $\bigvee$-prime *iff*

$$x \leq \bigvee_{t \in T} y_t \text{ implies } x \leq y_t \text{ for some } t \in T$$

(for each index set T *and all choices of* $y_t \in L$*).* ◇

This excludes the least element of the lattice.

Lemma 4. *A closed set X of a closure system* $\mathrm{Ext}(\mathbb{K})$ *on G is $\bigvee$-prime iff $X = \{x\}''$ for some base point x of the largest extent G.*

Proof. Let X be $\bigvee$-prime in $\mathrm{Ext}(\mathbb{K})$. Clearly

$$X \leq \bigvee_{x \in X} x'',$$

and thus $X \leq x''$ for some $x \in X$, which implies $X = x''$.

We claim that x is a base point of G. If not, then

$$x \in (G \setminus \{h \mid x'' \subseteq h''\})'',$$

which translates to

$$X = x'' \leq \bigvee_{x'' \nsubseteq h''} h''.$$

Since X is $\bigvee$-prime, we infer that

$$x'' \subseteq h'' \text{ for some } h \text{ with } x'' \nsubseteq h'',$$

a contradiction in itself.

Conversely if $X = x''$ for some base point x of G then if $X \leq \bigvee_{t \in T} Y_t$, we get $x \in (\bigcup Y_t)''$ and thus

$$\bigcup Y_t \cap \{h \mid x'' \subseteq h''\} \neq \emptyset.$$

Thus there must be some $t \in T$ and some $h \in Y_t$ such that $x'' \subseteq h''$. But then $X = x'' \subseteq Y_t$, which shows that X is $\bigvee$-prime. □

Corollary 2. *If X and Y are closed sets of* $\mathrm{Ext}(\mathbb{K})$ *with $X \subseteq Y$, then X is $\bigvee$-prime in the interval $[\emptyset'', Y]$ iff $X = b''$ for some base point b of Y.*

A reformulation of Definition 2 in terms of implications is sometimes helpful:

Proposition 2. *Let A be a closed set and $g \in G$. Then*

- *g is an extremal point of A iff the following condition holds:*
  If $X \to g$ for some $X \subseteq A$, then $g \in X$.
- *g is a base point of A iff the following condition holds:*
  If $X \to g$ for some $X \subseteq A$, then $x \to g$ for some $x \in X$.

The following proposition is now evident:

Proposition 3. *If A is 1-closed, then all base points of A'' are in A.*

As a special case, we get

Proposition 4. *If A and B are closed, then all base points of $(A \cup B)''$ are in $A \cup B$. More generally, if A_t is closed for all $t \in T$, then all base points of $\left(\bigcup_{t \in T} A_t\right)''$ are in $\bigcup_{t \in T} A_t$.*

This finally yields

Proposition 5. *If A, B, and C are closed and $(A \cup B)''$ is the closure of its base points, then $(A \cup B)'' = (A \cup C)''$ implies $(A \cup B)'' = (A \cup (B \cap C))''$.*

Proof. All base points of $(A \cup B)''$ which are not in A belong to B, and also to C, thus to $B \cap C$. □

Closure systems in which every closed set is the closure of its extremal points (or, as a weaker condition, of its base points) are well studied, see the literature cited for [GW99, Theorem 43, Theorem 44]. Here we restrict these results to the finite case. Assuming finiteness allows us to restrict to the maximal base points: If a f inite extent is the closure of its base points, then it is also the closure of its maximal base points. The representation of a closed set X by the set

$$\{g'' \mid g \text{ maximal base point of } X\}$$

is called the *canonical representation.*

Theorem 3. *Let $\mathbb{K} = (G, M, I)$ be a finite context.*

1. *Every element of* $\operatorname{Ext}(\mathbb{K})$ *is closure of its (maximal) base points if and only if* $(\operatorname{Ext}(\mathbb{K}), \subseteq)$ *is a join-semidistributive lattice, i.e. iff for all $A, B, C \in \operatorname{Ext}(\mathbb{K})$:*

$$(A \cup B)'' = (A \cup C)'' \Rightarrow (A \cup B)'' = (A \cup (B \cap C))''.$$

2. *Every element of* $\operatorname{Ext}(\mathbb{K})$ *is closure of its extremal points if and only if in the lattice* $(\operatorname{Ext}(\mathbb{K}), \subseteq)$ *the anti-exchange property holds.*

4 Convex Geometry Refinements

The problem whether every finite join-semidistributive lattice has a refinement that is atomistic and join-semidistributive as well was affirmatively answered by Adaricheva, Gorbunov, Tumanov [AGT03]. Later Adaricheva and Nation [AN03] simplified the construction and showed that it refines each (finite) convex geometry to the largest possible convex geometry refinement. We repeat their results here because they give some idea of the role of the non-unit implications.

Let $(\operatorname{Ext}(\mathbb{K}), \subseteq)$ be a closure system on the finite set G. For $A \in \operatorname{Ext}(\mathbb{K})$ let $\beta(A)$ denote the set of base points of A.

Proposition 6. *If $A_1 \subseteq A_2$ are closed, then $\beta(A_2) \cap A_1 \subseteq \beta(A_1)$.*

Lemma 5 (Adaricheva and Nation). *Let $(\operatorname{Ext}(\mathbb{K}), \subseteq)$ be a closure system on G in which every closed set is the closure of its base points. Define for all $S \subseteq G$*

$$\varepsilon(S) := S \cup \bigcup\{A \in \operatorname{Ext}(\mathbb{K}) \mid \beta(A) \subseteq S\}.$$

Then ε is a closure operator on G.

Proof. We show that for $C \in \text{Ext}(\mathbb{K})$ always

$$\beta(C) \subseteq \varepsilon(S) \quad \text{implies} \quad C \subseteq \varepsilon(S).$$

In fact, $\beta(C) \subseteq \varepsilon(S)$ implies $\beta(C) \subseteq S \cup \bigcup_{\beta(A) \subseteq S} A$. Thus for each $p \in C$ we have $p \in S$ or $p \in A_p$ for some closed set A_p with $\beta(A_p) \subseteq S$. Now let

$$B := \left(C \cup \bigcup_{p \in \beta(C) \setminus S} A_p \right)''.$$

Clearly B is a closed set containing C and $\beta(B) \subseteq C \cup \bigcup \beta(A_p)$ by Proposition 4. So if $z \in \beta(B)$, then either $z \in C$ or $z \in A_p$ for some A_p with $\beta(A_p) \subseteq S$. In the first case, if $z \in C$, we get from Proposition 6 with $z \in \beta(B)$ that

$$z \in \beta(B) \cap C \subseteq \beta(C) \subseteq \varepsilon(S),$$

and therefore that $z \in A_z$ for some closed set A_z with $\beta(A_z) \subseteq S$, as in the second case.

In the second case, where $z \in A_p \subseteq B$ for some A_p with $\beta(A_p) \subseteq S$, we infer from Proposition 6 that

$$z \in \beta(B) \cap A_p \subseteq \beta(A_p) \subseteq S,$$

a contradiction. □

Note that the closure system described in Lemma 5 can be given in terms of object implications as follows: It consists of those sets respecting all implications in

$$\mathcal{AN}(\mathbb{K}) := \{\beta(A) \to A \mid A \in \text{Ext}(\mathbb{K})\}.$$

It was assumed in Lemma 5 that every extent is the closure of its base points, and it therefore might be surprising that $\mathcal{AN}(\mathbb{K})$ does not generate everything. But note that for a one-generated extent g'' we always have

$$\beta(g'') = \beta(g'' \setminus \{h \mid h'' = g''\}) \cup \{h \mid h'' = g''\}.$$

Therefore, $\mathcal{AN}(\mathbb{K})$ contains no nontrivial implications with one-element premise, and the derived closure system is elementary.

The following lemmas are also due to Adaricheva and Nation [AN03].

Lemma 6. *The closure system described in Lemma 5 is a complete lattice refinement of* $\text{Ext}(\mathbb{K})$.

Proof. It suffices to show that $\mathcal{N}$ follows from $\mathcal{AN}$. Let A be 1-closed. According to Proposition 3, then $\beta(A'') \subseteq A$, and the implication $A \to A'' \in \mathcal{N}$ follows from $\beta(A'') \to A'' \in \mathcal{AN}$. □

Lemma 7. *The closure system described in Lemma 5 has the anti-exchange property.*

Proof. Let C be ε-closed, let $p \neq q \in G$ such that $p \in \varepsilon(C \cup \{q\})$, $p \notin \varepsilon(C)$, and assume $q \in \varepsilon(C \cup \{p\})$. It was shown in the proof of Lemma 5 that there must be closed sets A_p, $A_q \in \text{Ext}(\mathbb{K})$ with $p \in A_p$, $q \in A_q$, $\beta(A_p) \subseteq C \cup \{q\}$ and $\beta(A_q) \subseteq C \cup \{p\}$. Since $A_p \cup A_q \subseteq \varepsilon(C \cup \{p, q\})$, we know that $\beta((A_p \cup A_q)'') \subseteq C \cup \{p, q\}$. But $p \notin \beta(A_p)$, and by Proposition 6 we conclude that $p \notin \beta((A_p \cup A_q)'')$. With the same argument, $q \notin \beta((A_p \cup A_q)'')$, so $\beta((A_p \cup A_q)'') \subseteq C$, a contradiction. □

Example 4. Figure 2 shows an example of this construction. The smaller lattice is join-semidistributive, so that each extent is the closure of its base points. There are only two extents which are not equal to the set of their base points, namely $\beta(G) = \{a, b, c\}$ and $\beta(\{a, c, d, e\}) = \{a, c, e\}$. In both cases the set of base points is 1-closed. Hence the convex geometry refinement equals the largest possible refinement. But that is an exceptional case.

5 Conclusion

The "non-unit implications" of a formal context (with $\emptyset'' = \emptyset$) are those implications in the stem base which not having a one-element premise. They have the same consequences as the set

$$\mathcal{N} := \{A \to B \mid A \text{ 1-closed, but not closed}, B \subseteq A^{II}\}.$$

The closure system of all sets respecting $\mathcal{N}$ is the largest lattice refinement of the original one. In fact, the family of all complete lattice refinements forms a lattice itself, and a context for this lattice can be given.

In case of a semidistributive closure system, a similar construction due to Adaricheva and Nation leads to the largest convex geometry refinement.

References

[AGT03] K. V. Adaricheva, V. A. Gorbunov, V. I. Tumanov: *Join-semidistributive lattices and convex geometries*, Advances in Mathematics **173**: pp. 1–49, 2003.

[AN03] K. V. Adaricheva, J. B. Nation: *Largest extension of a finite convex geometry*, Algebra Universalis **52**: pp. 185–195, 2004.

[EJ85] P. H. Edelmann, R. Jamison: *The theory of convex geometries*, Geom. Dedicata **19**: pp. 247–274, 1985.

[FJN91] R. Freese, J. Ježek, J. B. Nation: *Free Lattices*, Mathematical Surveys and Monographs **42**, AMS 1991.

[GW99] B. Ganter, R. Wille: *Formal Concept Analysis – Mathematical Foundations*, Springer, 1999.

[GN06] A. Gély and L. Nourine: *About the Family of Closure Systems Preserving Non-unit Implications in the Guigues-Duquenne Base*, Formal Concept Analysis, Proceedings of ICFCA06, LNAI **3874**: pp. 191–204, 2006.

[N04] J. B. Nation: *Closure operators and lattice extensions*, Order **21**: pp. 43–48, 2004.

Lattices of Relatively Axiomatizable Classes*

Dmitry E. Pal'chunov

Institute of Mathematics,
Siberian Branch of Russian Acad. Sci., 630090, Novosibirsk, Russia
palch@math.nsc.ru

Abstract. In the paper we study lattices of axiomatizable classes and relatively axiomatizable classes. This study is based on Formal Concept Analysis [4,5]. The notion of a relatively axiomatizable class is a generalization of such concepts as variety, quasivariety, $\forall$-axiomatizable class, $\exists$-axiomatizable class, Π_n^0-axiomatizable class, Σ_n^0-axiomatizable class and so on. Relatively axiomatizable classes were studied in [10]. It is proved in the paper that any finite lattice may be represented as the lattice of all relatively axiomatizable subclasses of the class of all models of a one-element signature with respect to some set of sentences. Also we prove that any finite or countable complete lattice is isomorphic to the lattice of all relatively axiomatizable subclasses of some class of models with respect to a proper set of sentences.

Keywords: lattice, axiomatizable class, relatively axiomatizable class.

1 Introduction

The main goal of the present paper is to investigate the structure of various lattices of relatively axiomatizable classes.

We consider classes of algebraic systems of finite or countable signature σ. The sets of formulas and sentences of this signature are countable. Therefore in this case the lattices of axiomatizable classes have power which is less or equal to the continuum.

We show that any finite lattice may be represented as the lattice of all relatively axiomatizable subclasses of the class of all models of the signature consisting of one binary predicate symbol with respect to some subset of the set of all sentences over the given signature. Also we prove that any finite or countable complete lattice is isomorphic to the lattice of all relatively axiomatizable subclasses of some class of models of the given signature with respect to a proper set of sentences over the given signature.

2 Preliminaries in Model Theory

The aim of this section is to introduce concepts, definitions and facts on the model theory which are necessary for understanding of the proofs below. An expert in the model theory may skip this section.

* Supported by RFBR grant N 05-01-04003-NNIO-a (DFG project COMO, GZ: 436 RUS 113/829/0-1).

S.O. Kuznetsov and S. Schmidt (Eds.): ICFCA 2007, LNAI 4390, pp. 221–239, 2007.

2.1 Semantics of First Order Predicate Logic

First we give some basic definitions of the first order predicate logic.

Definition 1. ***An algebraic system (a model)*** *is a tuple*

$$\mathfrak{A} =< A; P_1, \ldots, P_n, f_1, \ldots, f_m, c_1, \ldots, c_k >,$$

where the set A is called universe, $P_1, \ldots, P_n$ are predicates defined on the set A, $f_1, \ldots, f_m$ are functions defined on the set A and $c_1, \ldots, c_k$ are constants, i.e. names of some (distinguished) elements of the set A. Usually the universe of the algebraic system $\mathfrak{A}$ is denoted by $|\mathfrak{A}|$, i.e. $|\mathfrak{A}| := A$.

The set $\sigma =< P_1, \ldots, P_n, f_1, \ldots, f_m, c_1, \ldots, c_k >$ is called signature of the algebraic system $\mathfrak{A}$.

Definition 2. *Consider a signature $\sigma =< P_1, \ldots, P_n, f_1, \ldots, f_m, c_1, \ldots, c_k >$. We give a definition of a* **term** *of the signature σ by induction:*

1. *The constants $c_1, ..., c_k \in \sigma$ are terms and the variables $x_1, x_2, ...$ are terms.*
2. *If $t_1, \ldots, t_n$ are terms, $f \in \sigma$ and f is a symbol of function then $f(t_1, \ldots, t_n)$ is a term.*

Definition 3. *We give also an inductive definition of a* **formula** *of the signature σ:*

1. *If t_1 and t_2 are terms then $t_1 = t_2$ is a formula; if $t_1, \ldots, t_n$ are terms and $P^n \in \sigma$ is a predicate symbol then $P(t_1, \ldots, t_n)$ is a formula.*
2. *If ϕ, ψ are formulas then $(\phi \vee \psi)$, $(\phi \& \psi)$, $(\phi \to \psi)$, $\neg\phi$, $\forall x\, \phi$ and $\exists x\, \phi$ are formulas.*

Recall that an occurrence of a variable in a formula is called **free** if it does not belong to the scope of a quantifier over this variable. A variable which has at least one free occurrence in a formula is called **free variable of the formula.** Denote by $FV(\varphi)$ the set of all free variables of a formula φ. A formula having no free variables is called **sentence**.

Definition 4. *For a signature σ we denote:*
$T(\sigma) := \{t \mid t\ is\ a\ term\ of\ the\ signature\ \sigma\}$,
$F(\sigma) := \{\varphi \mid \varphi\ is\ a\ formula\ of\ the\ signature\ \sigma\}$,
$S(\sigma) := \{\varphi \mid \varphi\ is\ a\ sentence\ of\ the\ signature\ \sigma\}$ *and*
$K(\sigma) := \{\mathfrak{A} \mid \mathfrak{A}\ is\ a\ model\ of\ the\ signature\ \sigma\}$.

Remark 1. *If $\sigma_1 \subseteq \sigma_2$ then*

1. $T(\sigma_1) \subseteq T(\sigma_2)$,
2. $F(\sigma_1) \subseteq F(\sigma_2)$,
3. $S(\sigma_1) \subseteq S(\sigma_2)$,
4. *if $\sigma_1 \neq \sigma_2$, then $K(\sigma_1) \cap K(\sigma_2) = \emptyset$.*

Let us present some basic notions on cardinal numbers. Two sets have equal cardinality (are equinumerous), denoted by $||A|| = ||B||$, if there is a bijective mapping $f : A \to B$. A set A is called countable if $||A|| = ||\mathbb{N}||$, A has the power of the continuum if $||A|| = ||\mathbb{R}||$.

Proposition 1. $||\mathbb{R}|| = ||\wp(\mathbb{N})||$, *that is the cardinal number of the power set of the set of natural numbers is continuum. The cardinal number of the power set of a countable set is continuum.*

Proposition 2. *If a set A is countable, then the set $\bigcup\limits_{n\in\mathbb{N}} A^n$ of all finite words of the language A is countable too.*

Corollary 1. *The set of all formulas (as well as the set of all sentences) of finite or countable signature is countable.*

Now we give some necessary definitions and facts on model theory.

Definition 5. *(The value of a term in a model.)*

Consider a signature σ and a model $\mathfrak{A} \in K(\sigma)$. Let X be a set of variables. A mapping $\gamma : X \to |\mathfrak{A}|$ is called ***interpretation (of the variables from the set X in the model $\mathfrak{A}$).***

Let $FV(t) \subseteq X$ for a term $t \in T(\sigma)$.

We define ***the value of the term*** *t in the model $\mathfrak{A}$ under the interpretation γ (denoted by $t^{\mathfrak{A}}[\gamma]$) in the following way:*

1. *If $t = c$ then $t^{\mathfrak{A}}[\gamma] = c^{\mathfrak{A}}$; if $t = x$ then $t^{\mathfrak{A}}[\gamma] = \gamma(x)$.*
2. *If $t = f(t_1, ..., t_n)$, where $f \in \sigma$ and $t_1, ..., t_n \in T(\sigma)$, then*

$$t^{\mathfrak{A}}[\gamma] = f^{\mathfrak{A}}(t_1^{\mathfrak{A}}[\gamma], ..., t_n^{\mathfrak{A}}[\gamma]).$$

Definition 6. *(The truth value of a formula in a model)*

Consider a signature σ and a model $\mathfrak{A} \in K(\sigma)$. Let $\varphi \in F(\sigma)$, $FV(\varphi) \subseteq X$ and a mapping $\gamma : X \to |\mathfrak{A}|$ be an ***interpretation.***

We denote $\mathfrak{A} \models \varphi[\gamma]$ if the formula φ is true in the model $\mathfrak{A}$ under the interpretation γ and denote $\mathfrak{A} \not\models \varphi[\gamma]$ if φ is false.

We define the relation $\mathfrak{A} \models \varphi[\gamma]$ by induction in the following way:

$\mathfrak{A} \models (t_1 = t_2)[\gamma] \Leftrightarrow \mathfrak{A} \models (t_1^{\mathfrak{A}}[\gamma] = t_2^{\mathfrak{A}}[\gamma])$;

$\mathfrak{A} \models P(t_1, ..., t_n)[\gamma] \Leftrightarrow \mathfrak{A} \models P^{\mathfrak{A}}(t_1^{\mathfrak{A}}[\gamma], ..., t_n^{\mathfrak{A}}[\gamma])$;

$\mathfrak{A} \models (\varphi_1 \vee \varphi_2)[\gamma] \Leftrightarrow \mathfrak{A} \models \varphi_1[\gamma]$ *or* $\mathfrak{A} \models \varphi_2[\gamma]$;

$\mathfrak{A} \models (\varphi_1 \& \varphi_2)[\gamma] \Leftrightarrow \mathfrak{A} \models \varphi_1[\gamma]$ *and* $\mathfrak{A} \models \varphi_2[\gamma]$;

$\mathfrak{A} \models (\varphi_1 \to \varphi_2)[\gamma] \Leftrightarrow \mathfrak{A} \not\models \varphi_1[\gamma]$ *or* $\mathfrak{A} \models \varphi_2[\gamma]$;

$\mathfrak{A} \models \neg\varphi[\gamma] \Leftrightarrow \mathfrak{A} \not\models \varphi$;

$\mathfrak{A} \models \forall x \varphi(x)[\gamma] \Leftrightarrow \mathfrak{A} \models \varphi(a)[\gamma]$ *for any* $a \in \mathfrak{A}$;

$\mathfrak{A} \models \exists x \varphi(x)[\gamma] \Leftrightarrow$ *there exists* $a \in \mathfrak{A}$ *such that* $\mathfrak{A} \models \varphi(a)[\gamma]$.

A formula is called **identically true**, if it is true in any model under any interpretation, it is called **identically false**, if it is false in any model under any interpretation.

Remark 2. *A formula φ is identically true if and only if the formula $\neg\varphi$ is identically false.*

Definition 7. *We say that formulas φ and ψ are* **semantically equivalent** *(and denote $\varphi \sim \psi$) if*

$$\mathfrak{A} \models \varphi[\gamma] \;\Leftrightarrow\; \mathfrak{A} \models \psi[\gamma].$$

for any model $\mathfrak{A} \in K(\sigma(\varphi) \cup \sigma(\psi))$ and any interpretation $\gamma : FV(\varphi) \cup FV(\psi) \to |\mathfrak{A}|$.

2.2 Predicate Calculus

Above we have presented a semantical approach to logic of predicates, which is called Tarski-style truth definition.

The predicate calculus is a syntactic approach to first order logic. Below we give some necessary definitions and statements.

Definition 8. *A sequence of formulas $\varphi_1, ..., \varphi_n$ is called* **proof** *if for any $i \leq n$ the formula φ_i either is an axiom, or is derived from the previous ones using a rule of inference.*

A formula φ is called **provable** *if there exists a proof $\varphi_1, ..., \varphi_n = \varphi$.*

The following statements show the equivalence between syntactical and semantical approaches.

Theorem 1. *(Soundness Theorem) If a formula is provable then it is identically true.*

Theorem 2. *(Goedel's Completeness Theorem) Any identically true formula is provable.*

Corollary 2. *A formula φ is provable if and only if φ it is identically true.*

Definition 9. *Let $\Gamma \subseteq F(\sigma)$ and $\varphi \in F(\sigma)$ for a signature σ. We denote:*

1. *$\Gamma \vdash \varphi$ if there exist formulas $\varphi_1, ..., \varphi_n \in \Gamma$ such that the formula $(\varphi_1 \& ... \& \varphi_n \to \varphi)$ is provable;*
2. *$\Gamma \vdash$ if there exist formulas $\varphi_1, ..., \varphi_n \in \Gamma$ such that the formula $\neg(\varphi_1 \& ... \& \varphi_n)$ is provable; in this case we say that the set of formulas Γ is* **inconsistent;**
3. *$\Gamma \nvdash$ if the set Γ is not inconsistent; we say that such set Γ is* **consistent.**

Remark 3. *For any set of sentences $T \subseteq S(\sigma)$ the following statements are equivalent:*

1. *The set T is inconsistent.*
2. *$T \vdash \varphi$ holds for any $\varphi \in S(\sigma)$.*
3. *There is $\varphi \in S(\sigma)$ such that $T \vdash \varphi$ and $T \vdash \neg\varphi$.*

Definition 10. *Let $\Gamma \subseteq F(\sigma)$ and $T \subseteq S(\sigma)$. The set formulas Γ* ***is satisfiable*** *if there is a model $\mathfrak{A} \in K(\sigma)$ and an interpretation $\gamma : FV(\Gamma) \to |\mathfrak{A}|$ such that $\mathfrak{A} \models \Gamma[\gamma]$. We denote $\mathfrak{A} \models \Gamma[\gamma]$ if*

$$\mathfrak{A} \models \varphi[\gamma] \;\; for \; any \; formula \;\; \varphi \in \Gamma.$$

We say that $\mathfrak{A}$ is a ***model of the set of sentences*** *T and denote $\mathfrak{A} \models T$ if*

$$\mathfrak{A} \models \varphi \;\; for \; any \;\; \varphi \in T.$$

Theorem 3. *(Henkin's Model Existence Theorem) Any consistent set of formulas is satisfiable, i. e. has a model:*

if $\Gamma \nvdash$ then there exist $\mathfrak{A} \in K(\sigma(\Gamma))$ and $\gamma : FV(\Gamma) \to |\mathfrak{A}|$ such that $\mathfrak{A} \models \Gamma\,[\gamma]$.

For the further consideration we need the notion of first order theory.

Definition 11. *A set of sentences $T \subseteq S(\sigma)$ is called* ***(first order) theory*** *of a signature σ if*

$$T \vdash \varphi \;\; implies \;\; \varphi \in T \;\; for \; any \;\; \varphi \in S(\sigma).$$

i. e. the set of sentence T is deductively closed.

Remark 4. *A set of sentences $T \subseteq S(\sigma)$ is a theory of a signature σ if and only if*

$$T \vdash \varphi \;\; is \; equivalent \; to \;\; \varphi \in T \;\; for \; any \;\; \varphi \in S(\sigma).$$

Corollary 3. *Let T be a theory of a signature σ. Then $T \vdash$ if and only if $T = S(\sigma)$.*

Recall that for a set of formulas (or sentences) Γ we denote by $\sigma(T)$ the set of all signature symbols occurring in Γ.

Remark 5. *Let T be a theory, $\sigma = \sigma(T)$, $\sigma \subseteq \sigma_1$ and $\sigma \neq \sigma_1$. Then T is not a theory of the signature σ_1.*

Definition 12. *For a model $\mathfrak{A} \in K(\sigma)$ the set*

$$Th(\mathfrak{A}) := \{\varphi \mid \varphi \in S(\sigma) \;\; and \;\; \mathfrak{A} \models \varphi\}$$

is called ***elementary theory*** *of the model $\mathfrak{A}$.*

Models $\mathfrak{A}, \mathfrak{B} \in K(\sigma)$ are called ***elementary equivalent*** *(denoted by $\mathfrak{A} \equiv \mathfrak{B}$) if $Th(\mathfrak{A}) = Th(\mathfrak{B})$, that is*

$$\mathfrak{A} \models \varphi \;\; iff \;\; \mathfrak{B} \models \varphi \;\; for \; any \;\; \varphi \in S(\sigma).$$

Thus elementary equivalent models are indistinguishable by first order predicate logic.

Roughly speaking, algebra studies algebraic systems up to isomorphism as well as logic investigates these systems up to elementary equivalence.

Remark 6. *The elementary theory of a model $\mathfrak{A}$ is a consistent theory of the signature $\sigma(\mathfrak{A})$.*

We have presented almost all necessary definitions on model theory. Further information on first order predicate logic and model theory can be found in [2,3]. For definitions and results in lattice theory we refer to [1,8,4].

3 Axiomatizable Classes

In this section we discuss the structure of lattices of axiomatizable classes.

3.1 Lattices of Axiomatizable Classes

First we give some necessary definitions.

Definition 13. *Let $K \subseteq K(\sigma)$. For a sentence $\varphi \in S(\sigma)$ we denote $K \models \varphi$ if*

$$\mathfrak{A} \models \varphi \ \ for\ any\ \ \mathfrak{A} \in K.$$

For a set of sentences $\Gamma \subseteq S(\sigma)$ we denote $K \models \Gamma$ if

$$\mathfrak{A} \models \varphi \ \ for\ any\ \ \mathfrak{A} \in K \ \ and\ \ \varphi \in \Gamma.$$

The set of sentences

$$ThK := \{\varphi \in S(\sigma) \mid \ K \models \varphi\}$$

is called ***theory of the class*** *K.*
For a set of sentences $\Gamma \subseteq S(\sigma)$ the class

$$K(\Gamma) := K_\sigma(\Gamma) := \{\mathfrak{A} \in K(\sigma) \mid \mathfrak{A} \models \varphi \ \ for\ any\ \ \varphi \in \Gamma\}$$

is called ***class axiomatized by Γ (in the signature σ).***

Remark 7. *A set of sentences Γ axiomatizes different classes if different signatures are treated. It means that if $\Gamma \subseteq S(\sigma_1)$, $\Gamma \subseteq S(\sigma_2)$ and $\sigma_1 \neq \sigma_2$ then $K_{\sigma_1}(\Gamma) \neq K_{\sigma_2}(\Gamma)$. Moreover, clearly if $\sigma_1 \neq \sigma_2$ then $K_{\sigma_1}(\Gamma) \cap K_{\sigma_2}(\Gamma) = \emptyset$.*

However the notation $K(\Gamma)$ is often used instead of $K_\sigma(\Gamma)$ if signature σ is fixed or is not important.

Definition 14. *A class K is called* ***axiomatizable*** *if $K = K(\Gamma)$ for some set of sentences Γ (that is $K = K_\sigma(\Gamma)$ for a proper signature σ).*
In this case Γ is called ***set of axioms*** *for the class K.*

Proposition 3. *Each axiomatizable class is closed with respect to elementary equivalence, i.e. if K is an axiomatizable class, $\mathfrak{A} \in K$ and $\mathfrak{B} \equiv \mathfrak{A}$ then $\mathfrak{B} \in K$.*

Proposition 4. *$K \subseteq K(Th(K))$.*

Proposition 5. *If $K = K(\Gamma)$ then $\Gamma \subseteq ThK$, i.e. $\Gamma \subseteq ThK(\Gamma)$*

Proposition 6. *A class K is axiomatizable if and only if $K = K(Th(K))$.*

Corollary 4. *For any axiomatizable class K there is the largest (w.r.t. inclusion) set of axioms, which is exactly $Th(K)$.*

Proposition 7. *A set of sentences* Γ *is a theory if and only if* $\Gamma = ThK(\Gamma)$.

Corollary 5. *The mappings* $K \to ThK$ *and* $T \to K(T)$ *define a bijection between axiomatizable classes and theories (of the same signatures).*

Proposition 8. *Let* $\Gamma_1, \Gamma_2 \subseteq S(\sigma)$ *and* $K_1, K_2 \subseteq K(\sigma)$.

1. *If* $\Gamma_1 \subseteq \Gamma_2$ *then* $K(\Gamma_2) \subseteq K(\Gamma_1)$.
2. *If* $K_1 \subseteq K_2$ *then* $Th(K_2) \subseteq Th(K_1)$.

Corollary 6. *The pair of mappings* $K \to ThK$ *and* $\Gamma \to K(\Gamma)$ *is a* ***Galois connection*** *between subsets of* $K(\sigma)$ *and subsets of* $S(\sigma)$*, ordered by inclusion.*

Remark 8. *In the general case the following statements do not hold:*

1. $K = K(Th(K))$.
2. $\Gamma = Th(K(\Gamma))$.

Corollary 7. *There are classes which are not axiomatizable.*

We will present simple examples of non-axiomatizable classes. The following theorem is very useful for proving that classes are not axiomatizable.

Theorem 4. *(Maltsev's Compactness Theorem) A set of sentences* Γ *is satisfiable if and only if it is locally satisfiable, i.e. every finite subset of* Γ *is satisfiable.*

The next remark gives an example of application of the Maltsev's Compactness Theorem.

Remark 9.
1. *The class* $K_f(\sigma)$ *of the finite models of a signature* σ *is not axiomatizable.*
2. *Suppose that a class* K *consists of finite models and for any natural number* n *there exists a model from* K *which has more than* n *elements. Then the class* K *is not axiomatizable.*

Proof. Statement (1) is an immediate consequence of statement (2). Hence it is sufficient to prove (2).

Suppose that a class $K \subseteq K(\sigma)$ is axiomatizable, so $K = K(\Gamma)$ for a proper set of sentences $\Gamma \subseteq S(\sigma)$. For each natural number n consider a sentence $\varphi_n \in S(\emptyset)$ which holds true in a model iff it has at least n elements. Denote $\Delta := \{ \varphi_n \mid n \in \mathbb{N} \}$ and $\Lambda = \Gamma \cup \Delta$. It is not difficult to prove that every finite subset of Λ is satisfiable. Hence, by the Maltsev's Compactness Theorem, the set Λ is satisfiable itself. Then there is a model $\mathfrak{A} \in K(\sigma)$ such that $\mathfrak{A} \models \Lambda$. Therefore, $\mathfrak{A} \models \Gamma$ and $\mathfrak{A} \models \Delta$, so $\mathfrak{A}$ belongs to the class K and $\mathfrak{A}$ is infinite, which contradicts our assumption.

Thus the class K is not axiomatizable.

Corollary 8. *Classes of finite groups, finite rings and finite fields are not axiomatizable.*

Let us consider a question: what is a structure of the set of all axiomatizable classes? First, is it a lattice?

Proposition 9. *If classes K_1 and K_2 are axiomatizable then the classes $K_1 \cap K_2$ and $K_1 \cup K_2$ are axiomatizable as well.*

Proof. Assume that $K_1 = K(\Gamma_1)$ and $K_2 = K(\Gamma_2)$. Denote $\Gamma' := \Gamma_1 \cup \Gamma_2$ and $\Gamma'' := \{\varphi \vee \psi \mid \varphi \in \Gamma_1 \ \ and \ \ \psi \in \Gamma_2\}$. Then $K_1 \cap K_2 = K(\Gamma')$ and $K_1 \cup K_2 = K(\Gamma'')$. Thus the classes $K_1 \cap K_2$ and $K_1 \cup K_2$ are axiomatizable.

Corollary 9. 1. *For any signature the set of all axiomatizable classes of this signature is a distributive lattice (w.r.t. inclusion).*
2. *For any class K of algebraic systems the set of all axiomatizable subclasses of K is a distributive lattice.*

3.2 Finitely Axiomatizable Classes

Now we pass on to a very important special case of axiomatizable classes — finitely axiomatizable classes.

Definition 15. *A class K is called finitely axiomatizable if there exists a finite set of sentences Γ such that $K = K(\Gamma)$.*

Remark 10. 1. *A class K is finitely axiomatizable if and only if $K = K(\{\varphi\})$ for a proper sentence $\varphi \in S(\sigma)$.*
2. *If $K = K(\{\varphi\})$ then $\overline{K} = K(\neg\varphi)$, where $\overline{K} := K(\sigma) \setminus K$.*
3. *A class K is finitely axiomatizable if and only if the class $\overline{K}$ is finitely axiomatizable.*

Theorem 5. *A class K is finitely axiomatizable if and only if the classes K and $\overline{K}$ are axiomatizable.*

Corollary 10. *Classes of infinite groups, infinite rings and infinite fields are not finitely axiomatizable.*

The following remark is obviously true.

Remark 11. *If classes K_1 and K_2 are finitely axiomatizable then the classes $K_1 \cap K_2$ and $K_1 \cup K_2$ are finitely axiomatizable too.*

Corollary 11. 1. *For any signature the set of all finitely axiomatizable classes of this signature forms a distributive lattice.*
2. *For any class K the set of all finitely axiomatizable subclasses of K is a distributive lattice.*

4 Classes Axiomatizable by Formulas of Special Kinds

In this section we consider axiomatizable classes having axioms from some restricted sets of sentences. Results on different restrictions on sets of axioms are given in [12].

4.1 Universally and Existentially Axiomatizable Classes

First we consider universally and existentially axiomatizable classes. These classes play an important role in the model theory.

Definition 16. *If a formula $\psi(\overline{x}, \overline{y}) \in F(\sigma)$ is quantifier-free, $\overline{x} = (x_1, ..., x_n)$ and $\overline{y} = (y_1, ..., y_m)$, then*
the formula $\exists x_1 ... \exists x_n \psi(\overline{x}, \overline{y})$ is called ***∃-formula*** *and*
the formula $\forall x_1 ... \forall x_n \psi(\overline{x}, \overline{y})$ is called ***∀-formula****.*
A sentence φ is called ***∃-sentence*** *if it is an ∃-formula*
and it is called ***∀-sentence*** *if it is an ∀-formula.*
A class K is called ∃-axiomatizable if $K = K(\Gamma)$ for some $\Gamma \subseteq S(\sigma)$ such that φ is an ∃-sentence for any $\varphi \in \Gamma$.
A class K is called ∀-axiomatizable if $K = K(\Gamma)$ for a certain $\Gamma \subseteq S(\sigma)$ such that φ is an ∀-sentence for any $\varphi \in \Gamma$.

Definition 17. *For a class $K \subseteq K(\sigma)$ the set of sentences*

$$Th_{\exists}(K) := \{\, \varphi \in ThK \mid \varphi \text{ is an } \exists\text{-sentence} \,\} =$$

$$= \{\, \varphi \in S(\sigma) \mid K \models \varphi \text{ and } \varphi \text{ is an } \exists\text{-sentence} \,\}$$

is called ***∃-theory of*** *K and the set of sentences*

$$Th_{\forall}(K) := \{\, \varphi \in ThK \mid \varphi \text{ is an } \forall\text{-sentence} \,\} =$$

$$= \{\, \varphi \in S(\sigma) \mid K \models \varphi \text{ and } \varphi \text{ is an } \forall\text{-sentence} \,\}$$

is ***∀-theory of*** *K.*

∃-theory $Th_{\exists}(K)$ and ∀-theory $Th_{\forall}(K)$ of a class K have quite similar properties as the theory $Th(K)$ of the class K has.

Proposition 10. *1. $K \subseteq K(Th_{\forall}(K))$.*
2. $K \subseteq K(Th_{\exists}(K))$.

Proposition 11. *1. A class K is ∀-axiomatizable if and only if $K = K(Th_{\forall}(K))$.*
2. A class K is ∃-axiomatizable if and only if $K = K(Th_{\exists}(K))$.

The ∀-axiomatizable and ∃-axiomatizable classes have a good algebraic description in terms of subsystems and supersystems. Denote $\mathfrak{A} \leq \mathfrak{B}$ if an algebraic system $\mathfrak{A}$ is a subsystem of $\mathfrak{B}$. In this case we say that the system $\mathfrak{B}$ is a supersystem of the system $\mathfrak{A}$.

Proposition 12. *Let $\mathfrak{A}, \mathfrak{B} \in K(\sigma)$, $\mathfrak{A} \leq \mathfrak{B}$, $a_1, ..., a_n \in |\mathfrak{A}|$ and $\varphi(x_1, ..., x_n)$ be a quantifier-free formula of a signature σ. Then*

$$\mathfrak{A} \models \varphi(a_1, ..., a_n) \text{ if and only if } \mathfrak{B} \models \varphi(a_1, ..., a_n).$$

Thus the quantifier-free formulas keep truth values when we go from a system to a subsystem and back.

Definition 18. *We say that a class K* **is closed under subsystems** *if it contains all subsystems of each system from K, i.e. $\mathfrak{B} \in K$ for any $\mathfrak{A} \in K$ and $\mathfrak{B} \leq \mathfrak{A}$.*

We say that a class K **is closed under supersystems** *if it contains all supersystems of each system from K, i.e. $\mathfrak{B} \in K$ for any $\mathfrak{A} \in K$ and $\mathfrak{B} \geq \mathfrak{A}$.*

Now we present an algebraic description of $\forall$-axiomatizable and $\exists$-axiomatizable classes.

Theorem 6. *Let a class K be axiomatizable.*

1. *K is $\forall$-axiomatizable if and only if K is closed under subsystems.*
2. *K is $\exists$-axiomatizable if and only if K is closed under supersystems.*

4.2 Σ_n^0- and Π_n^0-Axiomatizable Classes

Before studying lattices of $\exists$-axiomatizable and $\forall$-axiomatizable classes we present a very important generalization of concepts of $\exists$-formula and $\forall$-formula — notions of Σ_n^0-formula and Π_n^0-formula respectively. These notions are useful in model theory [2] and computability theory [6]. A similar notion of equivalence modulo sentences with fixed number of quantifiers was studied in [11].

Definition 19. *We define Σ_n^0-formulas and Π_n^0-formulas by induction over n.*

1. *A quantifier-free formula is a Σ_0^0-formula and a Π_0^0-formula.*
2. *If ψ is a Π_n^0-formula then ψ itself and $\varphi = \exists x_1 ... \exists x_m \psi$ are Σ_{n+1}^0-formulas.*
3. *If ψ is a Σ_n^0-formula then ψ itself and $\varphi = \forall x_1 ... \forall x_m \psi$ are Π_{n+1}^0-formulas.*

Denote

$$\Sigma_n := \{ \varphi \in S(\sigma) \mid \textit{the sentence } \varphi \textit{ is a } \Sigma_n^0\textit{-formula} \} \textit{ and}$$

$$\Pi_n := \{ \varphi \in S(\sigma) \mid \textit{the sentence } \varphi \textit{ is a } \Pi_n^0\textit{-formula} \}.$$

The following remark shows that $\exists$-formulas and $\forall$-formulas are special cases of Σ_n^0-formulas and Π_n^0-formulas respectively.

Remark 12. *1. $\Sigma_1 = \{ \varphi \mid \varphi$ is a $\exists$-sentence $\}$, $\Pi_1 = \{ \varphi \mid \varphi$ is a $\forall$-sentence $\}$.*
2. $\{ \varphi \mid \varphi$ is a $\exists$-formula $\} = \{ \varphi \mid \varphi$ is a Σ_1^0-formula $\}$,
$\{ \varphi \mid \varphi$ is a $\forall$-formula $\} = \{ \varphi \mid \varphi$ is a Π_1^0-formula $\}$.
3. A class K is $\forall$-axiomatizable if and only if K is Π_1-axiomatizable.
A class K is $\exists$-axiomatizable if and only if K is Σ_1-axiomatizable.

Further we will treat all types of axiomatizable classes mentioned above together. To do it we introduce a general notion of Δ-axiomatizable class.

Definition 20. *A class K is called Δ-axiomatizable if $K = K(\Gamma)$ and $\Gamma \subseteq \Delta$.*

Definition 21. *We say that a set $\Delta \subseteq S(\sigma)$ is closed under disjunction modulo equivalence (of formulas) if for any $\varphi, \psi \in \Delta$ there is a sentence $\xi \in \Delta$ such that $\xi \sim (\varphi \vee \psi)$.*

Proposition 13. *Suppose that a set of sentences Δ is closed under disjunction modulo equivalence. Then finite unions and intersections of Δ-axiomatizable classes are Δ-axiomatizable: if classes K_1 and K_2 are Δ-axiomatizable then the classes $K_1 \cap K_2$ and $K_1 \cup K_2$ are Δ-axiomatizable too.*

Proof. Assume that $\Delta \subseteq S(\sigma)$, $\Gamma_1, \Gamma_2 \subseteq \Delta$, $K_1, K_2 \subseteq K(\sigma)$, $K_1 = K(\Gamma_1)$ and $K_2 = K(\Gamma_2)$. Denote $\Gamma' := \Gamma_1 \cup \Gamma_2$ and

$$\Gamma'' := \{\, \xi \in \Delta \mid \xi \sim (\varphi \vee \psi) \textit{ for some } \varphi \in \Gamma_1 \textit{ and } \psi \in \Gamma_2 \}.$$

Then $K_1 \cap K_2 = K(\Gamma')$ and $K_1 \cup K_2 = K(\Gamma'')$, moreover, $\Gamma' \subseteq \Delta$ and $\Gamma'' \subseteq \Delta$. Therefore, the classes $K_1 \cap K_2$ and $K_1 \cup K_2$ are Δ-axiomatizable.

Corollary 12. *If a set of sentences Δ is closed under disjunction modulo equivalence then the Δ-axiomatizable classes form a distributive lattice.*

Proposition 14. *For any n the sets Σ_n and Π_n are closed under disjunction modulo equivalence.*

Proof. Consider sentences $\varphi, \psi \in \Sigma_n$. First we rename variables of φ such that the resulting sentence $\varphi_1 \sim \varphi$ would not have common variables with ψ. After that we move all existential quantifiers outside the disjunction $(\varphi_1 \vee \psi)$, which may be moved without touching universal quantifiers. Then we move all universal quantifiers outside the disjunction, not touching existential quantifiers. Next we move all existential quantifiers outside the disjunction again, and so on. After a finite number of such steps we obtain a Σ_n^0-sentence ξ which is equivalent to the initial disjunction: $\xi \sim (\varphi \vee \psi)$. So $\xi \in \Sigma_n$. Therefore, the class Σ_n is closed under disjunction modulo equivalence.

The fact that the class Π_n is closed under disjunction modulo equivalence may be proved in a similar way.

Corollary 13. *For any n the set of all Σ_n-axiomatizable classes, as well as the set of all Π_n-axiomatizable classes, is a distributive lattice.*

4.3 Varieties and Quasivarieties

In this section we consider lattices of varieties and quasivarieties. A variety (quasivariety) is the class of all algebraic structures of a given signature satisfying a given set of identities (respectively, quasiidentities).

Definition 22. *(**Atomic formulas** of the signature σ)*

1. *If t_1 and t_2 are terms then $t_1 = t_2$ is an atomic formula.*
2. *If $t_1, \ldots, t_n$ are terms and $P^n \in \sigma$ is a predicate symbol then $P(t_1, \ldots, t_n)$ is an atomic formula.*

Definition 23. *Let $\psi, \varphi_1, ..., \varphi_n$ be atomic formulas.*

1. *A sentence of the form $\forall x_1 ... \forall x_n(\varphi_1 \& ... \& \varphi_n)$ is called* ***identity****.*
2. *A sentence of the form $\forall x_1 ... \forall x_n((\varphi_1 \& ... \& \varphi_n) \to \psi)$ is called* ***quasiidentity****.*

Definition 24. *1. A class V is called* ***variety*** *if $V = K(\Gamma)$ for some set of identities Γ.*
2. A class Q is called ***quasivariety*** *if $Q = K(\Gamma)$ for some set of quasiidentities Γ.*

Definition 25. *Let $K \subseteq K(\sigma)$.*

1. *A class $V \subseteq K$ is called* ***subvariety*** *of K if $V = \{\mathfrak{A} \in K \mid \mathfrak{A} \models \Gamma\}$ for some set of identities Γ.*
2. *A class $Q \subseteq K$ is called* ***subquasivariety*** *of K if $Q = \{\mathfrak{A} \in K \mid \mathfrak{A} \models \Gamma\}$ for some set of quasiidentities Γ.*

Lattices of varieties and quasivarieties are very complicated. In the general case these lattices are not distributive.

Problem. (*G.Birkhoff, 1945, A.I.Maltsev, 1966) To describe lattices of subquasivarieties for different quasivarieties.*

Many results on lattices of varieties and quasivarieties are presented in [7,8,9].

5 Relatively Axiomatizable Classes

In this section we pass on to the main subject of the paper, the lattices of relatively axiomatizable classes, which were studied in [10]. In this paper the study of axiomatizable classes is based on Formal Concept Analysis [4,5].

5.1 Relatively Axiomatizable Classes and FCA

Definition 26. *Let $K_0, K_1 \subseteq K(\sigma)$ and $\Delta \subseteq S(\sigma)$. We say that the class K_0 is axiomatizable in the class K_1 relatively to the set of sentences Δ if there exists a set $\Gamma \subseteq \Delta$ such that $K_0 = \{\mathfrak{A} \in K_1 \mid \mathfrak{A} \models \Gamma\}$.*

Recall that a class $K \subseteq K(\sigma)$ is axiomatizable if there exists a set $\Gamma \subseteq S(\sigma)$ such that $K = \{\mathfrak{A} \in K(\sigma) \mid \mathfrak{A} \models \Gamma\}$. The notion of relatively axiomatizable class is a generalization of the concept of axiomatizable class as well as of such concepts as Σ_n-axiomatizable class, Π_n-axiomatizable class, variety and quasivariety.

Remark 13. *Let $K \subseteq K(\sigma)$. The class K is axiomatizable if and only if K is axiomatizable in the class $K_1 = K(\sigma)$ relatively to the set of sentences $\Delta = S(\sigma)$.*

Remark 14. *Let $K \subseteq K(\sigma)$.*

1. *The class K is a variety if and only if K is axiomatizable in the class $K(\sigma)$ relatively to the set of sentences $\Delta = \{\varphi \in S(\sigma) \mid \varphi$ is an identity$\}$.*
2. *The class K is a quasivariety if and only if K is axiomatizable in the class $K(\sigma)$ relatively to the set of sentences $\Delta = \{\varphi \in S(\sigma) \mid \varphi$ is a quasiidentity$\}$.*
3. *The class K is $\forall$-axiomatizable if and only if K is axiomatizable in the class $K(\sigma)$ relatively to the set of sentences $\Delta = \{\varphi \in S(\sigma) \mid \varphi$ is an $\forall$-formula, i.e. $\varphi = \forall x_1 ... \forall x_n \psi(\overline{x})$, where ψ is quantifier-free$\}$.*
4. *The class K is $\exists$-axiomatizable if and only if K is axiomatizable in the class $K(\sigma)$ relatively to the set of sentences $\Delta = \{\varphi \in S(\sigma) \mid \varphi$ is an $\exists$-formula, i.e. $\varphi = \exists x_1 ... \exists x_n \psi(\overline{x})$, where ψ is quantifier-free$\}$.*
5. *The class K is Π_n-axiomatizable if and only if K is axiomatizable in the class $K(\sigma)$ relatively to the set of sentences $\Delta = \Pi_n$.*
6. *The class K is Σ_n-axiomatizable if and only if K is axiomatizable in the class $K(\sigma)$ relatively to the set of sentences $\Delta = \Sigma_n$.*

Definition 27. *For $K \subseteq K(\sigma)$ and $\Delta \subseteq S(\sigma)$ we denote*

$$\mathbb{A}(K, \Delta) := \{K_0 \mid K_0 \text{ is axiomatizable in } K \text{ relatively to } \Delta\} \text{ and}$$

$$Th_\Delta(K) := \{\varphi \in \Delta \mid K \models \varphi\}.$$

We consider $\mathbb{A}(K, \Delta)$ as a set ordered by inclusion $\subseteq$.

Remark 15. *1. A class K is Δ-axiomatizable if and only if $K = K(Th_\Delta(K))$.*
2. For $K, K' \in K(\sigma)$ we have $K \in \mathbb{A}(K', \Delta)$ if and only if

$$K = \{\mathfrak{A} \in K' \mid \mathfrak{A} \models Th_\Delta(K)\}.$$

For each class $K \subseteq K(\sigma)$ and set $\Delta \subseteq S(\sigma)$ we consider the formal context

$$(K, \Delta, \models),$$

with derivation operator denoted by $()'$ as usual [4].

The following statement is an immediate consequence of the definitions of formal context and formal concept [4].

Remark 16. *Let $K \subseteq K(\sigma)$ and $\Delta \subseteq S(\sigma)$.*

1. *If $A \subseteq K$ then $A' = Th_\Delta(A)$.*
2. *If $B \subseteq \Delta$ then $B' = \{\mathfrak{C} \in K \mid \mathfrak{C} \models B\}$.*

Recall that for a formal context (G, M, I) by $\underline{\mathfrak{B}}(G, M, I)$ we denote the lattice of all formal concepts of the formal context (G, M, I).

Theorem 7. *Let $K \subseteq K(\sigma)$ and $\Delta \subseteq S(\sigma)$. Then $K_0 \in \mathbb{A}(K, \Delta)$ if and only if*

$$(K_0, Th_\Delta K_0) \in |\underline{\mathfrak{B}}(K, \Delta, \models)|.$$

Proof. ($\Rightarrow$) Suppose that $K_0 \in \mathbb{A}(K,\Delta)$.

Then there exists $\Gamma \subseteq \Delta$ such that $K_0 = \{\mathfrak{A} \in K \mid \mathfrak{A} \models \Gamma\}$. Then $\Gamma \subseteq Th_\Delta K_0$.

By the definition of $Th_\Delta K_0$ we have $\mathfrak{A} \models \varphi$ for any $\mathfrak{A} \in K_0$ and $\varphi \in Th_\Delta K_0$.

Suppose that $\mathfrak{A} \in K$ and $\mathfrak{A} \models \varphi$ for any $\varphi \in Th_\Delta K_0$. We have $\Gamma \subseteq Th_\Delta K_0$, so $\mathfrak{A} \models \Gamma$ and then $\mathfrak{A} \in K_0$. Thus $K_0 = (Th_\Delta K_0)'$.

Assume that $\varphi \in \Delta$ and $\mathfrak{A} \models \varphi$ for any $\mathfrak{A} \in K_0$. Then $\varphi \in Th_\Delta K_0$, so $Th_\Delta K_0 = (K_0)'$. Therefore, $(K_0, Th_\Delta K_0) \in \underline{\mathfrak{B}}(K,\Delta,\models)$.

($\Leftarrow$) Let $(K_0, Th_\Delta K_0) \in \underline{\mathfrak{B}}(K,\Delta,\models)$. Then $K_0 \subseteq K$. We have $K_0 = (Th_\Delta K_0)'$, hence $K_0 = \{\mathfrak{A} \mid \mathfrak{A} \models Th_\Delta K_0\}$. By definition, $Th_\Delta K_0 \subseteq \Delta$, so $K_0 \in \mathbb{A}(K,\Delta)$.

Theorem 7 is proved.

Corollary 14. *Let $K \subseteq K(\sigma)$ and $\Delta \subseteq S(\sigma)$. Then $K_0 \in \mathbb{A}(K,\Delta)$ if and only if $(K_0,\Gamma) \in \underline{\mathfrak{B}}(K,\Delta,\models)$ for some $\Gamma \subseteq \Delta$.*

Proof. ($\Rightarrow$) is an obvious consequence of Theorem 7.

($\Leftarrow$) Let $(K_0,\Gamma) \in \underline{\mathfrak{B}}(K,\Delta,\models)$ and $\Gamma \subseteq \Delta$. Then $\Gamma = \{\varphi \in \Delta \mid \mathfrak{A} \models \varphi \textit{ for any } \mathfrak{A} \in K_0\}$, so $\Gamma = Th_\Delta K_0$. Hence $(K_0, Th_\Delta K_0) \in \underline{\mathfrak{B}}(K,\Delta,\models)$ and $K_0 \in \mathbb{A}(K,\Delta)$, which completes the proof of Corollary 14.

Corollary 15. $|\underline{\mathfrak{B}}(K,\Delta,\models)| = \{(K_0, Th_\Delta K_0) \mid K_0 \in \mathbb{A}(K,\Delta)\}$.

Therefore, the classes which are axiomatizable in a class K relatively to a set of sentences Δ are exactly extents of the formal concepts of the formal context $(K,\Delta,\models)$.

Corollary 16. *Let $K \subseteq K(\sigma)$, $\Delta \subseteq S(\sigma)$, $A \subseteq K$ and $B \subseteq \Delta$.*

1. *$A = A''$ if and only if A is axiomatizable in the class K relatively to the set of sentences Δ.*
2. *$B = B''$ if and only if $B = Th_\Delta(\{\mathfrak{C} \in K \mid \mathfrak{C} \models B\})$.*

This Corollary is similar to Propositions 6 and 7.

Corollary 17. *$\mathbb{A}(K,\Delta) \cong \underline{\mathfrak{B}}(K,\Delta,\models)$ as ordered sets for any $K \subseteq K(\sigma)$ and $\Delta \subseteq S(\sigma)$.*

Therefore, using properties of lattices of formal concepts [4] we have

Proposition 15. *For $K \subseteq K(\sigma)$ and $\Delta \subseteq S(\sigma)$ the set $\mathbb{A}(K,\Delta)$ is a complete lattice.*

Corollary 18. *Let us fix a signature σ.*

1. *If Δ is closed under disjunction modulo equivalence then the set of all Δ-axiomatizable classes is a distributive complete lattice.*
2. *For any n the set of all Σ_n-axiomatizable classes is a distributive complete lattice.*
3. *For any n the set of all Π_n-axiomatizable classes is a distributive complete lattice.*

5.2 Lattices of Relatively Axiomatizable Classes

For our further consideration we fix a signature $\sigma = \{R^2\}$. It means that σ consists of one symbol of a binary predicate.

Recall that for two formal contexts $C_1 = (G_1, M_1, I_1)$ and $C_2 = (G_2, M_2, I_2)$ we say that C_1 is isomorphic to C_2 and denote $C_1 \cong C_2$ if there exist bijective mappings $f : G_1 \to G_2$ and $h : M_1 \to M_2$ such that gI_1m is equivalent to $f(g)I_2h(m)$ for any $g \in G_1$ and $m \in M_1$.

Proposition 16. *Consider a formal context $C = (G, M, I)$ such that the cardinality of the set G is less or equal to continuum and the cardinality of the set M is less or equal to countable. There exist $K \subseteq K(\sigma)$ and $\Delta \subseteq S(\sigma)$ such that $C \cong (K, \Delta, \models)$.*

Proof. Let $C = (G, M, I)$ be an arbitrary formal context with cardinality conditions mentioned above. Consider the class of all ordered sets POS in the signature σ. For convenience for any $\mathfrak{A} \in POS$ and every $a, b \in \mathfrak{A}$ we write $a \leq b$ instead of $R(a, b)$.

For each $n \in \mathbb{N}$ consider a sentence

$$\varphi_n := \exists x_1 ... \exists x_n (\bigwedge_{i \neq j} (x_i \neq x_j) \ \& \bigwedge_{i \leq j} (x_i \leq x_j) \ \& \ \forall y (\bigwedge_{i} (y \neq x_i) \to$$

$$\to \bigwedge_{i} (\neg (x_i \leq y) \& \neg (y \leq x_i))))$$

stating that there is a maximal w.r.t. inclusion linear ordered subset the cardinality of which is equal to n, such that all other elements are not compatible with the elements from this subset.

The set M is finite or countable, so there exists an injective mapping $\nu : M \to \mathbb{N}$ and a mapping $h : M \to \{\varphi_n \mid n \in \mathbb{N}\}$ such that $h(m) = \varphi_{\nu(m)}$ for any $m \in M$.

Denote $\Delta := h(M) = \{h(m) \mid m \in M\}$. It is obvious that the mapping $h : M \to \Delta$ is bijective.

For any $n \in \mathbb{N}$ let $\mathfrak{L}_n$ be a linear ordering consisting of n elements. Denote by $\mathbb{B}$ a set of countable Boolean algebras with universe $\mathbb{N}$.

The cardinality of isomorphism types of models from the set $\mathbb{B}$ is continuum. So there exists a mapping $f_0 : G \to \mathbb{B}$ such that if $g_1 \neq g_2$ then $f_0(g_1) \ncong f_0(g_2)$.

For each $g \in G$ we construct a model $\mathfrak{A}_g \in POS$ in the following way. Denote $M(g) := \{m \in M \mid gIm\}$ and put $\mathfrak{A}_g := f_0(g) \cup \bigcup_{m \in M(g)} \mathfrak{L}_{\nu(m)}$. It means that for the universe of this model we have $|\mathfrak{A}_g| = |f_0(g)| \cup \bigcup_{m \in M(g)} |\mathfrak{L}_{\nu(m)}|$ and the predicate is defined as follows: $R^{\mathfrak{A}_g} = R^{f_0(g)} \cup \bigcup_{m \in M(g)} R^{\mathfrak{L}_{\nu(m)}}$ (here $R^{\mathfrak{L}}$ denotes the predicate R in the model $\mathfrak{L}$).

It is not difficult to prove the following

Lemma 1. *1. If $g_1, g_2 \in G$ and $g_1 \neq g_2$ then $\mathfrak{A}_{g_1} \ncong \mathfrak{A}_{g_2}$.*

2. If $g \in G$ and $m \in M$ then gIm if and only if $\mathfrak{A}_g \models \varphi_{\nu(m)}$.

Consider a mapping $f : G \to POS$ such that $f(g) = \mathfrak{A}_g$ for any $g \in G$. Denote $K := f(G) = \{\mathfrak{A}_g \mid g \in G\}$.

Then for any $g \in G$ and $m \in M$ we have gIm if and only if $f(g) \models h(m)$. Mappings g and h are bijective, so $C \cong (K, \Delta, \models)$.

Proposition 16 is proved.

Theorem 8. *For any finite or countable complete lattice L there exists a countable $K \subseteq K(\sigma)$ and a countable $\Delta \subseteq S(\sigma)$ such that $L \cong \mathbb{A}(K, \Delta)$.*

Proof. Consider a complete lattice L. Suppose that L is countable or finite. Put $G := L$, $M := L$ and $I :=\leq$. By the Basic Theorem on Concept Lattices [4] $L \cong \underline{\mathfrak{B}}(G, M, I)$. Denote $C := (G, M, I)$. Then G and M are finite or countable sets. Then by Proposition 16 there exist $K \subseteq K(\sigma)$ and $\Delta \subseteq S(\sigma)$ such that $C \cong (K, \Delta, \models)$.

Therefore $L \cong B(K, \Delta, \models)$ and, by Corollary 17, $L \cong \mathbb{A}(K, \Delta)$.

Theorem 8 is proved.

Thus each countable complete lattice may by represented as the lattice of relatively axiomatizable subclasses of some class $K \subseteq K(\sigma)$ for a proper $\Delta \subseteq S(\sigma)$.

However, most interesting are classes which are axiomatizable in $K(\sigma)$ relatively to some $\Delta \subseteq S(\sigma)$. Varieties, quasivarieties, $\forall$- and $\exists$-axiomatizable classes are such examples.

Definition 28. *For $\mathfrak{A}, \mathfrak{B} \in K(\sigma)$ and $\Delta \subseteq S(\sigma)$ we denote $\mathfrak{A} \equiv_\Delta \mathfrak{B}$ if*

$$\{\varphi \in \Delta \mid \mathfrak{A} \models \varphi\} = \{\varphi \in \Delta \mid \mathfrak{B} \models \varphi\}$$

and $[\mathfrak{A}]_\Delta := \{\mathfrak{B} \in K(\varphi) \mid \mathfrak{A} \equiv_\Delta \mathfrak{B}\}$. For $K \subseteq K(\sigma)$ denote

$$K/\Delta := \{[\mathfrak{A}]_\Delta | \mathfrak{A} \in K\}.$$

For $\varphi \in \Delta$ we denote $[\mathfrak{A}]_\Delta \models \varphi$ if $\mathfrak{A} \models \varphi$.

It is obvious that Definition 28 is correct, i.e. does not depend on the choice of $\mathfrak{B} \in [\mathfrak{A}]_\Delta$.

Proposition 17. *For any finite object-clarified formal context C there exists $\Delta \subseteq S(\sigma)$ such that $C \cong (K(\sigma)/\Delta, \Delta, \models)$.*

Proof. Let $C = (G, M, I)$ be a finite object-clarified formal context.

Case 1. Assume that there exists a formal object $g \in G$ such that there is no formal attribute $m \in M$ with gIm.

Consider a sentence

$$\psi = \forall x \forall y \forall z ((x \leq x) \;\&\; (((x \leq y \;\&\; y \leq x) \;\to\; x = y) \;\&$$

$$\& \; ((x \leq y \;\&\; y \leq z) \;\to\; x \leq z)))$$

stating that $\leq$ is a (partial) order.

The sets G and M are finite. Suppose that $G=\{g_1, ..., g_n\}$, $M=\{m_1, ..., m_k\}$, $g_i \neq g_j$ for any $i \neq j$, and $m_i \neq m_j$ for any $i \neq j$.

For any $i \leq n$ we define $\varepsilon^i = (\varepsilon^i_1, ..., \varepsilon^i_k)$ in the following way:

$$\varepsilon^i_j := \begin{cases} 1, & if\ g_i I m_j; \\ 0, & otherwise. \end{cases}$$

Denote $\xi := \bigvee\limits_{i\leq n} (\bigwedge\limits_{j\leq k} \varphi_j^{\varepsilon^i_j})$, where $\varphi^0 := \neg\varphi$ and $\varphi^1 := \varphi$.

Denote $\psi_i := (\varphi_i \& \psi \& \xi)$ for any $i \leq k$.

Note,that for any $\mathfrak{A} \in K(\sigma)$ and $i \leq k$ we have:

– if $\mathfrak{A}$ is not an ordered set then $\mathfrak{A} \not\models \psi_i$;

– if $\mathfrak{A}$ is an ordered set, $\mathfrak{A} \models \varphi_i$ and there exists a formal object $g \in G$ such that $\mathfrak{A} \models \varphi_j \ \Leftrightarrow\ gIm_j$ then $\mathfrak{A} \models \psi_i$;

– $\mathfrak{A} \not\models \psi_i$ otherwise.

Thus for any $\mathfrak{A} \in K(\sigma)$ we have only two possibilities:

– $\mathfrak{A} \not\models \psi_i$ for any $i \leq k$;

– there exists $g \in G$ such that $\mathfrak{A} \models \psi_i$ if and only if gIm_i for any $i \leq k$.

Recall that G contains a formal object $g \in G$ such that gIm is not true for any $m \in M$. Therefore for any $\mathfrak{A} \in K(\sigma)$ there is $g \in G$ such that $\mathfrak{A} \models \psi_i \ \Leftrightarrow\ gIm_i$ for each $i \leq k$.

Moreover, for any $g \in G$ there is a model $\mathfrak{A} \in K(\sigma)$ such that $\mathfrak{A} \models \psi_i$ if and only if gIm_i for any $i \leq k$.

For convenience of further consideration we denote $\xi_i := \psi_i$ for any $i \leq k$.

Put $\Delta := \{\xi_1, ..., \xi_k\}$.

Case 2. For any $g \in G$ there is $m \in M$ such that gIm.

Denote

$$\xi_i := \begin{cases} ((\psi \ \&\ \xi) \ \rightarrow\ \psi_i) & if\ \varepsilon^1_i = 1; \\ \psi_i, & if\ \varepsilon^1_i = 0. \end{cases}$$

Put $\Delta := \{\xi_1, ..., \xi_k\}$.

It is easy to prove the following

Lemma 2. *Let* $\mathfrak{A} \in K(\sigma)$.

a) If $\mathfrak{A} \models (\psi \ \&\ \xi)$ *then* $\mathfrak{A} \models \xi_j$ *if and only if* $\mathfrak{A} \models \psi_j$ *for any* $j \leq k$;

b) If $\mathfrak{A} \models (\psi \ \&\ \xi)$ *then there is* $i \leq n$ *such that* ($\mathfrak{A} \models \xi_j$ *if and only if* $g_i I m_j$) *for any* $j \leq k$;

c) If $\mathfrak{A} \models \neg(\psi \ \&\ \xi)$ *then* ($\mathfrak{A} \models \xi_j$ *if and only if* $\varepsilon^1_j = 1$) *for any* $j \leq k$;

d) There is $g \in G$ *such that* ($\mathfrak{A} \models \xi_j$ *if and only if* gIm_j) *for any* $j \leq k$;

e) For each $g \in G$ *there is a model* $\mathfrak{L} \in K(\sigma)$ *such that* ($\mathfrak{L} \models \xi_j$ *if and only if* gIm_j) *for any* $j \leq k$.

Further we treat Case 1 and Case 2 simultaneously.

We consider the formal context $C_\Delta := (K(\sigma)/_\Delta, \Delta, \models)$. By the definition of C_Δ this formal context is object-clarified. Recall that the formal context C is object-clarified too. Then, in virtue of Lemma 2 (d,e), there exists a bijective mapping $f : G \to K(\sigma)/_\Delta$ such that for each $g \in G$ we have (gIm_i if and only if $f(g) \models \xi_i$) for any $i \leq k$.

Consider a mapping $h : M \to \Delta$ defined as follows: $h(m_i) = \xi_i$ for any $i \leq k$. Obviously h is a bijective mapping.

Thus for any $g \in G$ and $i \leq k$ we have gIm_i if and only if $f(g) \models h(m_i)$. Therefore, $C \cong (K(\sigma)/_\Delta, \Delta, \models)$.

Proposition 17 is proved.

Theorem 9. *For any finite lattice L there exists* $\Delta \subseteq S(\sigma)$ *such that*

$$L \cong \mathbb{A}(K(\sigma), \Delta).$$

Proof. Let L be a finite lattice. By the Basic Theorem on Concept Lattices [4] $L \cong \mathfrak{B}(G, M, I)$, where $G := L$, $M := L$ and $I := \leq$.

So the sets G and M are finite. Then, by Proposition 17, there exists $\Delta \subseteq S(\sigma)$ such that $(G, M, I) \cong (K(\sigma)/_\Delta, \Delta, \models)$. Therefore

$$L \cong \mathfrak{B}(K(\sigma)/_\Delta, \Delta, \models)\,.$$

The formal context $(K(\sigma)/_\Delta, \Delta, \models)$ is the object clarification of the formal context $(K(\sigma), \Delta, \models)$. Then $\mathfrak{B}(K(\sigma)/_\Delta, \Delta, \models) \cong \mathfrak{B}(K(\sigma), \Delta, \models)$. By Corollary 17 $\mathbb{A}(K(\sigma), \Delta) \cong \mathfrak{B}(K(\sigma), \Delta, \models)$. Therefore $L \cong \mathbb{A}(K(\sigma), \Delta)$.

Theorem 9 is proved.

Thus each finite lattice may be represented as a lattice of all relatively axiomatizable subclasses of $K(\sigma)$ for a proper set $\Delta \subseteq S(\sigma)$.

Problem 1. *What are the lattices of all relatively axiomatizable subclasses of* $K(\sigma)$ *for various sets* $\Delta \subseteq S(\sigma)$*? Is it the class of all complete lattices?*

Acknowledgements. I am very grateful to Rudolf Wille, Karl Erich Wolff and Peter Burmeister for interesting discussions which have inspired me to make this investigation.

References

1. G.Birkhoff. Lattice theory, 3rd ed. Vol. 25 of American Mathematical Society Colloquium Publications. American Mathematical Society, 1973.
2. C.C.Chang, H.J.Keisler. Model theory. North-Holland Publishing Company, New York, 1973.
3. Yu.L.Ershov, E.A.Palutin. Mathematical logic. Walter de Gruyter, Berlin and New York, 1989.
4. B.Ganter, R.Wille. Formal Concept Analysis. Mathematical foundations. Springer-Verlag, Berlin, Heidelberg, 1999.
5. B.Ganter, G.Stumme, R.Wille. Formal Concept Analysis. Foundations and Applications. Springer-Verlag, Berlin, Heidelberg, 2005.
6. S.S.Goncharov. Countable Boolean Algebras and Decidability. Siberian School of Algebra and Logic. Consultants Bureau, New York,London and Moscow, 1997.

7. V.A.Gorbunov. Algebraic theory of quasivarieties. Siberian School of Algebra and Logic. New York: Consultants Bureau, 1998.
8. G.Gratzer. General lattice theory. Akademie-Verlag, Berlin, 1978.
9. P.Jipsen, H.Rose. Varieties of Lattices. Lecture Notes in Mathematics 1533, Springer Verlag, 1992.
10. D.E.Pal'chunov. On some logic calculus describing calculations on computer. The International conference on computer logic COLOG-88, Part 1, Tallinn, 1988, p. 76 - 79.
11. D.E.Pal'chunov. Finitely axiomatizable Boolean algebras with distinguished ideals. Algebra and Logic, vol. 26, N 4, 1987, p. 435 - 455.
12. A.G.Pinus, Yu.M.Vazhenin. Elementary classification and decidability of theories of derived structures. RUSS MATH SURV, 60 (3), 2005, p. 395-432.

A Solution of the Word Problem for Free Double Boolean Algebras

Björn Vormbrock

Technische Universität Darmstadt, Fachbereich Mathematik
Schloßgartenstr. 7, D-64289 Darmstadt,
vormbrock@mathematik.tu-darmstadt.de

Abstract. Double Boolean algebras were introduced in [Wi00a] as a variety fundamental for Boolean Concept Logic, an extension of Formal Concept Analysis allowing negations of formal concepts. In this paper, the free double Boolean algebra generated by the constants is described. Moreover, we show that every free double Boolean algebra with at least one generator is infinite. A measure of the complexity of terms specific for double Boolean algebras is introduced. This, together with a modification of the algorithm for protoconcept exploration (cf. [Vo04]) yields double Boolean algebras containing a counterexample to every term identity up to a given complexity if the identity does not hold in general. These algebras can be constructed automatically, thus the word problem for free double Boolean algebras is solved.

1 Introduction

Double Boolean algebras form the variety generated by protoconcept algebras. Protoconcept algebras were defined in order to introduce negations of concepts in Formal Concept Analysis and to develop a Boolean Concept Logic in the framework of Contextual Logic (see [Wi00b] for an introduction to Contextual Logic, [Wi00a] for an introduction to Boolean Concept Logic). For the development of Boolean Concept Logic, a solution to the word problem for the free double Boolean algebras is essential. In[HLSW00] Herrmann et al. show that double Boolean algebras have the finite embedding property. Since double Boolean algebras are finitely axiomatized it follows that their universal theory is decidable. Moreover, as every Boolean algebra may be regarded as a double Boolean algebra, this decision problem is $\mathcal{NP}$-complete.

In this paper, a measure for the compelxity of terms appropiate for double Boolean algebras is developed and an upper bound for the size of a minimal counterexample to an invalid term identity $s \sim t$ depending of the complexity of s and t is derived. Moreover, a class of protoconcept algebras containing all counterexamples for invalid term identities up to a given level of complexity is described. An algorithm for the automated construction of their underlying contexts is given. This provides a semantic solution of the word problem for free double Boolean algebras where the algorithm depends only of the complexity of the terms.

S.O. Kuznetsov and S. Schmidt (Eds.): ICFCA 2007, LNAI 4390, pp. 240–270, 2007.

In Section 1, the basic definitions for double Boolean algebras and protoconcept algebras are introduced. Moreover, the free double Boolean algebra generated by the constants is described. In Section 2, an algorithm for the stepwise exploration of double Boolean algebras is developed. The application of this algorithm to free double Boolean algebras in Section 3 yields a class of counterexamples to invalid term identities depending on the complexity of the terms and the number of variables. An upper bound for the size of a minimal counterexample is derived in Corollary 4. Section 4 is dedicated to the automated generation of the contexts underlying the described class of counterexamples.

1.1 Double Boolean Algebras and Protoconcept Algebras

Double Boolean Algebras

Definition 1. *A* double Boolean algebra *is an algebra* $\underline{D} := (D, \sqcap, \sqcup, \neg, \lrcorner, \bot, \top)$ *of type (2,2,1,1,0,0), satisfying the equations*

1a) $(x \sqcap x) \sqcap y = x \sqcap y$	*1b)* $(x \sqcup x) \sqcup y = x \sqcup y$
2a) $x \sqcap y = y \sqcap x$	*2b)* $x \sqcup y = y \sqcup x$
3a) $x \sqcap (y \sqcap z) = (x \sqcap y) \sqcap z$	*3b)* $x \sqcup (y \sqcup z) = (x \sqcup y) \sqcup z$
4a) $x \sqcap (x \sqcup y) = x \sqcap x$	*4b)* $x \sqcup (x \sqcap y) = x \sqcup x$
5a) $x \sqcap (x \veebar y) = x \sqcap x$	*5b)* $x \sqcup (x \barwedge y) = x \sqcup x$
6a) $x \sqcap (y \veebar z) = (x \sqcap y) \veebar (x \sqcap z)$	*6b)* $x \sqcup (y \barwedge z) = (x \sqcup y) \barwedge (x \sqcup z)$
7a) $\neg\neg(x \sqcap y) = x \sqcap y$	*7b)* $\lrcorner\lrcorner(x \sqcup y) = x \sqcup y$
8a) $\neg(x \sqcap x) = \neg x$	*8b)* $\lrcorner(x \sqcup x) = \lrcorner x$
9a) $x \sqcap \neg x = \bot$	*9b)* $x \sqcup \lrcorner x = \top$
10a) $\neg\bot = \top \sqcap \top$	*10b)* $\lrcorner\top = \bot \sqcup \bot$
11a) $\neg\top = \bot$	*11b)* $\lrcorner\bot = \top$
12) $(x \sqcap x) \sqcup (x \sqcap x) = (x \sqcup x) \sqcap (x \sqcup x)$	

with the operations $\veebar, \barwedge, \overline{\top}, \underline{\bot}$ *defined by*

$$x \veebar y := \neg(\neg x \sqcap \neg y)$$
$$x \barwedge y := \lrcorner(\lrcorner x \sqcup \lrcorner y)$$
$$\overline{\top} := \neg\bot$$
$$\underline{\bot} := \lrcorner\top$$

A pure double Boolean algebra *is a double Boolean algebra that satisfies the additional condition*

$$13)\ x = x \sqcap x \text{ or } x = x \sqcup x.$$

To shorten notation we write $x_\sqcap$ for $x \sqcap x$ and $x_\sqcup$ for $x \sqcup x$, and define $D_\sqcap := \{x_\sqcap \mid x \in D\}$, $D_\sqcup := \{x_\sqcup \mid x \in D\}$ and $D_p := D_\sqcap \cup D_\sqcup$. The restriction of $\underline{D}$ to D_p is a pure subalgebra of $\underline{D}$.

On double Boolean algebras we define a quasi-order $\sqsubseteq$ by:

$$x \sqsubseteq y :\Leftrightarrow x \sqcap y = x_\sqcap \text{ and } x \sqcup y = y_\sqcup.$$

Some basic properties of double Boolean algebras were discussed in [HLSW00]:

Theorem 1. *Let* $\underline{D} := (D, \sqcap, \sqcup, \neg, \lrcorner, \bot, \top)$ *be a double Boolean algebra. Then the following conditions are satisfied:*

(1) $(D, \sqsubseteq)$ *is a quasi-ordered set.*
(2) $\underline{D}_\sqcap := (D_\sqcap, \sqcap, \veebar, \neg, \bot, \overline{\top})$ *is a Boolean algebra whose order relation is the restriction of* $\sqsubseteq$ *to* $D_\sqcap$.
(3) $\underline{D}_\sqcup := (D_\sqcup, \overline{\sqcap}, \sqcup, \lrcorner, \underline{\bot}, \top)$ *is a Boolean algebra whose order relation is the restriction of* $\sqsubseteq$ *to* $D_\sqcup$.
(4) $y \sqsubseteq x_\sqcap \Leftrightarrow y \sqsubseteq x$ *for* $x \in D$ *and* $y \in D_\sqcap$.
(5) $x_\sqcup \sqsubseteq y \Leftrightarrow x \sqsubseteq y$ *for* $x \in D$ *and* $y \in D_\sqcup$.
(6) $x \sqsubseteq y \Leftrightarrow x_\sqcap \sqsubseteq y_\sqcap$ *and* $x_\sqcup \sqsubseteq y_\sqcup$ *for* $x, y \in D$.

Protoconcept algebras. Protoconcepts were defined in order to introduce negations on formal concepts. We briefly repeat the basic definitions since protoconcept algebras form an important class of examples for double Boolean algebras. We assume that the reader is familiar with Formal Concept Analysis, for a textbook we refer to [GW99]. An analysis of the meaning of protoconcepts and the two operations "negation" ($\neg$) and "opposition" ($\lrcorner$) is presented in [Wi00a] and [VW03].

Definition 2. *A* protoconcept *of a formal context* $\mathbb{K} := (G, M, I)$ *is a pair* (A, B) *with* $A \subseteq G$ *and* $B \subseteq M$ *such that* $A' = B''$ *or, equivalently,* $A'' = B'$. *We denote the set of all protoconcepts of a context* $\mathbb{K}$ *by* $\mathfrak{P}(\mathbb{K})$ *and define on* $\mathfrak{P}(\mathbb{K})$ *operations* $\sqcap$, $\sqcup$, $\lrcorner$, $\neg$, $\top$ *and* $\bot$ *by:*

$$\begin{aligned}(A_1, B_1) \sqcap (A_2, B_2) &:= (A_1 \cap A_2, (A_1 \cap A_2)')\\ (A_1, B_1) \sqcup (A_2, B_2) &:= ((B_1 \cap B_2)', B_1 \cap B_2)\\ \neg(A, B) &:= (G \setminus A, (G \setminus A)')\\ \lrcorner(A, B) &:= ((M \setminus B)', M \setminus B)\\ \top &:= (G, \emptyset)\\ \bot &:= (\emptyset, M)\end{aligned}$$

The set of all protoconcepts of a context $\mathbb{K}$ *together with these operations is called the* protoconcept algebra *of* $\mathbb{K}$ *and denoted by* $\underline{\mathfrak{P}}(\mathbb{K})$.

On protoconcept algebras the quasi-order $\sqsubseteq$ is an order relation and

$$(A_1, B_1) \sqsubseteq (A_2, B_2) \Leftrightarrow A_1 \subseteq A_2 \text{ and } B_1 \supseteq B_2$$

Note that the result of any operation in a protoconcept algebra is a protoconcept of the form (A, A') or (B', B). These protoconcepts are called $\sqcap$-*semiconcepts* or $\sqcup$-*semiconcepts*, respectively. The set of all $\sqcap$-semiconcepts of a protoconcept algebra $\underline{\mathfrak{P}}(\mathbb{K}) (= \underline{D})$ is denoted by $\mathfrak{P}(\mathbb{K})_\sqcap$ $(= D_\sqcap)$ and the set of all $\sqcup$-semiconcepts by $\mathfrak{P}(\mathbb{K})_\sqcup$ $(= D_\sqcup)$. As before, the set $\mathfrak{H}(\mathbb{K}) := \mathfrak{P}(\mathbb{K})_\sqcap \cup \mathfrak{P}(\mathbb{K})_\sqcup$

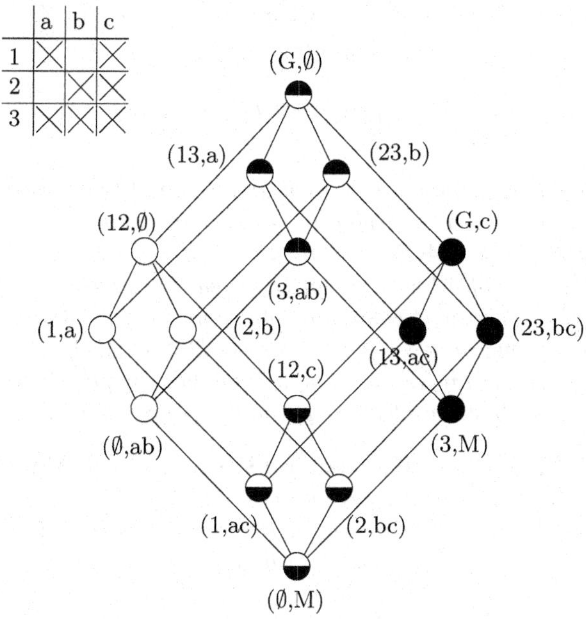

Fig. 1. A context and its protoconcept algebra

of all semiconcepts of $\mathbb{K}$, together with the operations of $\underline{\mathfrak{P}}(\mathbb{K})$ is a subalgebra of $\underline{\mathfrak{P}}(\mathbb{K})$. We call this subalgebra the *semiconcept algebra* of the context $\mathbb{K}$.

Note that the formal concepts of a context are those protoconcepts that are both $\sqcap$- and $\sqcup$-semiconcepts.

Example 1. Figure 1 depicts a context and its protoconcept algebra. The elements represented by filled circles are formal concepts. The circles with the upper half filled represent $\sqcup$-semiconcepts, those with the lower half filled represent $\sqcap$-semiconcepts.

In [Wi00a] it was shown that the axioms of double Boolean algebras generate the equational theory of protoconcept algebras.

Basic Theorems for semiconcept algebras and protoconcept algebras. In [VW03] Basic Theorems for semiconcept algebras and protoconcept algebras were shown. In order to quote them here we have to introduce the notions of contextual, fully contextual and complete double Boolean algebras: A double Boolean algebra $\underline{D}$ is called *contextual* if its quasiorder $\sqsubseteq$ is antisymmetric, i.e. the relation $\sqsubseteq$ is an order on $\underline{D}$. A contextual double Boolean algebra $\underline{D}$ is said to be *fully contextual* if, in addition, for each $x \in \underline{D}_\sqcap$ and $y \in \underline{D}_\sqcup$ with $x_\sqcup = y_\sqcap$ there is a unique $z \in \underline{D}$ with $z_\sqcap = x$ and $z_\sqcup = y$. The double Boolean algebra $\underline{D}$ is called complete if and only if its Boolean algebras $\underline{D}_\sqcap$ and $\underline{D}_\sqcup$ are complete.

Theorem 2 (The Basic Theorem on Semiconcept Algebras). *For a context $\mathbb{K} := (G, M, I)$, the semiconcept algebra $\underline{\mathfrak{H}}(\mathbb{K})$ is a complete pure double*

Boolean algebra whose Boolean algebras $\underline{\mathfrak{H}}_\sqcap(\mathbb{K})$ *and* $\underline{\mathfrak{H}}_\sqcup(\mathbb{K})$ *are atomic. The (arbitrary) meet and join of* $\underline{\mathfrak{H}}(\mathbb{K})$ *are given by*

$$\bigsqcap_{t\in T}(A_t,B_t) = (\bigcap_{t\in T} A_t, (\bigcap_{t\in T} A_t)') \quad \text{and} \quad \bigsqcup_{t\in T}(A_t,B_t) = ((\bigcap_{t\in T} B_t)', \bigcap_{t\in T} B_t).$$

In general, a complete pure double Boolean algebra $\underline{D}$ *whose Boolean algebras* $\underline{D}_\sqcap$ *and* $\underline{D}_\sqcup$ *are atomic, is isomorphic to* $\underline{\mathfrak{H}}(\mathbb{K})$ *if and only if there exist a bijection* $\tilde{\gamma}$ *from* G *onto the set* $\mathcal{A}(\underline{D}_\sqcap)$ *of all atoms of* $\underline{D}_\sqcap$ *and a bijection* $\tilde{\mu}$ *from* M *onto the set* $\mathcal{C}(\underline{D}_\sqcup)$ *of all coatoms of* $\underline{D}_\sqcup$ *such that* $gIm \Leftrightarrow \tilde{\gamma}(g) \sqsubseteq \tilde{\mu}(m)$ *for all* $g \in G$ *and* $m \in M$. *In particular, for any complete pure double Boolean algebra* $\underline{D}$ *whose Boolean algebras are atomic, we get* $\underline{D} \cong \underline{\mathfrak{H}}(\mathcal{A}(\underline{D}_\sqcap), \mathcal{C}(\underline{D}_\sqcup), \sqsubseteq)$, *i.e., the semiconcept algebras are up to isomorphism the complete pure double Boolean algebras* $\underline{D}$ *whose Boolean algebras* $\underline{D}_\sqcap$ *and* $\underline{D}_\sqcup$ *are atomic.*

Theorem 3 (The Basic Theorem on Protoconcept Algebras). *For a context* $\mathbb{K} := (G, M, I)$, *the protoconcept algebra* $\underline{\mathfrak{P}}(\mathbb{K})$ *of* $\mathbb{K}$ *is a complete fully contextual double Boolean algebra whose Boolean algebras* $\underline{\mathfrak{H}}_\sqcap(\mathbb{K})$ *and* $\underline{\mathfrak{H}}_\sqcup(\mathbb{K})$ *are atomic. The (arbitrary) meet and join of* $\underline{\mathfrak{P}}(\mathbb{K})$ *are given by*

$$\bigsqcap_{t\in T}(A_t,B_t) = (\bigcap_{t\in T} A_t, (\bigcap_{t\in T} A_t)') \quad \text{and} \quad \bigsqcup_{t\in T}(A_t,B_t) = ((\bigcap_{t\in T} B_t)', \bigcap_{t\in T} B_t).$$

In general, a complete fully contextual double Boolean algebra $\underline{D}$ *whose Boolean algebras* $\underline{D}_\sqcap$ *and* $\underline{D}_\sqcup$ *are atomic, is isomorphic to* $\underline{\mathfrak{P}}(\mathbb{K})$ *if and only if there exist a bijection* $\tilde{\gamma}$ *from* G *onto the set* $\mathcal{A}(\underline{D}_\sqcap)$ *of all atoms of* $\underline{D}_\sqcap$ *and a bijection* $\tilde{\mu}$ *from* M *onto the set* $\mathcal{C}(\underline{D}_\sqcup)$ *of all coatoms of* $\underline{D}_\sqcup$ *such that* $gIm \Leftrightarrow \tilde{\gamma}(g) \sqsubseteq \tilde{\mu}(m)$ *for all* $g \in G$ *and* $m \in M$. *In particular, for any complete fully contextual double Boolean algebra* $\underline{D}$ *whose Boolean algebras are atomic, we get* $\underline{D} \cong \underline{\mathfrak{P}}(\mathcal{A}(\underline{D}_\sqcap), \mathcal{C}(\underline{D}_\sqcup), \sqsubseteq)$, *i.e., the protoconcept algebras are up to isomorphism the complete fully contextual double Boolean algebras* $\underline{D}$ *whose Boolean algebras* $\underline{D}_\sqcap$ *and* $\underline{D}_\sqcup$ *are atomic.*

1.2 Terms and Free Double Boolean Algebras

In this subsection terms and free double Boolean algebras are introduced in accordance with the standard definitions of universal algebra (cp. [BS00] for example). Moreover, a notion of the complexity of terms which is specific for double Boolean algebras is defined. The free double Boolean algebra generated by the constants is described and, finally, it is shown that every free double Boolean algebra with at least one generator is infinite.

Definition 3. *Let* X *be a set of variables with* $\{\bot, \top, \neg, \lrcorner, \sqcap, \sqcup\} \cap X = \emptyset$. *The set* $\mathcal{T}(X)$ *of all terms over* X *is defined recursively by:*

1. $X \cup \{\bot, \top\} \subseteq \mathcal{T}(X)$
2. $s, t \in \mathcal{T}(X) \Rightarrow \{(\neg s), (\lrcorner s), (s \sqcap t), (s \sqcup t)\} \subseteq \mathcal{T}(X)$

As before we write $\overline{\top}$ for $(\top \sqcap \top)$, $\underline{\bot}$ for $(\bot \sqcup \bot)$, $(s \mathbin{\underline{\sqcup}} t)$ for $(\neg((\neg s) \sqcap (\neg t)))$, $(s \mathbin{\bar{\sqcap}} t)$ for $(\lrcorner((\lrcorner s) \sqcup (\lrcorner t)))$, $t_\sqcap$ for $(t \sqcap t)$ and $t_\sqcup$ for $(t \sqcup t)$. In order to avoid unnecessary parentheses we write $x_1 f_1 x_2 f_2 x_3 \dots f_{n-1} x_n$ for $(\dots(x_1 f_1 x_2) f_2 x_3) \dots) f_{n-1} x_n)$ with $f_1, \dots, f_n \in \{\sqcap, \sqcup, \bar{\sqcap}, \underline{\sqcup}\}$ and $x_1, \dots, x_n \in X$.

For $A := \{a_1, \dots, a_n\}$ we set $\bigsqcap A := a_1 \sqcap a_2 \sqcap a_3 \cdots \sqcap a_n$ and define likewise $\bigsqcup A$, $\bar{\bigsqcap} A$ and $\underline{\bigsqcup} A$.

Term functions are a common concept in universal algebra. They are used throughout the paper, therefore we give the definition briefly:

Definition 4. *Given a term $p(x_1, \dots x_n)$ and a double Boolean algebra $\underline{D}$ we obtain a function $p^{\underline{D}} : D^n \to D$ as follows:*

1. *If p is a variable x_i, then*

$$p^{\underline{D}}(a_1, \dots, a_n) := a_i$$

   for $a_1, \dots, a_n \in D$.
2. *Otherwise we set*
   (a) $p^{\underline{D}}(a_1, \dots, a_n) := (\neg p_1^{\underline{D}}(a_1, \dots, a_n))$ if $p(x_1, \dots, x_n) = (\neg p_1(x_1, \dots, x_n))$
   (b) $p^{\underline{D}}(a_1, \dots, a_n) := (\lrcorner p_1^{\underline{D}}(a_1, \dots, a_n))$ if $p(x_1, \dots, x_n) = (\lrcorner p_1(x_1, \dots, x_n))$
   (c) $p^{\underline{D}}(a_1, \dots, a_n) := (p_1(a_1, \dots, a_n) \sqcap p_2(a_1, \dots, a_n)$ if $p(x_1, \dots, x_n) = (p_1(x_1, \dots, x_n) \sqcap p_2(x_1, \dots, x_n))$
   (d) $p^{\underline{D}}(a_1, \dots, a_n) := (p_1(a_1, \dots, a_n) \sqcup p_2(a_1, \dots, a_n)$ if $p(x_1, \dots, x_n) = (p_1(x_1, \dots, x_n) \sqcup p_2(x_1, \dots, x_n))$
   for $a_1, \dots, a_n \in D$.

The definition of free double Boolean algebras follows [BS00].

Definition 5. *Let $\mathfrak{D}$ be the variety of double Boolean algebras. Given a set X of variables we define a congruence relation $\approx$ on $\underline{T}(X)$ by*

$$s \approx t :\Leftrightarrow (s, t) \in \bigcap \Phi_{\mathfrak{D}}(X),$$

where

$$\Phi_{\mathfrak{D}}(X) := \{\phi \in Con\, \underline{T}(X) \mid \underline{T}(X)/\phi \in \mathfrak{D}\};$$

and then define $\mathcal{D}(X)$, the free double Boolean algebra over X, by

$$\mathcal{D}(X) := \underline{T}(X)/\approx .$$

The $\approx$-congruence class of a term t is denoted with $\bar{t}$. Note that for $x, y \in X$ we have $\bar{x} \neq \bar{y}$ if $x \neq y$ since $\mathfrak{D}$ contains nontrivial algebras. It is well known that $\mathcal{D}(X) \in \mathfrak{D}$ and that for $s(x_1, \dots x_n), t(x_1, \dots x_n) \in \mathcal{T}(X)$ holds

$$s \approx t \Leftrightarrow s^{\underline{D}} = t^{\underline{D}} \text{ in every double Boolean algebra } \underline{D}.$$

Example 2. Fig. 2 depicts the free double Boolean algebra generated by the empty set (cp. Proposition 1). Since it is a pure, finite double Boolean algebra, it is isomorphic to the semiconcept algebra of the given context.

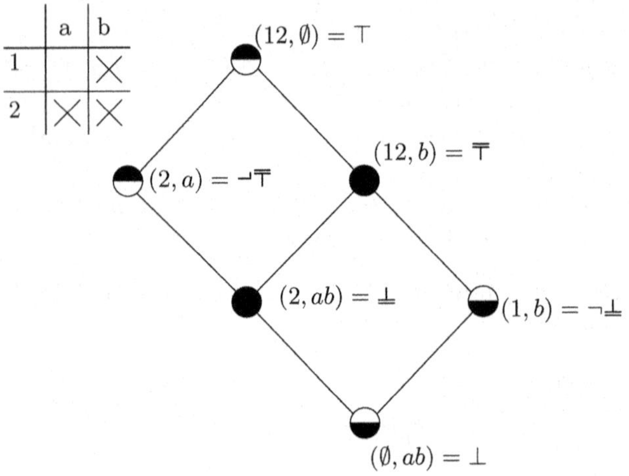

Fig. 2. A protoconcept algebra isomorphic to the free double Boolean algebra generated by the empty set

Example 3. There are infinite double Boolean algebras generated by a single element: Let $G := \{g_1, g_2, \ldots\}$ be a countable set of objects, let $M := \{m_1, m_2, \ldots\}$ be a countable set of attributes and define an incidence relation I by

$$g_i \, I \, m_j :\Leftrightarrow j \notin \{i-1, i\}.$$

Then with $\mathbb{K} := (G, M, I)$ (cp. Fig. 3) the element $\mathfrak{p} \in \mathfrak{H}(\mathbb{K})$, $\mathfrak{p} := (\{g_1\}, \{g_1\}')$ generates an infinite subalgebra of $\mathfrak{H}(\mathbb{K})$: We obtain the elements $(\{m_1\}', \{m_1\}) = \lrcorner\mathfrak{p}$, $(\{g_2\}, \{g_2\}') = \neg\lrcorner\mathfrak{p} \sqcap \neg\mathfrak{p}$ and $(\{m_2\}', \{m_2\}) = \lrcorner(\neg\lrcorner\mathfrak{p} \sqcap \neg\mathfrak{p}) \sqcup \mathfrak{p}$ from $\mathfrak{p}$. For $i \geq 2$ the elements $\{(\{g_1\}, \{g_1\}'), \ldots, (\{g_i\}, \{g_i\}'), (\{m_1\}', \{m_1\}), \ldots, (\{m_i\}', \{m_i\})\}$ generate

$$(\{g_{i+1}\}, \{g_{i+1}\}') = \neg(\{m_i\}', \{m_i\}) \sqcap \neg(\{g_i\}, \{g_i\}')$$

and

$$(\{m_{i+1}\}', \{m_{i+1}\}) = \lrcorner(\{g_{i+1}\}, \{g_{i+1}\}') \sqcup \lrcorner(\{m_i\}', \{m_i\}).$$

For our investigation, an appropiate measure of the complexity of terms for double Boolean algebras is needed. In Example 3 an infinite subalgebra is generated through permanent switches from $\sqcap$-semiconcepts to $\sqcup$-semiconcepts and back. This suggests to take the number of switches between $\underline{D}_\sqcap$ and $\underline{D}_\sqcup$ as a measure for the complexity.

Definition 6. *For a finite set X of variables we set:*

$$\mathcal{T}_\sqcap(X) := \{t \in \mathcal{T}(X) \mid t = (t_1 \sqcap t_2) \text{ or } t = (\neg s) \text{ for } t_1, t_2, s \in \mathcal{T}(X)\} \cup \{\bot\}$$
$$\mathcal{T}_\sqcup(X) := \{t \in \mathcal{T}(X) \mid t = (t_1 \sqcup t_2) \text{ or } t = (\lrcorner s) \text{ for } t_1, t_2, s \in \mathcal{T}(X)\} \cup \{\top\}.$$

	m_1	m_2	m_3	m_4	m_5	...
g_1		X	X	X	X	...
g_2			X	X	X	...
g_3	X			X	X	...
g_4	X	X			X	...
g_5	X	X	X			...
...	...	...	...	...	...	...

Fig. 3. The context of Example 3

We define the complexity $k(t)$ *of a term* t *inductively by:*

$k(x) := 0$ *for* $x \in X$,
$k(\bot) := k(\top) := 1$ *and*

$$k(t_1 \sqcap t_2) := max(\{k(t_i) \mid t_i \in \mathcal{T}_\sqcap(X), i \in \{1,2\}\} \cup \{1 + k(t_i) \mid t_i \notin \mathcal{T}_\sqcap(X), i \in \{1,2\}\})$$

$$k(t_1 \sqcup t_2) := max(\{k(t_i) \mid t_i \in \mathcal{T}_\sqcup(X), i \in \{1,2\}\} \cup \{1 + k(t_i) \mid t_i \notin \mathcal{T}_\sqcup(X), i \in \{1,2\}\})$$

$$k(\neg t) = \begin{cases} k(t) & for\ t \in \mathcal{T}_\sqcap(X) \\ 1 + k(t) & for\ t \notin \mathcal{T}_\sqcap(X) \end{cases}$$

$$k(\lrcorner t) = \begin{cases} k(t) & for\ t \in \mathcal{T}_\sqcup(X) \\ 1 + k(t) & for\ t \notin \mathcal{T}_\sqcup(X) \end{cases}.$$

In the following we will often need the set of all terms over a given set of variables with complexity less or equal to a given natural number. Therefore, we define:

Definition 7. *For a finite set of variables* X *we set*

$$\mathcal{T}^i(X) := \{t \in \mathcal{T}(X) \mid k(t) \leq i\}$$
$$\mathcal{T}^i_\sqcap(X) := \{t \in \mathcal{T}_\sqcap(X) \mid k(t) \leq i\}$$
$$\mathcal{T}^i_\sqcup(X) := \{t \in \mathcal{T}_\sqcup(X) \mid k(t) \leq i\}.$$

The free double Boolean algebra $\mathcal{D}(\emptyset)$. The free double Boolean algebra generated by the empty set was first described in [Vo02]. We briefly repeat the results given there.

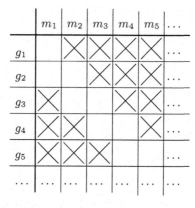

Lemma 1. *Let $\underline{D}$ be a double Boolean algebra. then*

$$1)\ x, y \in D_{\sqcup} \Rightarrow x \sqcap y \sqsubseteq x \mathbin{\overline{\sqcap}} y$$
$$2)\ x, y \in D_{\sqcap} \Rightarrow x \mathbin{\underline{\sqcup}} y \sqsubseteq x \sqcup y.$$

Proof. From Theorem 1.5 we obtain $x \sqcap y \sqsubseteq x \mathbin{\overline{\sqcap}} y \Leftrightarrow (x \sqcap y)_{\sqcup} \sqsubseteq x \mathbin{\overline{\sqcap}} y$. Since the map $a \mapsto a \sqcup b$ preserves $\sqsubseteq$ and $\sqcup$ (cp. [Wi00a]) follows $x \sqcap y \sqsubseteq x \Rightarrow (x \sqcap y)_{\sqcup} \sqsubseteq x \sqcup (x \sqcap y) = x \sqcup x = x$. Analogously we obtain $(x \sqcap y)_{\sqcup} \sqsubseteq y$. From $(x \sqcap y)_{\sqcup} \in \underline{D}_{\sqcup}$ follows $(x \sqcap y)_{\sqcup} \sqsubseteq x \mathbin{\overline{\sqcap}} y$ and thus 1). Dually, 2) is obtained. □

Lemma 2. *Let $\underline{D}$ be a double Boolean algebra and $x \in D_{\sqcup}$, $y \in D_{\sqcap}$. Then $x \sqcap \lrcorner x = \underline{\bot}$ and $y \sqcup \neg y = \overline{\top}$.*

Proof. We show $x \sqcap \lrcorner x \sqsubseteq \underline{\bot}$ and $\underline{\bot} \sqsubseteq x \sqcap \lrcorner x$ and use that $\sqsubseteq$ is an order on $\underline{D}_p$. The previous lemma yields $x \sqcap \lrcorner x \sqsubseteq x \mathbin{\overline{\sqcap}} \lrcorner x = \underline{\bot}$. Conversely, $\underline{\bot} \sqsubseteq x \sqcap \lrcorner x$, since $(x \sqcap \lrcorner x) \sqcup \underline{\bot} = (x \sqcap \lrcorner x) \sqcup \bot = (x \sqcap \lrcorner x)_{\sqcup}$ and $(x \sqcap \lrcorner x) \sqcap \underline{\bot} = (x \sqcap \underline{\bot}) \sqcap (\lrcorner x \sqcap \underline{\bot})$. Since $x \in D_{\sqcup}$ we have $\underline{\bot} \sqsubseteq x$ and $\underline{\bot} \sqsubseteq \lrcorner x$. This yields $\underline{\bot} = \underline{\bot} \sqcap \underline{\bot} \sqsubseteq (x \sqcap \underline{\bot})$ and $\underline{\bot} \sqsubseteq (\lrcorner x \sqcap \underline{\bot})$. Thus $(x \sqcap \underline{\bot}) \sqcap (\lrcorner x \sqcap \underline{\bot}) = (x \sqcap \underline{\bot}) \sqcap \underline{\bot} \sqcap (\lrcorner x \sqcap \underline{\bot}) \sqcap \underline{\bot} = \underline{\bot} \sqcap \underline{\bot}$. □

Proposition 1. *The free double Boolean algebra $\mathcal{D}(\emptyset)$ is the double Boolean algebra depicted in Fig. 2.*

Proof. The following tables show that the set $\{\top, \bot, \overline{\top}, \underline{\bot}, \lrcorner\overline{\top}, \neg\underline{\bot}\}$ is closed under the operations $\lrcorner, \neg, \sqcap$ und $\sqcup$.

	$\neg$	$\lrcorner$
$\top$	$\bot$	$\underline{\bot}$
$\bot$	$\overline{\top}$	$\top$
$\overline{\top}$	$\bot$	$\lrcorner\overline{\top}$
$\underline{\bot}$	$\neg\underline{\bot}$	$\top$
$\neg\underline{\bot}$	$\underline{\bot}$	$\lrcorner\overline{\top}$
$\lrcorner\overline{\top}$	$\neg\underline{\bot}$	$\overline{\top}$

$\sqcap$	$\top$	$\bot$	$\overline{\top}$	$\underline{\bot}$	$\neg\underline{\bot}$	$\lrcorner\overline{\top}$
$\top$	$\overline{\top}$	$\bot$	$\overline{\top}$	$\underline{\bot}$	$\neg\underline{\bot}$	$\underline{\bot}$
$\bot$	$\bot$	$\bot$	$\bot$	$\bot$	$\bot$	$\bot$
$\overline{\top}$	$\overline{\top}$	$\bot$	$\overline{\top}$	$\underline{\bot}$	$\neg\underline{\bot}$	$\underline{\bot}$
$\underline{\bot}$	$\underline{\bot}$	$\bot$	$\underline{\bot}$	$\underline{\bot}$	$\bot$	$\underline{\bot}$
$\neg\underline{\bot}$	$\neg\underline{\bot}$	$\bot$	$\neg\underline{\bot}$	$\bot$	$\neg\underline{\bot}$	$\bot$
$\lrcorner\overline{\top}$	$\underline{\bot}$	$\bot$	$\underline{\bot}$	$\underline{\bot}$	$\bot$	$\underline{\bot}$

$\sqcup$	$\top$	$\bot$	$\overline{\top}$	$\underline{\bot}$	$\neg\underline{\bot}$	$\lrcorner\overline{\top}$
$\top$	$\top$	$\top$	$\top$	$\top$	$\top$	$\top$
$\bot$	$\top$	$\underline{\bot}$	$\overline{\top}$	$\underline{\bot}$	$\overline{\top}$	$\lrcorner\overline{\top}$
$\overline{\top}$	$\top$	$\overline{\top}$	$\overline{\top}$	$\overline{\top}$	$\overline{\top}$	$\top$
$\underline{\bot}$	$\top$	$\underline{\bot}$	$\overline{\top}$	$\underline{\bot}$	$\overline{\top}$	$\lrcorner\overline{\top}$
$\neg\underline{\bot}$	$\top$	$\overline{\top}$	$\overline{\top}$	$\overline{\top}$	$\overline{\top}$	$\top$
$\lrcorner\overline{\top}$	$\top$	$\lrcorner\overline{\top}$	$\top$	$\lrcorner\overline{\top}$	$\top$	$\lrcorner\overline{\top}$

Most of these equations follow easily from the axioms. Lemma 2 is needed to determine $\lrcorner\overline{\top} \sqcap \overline{\top} = \underline{\bot}$ and $\neg\underline{\bot} \sqcup \underline{\bot} = \overline{\top}$. Then $\neg\underline{\bot} \sqcup \neg\underline{\bot} = \neg\underline{\bot} \sqcup \underline{\bot} = \overline{\top}$ is obtained from $x \sqcup \underline{\bot} = x \sqcup x$. Dually follows $\lrcorner\overline{\top} \sqcap \lrcorner\overline{\top} = \underline{\bot}$. This implies $\lrcorner\neg\underline{\bot} = \lrcorner(\neg\underline{\bot} \sqcup \neg\underline{\bot}) = \lrcorner\overline{\top}$ and, dually, $\neg\lrcorner\overline{\top} = \neg\underline{\bot}$.

2 An Algorithm for the Stepwise Exploration of Double Boolean Algebras

In this section an algorithm for the stepwise exploration of double Boolean algebras is developed. It is a generalization of the algorithm for protoconcept exploration as introduced in [Vo04]. The basic idea of the exploration approach as a knowledge acquisition tool was formulated early in the development of Formal Concept Analysis (cp. [Wi82]) and led to exporation algorithms for attributes (cp. [GW99], [Bu00]) and concepts (cp. [St97]). The common assumption is that in applications of Formal Concept Analysis we may have only implicit knowledge of a domain of interest and it may be impossible to write down explicitly all objects, all attributes and their incidence relation in a context, although some attributes, concepts or protoconcepts may be 'fairly clear'. The aim of protoconcept exploration can be stated as follows: We assume that there exists an underlying protoconcept algebra $\underline{\mathfrak{P}}_u = \mathfrak{P}(G_u, M_u, I_u)$ which may be only 'vaguely known' by the investigator. For a finite subset $\mathcal{B} = \{\mathfrak{p}_1, \dots \mathfrak{p}_n\}$ of $\underline{\mathfrak{P}}_u$, the set of basic protoconcepts, protoconcept exploration shall support him in determining the structure of the subalgebra $\underline{\mathfrak{P}}_g$ of $\underline{\mathfrak{P}}_u$ which is generated by $\mathcal{B}$. Therefore, the algorithm generates questions about implications between the identified protoconcepts. The answers given by the user are then used to find more protoconcepts.

In this paper we want to use a similar approach to explore free double Boolean algebras, which in general are not protoconcept algebras. Therefore, we have to adjust slightly the notions and proofs introduced in [Vo04] to the exploration of an arbitrary double Boolean algebra $\underline{\mathcal{D}}_g$ generated by elements $\mathfrak{g}_1, \dots, \mathfrak{g}_n$ in an underlying double Boolean algebra $\underline{\mathcal{D}}_u$. Here we still assume the existence of a domain expert that answers the questions generated by the algorithm. In Section 4 algorithms are developed that find the answers automatically for the case of free double Boolean algebras.

In this section we will always consider a finite set $\{\mathfrak{g}_1, \dots \mathfrak{g}_n\}$ of generators and a set of variables $X := \{x_1, \dots x_n\}$ of equal cardinality. We set $\vec{\mathfrak{g}} := (\mathfrak{g}_1, \dots \mathfrak{g}_n)$.

Definition 8. *For a set X of variables and sets of terms $S \subseteq \mathcal{T}_{\sqcup}(X)$ and $T \subseteq \mathcal{T}_{\sqcap}(X)$ we say that:*

1. *The pair $(\mathfrak{A}, \mathfrak{B}) \in \mathfrak{P}(S) \times \mathfrak{P}(S)$ is a* $\sqcup$-sequent over S *if $\mathfrak{A} \cap \mathfrak{B} = \emptyset$.*
2. *Dually, the pair $(\mathfrak{A}, \mathfrak{B}) \in \mathfrak{P}(T) \times \mathfrak{P}(T)$ is a* $\sqcap$-sequent *over T if $\mathfrak{A} \cap \mathfrak{B} = \emptyset$,*
3. *A $\sqcup$-sequent over S ($\sqcap$-sequent over T) is* full *iff $\mathfrak{A} \cup \mathfrak{B} = S$ ($\mathfrak{A} \cup \mathfrak{B} = T$).*

On sequents of the same type we define an order by $(\mathfrak{A}_1, \mathfrak{B}_1) \leq (\mathfrak{A}_2, \mathfrak{B}_2) :\Leftrightarrow \mathfrak{A}_1 \subseteq \mathfrak{A}_2$ and $\mathfrak{B}_1 \subseteq \mathfrak{B}_2$.

To shorten notation, for $\sqcap$-sequents $(\mathfrak{A}, \mathfrak{B})$, we set

$$m(\mathfrak{A}, \mathfrak{B}) := ((\bigsqcap \mathfrak{A}) \sqcap (\neg \bigsqcup\!\!\!\!\!\!- \ \mathfrak{B}))$$

and, dually for $\sqcup$-sequents $(\mathfrak{A}, \mathfrak{B})$,

$$j(\mathfrak{A}, \mathfrak{B}) := ((\lrcorner \overline{\bigsqcap} \mathfrak{A}) \sqcup (\bigsqcup \mathfrak{B})).$$

In the sense of Definition 4, we denote the evaluation of a term t in the algebra $\mathcal{D}_u$ assigning to every variable x_i the generator $\mathfrak{g}_i$ with $(t)^{\mathcal{D}_u}(\vec{\mathfrak{g}})$. Note that, for n generators $\mathfrak{g}_1, \dots \mathfrak{g}_n$ of $\mathcal{D}_g$, $X := \{x_1, \dots, x_n\}$ and a $\sqcap$-sequent $(\mathfrak{A}, \mathfrak{B})$ over $T \subseteq \mathcal{T}_{\sqcap}(X)$ we have

$$(\bigsqcap \mathfrak{A})^{\mathcal{D}_u}(\vec{\mathfrak{g}}) \sqsubseteq (\bigsqcup\!\!\!\!\!\!- \ \mathfrak{B})^{\mathcal{D}_u}(\vec{\mathfrak{g}}) \Leftrightarrow m(\mathfrak{A}, \mathfrak{B})^{\mathcal{D}_u}(\vec{\mathfrak{g}}) = \bot,$$

and, dually for a $\sqcup$-sequent $(\mathfrak{A}, \mathfrak{B})$ over $S \subseteq \mathcal{T}_{\sqcup}(X)$, we have

$$(\bigsqcap\mathfrak{A})^{\mathcal{D}_u}(\vec{\mathfrak{g}}) \sqsubseteq (\bigsqcup\mathfrak{B})^{\mathcal{D}_u}(\vec{\mathfrak{g}}) \Leftrightarrow j(\mathfrak{A},\mathfrak{B})^{\mathcal{D}_u}(\vec{\mathfrak{g}}) = \top.$$

The exploration algorithm uses that for $i \in \mathbb{N}$ the set $\mathcal{B}^i_\sqcap := \{\mathfrak{p} \in \mathcal{D}_g \mid \exists t \in \mathcal{T}^i_\sqcap(X). \mathfrak{p} = (t)^{\mathcal{D}_u}(\vec{\mathfrak{g}})\}$ is the universe of a finite subalgebra $\underline{\mathcal{B}}^i_\sqcap := (\mathcal{B}^i_\sqcap, \sqcap, \uplus, \neg, \bot, \top)$ of the Boolean algebra $\underline{\mathcal{D}}_{g\sqcap}$. Dually, $\mathcal{B}^i_\sqcup := \{\mathfrak{p} \in \mathcal{D}_g \mid \exists t \in \mathcal{T}^i_\sqcup(X).\mathfrak{p} = (t)^{\mathcal{D}_u}(\vec{\mathfrak{g}})\}$ is the universe of a finite subalgebra $\underline{\mathcal{B}}^i_\sqcup := (\mathcal{B}^i_\sqcup, \bar{\sqcap}, \sqcup, \lrcorner, \bot\!\!\!\bot, \top)$ of the Boolean algebra $\underline{\mathcal{D}}_{g\sqcup}$ (cp. Corollary 2). In the i-th iteration, the exploration algorithm generates sets $S^i_\sqcap$ and $S^i_\sqcup$ of sequents that describe the atoms resp. coatoms of $\mathcal{B}^i_\sqcap$ resp. $\mathcal{B}^i_\sqcup$. The algorithm has four steps:

Exploration algorithm

1. For n generators we take $X := \{x_1, \ldots x_n\}$ as set of variables and set $\tilde{S}^1_\sqcup$: $\{x_{1\sqcup}, x_{2\sqcup}, \ldots x_{n\sqcup}\}$, $\tilde{S}^1_\sqcap := \{x_{1\sqcap}, x_{2\sqcap}, \ldots x_{n\sqcap}\}$ and $S^0_\sqcup := S^0_\sqcap := \emptyset$.
2. (a) For $i \geq 1$ we determine the set

   $$S^i_\sqcup := \{(\mathfrak{A},\mathfrak{B}) \mid (\mathfrak{A},\mathfrak{B}) \text{ is a full } \sqcup\text{-sequent over } \tilde{S}^i_\sqcup \text{ and } j(\mathfrak{A},\mathfrak{B})^{\mathcal{D}_u}(\vec{\mathfrak{g}}) \neq \top\}$$

   in interaction with the user.
   (b) Dually, we determine for $i \geq 1$ the set

   $$S^i_\sqcap := \{(\mathfrak{A},\mathfrak{B}) \mid (\mathfrak{A},\mathfrak{B}) \text{ is a full } \sqcap\text{-sequent over } \tilde{S}^i_\sqcap \text{ and } m(\mathfrak{A},\mathfrak{B})^{\mathcal{D}_u}(\vec{\mathfrak{g}}) \neq \bot\}.$$

3. We set $G_i := \{(m(\mathfrak{A},\mathfrak{B}))^{\mathcal{D}_u}(\vec{\mathfrak{g}}) \mid (\mathfrak{A},\mathfrak{B}) \in S^i_\sqcap\}$, $M_i := \{(j(\mathfrak{A},\mathfrak{B}))^{\mathcal{D}_u}(\vec{\mathfrak{g}}) \mid (\mathfrak{A},\mathfrak{B}) \in S^i_\sqcup\}$ and $\mathbb{K}_i := (G_i, M_i, \sqsubseteq)$ where we determine the relation $\sqsubseteq$ with the aid of the expert.
4. We stop if $|S^{i-1}_\sqcup| = |S^i_\sqcup|$ and $|S^{i-1}_\sqcap| = |S^i_\sqcap|$, otherwise we set $\tilde{S}^{i+1}_\sqcap := \tilde{S}^i_\sqcap \cup \{(j(\mathfrak{A},\mathfrak{B}) \sqcap j(\mathfrak{A},\mathfrak{B}) \mid (\mathfrak{A},\mathfrak{B}) \in S^i_\sqcup\}$ and $\tilde{S}^{i+1}_\sqcup := \tilde{S}^i_\sqcup \cup \{(m(\mathfrak{A},\mathfrak{B}) \sqcup m(\mathfrak{A},\mathfrak{B}) \mid (\mathfrak{A},\mathfrak{B}) \in S^i_\sqcap\}$ and continue with 2).

The rest of this section is used to prove that the sets $S^i_\sqcap$ and $S^i_\sqcup$ of sequents found by th exploration algorithm indeed describe the atoms resp. coatoms of $\mathcal{B}^i_\sqcap$ resp. $\mathcal{B}^i_\sqcup$. This follows from Theorem 4 which states that every element $\mathfrak{a}$ of the pure double Boolean $\mathcal{D}_{g\,p}$ that can be generated as $\mathfrak{a} = t^{\mathcal{D}_u}(\vec{\mathfrak{g}})$ with $k(t) \leq i$ may be represented as a set of objects or attributes of the context $\mathbb{K}_i$ generated in the i-th iteration of the exploration algorithm. In the next section, this allows us to map these generated elements to semiconcepts of $\mathbb{K}_i$.

Proposition 2. *1) Let $i \geq 1$ and let $(\mathfrak{A}_1,\mathfrak{B}_1)$, $(\mathfrak{A}_2,\mathfrak{B}_2) \in S^i_\sqcap$ with $(\mathfrak{A}_1,\mathfrak{B}_1) \neq (\mathfrak{A}_2,\mathfrak{B}_2)$. Then*

$$(m(\mathfrak{A}_1,\mathfrak{B}_1) \sqcap m(\mathfrak{A}_2,\mathfrak{B}_2))^{\mathcal{D}_u}(\vec{\mathfrak{g}}) = \bot.$$

2)Dually, if $(\mathfrak{A}_1,\mathfrak{B}_1)$, $(\mathfrak{A}_2,\mathfrak{B}_2) \in S^i_\sqcup$ with $(\mathfrak{A}_1,\mathfrak{B}_1) \neq (\mathfrak{A}_2,\mathfrak{B}_2)$ then

$$(j(\mathfrak{A}_1,\mathfrak{B}_1) \sqcup j(\mathfrak{A}_2,\mathfrak{B}_2))^{\mathcal{D}_u}(\vec{\mathfrak{g}}) = \top.$$

Proof. 1) From $(\mathfrak{A}_1,\mathfrak{B}_1) \neq (\mathfrak{A}_2,\mathfrak{B}_2)$ it follows $\mathfrak{A}_1 \neq \mathfrak{A}_2$ since the sequents are full. Thus there is either a $t \in \tilde{S}^i_\sqcap$ with $t \in \mathfrak{A}_1$, $t \notin \mathfrak{A}_2$ or vice versa. We assume w.l.o.g. the first case. Then

$$\begin{aligned} m(\mathfrak{A}_1, \mathfrak{B}_1) \sqcap m(\mathfrak{A}_2, \mathfrak{B}_2) &= ((\bigsqcap \mathfrak{A}_1) \sqcap (\neg \overline{\bigsqcap} \mathfrak{B}_1)) \sqcap ((\bigsqcap \mathfrak{A}_2) \sqcap (\neg \overline{\bigsqcap} \mathfrak{B}_2)) \\ &\approx ((\bigsqcap (\mathfrak{A}_1 \setminus \{t\})) \sqcap (\neg \overline{\bigsqcap} \mathfrak{B}_1) \sqcap t) \\ &\quad \sqcap ((\bigsqcap \mathfrak{A}_2) \sqcap (\neg \overline{\bigsqcap} (\mathfrak{B}_2 \setminus \{t\})) \sqcap \neg t) \\ &\approx (\bigsqcap (\mathfrak{A}_1 \setminus \{t\})) \sqcap (\neg \overline{\bigsqcap} \mathfrak{B}_1) \sqcap (\bigsqcap \mathfrak{A}_2) \\ &\quad \sqcap (\neg \overline{\bigsqcap} (\mathfrak{B}_2 \setminus \{t\})) \sqcap (t \sqcap \neg t) \\ &\approx \bot. \end{aligned}$$

2) Follows dually. □

The maps m and j assigning terms to sequents extend to sets of sequents: For a set $\mathcal{A}$ of $\sqcap$-sequents over a given set of terms we define

$$m(\mathcal{A}) := \{m(\mathfrak{A}, \mathfrak{B}) \in \mathcal{T}_\sqcap(X) \mid (\mathfrak{A}, \mathfrak{B}) \in \mathcal{A}\}$$

and, dually for a set $\mathcal{B}$ of $\sqcup$-sequents over a given set of terms we define

$$j(\mathcal{B}) := \{j(\mathfrak{A}, \mathfrak{B}) \in \mathcal{T}_\sqcup(X) \mid (\mathfrak{A}, \mathfrak{B}) \in \mathcal{B}\}.$$

Corollary 1. *1) Let $i \geq 1$ and $\mathcal{A}_1, \mathcal{A}_2 \subseteq S^i_\sqcap$. Then*

$$(\underline{\bigsqcup} m(\mathcal{A}_1) \sqcap \underline{\bigsqcup} m(\mathcal{A}_2))^{\mathcal{D}_u}(\overrightarrow{\mathfrak{g}}) = (\underline{\bigsqcup} m(\mathcal{A}_1 \cap \mathcal{A}_2))^{\mathcal{D}_u}(\overrightarrow{\mathfrak{g}}).$$

2) Dually, if $\mathcal{A}_1, \mathcal{A}_2 \subseteq S^i_\sqcup$. Then

$$(\overline{\bigsqcap} j(\mathcal{A}_1) \sqcup \overline{\bigsqcap} j(\mathcal{A}_2))^{\mathcal{D}_u}(\overrightarrow{\mathfrak{g}}) = (\overline{\bigsqcap} j(\mathcal{A}_1 \cap \mathcal{A}_2))^{\mathcal{D}_u}(\overrightarrow{\mathfrak{g}}).$$

Lemma 3. *1) Let $(\mathfrak{A}, \mathfrak{B})$ be a $\sqcup$-sequent over a finite set $S \subseteq \mathcal{T}_\sqcup(X)$. Then*

$$\begin{aligned} j(\mathfrak{A}, \mathfrak{B})^{\mathcal{D}_u}(\overrightarrow{\mathfrak{g}}) = (\overline{\bigsqcap}\{j(\mathfrak{C}, \mathfrak{D}) \mid (\mathfrak{C}, \mathfrak{D}) \textit{ is a full } \sqcup\textit{-sequent over } S \textit{ with} \\ j(\mathfrak{C}, \mathfrak{D})^{\mathcal{D}_u}(\overrightarrow{\mathfrak{g}}) \neq \top \textit{ and } (\mathfrak{A}, \mathfrak{B}) \leq (\mathfrak{C}, \mathfrak{D})\})^{\mathcal{D}_u}(\overrightarrow{\mathfrak{g}}). \end{aligned}$$

2) Dually, let $(\mathfrak{A}, \mathfrak{B})$ be a $\sqcap$-sequent over a set $S \subseteq \mathcal{T}_\sqcap(X)$. Then

$$\begin{aligned} m(\mathfrak{A}, \mathfrak{B})^{\mathcal{D}_u}(\overrightarrow{\mathfrak{g}}) = (\underline{\bigsqcup}\{m(\mathfrak{C}, \mathfrak{D}) \mid (\mathfrak{C}, \mathfrak{D}) \textit{ is a full } \sqcap\textit{-sequent over } S \textit{ with} \\ m(\mathfrak{C}, \mathfrak{D})^{\mathcal{D}_u}(\overrightarrow{\mathfrak{g}}) \neq \bot \textit{ and } (\mathfrak{A}, \mathfrak{B}) \leq (\mathfrak{C}, \mathfrak{D})\})^{\mathcal{D}_u}(\overrightarrow{\mathfrak{g}}). \end{aligned}$$

Proof. 1) Note that

$$\begin{aligned} &\overline{\bigsqcap} \quad \{j(\mathfrak{C}, \mathfrak{D}) \mid (\mathfrak{C}, \mathfrak{D}) \text{ is a full } \sqcup\text{-sequent over } S \text{ with } j(\mathfrak{C}, \mathfrak{D})^{\mathcal{D}_u}(\overrightarrow{\mathfrak{g}}) \neq \top \\ &\qquad\qquad \text{and } (\mathfrak{A}, \mathfrak{B}) \leq (\mathfrak{C}, \mathfrak{D})\}^{\mathcal{D}_u}(\overrightarrow{\mathfrak{g}}) \\ &= \overline{\bigsqcap} \{j(\mathfrak{C}, \mathfrak{D}) \mid (\mathfrak{C}, \mathfrak{D}) \text{ is a full } \sqcup\text{-sequent over } S \text{ and } (\mathfrak{A}, \mathfrak{B}) \leq (\mathfrak{C}, \mathfrak{D})\}^{\mathcal{D}_u}(\overrightarrow{\mathfrak{g}}) \end{aligned}$$

since $\top \overline{\sqcap} (a \sqcup a) \approx (a \sqcup a)$. For every $t \in S \setminus (\mathfrak{A} \cup \mathfrak{B})$ and for every full $\sqcup$-sequent $(\mathfrak{C}, \mathfrak{D})$ over $S \setminus \{t\}$, the sequents $(\mathfrak{C} \cup \{t\}, \mathfrak{D})$ and $(\mathfrak{C}, \mathfrak{D} \cup \{t\})$ are full $\sqcup$-sequents over S and every full $\sqcup$-sequent over S is of that form. From

$$\begin{aligned} j(\mathfrak{C}\cup\{t\},\mathfrak{D})\sqcap j(\mathfrak{C},\mathfrak{D}\cup\{t\}) &= (\neg\bigsqcap(\mathfrak{C}\cup\{t\})\sqcup\bigsqcup\mathfrak{D})\sqcap(\neg\bigsqcap\mathfrak{C}\sqcup\bigsqcup(\mathfrak{D}\cup\{t\}))\\ &\approx(\neg\bigsqcap\mathfrak{C}\sqcup\bigsqcup\mathfrak{D}\sqcup\neg t)\sqcap(\neg\bigsqcap\mathfrak{C}\sqcup\bigsqcup\mathfrak{D}\sqcup t)\\ &\approx((\neg\bigsqcap\mathfrak{C}\sqcup\bigsqcup\mathfrak{D}\sqcup\neg t)\sqcap(\neg\bigsqcap\mathfrak{C}\sqcup\bigsqcup\mathfrak{D}))\\ &\quad\sqcup((\neg\bigsqcap\mathfrak{C}\sqcup\bigsqcup\mathfrak{D}\sqcup\neg t)\sqcap t)\end{aligned}$$

it follows with

$$(\neg\bigsqcap\mathfrak{C}\sqcup\bigsqcup\mathfrak{D}\sqcup\neg t)\sqcap(\neg\bigsqcap\mathfrak{C}\sqcup\bigsqcup\mathfrak{D})\approx\neg\bigsqcap\mathfrak{C}\sqcup\bigsqcup\mathfrak{D}$$

and $(\neg\bigsqcap\mathfrak{C}\sqcup\bigsqcup\mathfrak{D}\sqcup\neg t)\sqcap t\approx(\neg\bigsqcap\mathfrak{C}\sqcup\bigsqcup\mathfrak{D})\sqcap t$ that $j(\mathfrak{C}\cup\{t\},\mathfrak{D})\sqcap j(\mathfrak{C},\mathfrak{D}\cup\{t\})\approx j(\mathfrak{C},\mathfrak{D})$. Thus

$$\begin{aligned} &\bigsqcap\{j(\mathfrak{C},\mathfrak{D})\mid(\mathfrak{C},\mathfrak{D})\text{ is a full }\sqcup\text{-sequent over }S\text{ and }(\mathfrak{A},\mathfrak{B})\leq(\mathfrak{C},\mathfrak{D})\}\\ \approx&\bigsqcap\{j(\mathfrak{C},\mathfrak{D})\mid(\mathfrak{C},\mathfrak{D})\text{ is a full }\sqcup\text{-sequent over }S\setminus\{t\}\text{ and }(\mathfrak{A},\mathfrak{B})\leq(\mathfrak{C},\mathfrak{D})\}.\end{aligned}$$

2) follows dually. □

Lemma 4. *1) Let $t\in\widetilde{S}^i_\sqcup$. Then*

$$(t_\sqcup)^{\mathcal{D}_u}(\overrightarrow{\mathfrak{g}})=(\bigsqcap\{j(\mathfrak{A},\mathfrak{B})\mid(\mathfrak{A},\mathfrak{B})\in S^i_\sqcup\textit{ and }t\in\mathfrak{B}\})^{\mathcal{D}_u}(\overrightarrow{\mathfrak{g}})$$

and

$$(\neg t_\sqcup)^{\mathcal{D}_u}(\overrightarrow{\mathfrak{g}})=(\bigsqcap\{j(\mathfrak{A},\mathfrak{B})\mid(\mathfrak{A},\mathfrak{B})\in S^i_\sqcup\textit{ and }t\in\mathfrak{A}\})^{\mathcal{D}_u}(\overrightarrow{\mathfrak{g}})$$

2) Dually, let $t\in\widetilde{S}^i_\sqcap$. Then

$$(t_\sqcap)^{\mathcal{D}_u}(\overrightarrow{\mathfrak{g}})=(\bigsqcup\{m(\mathfrak{A},\mathfrak{B})\mid(\mathfrak{A},\mathfrak{B})\in S^i_\sqcap\textit{ and }t\in\mathfrak{A}\})^{\mathcal{D}_u}(\overrightarrow{\mathfrak{g}})$$

and

$$(\neg t_\sqcap)=(\bigsqcup\{m(\mathfrak{A},\mathfrak{B})\mid(\mathfrak{A},\mathfrak{B})\in S^i_\sqcap\textit{ and }t\in\mathfrak{B}\})^{\mathcal{D}_u}(\overrightarrow{\mathfrak{g}})$$

Proof. This is a consequence of Lemma 3. For 1) we have that $(t\sqcup t)\approx j(\emptyset,\{t\})$ and thus

$$\begin{aligned}(t_\sqcup)^{\mathcal{D}_u}(\overrightarrow{\mathfrak{g}})&=(\bigsqcap\{j(\mathfrak{A},\mathfrak{B})\mid(\mathfrak{A},\mathfrak{B})\in S^i_\sqcup\text{ and }(\emptyset,\{t\})\leq(\mathfrak{A},\mathfrak{B})\})^{\mathcal{D}_u}(\overrightarrow{\mathfrak{g}})\\ &=(\bigsqcap\{j(\mathfrak{A},\mathfrak{B})\mid(\mathfrak{A},\mathfrak{B})\in S^i_\sqcup\text{ and }t\in\mathfrak{B}\})^{\mathcal{D}_u}(\overrightarrow{\mathfrak{g}}).\end{aligned}$$

Analogously it follows from $(\neg t_\sqcup)\approx j(\{t\},\emptyset)$ that $(\neg t_\sqcup))^{\mathcal{D}_u}(\overrightarrow{\mathfrak{g}})=(\bigsqcap\{j(\mathfrak{A},\mathfrak{B})\mid(\mathfrak{A},\mathfrak{B})\in S^i_\sqcup\text{ and }t\in\mathfrak{A}\})^{\mathcal{D}_u}(\overrightarrow{\mathfrak{g}})$. Dually we obtain 2). □

Lemma 5. *Let $i\geq 1$, $\mathcal{A}\subseteq S^i_\sqcap$ and $\mathcal{B}\subseteq S^i_\sqcup$. Then*

$$(\neg\bigsqcup m(\mathcal{A}))^{\mathcal{D}_u}(\overrightarrow{\mathfrak{g}})=(\bigsqcup m(S^i_\sqcap\setminus\mathcal{A}))^{\mathcal{D}_u}(\overrightarrow{\mathfrak{g}}).$$

and

$$(\neg\bigsqcap j(\mathcal{B}))^{\mathcal{D}_u}(\overrightarrow{\mathfrak{g}})=(\bigsqcap j(S^i_\sqcup\setminus\mathcal{B}))^{\mathcal{D}_u}(\overrightarrow{\mathfrak{g}}).$$

Proof. From Corollary 1 we obtain immediately

$$(\biguplus m(\mathcal{A}))^{\mathcal{D}_u}(\vec{\mathfrak{g}}) \sqcap (\biguplus m(S^i_\sqcap \setminus \mathcal{A}))^{\mathcal{D}_u}(\vec{\mathfrak{g}}) = (\biguplus m(\mathcal{A}) \sqcap \biguplus m(S^i_\sqcap \setminus \mathcal{A}))^{\mathcal{D}_u}(\vec{\mathfrak{g}})$$
$$= (\biguplus \emptyset)^{\mathcal{D}_u}(\vec{\mathfrak{g}}) = \bot.$$

Moreover, Lemma 3 yields

$$(\biguplus m(\mathcal{A}))^{\mathcal{D}_u}(\vec{\mathfrak{g}}) \uplus (\biguplus (S^i_\sqcap \setminus \mathcal{A}))^{\mathcal{D}_u}(\vec{\mathfrak{g}}) = (\biguplus m(S^i_\sqcap))^{\mathcal{D}_u}(\vec{\mathfrak{g}})$$
$$= (\biguplus m(\{(\mathfrak{A},\mathfrak{B}) \setminus (\mathfrak{A},\mathfrak{B}) \text{ is a full } \sqcap\text{-sequent over } \tilde{S}^i_\sqcap \text{ and } (\emptyset,\emptyset) \leq (\mathfrak{A},\mathfrak{B})\}))^{\mathcal{D}_u}(\vec{\mathfrak{g}})$$
$$= (\bigsqcap \emptyset \sqcap \neg \biguplus \emptyset)^{\mathcal{D}_u}(\vec{\mathfrak{g}})$$
$$= \top,$$

thus $(\biguplus m(S^i_\sqcap \setminus \mathcal{A}))^{\mathcal{D}_u}(\vec{\mathfrak{g}})$ is the negation of $(\biguplus m(\mathcal{A}))^{\mathcal{D}_u}(\vec{\mathfrak{g}})$ in the Boolean algebra $\mathfrak{P}_{u\sqcap}$. Dually we obtain $(\lrcorner \bigsqcap j(\mathcal{B}))^{\mathcal{D}_u}(\vec{\mathfrak{g}}) = (\bigsqcap j(S^i_\sqcup \setminus \mathcal{B}))^{\mathcal{D}_u}(\vec{\mathfrak{g}})$. □

Lemma 6. *For every set $A \subseteq S^i_\sqcap$ exists a set $A^* \subseteq S^{i+1}_\sqcap$ such that*

$$(\biguplus m(A))^{\mathcal{D}_u}(\vec{\mathfrak{g}}) = (\biguplus m(A^*))^{\mathcal{D}_u}(\vec{\mathfrak{g}}).$$

Dually, for $B \subseteq S^i_\sqcup$ there exists a set $B^* \subseteq S^{i+1}_\sqcup$ such that

$$(\bigsqcap j(B))^{\mathcal{D}_u}(\vec{\mathfrak{g}}) = (\bigsqcap j(B^*))^{\mathcal{D}_u}(\vec{\mathfrak{g}}).$$

Proof. This follows from $\tilde{S}^i_\sqcap \subseteq \tilde{S}^{i+1}_\sqcap$ resp. $\tilde{S}^i_\sqcup \subseteq \tilde{S}^{i+1}_\sqcup$ and Lemma 3: We find that every $\sqcap$-sequent over $\tilde{S}^i_\sqcap$ is a $\sqcap$-sequent over $\tilde{S}^{i+1}_\sqcap$ and thus

$$(\biguplus m(A))^{\mathcal{D}_u}(\vec{\mathfrak{g}}) = (\biguplus \{m(\mathfrak{C},\mathfrak{D}) \mid (\mathfrak{C},\mathfrak{D}) \text{ is a full } \sqcap\text{-sequent over } \tilde{S}^{i+1}_\sqcap \text{ and } (\mathfrak{A},\mathfrak{B}) \leq (\mathfrak{C},\mathfrak{D}) \text{ for some } (\mathfrak{A},\mathfrak{B}) \in A\})^{\mathcal{D}_u}(\vec{\mathfrak{g}}).$$

Dually, the second claim is obtained. □

Proposition 3. *1) Let $t \in T^i_\sqcap(X)$ and $A \subseteq S^i_\sqcap$ with $t^{\mathcal{D}_u}(\vec{\mathfrak{g}}) = (\biguplus m(A))^{\mathcal{D}_u}(\vec{\mathfrak{g}})$. Then there exists a subset $B \subseteq S^{i+1}_\sqcup$ such that $(t \sqcup t)^{\mathcal{D}_u}(\vec{\mathfrak{g}}) = (\bigsqcap j(B))^{\mathcal{D}_u}(\vec{\mathfrak{g}})$. 2) Dually, let $t \in T^i_\sqcup(X)$ and $A \subseteq S^i_\sqcup$ with $t^{\mathcal{D}_u}(\vec{\mathfrak{g}}) = (\bigsqcap j(A))^{\mathcal{D}_u}(\vec{\mathfrak{g}})$. Then there exists a subset $B \subseteq S^{i+1}_\sqcap$ such that $(t \sqcap t)^{\mathcal{D}_u}(\vec{\mathfrak{g}}) = (\biguplus m(B))^{\mathcal{D}_u}(\vec{\mathfrak{g}})$.*

Proof. 1) Since for elements $\mathfrak{a}, \mathfrak{b} \in \mathfrak{D}_{u\sqcap}$ holds $(\mathfrak{a} \uplus \mathfrak{b}) \sqcup (\mathfrak{a} \uplus \mathfrak{b}) = \mathfrak{a} \sqcup \mathfrak{b}$ (cf.[Vo03]), it follows from $t^{\mathcal{D}_u}(\vec{\mathfrak{g}}) = (\biguplus m(A))^{\mathcal{D}_u}(\vec{\mathfrak{g}})$ that

$$(t \sqcup t)^{\mathcal{D}_u}(\vec{\mathfrak{g}}) = \bigsqcup (\{m(\mathfrak{A},\mathfrak{B}) \mid (\mathfrak{A},\mathfrak{B}) \in A\})^{\mathcal{D}_u}(\vec{\mathfrak{g}})$$
$$= \bigsqcup (\{m(\mathfrak{A},\mathfrak{B}) \sqcup m(\mathfrak{A},\mathfrak{B}) \mid (\mathfrak{A},\mathfrak{B}) \in A\})^{\mathcal{D}_u}(\vec{\mathfrak{g}}).$$

Since $(m(\mathfrak{A},\mathfrak{B}) \sqcup m(\mathfrak{A},\mathfrak{B})) \in \tilde{S}^{i+1}_\sqcup$ for $(\mathfrak{A},\mathfrak{B}) \in S^i_\sqcap \supseteq A$, Lemma 3 yields

$$(m(\mathfrak{A},\mathfrak{B}) \sqcup m(\mathfrak{A},\mathfrak{B}))^{\mathcal{D}_u}(\vec{\mathfrak{g}}) = j(\emptyset, \{(m(\mathfrak{A},\mathfrak{B}) \sqcup m(\mathfrak{A},\mathfrak{B}))\})^{\mathcal{D}_u}(\vec{\mathfrak{g}})$$
$$= \bigsqcap \{j(\mathfrak{C},\mathfrak{D}) \mid (\mathfrak{C},\mathfrak{D}) \in S^{i+1}_\sqcup \text{ and } (m(\mathfrak{A},\mathfrak{B}))_\sqcup \in \mathfrak{D}\}.$$

We conclude from Corollary 1 that

$$(t \sqcup t)^{\mathcal{D}_u}(\vec{\mathfrak{g}}) = (\bigsqcup_{(\mathfrak{A},\mathfrak{B})\in A} \bigsqcap\{j(\mathfrak{C},\mathfrak{D}) \mid (\mathfrak{C},\mathfrak{D}) \in S_{\sqcup}^{i+1} \text{ and } (m(\mathfrak{A},\mathfrak{B}) \sqcup m(\mathfrak{A},\mathfrak{B})) \in \mathfrak{D}\})^{\mathcal{D}_u}(\vec{\mathfrak{g}})$$
$$= (\bigsqcap\{j(\mathfrak{C},\mathfrak{D}) \mid (\mathfrak{C},\mathfrak{D}) \in S_{\sqcup}^{i+1} \text{ and } (m(\mathfrak{A},\mathfrak{B}) \sqcup m(\mathfrak{A},\mathfrak{B})) \in \mathfrak{D} \text{ for every } (\mathfrak{A},\mathfrak{B}) \in A\})^{\mathcal{D}_u}(\vec{\mathfrak{g}}).$$

2) follows dually. □

The following theorem is the main result of this section. It assigns to every term $t \in T_{\sqcap}^i$ a subset A_t of $S_{\sqcap}^i$ such that the evaluation of t and $\bigsqcap m(A_t)$ in $\mathcal{D}_u$ yield the same element of $\mathcal{D}_g$. If we say that an element $\mathfrak{a}$ of $\mathcal{D}_g$ is generated in i steps from $\mathfrak{g}_1, \dots \mathfrak{g}_n$ if there exists a term t with $k(t) \leq i$ such that $\mathfrak{a} = (t)^{\mathcal{D}_u}(\vec{\mathfrak{g}})$ then Theorem 4 ensures that every element of $\mathcal{D}_{g\sqcap}$ generated in i steps from $\mathfrak{g}_1, \dots \mathfrak{g}_n$ is obtained from $S_{\sqcap}^i$. Dually, every element of $\mathcal{D}_{g\sqcup}$ generated in i steps from $\mathfrak{g}_1, \dots \mathfrak{g}_n$ is obtained from $S_{\sqcup}^i$. This yields the desired Corollary 2. Moreover, in the next section we obtain for the case of free double Boolean algebra a map of equivalence classes of $T^i(X)$ to semiconcepts of $\mathbb{K}_i$ and show that $\mathfrak{P}(\mathbb{K}_i)$ is a counterexample to all invalid term identities $s \sim t$ for $k(s), k(t) < i$.

Theorem 4. *Let $t \in T_{\sqcap}^i(X)$. We set*

$$A_t := \{(\mathfrak{A},\mathfrak{B}) \in S_{\sqcap}^i \mid m(\mathfrak{A},\mathfrak{B})^{\mathcal{D}_u}(\vec{\mathfrak{g}}) \sqsubseteq t^{\mathcal{D}_u}(\vec{\mathfrak{g}})\}.$$

Then $t^{\mathcal{D}_u}(\vec{\mathfrak{g}}) = (\bigsqcup m(A_t))^{\mathcal{D}_u}(\vec{\mathfrak{g}})$. Dually, for $t \in T_{\sqcup}^i(X)$ we set

$$A_t := \{(\mathfrak{A},\mathfrak{B}) \in S_{\sqcup}^i \mid j(\mathfrak{A},\mathfrak{B})^{\mathcal{D}_u}(\vec{\mathfrak{g}}) \sqsupseteq t^{\mathcal{D}_u}(\vec{\mathfrak{g}})\}.$$

Then $t^{\mathcal{D}_u}(\vec{\mathfrak{g}}) = (\bigsqcap j(A_t))^{\mathcal{D}_u}(\vec{\mathfrak{g}})$.

Proof. First, we show the existence of sets $A \subseteq S_{\sqcap}^i$ and $B \subseteq S_{\sqcup}^i$ with $t^{\mathcal{D}_u}(\vec{\mathfrak{g}}) = (\bigsqcup m(A))^{\mathcal{D}_u}(\vec{\mathfrak{g}})$ resp. $t^{\mathcal{D}_u}(\vec{\mathfrak{g}}) = (\bigsqcap j(B))^{\mathcal{D}_u}(\vec{\mathfrak{g}})$ through induction over i: For $i = 1$, $t \in T_{\sqcap}^1(X)$, t is generated from $\bot, x_1, \dots, x_n$ by applying the operations $\sqcap$ and $\neg$. Since $s_1 \sqcap s_2 \approx (s_1 \sqcap s_1) \sqcap (s_2 \sqcap s_2)$ and $\neg s_1 \approx \neg(s_1 \sqcap s_1)$ we may replace any occurence of x_j by $x_{j\sqcap}$. For each of these generators g we find a set $A \subseteq G_1$ such that $g = (\bigsqcup m(A))^{\mathcal{D}_u}(\vec{\mathfrak{g}})$: We set $A := \emptyset$ for $g = \bot$, and for $g = (x_{j\sqcap})$ for some $j \in \{1, \dots, n\}$ Lemma 4 yields the corresponding set A.

Moreover, if we have $s, t \in T_{\sqcap}^1(X)$ and sets $A_s, A_t \subseteq G_1$ with $s^{\mathcal{D}_u}(\vec{\mathfrak{g}}) = (\bigsqcup A_s)^{\mathcal{D}_u}(\vec{\mathfrak{g}})$ and $t^{\mathcal{D}_u}(\vec{\mathfrak{g}}) = (\bigsqcup A_t)^{\mathcal{D}_u}(\vec{\mathfrak{g}})$ then Lemma 5 and Corollary 1 yield sets $B, C \subseteq G_1$ with $(\neg s)^{\mathcal{D}_u}(\vec{\mathfrak{g}}) = (\bigsqcup m(B))^{\mathcal{D}_u}(\vec{\mathfrak{g}})$ and $(s \sqcap t)^{\mathcal{D}_u}(\vec{\mathfrak{g}}) = (\bigsqcup C)^{\mathcal{D}_u}(\vec{\mathfrak{g}})$. Thus we find for every $t \in T_{\sqcap}^1(X)$ a subset of G_1 with the desired property. Analogously we conclude that there is a set $A \subseteq M_1$ for every $t \in T_{\sqcup}^1$ with $t^{\mathcal{D}_u}(\vec{\mathfrak{g}}) = (\bigsqcap j(A))^{\mathcal{D}_u}(\vec{\mathfrak{g}})$.

To conclude from i to $i+1$ let $t \in T_{\sqcap}^{i+1}(X)$. Then t is obtained from subterms $t_1, \dots, t_m \in T^i(X)$ by applying the operations $\sqcap$ and $\neg$. Again we may replace each of these subterms $t_j \notin \mathcal{T}^i_\sqcap$ by $t'_j := (t_j \sqcap t_j)$. By assumption we find a set $A_j \subseteq S^i_\sqcap$ with $(t_j)^{\mathcal{D}_u}(\overrightarrow{\mathfrak{g}}) = (\bigsqcup m(A_j))^{\mathcal{D}_u}(\overrightarrow{\mathfrak{g}})$ for every subterm $t_j \in \mathcal{T}^i_\sqcap(X)$ and thus by Lemma 6 we also find a set $A'_j \subseteq S^{i+1}_\sqcap$ with $t_j^{\mathcal{D}_u}(\overrightarrow{\mathfrak{g}}) = (\bigsqcup A'_j)^{\mathcal{D}_u}(\overrightarrow{\mathfrak{g}})$. Likewise we obtain sets $A_j \subseteq S^i_\sqcup$ with $(t_j)^{\mathcal{D}_u}(\overrightarrow{\mathfrak{g}}) = (\bigsqcap j(A_j))^{\mathcal{D}_u}(\overrightarrow{\mathfrak{g}})$ for every subterm $t_j \in \mathcal{T}^i_\sqcup$. Proposition 3 then yields a set $A'_j \subseteq S^{i+1}_\sqcap$ with $(t'_j)^{\mathcal{D}_u}(\overrightarrow{\mathfrak{g}}) = (\bigsqcup m(A'_j))^{\mathcal{D}_u}(\overrightarrow{\mathfrak{g}})$. If $t_j \in X$ or $t_j = \bot$ then we may again replace t_j by t'_j as above and since $i \geq 1$, $t_j \in \mathcal{T}^i_\sqcap(X)$ the first case applies. If $t_j = \top$, then $t_j \approx \top \sqcup \top \in \mathcal{T}^1_\sqcup(X)$. Thus, replacing $\top$ by $\top \sqcup \top$ leads back to the second case. As before, we find the desired subset $A \subseteq S^{i+1}_\sqcap$ by intersecting and taking complements of the respective sets existing for the subterms $t_1, \ldots, t_m$. Analogously, we obtain a set $A \subseteq S^{i+1}_\sqcup$ with $t^{\mathcal{D}_u}(\overrightarrow{\mathfrak{g}}) = (\bigsqcap j(A))^{\mathcal{D}_u}(\overrightarrow{\mathfrak{g}})$ for every $t \in \mathcal{T}^{i+1}_\sqcup$.

Finally, we show that if $t^{\mathcal{D}_u}(\overrightarrow{\mathfrak{g}}) = (\bigsqcup m(A))^{\mathcal{D}_u}(\overrightarrow{\mathfrak{g}})$, $A \subseteq S^i_\sqcap$, then $A = A_t$. Since A may contain only sequents $(\mathfrak{A}, \mathfrak{B}) \in S^i_\sqcap$ with $(m(\mathfrak{A}, \mathfrak{B}))^{\mathcal{D}_u}(\overrightarrow{\mathfrak{g}}) \sqsubseteq t^{\mathcal{D}_u}(\overrightarrow{\mathfrak{g}})$ we obtain immediately $A \subseteq A_t$ and thus $(\bigsqcup m(A))^{\mathcal{D}_u}(\overrightarrow{\mathfrak{g}}) \sqsubseteq (\bigsqcup m(A_t))^{\mathcal{D}_u}(\overrightarrow{\mathfrak{g}}) = t^{\mathcal{D}_u}(\overrightarrow{\mathfrak{g}})$. If $(\mathfrak{C}, \mathfrak{D}) \in A_t \setminus A$ it follows from Proposition 2 that

$$(m(\mathfrak{C}, \mathfrak{D}) \sqcap (\bigsqcup m(A)))^{\mathcal{D}_u}(\overrightarrow{\mathfrak{g}}) = (\bigsqcup \{m(\mathfrak{A}, \mathfrak{B}) \sqcap m(\mathfrak{C}, \mathfrak{D}) \mid (\mathfrak{A}, \mathfrak{B}) \in A\})^{\mathcal{D}_u}(\overrightarrow{\mathfrak{g}}) \\ = (\bigsqcup \bot)^{\mathcal{D}_u}(\overrightarrow{\mathfrak{g}}) = \bot.$$

Thus $(m(\mathfrak{C}, \mathfrak{D}))^{\mathcal{D}_u}(\overrightarrow{\mathfrak{g}}) \not\sqsubseteq (\bigsqcup m(A))^{\mathcal{D}_u}(\overrightarrow{\mathfrak{g}})$ and therefore

$$(\bigsqcup m(A))^{\mathcal{D}_u}(\overrightarrow{\mathfrak{g}}) \sqsubset (\bigsqcup A \cup \{(\mathfrak{C}, \mathfrak{D})\})^{\mathcal{D}_u}(\overrightarrow{\mathfrak{g}}) \sqsubseteq (\bigsqcup m(A_t))^{\mathcal{D}_u}(\overrightarrow{\mathfrak{g}}) = t^{\mathcal{D}_u}(\overrightarrow{\mathfrak{g}})$$

in contradiction to the assumption $t^{\mathcal{D}_u}(\overrightarrow{\mathfrak{g}}) = (\bigsqcup m(A))^{\mathcal{D}_u}(\overrightarrow{\mathfrak{g}})$. Dually, we obtain the result for $A \subseteq S^i_\sqcup$ and $t^{\mathcal{D}_u}(\overrightarrow{\mathfrak{g}}) = (\bigsqcap j(A))^{\mathcal{D}_u}(\overrightarrow{\mathfrak{g}})$. □

Corollary 2. *For $i \in \mathbb{N}$ we set*

$$\mathcal{B}^i_\sqcap := \{\mathfrak{p} \in \mathcal{D}_g \mid \exists t \in \mathcal{T}^i_\sqcap(X).(t)^{\mathcal{D}_u}(\overrightarrow{\mathfrak{g}}) = \mathfrak{p}\}$$

and

$$\mathcal{B}^i_\sqcup := \{\mathfrak{p} \in \mathcal{D}_g \mid \exists t \in \mathcal{T}^i_\sqcup(X).(t)^{\mathcal{D}_u}(\overrightarrow{\mathfrak{g}}) = \mathfrak{p}\}.$$

Then $\underline{\mathcal{B}}^i_\sqcap := (\mathcal{B}^i_\sqcap, \sqcap, \veebar, \neg, \bot, \bar{\top})$ is a finite subalgebra of $\underline{\mathcal{D}}_{g\sqcap}$ and, dually $\underline{\mathcal{B}}^i_\sqcup := (\mathcal{B}^i_\sqcup, \bar{\sqcap}, \sqcup, \lrcorner, \bar{\bot}, \top)$ is a finite subalgebra of $\underline{\mathcal{D}}_{g\sqcup}$. Moreover, G_i is the set of atoms of $\underline{\mathcal{B}}^i_\sqcap$ and M_i is the set of coatoms of $\underline{\mathcal{B}}^i_\sqcup$.

Proof. Obviously, for every i the set $S^i_\sqcap$ consists of full $\sqcap$-sequents over a finite set, hence $S^i_\sqcap$ is finite. Theorem 4 states that every element of $\underline{\mathcal{B}}^i_\sqcap$ is obtained from a disjunction of terms in $S^i_\sqcap$, thus $\underline{\mathcal{B}}^i_\sqcap$ is finite. It is evident that $\underline{\mathcal{B}}^i_\sqcap$ is closed under the operations $\sqcap$, $\veebar$ and $\neg$. From $k(\bot) = 1$ and $\bar{\top} \approx \neg\bot$ follows that $\bot$, $\bar{\top} \in \mathcal{B}^i_\sqcap$ for every i, hence $\underline{\mathcal{B}}^i_\sqcap$ is a finite subalgebra of $\underline{\mathcal{D}}_{g\sqcap}$. Dually we obtain that $\underline{\mathcal{B}}^i_\sqcup$ is a finite subalgebra of $\underline{\mathcal{D}}_{g\sqcup}$.

As for every $\sqcap$-sequent $(\mathfrak{A}, \mathfrak{B}) \in S^i_\sqcap$ the complexity of the corresponding term $k(m(\mathfrak{A}, \mathfrak{B}))$ equals i, we find that G_i, defined in the exploration algorithm as $G_i := \{(m(\mathfrak{A}, \mathfrak{B})^{\mathcal{D}_u}(\overrightarrow{\mathfrak{g}}) \mid (\mathfrak{A}, \mathfrak{B}) \in S^i_\sqcap\}$, is a subset of $\mathcal{B}^i_\sqcap$. Proposition 2 yields

$\mathfrak{a} \sqcap \mathfrak{b} = \bot$ for $\mathfrak{a}, \mathfrak{b} \in G_i$ and from Theorem 4 we conclude that $\mathfrak{p} = \bigsqcup\{\mathfrak{a} \in G_i \mid \mathfrak{a} \sqsubseteq \mathfrak{p}\}$ for every $\mathfrak{p} \in \mathcal{B}^i_\sqcap$. Thus G_i is the set of atoms of $\mathcal{B}^i_\sqcap$. Dually we obtain that M_i, defined in the exploration algorithm as $M_i := \{(j(\mathfrak{A}, \mathfrak{B})^{\mathcal{D}_u}(\vec{\mathfrak{g}}) \mid (\mathfrak{A}, \mathfrak{B}) \in S^i_\sqcup\}$, is the set of coatoms of $\mathcal{B}^i_\sqcup$. □

From the Basic Theorem on Semiconcept Algebras we obtain immediately:

Corollary 3. *If the algorithm stops after i iterations, then $\underline{\mathfrak{H}}(\mathbb{K}_i)$ is isomorphic to the pure subalgebra $\underline{\mathcal{D}}_{gp}$ of $\mathcal{D}_u$.*

3 Exploring Free Double Boolean Algebras

As we have seen, the free double Boolean algebras with one or more generators are infinite. Thus, the described exploration algorithm will not terminate when exploring such a free algebra. Nevertheless, we gain insight from the $\mathbb{K}_i$ constructed during the exploration process as will be explained in this section.

For a given (finite) set of variables X we define $\mathbb{K}_i(X)$ to be the context $\mathbb{K}_i$ obtained when exploring the free double Boolean algebra $D(X)$. Theorem 4 motivates the following map:

Proposition 4. *Let $X := \{x_1, \ldots x_n\}$ be a set of variables and let Θ^i be the restriction of the term equivalence $\approx$ to $T^i(X) \times T^i(X)$. Then*

$$\begin{aligned}\phi_i(\bar{t}) := (\{\overline{m(\mathfrak{A}, \mathfrak{B})} \mid (\mathfrak{A}, \mathfrak{B}) \in S^i_\sqcap \text{ and } \overline{m(\mathfrak{A}, \mathfrak{B})} \sqsubseteq \bar{t}\},\\ \{\overline{j(\mathfrak{A}, \mathfrak{B})} \mid (\mathfrak{A}, \mathfrak{B}) \in S^i_\sqcup \text{ and } \overline{j(\mathfrak{A}, \mathfrak{B})} \sqsupseteq \bar{t}\})\end{aligned}$$

defines a map $\phi_i : T^i(X)/\Theta^i \to \mathfrak{P}(\mathbb{K}_i(X))$.

Proof. For any term $t \in \mathcal{T}(X)$ we find that $t^{D(X)}(\overline{x}_1, \ldots, \overline{x}_n) = \bar{t}$, thus ϕ_i assigns to each class of terms subsets of the object set and the attribute set of $\mathbb{K}_i(X)$. For $s \in \bar{t}$ obviously holds $\phi_i(\bar{s}) = \phi_i(\bar{t})$. In order to show that for $t \in \mathcal{T}^i(X)$ the image $\phi_i(\bar{t})$ is a protoconcept of $\mathbb{K}_i(X)$ we distinguish three cases:
1) $t \in \mathcal{T}^i_\sqcap(X)$. We set $(A_t, B_t) := \phi_i(\bar{t})$. Then Theorem 4 yields $\bar{t} = \bigsqcup A_t$ and for $\overline{j(\mathfrak{C}, \mathfrak{D})} \in B_t$ we obtain:

$$\begin{aligned}\overline{j(\mathfrak{C}, \mathfrak{D})} \in B_t &\Leftrightarrow \bar{t} \sqsubseteq \overline{j(\mathfrak{C}, \mathfrak{D})}\\ &\Leftrightarrow \bigsqcup A_t \sqsubseteq \overline{j(\mathfrak{C}, \mathfrak{D})}\\ &\Leftrightarrow \mathfrak{a} \sqsubseteq \overline{j(\mathfrak{C}, \mathfrak{D})} \text{ for every } \mathfrak{a} \in A_t\\ &\Leftrightarrow \overline{j(\mathfrak{C}, \mathfrak{D})} \in A_t'.\end{aligned}$$

Thus $B_t = A_t'$ and ϕ_i maps $\sqcap$-semiconcepts of $D(X)$ to $\sqcap$-semiconcepts of $\mathfrak{P}(\mathbb{K}_i(X))$.
2) $t \in \mathcal{T}^i_\sqcup(X)$. We obtain analogously that ϕ_i maps $\sqcup$-semiconcepts of $D(X)$ to $\sqcup$-semiconcepts of $\mathfrak{P}(\mathbb{K}_i(X))$.
3) $t = x_j$ for some $1 \leq j \leq n$. Again we set $(A_t, B_t) := \phi_i(\bar{t})$ and need to show $A_t' = B_t''$. First, note that $A_t \subseteq B_t'$ and, conversely, $B_t \subseteq A_t'$ as from

$\overline{m(\mathfrak{A},\mathfrak{B})} \in A_t$ and $\overline{j(\mathfrak{C},\mathfrak{D})} \in B_t$ it follows that $\overline{m(\mathfrak{A},\mathfrak{B})} \sqsubseteq \bar{t} \sqsubseteq \overline{j(\mathfrak{C},\mathfrak{D})}$ and thus $\overline{m(\mathfrak{A},\mathfrak{B})}\, I\, \overline{j(\mathfrak{C},\mathfrak{D})}$ in $\mathbb{K}_i(X)$. This yields $B_t'' \subseteq A_t'$ and $A_t'' \subseteq B_t'$.

Now assume $\overline{j(\mathfrak{C},\mathfrak{D})} \in A_t' \setminus B_t''$. Then we find $\overline{m(\mathfrak{A},\mathfrak{B})} \in B_t' \setminus A_t$ such that $\overline{m(\mathfrak{A},\mathfrak{B})} \not\sqsubseteq \overline{j(\mathfrak{C},\mathfrak{D})}$. But since $\overline{x}_{j\sqcup} = \bigsqcap B_t$ we obtain from $\overline{m(\mathfrak{A},\mathfrak{B})} \in B_t'$ that $\overline{m(\mathfrak{A},\mathfrak{B})} \sqsubseteq \overline{x}_{j\sqcup}$ and from $m(\mathfrak{A},\mathfrak{B}) \in T_\sqcap(X)$ it follows that $\overline{m(\mathfrak{A},\mathfrak{B})} \sqsubseteq (\overline{x}_{j\sqcup})_\sqcap = (\overline{x}_{j\sqcap})_\sqcup$. Likewise we find for $\overline{j(\mathfrak{C},\mathfrak{D})} \in A_t'$ that $(\overline{x}_{j\sqcap})_\sqcup \sqsubseteq \overline{j(\mathfrak{C},\mathfrak{D})}$ and therefore $\overline{m(\mathfrak{A},\mathfrak{B})} \sqsubseteq (\overline{x}_{j\sqcap})_\sqcup \sqsubseteq \overline{j(\mathfrak{C},\mathfrak{D})}$ in contradiction to our assumption. Thus $B_t'' = A_t'$. □

Proposition 5. *1) The restriction $\phi_{\sqcap i}$ of ϕ_i to $T^i_\sqcap(X)/\Theta^i$ is an isomorphism between the Boolean algebras $T^i_\sqcap(X)/\Theta^i$ and $\underline{\mathfrak{H}}_\sqcap(\mathbb{K}_i(X))$.*
2) Dually, the restriction $\phi_{\sqcup i}$ of ϕ_i to $T^i_\sqcup(X)/\Theta^i$ is an isomorphism between the Boolean algebras $T^i_\sqcup(X)/\Theta^i$ and $\underline{\mathfrak{H}}_\sqcup(\mathbb{K}_i(X))$.

Proof. This follows immediately from Corollary 2 and the definition of ϕ_i. □

Proposition 6 shows that the ϕ_i respect to a certain degree the operations of double Boolean algebras. This is essential for the proof of Theorem 5.

Proposition 6. *Let $s, t \in T^i(X)$. Then*

1. $\phi_i(\overline{s \sqcap t}) = \phi_i(\overline{s}) \sqcap \phi_i(\overline{t})$ *if* $k(s \sqcap t) \leq i$.
2. $\phi_i(\overline{s \sqcup t}) = \phi_i(\overline{s}) \sqcup \phi_i(\overline{t})$ *if* $k(s \sqcup t) \leq i$.
3. $\phi_i(\overline{\neg t}) = \neg\phi_i(\overline{t})$ *if* $k(\neg t) \leq i$.
4. $\phi_i(\overline{\lrcorner t}) = \lrcorner\phi_i(\overline{t})$ *if* $k(\lrcorner t) \leq i$.

Moreover, $\phi_i(\overline{\top}) = \top$ *and* $\phi_i(\overline{\bot}) = \bot$.

Proof. We show that for any term t with $k(t \sqcap t) \leq i$ holds $\phi_i(\bar{t}) \sqcap \phi_i(\bar{t}) = \phi_i(\overline{t \sqcap t})$: Let $(A, B) := \phi_i(\bar{t})$. For $(\mathfrak{A},\mathfrak{B}) \in S^i_\sqcap$ we have $m(\mathfrak{A},\mathfrak{B})^{D(X)}(\underline{\mathfrak{x}}) \sqsubseteq \bar{t} \Leftrightarrow \overline{m(\mathfrak{A},\mathfrak{B})} \sqsubseteq \overline{t \sqcap t}$. Since ϕ_i maps elements of $T^i_\sqcap(X)$ to $\sqcap$-semiconcepts of $\mathbb{K}_i(X)$, we obtain $\phi_i(\overline{t \sqcap t}) = (A, A') = \phi_i(\bar{t}) \sqcap \phi_i(\bar{t})$. Dually, we obtain for any term s with $k(s \sqcup s) \leq i$ that $\phi_i(\overline{s \sqcup s}) = \phi_i(\overline{s}) \sqcup \phi_i(\overline{s})$. This yields together with Proposition 5 the statement. □

Now we have the means to prove the main result of this paper. Two terms s, t with $k(s), k(t) < i$ are equivalent if and only if $\underline{\mathfrak{P}}(\mathbb{K}_i(X))$ is not a counterexample.

Theorem 5. *For $s, t \in T^i(X)$ the following conditions are equivalent:*

1. $s \approx t$
2. $\phi_{i+1}(\overline{s}) = \phi_{i+1}(\overline{t})$
3. $s^{\underline{\mathfrak{P}}(\mathbb{K}_{i+1}(X))}(\phi_{i+1}(\overline{x}_1), \ldots \phi_{i+1}(\overline{x}_n)) = t^{\underline{\mathfrak{P}}(\mathbb{K}_{i+1}(X))}(\phi_{i+1}(\overline{x}_1), \ldots \phi_{i+1}(\overline{x}_n))$.

Proof. 1) $\Leftrightarrow$ 2): For $s, t \in T^i_\sqcap(X)$ (resp. $s, t \in T^i_\sqcup(X)$) Proposition 5 and $T^i_\sqcap(X) \subseteq T^{i+1}_\sqcap(X)$ (resp. $T^i_\sqcup(X) \subseteq T^{i+1}_\sqcup(X)$) yield $s \approx t \Leftrightarrow \phi_i(\overline{s}) = \phi_i(\overline{t}) \Leftrightarrow \phi_{i+1}(\overline{s}) = \phi_{i+1}(\overline{t})$.

Now assume $s \in T^i_\sqcap(X)$, $t \in T^i_\sqcup(X)$. If $s \approx t$ we obtain immediately from the definition of ϕ_{i+1} that $\phi_{i+1}(\overline{s}) = \phi_{i+1}(\overline{t})$. If $s \not\approx t$ then $\overline{s} \neq \overline{t \sqcap t}$ or $\overline{s \sqcup s} \neq \overline{t}$ (otherwise $\overline{s} \sqsubseteq \overline{t}$ and $\overline{t} \sqsubseteq \overline{s}$ which implies $\overline{s} = \overline{t}$ since $\sqsubseteq$ is an order on $D_p(X)$).

In the first case, $\overline{s} \neq \overline{t_\sqcap}$, Proposition 5 yields $\phi_{\sqcap i+1}(\overline{s}) \neq \phi_{\sqcap i+1}(\overline{t_\sqcap})$ and thus
$\{\overline{m(\mathfrak{A},\mathfrak{B})} \mid (\mathfrak{A},\mathfrak{B}) \in S_\sqcap^{i+1} \text{ and } \overline{m(\mathfrak{A},\mathfrak{B})} \sqsubseteq \overline{s}\}$
$\neq \{\overline{m(\mathfrak{A},\mathfrak{B})} \mid (\mathfrak{A},\mathfrak{B}) \in S_\sqcap^{i+1} \text{ and } \overline{m(\mathfrak{A},\mathfrak{B})} \sqsubseteq \overline{t \sqcap t}\}$.

Since for $(\mathfrak{A},\mathfrak{B}) \in S_\sqcap^{i+1}$ we have $\overline{m(\mathfrak{A},\mathfrak{B})} \sqsubseteq \overline{t \sqcap t} \Leftrightarrow \overline{m(\mathfrak{A},\mathfrak{B})} \sqsubseteq \overline{t}$ we obtain $\phi_{i+1}(\overline{s}) \neq \phi_{i+1}(\overline{t})$. The second case is shown dually.

In the case $s \in X$, $s = x_j$ for some $1 \leq j \leq n$, we set $(A, B) := \phi_{i+1}(\overline{x}_j)$. Note that $\phi_{i+1}(\overline{x_{j\sqcap}}) = (A, A')$ and $\phi_{i+1}(\overline{x_{j\sqcup}}) = (B', B)$. If $t \in X$, $t \neq s$, then $\overline{x_{k\sqcap}} \neq \overline{x_{j\sqcap}}$ in $\mathcal{D}(X)$ and Proposition 5 yield that $\phi_1(\overline{t_\sqcap}) \neq \phi_1(\overline{s_\sqcap})$ and thus $\phi_1(\overline{t}) \neq \phi_1(\overline{s})$. This extends to ϕ_i for any i.

For the case $t \notin X$, $s \in X$ we show that $\phi_k(\overline{s}) \in \mathfrak{P}(\mathbb{K}_k(X)) \setminus \mathfrak{H}(\mathbb{K}_k(X))$ for $k \geq 2$. Since ϕ_k maps classes of terms of $\mathcal{T}_\sqcap^i(X)$ to $\sqcap$-semiconcepts and classes of terms of $\mathcal{T}_\sqcup^i(X)$ to $\sqcup$-semiconcepts this implies $\phi_k(\overline{s}) \neq \phi_k(\overline{t})$. In order to show $\phi_k(\overline{s}) \in \mathfrak{P}(\mathbb{K}_k(X)) \setminus \mathfrak{H}(\mathbb{K}_k(X))$ note that for $k \geq 2$ we find a set $S \subseteq M_k$ with $\overline{\bigsqcap j(S)} = \overline{x_{j\sqcap\sqcup}}$. Moreover, since $\overline{s_{j\sqcap}} \sqsubseteq \overline{x_{j\sqcap\sqcup}}$, the set $\{\overline{j(\mathfrak{A},\mathfrak{B})} \mid (\mathfrak{A},\mathfrak{B}) \in S\}$ is contained in A'. It is easy to check that $\overline{x_{j\sqcap\sqcup}}$ and $\overline{x_j}$ are incomparable in $D(X)$ and therefore there exists a sequent $(\mathfrak{A},\mathfrak{B}) \in S$ such that $\overline{j(\mathfrak{A},\mathfrak{B})}$ is not contained in B. Hence $\phi_k(\overline{x}_j)$ is not of the form (A, A').

Dually we obtain, that $\phi_k(\overline{x}_j)$ is not of the form (B', B), thus $\phi_k(\overline{x}_j)$ is not a semiconcept of $\mathbb{K}_k(X)$.

The equivalence 2) $\Leftrightarrow$ 3) follows directly from Proposition 6. □

If we know the $\mathbb{K}_i(X)$ then Theorem 5 yields an easy criterion to decide whether two terms are equivalent since the $\phi_i(\overline{x}_j)$ are computed easily:

$$\begin{aligned}\phi_i(\overline{x}_j) = (\{&\overline{m(\mathfrak{A},\mathfrak{B})} \mid (\mathfrak{A},\mathfrak{B}) \in S_\sqcap^i \text{ and } x_{j\sqcap} \in \mathfrak{A}\},\\ \{&\overline{j(\mathfrak{A},\mathfrak{B})} \mid (\mathfrak{A},\mathfrak{B}) \in S_\sqcup^i \text{ and } x_{j\sqcup} \in \mathfrak{B}\})\end{aligned}$$

Of course, the main problem remains to find the contexts $\mathbb{K}_i(X)$. But these results already allow to derive an upper bound for the size of a minimal counterexample to an identity $s = t$ of terms, if there exists one: For a given finite set of variables X with cardinality $|X| =: n$, we obtain in the first iteration of the exploration not more than 2^n full $\sqcap$-sequents and the same amount of full $\sqcup$-sequents. If at the end of the i-th iteration we have $|S_\sqcap^i| = |S_\sqcup^i| = x$, then we cannot generate more than $x2^x$ full $\sqcap$-sequents (and the same number of $\sqcup$-sequents) in the next step. This motivates the inductive definition of the following series:

Definition 9. *For $x \in \mathbb{N}$ we define inductively*

$$\begin{aligned}a_1^x &:= 2^x\\ a_{n+1}^x &:= a_n^x 2^{a_n^x}\end{aligned}$$

Corollary 4. *If X is a finite set of variables with $|X| = n$ and $s, t \in \mathcal{T}(X)$, $s \not\approx t$ with $k(s), k(t) \leq m$, then there exists a context $\mathbb{K} := (G, M, I)$ with $|G| = |M| \leq a_{m+1}^n$ such that $s^{\underline{\mathfrak{P}}(\mathbb{K})} \neq t^{\underline{\mathfrak{P}}(\mathbb{K})}$.*

This corollary gives only a very weak upper bound for the size of the $\mathbb{K}_i(X)$ as Figure 4 shows.

i	Number of variables $n := \|X\|$	value of a_i^n	$\|G_i\|$ of $\mathbb{K}_i(X) := (G_i, M_i, \sqsubseteq)$
1	1	2	2
2	1	8	6
3	1	2048	10
1	2	4	4
2	2	64	26

Fig. 4. Comparison of the upper bounds given in Corollary 4 and the sizes of the object sets of $\mathbb{K}_i(X)$

4 Automated Generation of $\mathbb{K}_i(X)$

In this section, the problem of stepwise generation of the contexts $\mathbb{K}_i(X)$ is attacked. If the context $\mathbb{K}_i(X)$ is known then the first task is to find the sets $S_\sqcap^{i+1}$ and $S_\sqcup^{i+1}$ which yield the object and the attribute set of the context $\mathbb{K}_{i+1}(X)$. This problem is dealt with in part 4.1. Once the object and the attribute set are generated, the next step is to determine the incidence relation $\sqsubseteq$ which will be treated in part 4.2. In both cases, our strategy is to reduce the problem to a relativey small number of possibilities which may then be checked automatically.

4.1 Generation of $S_\sqcap^{i+1}$ and $S_\sqcup^{i+1}$

In order to construct the contexts $\mathbb{K}_i(X)$ stepwise, the contexts $\mathbb{K}_1(X)$ are needed as a starting point. They are described in the following proposition:

Proposition 7. *Let $X := \{x_1, \dots x_n\}$ be a set of variables and let $(\mathfrak{A}, \mathfrak{B})$ be a full $\sqcap$-sequent over $\tilde{S}_\sqcap^1$ and $(\mathfrak{C}, \mathfrak{D})$ a full $\sqcup$-sequent over $\tilde{S}_\sqcup^1$. Then $m(\mathfrak{A}, \mathfrak{B}) \not\sqsubseteq \bot$ and $j(\mathfrak{C}, \mathfrak{D}) \not\sqsubseteq \top$ and*

$$\overline{m(\mathfrak{A}, \mathfrak{B})} \sqsubseteq \overline{j(\mathfrak{C}, \mathfrak{D})} \Leftrightarrow \exists x_j \in X : x_{j\sqcap} \in \mathfrak{A} \text{ and } x_{j\sqcup} \in \mathfrak{D}.$$

Proof. It is well-known that $l_1 \wedge \cdots \wedge l_n \neq 0$ for $l_i \in \{x_i, \neg x_i\}$ holds in the free Boolean algebra generated by X and, dually, $l_1 \vee \cdots \vee l_n \neq 1$ for $l_i \in \{x_i, \neg x_i\}$. Since every Boolean algebra $(\mathcal{B}, \wedge, \vee, \neg, 1, 0)$ can be considered as a double Boolean algebra $(\mathcal{B}, \wedge, \vee, \neg, \neg, 1, 0)$, this yields $l_1 \sqcap \cdots \sqcap l_n \not\sqsubseteq \bot$ and $l_1 \sqcup \cdots \sqcup l_n \not\sqsubseteq \top$ in double Boolean algebras.

From the definition of $m(\mathfrak{A}, \mathfrak{B})$ and $j(\mathfrak{C}, \mathfrak{D})$ we obtain immediately

$$\exists x_j \in X : x_{j\sqcap} \in \mathfrak{A} \text{ and } x_{j\sqcup} \in \mathfrak{D} \Rightarrow \overline{m(\mathfrak{A}, \mathfrak{B})} \sqsubseteq \overline{j(\mathfrak{C}, \mathfrak{D})}.$$

Now, consider $\mathbb{K} := (G_1, M_1, I)$ where I is defined as

$$\overline{m(\mathfrak{A}, \mathfrak{B})}\, I\, \overline{j(\mathfrak{C}, \mathfrak{D})} :\Leftrightarrow \exists x_j \in X : x_{j\sqcap} \in \mathfrak{A} \text{ and } x_{j\sqcup} \in \mathfrak{D}.$$

We will show that in this context $\mathbb{K}$ the equivalence holds which implies that it is true in $\mathbb{K}_1(X)$. We set

$$A_{x_j} := \{\overline{m(\mathfrak{A}, \mathfrak{B})} \in G_1 \mid x_{j\sqcap} \in \mathfrak{A}\}$$

and

$$B_{x_j} := \{\overline{j(\mathfrak{C},\mathfrak{D})} \in M_1 \mid x_{j\sqcup} \in \mathfrak{D}\}$$

and check that (A_{x_j}, B_{x_j}) is a protoconcept: Obviously $B_{x_j} \subseteq A_{x_j}^I$. If $\overline{j(\mathfrak{C},\mathfrak{D})} \in A_{x_j}^I \setminus B_{x_j}$ then we find $k \neq j$ such that $x_{k\sqcup} \in \mathfrak{D}$ and $x_{k\sqcap} \in \mathfrak{A}$ for all $\overline{m(\mathfrak{A},\mathfrak{B})} \in A_{x_j}$. This contradicts $\overline{m(\{x_{j\sqcap}\}, \tilde{S}^1_\sqcap \setminus \{x_{j\sqcap}\}} \in A_{x_j}$. Therefore, $B_{x_j} = A_{x_j}^I$ and (A_{x_j}, B_{x_j}) is a protoconcept of $\mathfrak{P}(\mathbb{K})$.

Finally, we show

$$m(\mathfrak{A},\mathfrak{B})^{\underline{\mathfrak{P}(\mathbb{K})}}((A_{x_1}, B_{x_1}), \ldots, (A_{x_n}, B_{x_n})) = (\{\overline{m(\mathfrak{A},\mathfrak{B})}\}, \{\overline{m(\mathfrak{A},\mathfrak{B})}\}^I)$$

and

$$j(\mathfrak{C},\mathfrak{D})^{\underline{\mathfrak{P}(\mathbb{K})}}((A_{x_1}, B_{x_1}), \ldots, (A_{x_n}, B_{x_n})) = (\{\overline{j(\mathfrak{C},\mathfrak{D})}\}^I, \{\overline{j(\mathfrak{C},\mathfrak{D})}\})$$

for $(\mathfrak{A},\mathfrak{B}) \in S^1_\sqcap$ and $(\mathfrak{C},\mathfrak{D}) \in S^1_\sqcup$:

$$\begin{aligned}
&m(\mathfrak{A},\mathfrak{B})^{\underline{\mathfrak{P}(\mathbb{K})}}((A_{x_1}, B_{x_1}), \ldots, (A_{x_n}, B_{x_n}))\\
&\quad = \bigsqcap\{(A_{x_j}, B_{x_j}) \mid x_{j\sqcap} \in \mathfrak{A}\} \sqcap \neg \bigsqcup\{(A_{x_j}, B_{x_j}) \mid x_{j\sqcap} \in \mathfrak{B}\}\\
&\quad = (\bigcap\{A_{x_j} \mid x_{j\sqcap} \in \mathfrak{A}\}, \bigcap\{A_{x_j} \mid x_{j\sqcap} \in \mathfrak{A}\}^I)\\
&\quad\quad \sqcap \neg(\bigcup\{A_{x_j} \mid x_{j\sqcap} \in \mathfrak{B}\}, \bigcap\{A_{x_j} \mid x_{j\sqcap} \in \mathfrak{B}\}^I)\\
&\quad = (\{\overline{m(\mathfrak{E},\mathfrak{F})} \in G_1 \mid \mathfrak{A} \subseteq \mathfrak{E}\} \setminus \{\overline{m(\mathfrak{E},\mathfrak{F})} \in G_1 \mid \mathfrak{E} \cap \mathfrak{B} \neq \emptyset\},\\
&\quad\quad (\{\overline{m(\mathfrak{E},\mathfrak{F})} \in G_1 \mid \mathfrak{A} \subseteq \mathfrak{E}\} \setminus \{\overline{m(\mathfrak{E},\mathfrak{F})} \in G_1 \mid \mathfrak{E} \cap \mathfrak{B} \neq \emptyset\})^I)\\
&\quad = (\{\overline{m(\mathfrak{A},\mathfrak{B})}\}, \{\overline{m(\mathfrak{A},\mathfrak{B})}\}^I).
\end{aligned}$$

Dually we obtain

$$j(\mathfrak{C},\mathfrak{D})^{\underline{\mathfrak{P}(\mathbb{K})}}((A_{x_1}, B_{x_1}), \ldots, (A_{x_n}, B_{x_n})) = (\{\overline{j(\mathfrak{C},\mathfrak{D})}\}^I, \{\overline{j(\mathfrak{C},\mathfrak{D})}\}).$$

If $(\overline{m(\mathfrak{A},\mathfrak{B})}, \overline{j(\mathfrak{C},\mathfrak{D})}) \notin I$ in $\mathbb{K}$ then $m(\mathfrak{A},\mathfrak{B})^{\underline{\mathfrak{P}(\mathbb{K})}}((A_{x_1}, B_{x_1}), \ldots, (A_{x_n}, B_{x_n})) \not\sqsubseteq j(\mathfrak{C},\mathfrak{D})^{\underline{\mathfrak{P}(\mathbb{K})}}((A_{x_1}, B_{x_1}), \ldots, (A_{x_n}, B_{x_n}))$ and thus $m(\mathfrak{A},\mathfrak{B}) \not\sqsubseteq j(\mathfrak{C},\mathfrak{D})$ in $\mathcal{D}(X)$. □

The following lemma describes the incidence relation in a special case. It is needed in the proof of Theorem 6.

Lemma 7. *Let* $(\mathfrak{A},\mathfrak{B}) \in S^i_\sqcap$, $(\mathfrak{C},\mathfrak{D}) \in S^i_\sqcup$ *with* $\overline{m(\mathfrak{A},\mathfrak{B})} \not\sqsubseteq \overline{j(\mathfrak{C},\mathfrak{D})}$. *Then*

1. $\overline{m(\mathfrak{A} \cup \{j(\mathfrak{C},\mathfrak{D})_\sqcap\}, \mathfrak{B})} \sqsubseteq \overline{j(\mathfrak{C},\mathfrak{D})}$ *and, if* $m(\mathfrak{A}, \mathfrak{B} \cup \{j(\mathfrak{C},\mathfrak{D})_\sqcap\}) \not\sim \bot$ *then* $\overline{m(\mathfrak{A}, \mathfrak{B} \cup \{j(\mathfrak{C},\mathfrak{D})_\sqcap\}) \not\sqsubseteq j(\mathfrak{C},\mathfrak{D})}$.
2. *Dually,* $\overline{m(\mathfrak{A},\mathfrak{B})} \sqsubseteq \overline{j(\mathfrak{C}, \mathfrak{D} \cup \{m(\mathfrak{A},\mathfrak{B})_\sqcup\})}$ *and, if* $j(\mathfrak{C} \cup \{m(\mathfrak{A},\mathfrak{B})_\sqcup\}, \mathfrak{D}) \not\sim \top$ *then* $\overline{m(\mathfrak{A},\mathfrak{B})} \not\sqsubseteq \overline{j(\mathfrak{C} \cup \{m(\mathfrak{A},\mathfrak{B})_\sqcup\}, \mathfrak{D})}$.

Proof. 1) From

$$\overline{m(\mathfrak{A} \cup \{j(\mathfrak{C},\mathfrak{D}))_\sqcap\}, \mathfrak{B})} = \overline{m(\mathfrak{A},\mathfrak{B}) \sqcap j(\mathfrak{C},\mathfrak{D})_\sqcap}$$

it follows immediately

$$\overline{m(\mathfrak{A} \cup \{j(\mathfrak{C},\mathfrak{D})_\sqcap\}, \mathfrak{B})} \sqsubseteq \overline{j(\mathfrak{C},\mathfrak{D})}.$$

From

$$m(\mathfrak{A}, \mathfrak{B} \cup \{j(\mathfrak{C}, \mathfrak{D})_\sqcap\}) \approx m(\mathfrak{A}, \mathfrak{B}) \sqcap \neg j(\mathfrak{C}, \mathfrak{D})_\sqcap$$

we obtain

$$\overline{m(\mathfrak{A}, \mathfrak{B} \cup \{j(\mathfrak{C}, \mathfrak{D})_\sqcap\})} \sqsubseteq \overline{\neg j(\mathfrak{C}, \mathfrak{D})_\sqcap}.$$

If $\overline{m(\mathfrak{A}, \mathfrak{B} \cup \{j(\mathfrak{C}, \mathfrak{D})_\sqcap\})} \sqsubseteq \overline{j(\mathfrak{C}, \mathfrak{D})_\sqcap}$ then

$$\overline{m(\mathfrak{A}, \mathfrak{B} \cup \{j(\mathfrak{C}, \mathfrak{D})_\sqcap\})} \sqsubseteq \overline{j(\mathfrak{C}, \mathfrak{D})_\sqcap} \sqcap \overline{\neg j(\mathfrak{C}, \mathfrak{D})_\sqcap} = \overline{\bot}.$$

2) follows dually. □

Now we have the means to prove the main result of this section. Theorem 6 shows that for a full $\sqcap$-sequent $(\mathfrak{A}, \mathfrak{B})$ over $\tilde{S}^{i+1}$ a counterexample to $m(\mathfrak{A}, \mathfrak{B}) \approx \bot$, if there exists any, can be found in an appropiate extension of $\mathbb{K}_i(X)$.

Theorem 6. *Let* $(\mathfrak{A}, \mathfrak{B}) \in S^i_\sqcap$ *and let*

$$T := \{j(\mathfrak{C}_1, \mathfrak{D}_1)_\sqcap, \dots, j(\mathfrak{C}_m, \mathfrak{D}_m)_\sqcap\} \subseteq \tilde{S}^{i+1}_\sqcap$$

be a family of terms such that $\overline{m(\mathfrak{A}, \mathfrak{B})} \not\sqsubseteq \overline{j(\mathfrak{C}_j, \mathfrak{D}_j)}$ *for* $1 \leq j \leq m$.
For subsets $Y_1, \dots, Y_k$ *of* T *with*

$$m(\mathfrak{A} \cup Y_j, \mathfrak{B} \cup (T \setminus Y_j)) \not\approx \bot$$

and

$$\bigsqcup_{1 \leq j \leq k} m(\mathfrak{A} \cup Y_j, \mathfrak{B} \cup (T \setminus Y_j)) \approx m(\mathfrak{A}, \mathfrak{B})$$

we modify the object set G_i *of* $\mathbb{K}_i(X)$ *by setting*

$$G := \{\overline{m(\mathfrak{E}, \mathfrak{F})} \in G_i \mid (\mathfrak{E}, \mathfrak{F}) \neq (\mathfrak{A}, \mathfrak{B})\} \cup \{\overline{m(\mathfrak{A} \cup Y_j, \mathfrak{B} \cup (T \setminus Y_j))} \mid k = 1, \dots, m\}.$$

We define a new context $\mathbb{K}$ *with this modified object set as* $\mathbb{K} := (G, M_i, \sqsubseteq)$ *and set*

$$A_{x_j} := \{\overline{m(\mathfrak{E}, \mathfrak{F})} \in G \mid x_{j\sqcap} \in \mathfrak{E}\},$$
$$B_{x_j} := \{\overline{j(\mathfrak{E}, \mathfrak{F})} \in M_i \mid x_{j\sqcup} \in \mathfrak{F}\}.$$

Then

$$m(\mathfrak{A} \cup Y_j, \mathfrak{B} \cup (T \setminus Y_j))^{\mathfrak{P}(\mathbb{K})}((A_{x_1}, B_{x_1}), \dots, (A_{x_n}, B_{x_n})) \neq (\emptyset, M_i).$$

Proof. First we prove that $(A_{x_j}, B_{x_j}) \in \mathfrak{P}(\mathbb{K})$ for every $x_j \in X$. Note that, for $l \in 1, \dots, k$, $x_{j\sqcap} \in \mathfrak{A} \cup Y_l \Leftrightarrow x_{j\sqcap} \in \mathfrak{A}$ since $x_{j\sqcap} \in \tilde{S}^1_\sqcap$. Thus

$$\overline{m(\mathfrak{A} \cup Y_l, \mathfrak{B} \cup (T \setminus Y_l))} \in A_{x_j} \Leftrightarrow \overline{m(\mathfrak{A} \cup Y_h, \mathfrak{B} \cup (T \setminus Y_h))} \in A_{x_j}$$

for all $h \in 1, \dots, k$.
From $\overline{x_{j\sqcap}} = \bigsqcup\{\overline{m(\mathfrak{E}, \mathfrak{F})} \in G_i \mid \overline{x_{j\sqcap}} \in \mathfrak{E}\}$ and $\bigsqcup_{1 \leq l \leq k} m(\mathfrak{A} \cup Y_l, \mathfrak{B} \cup (T \setminus Y_l)) \approx m(\mathfrak{A}, \mathfrak{B})$ we obtain $\bigsqcup A_{x_j} = \overline{x_{j\sqcap}}$. Likewise follows $\overline{\bigsqcap} x_j = \overline{x_{j\sqcup}}$, thus $A_{x_j} \, I \, B_{x_j}$.

In the case $i = 1$, we show that $A'_{x_j} = B_{x_j}$: Assume $\overline{j(\mathfrak{E}, \mathfrak{F})} \in A'_{x_j} \setminus B_{x_j}$. Then $x_{j\sqcup} \notin \mathfrak{F}$, but Proposition 7 yields the existence of $x_l \neq x_j$ with $x_{l\sqcap} \in \mathfrak{G}$ for all $\overline{m(\mathfrak{G}, \mathfrak{H})} \in A_{x_j}$ and $x_{l\sqcup} \in \mathfrak{F}$ in contradiction to $\overline{m(\{x_{j\sqcap}\}, \tilde{S}^1_\sqcap \setminus \{x_{j\sqcap}\})} \in A_{x_j}$ if $(\mathfrak{A}, \mathfrak{B}) \neq (\{x_{j\sqcap}\}, \tilde{S}^1_\sqcap \setminus \{x_j\})$ or $\overline{m(\mathfrak{A} \cup Y_1, \mathfrak{B} \cup (T \setminus Y_1))} \in A_{x_j}$ otherwise.

In the case $i \geq 2$,

$$\overline{j(\mathfrak{E}, \mathfrak{F})} \in A'_{x_j} \Leftrightarrow \bigsqcup A_{x_j} = \overline{x_{j\sqcap}} \sqsubseteq \overline{j(\mathfrak{E}, \mathfrak{F})} \Leftrightarrow \overline{x_{j\sqcap\sqcup}} \sqsubseteq \overline{j(\mathfrak{E}, \mathfrak{F})}.$$

Moreover, since $i \geq 2$ and $k(x_{j\sqcap\sqcup}) = 2$, Theorem 4 yields

$$\bigsqcap A'_{x_j} = \overline{x_{j\sqcap\sqcup}} = \overline{x_{j\sqcup\sqcap}} = \bigsqcup A''_{x_j}.$$

Likewise we obtain $\bigsqcup B'_{x_j} = \overline{x_{j\sqcup\sqcap}}$ and thus $A''_{x_j} = B'_{x_j}$. We conclude $(A_{x_j}, B_{x_j}) \in \mathfrak{P}(\mathbb{K})$.

In the next step we show for $(\mathfrak{E}_1, \mathfrak{F}_1) \in S^j_\sqcap$ and $(\mathfrak{E}_2, \mathfrak{F}_2) \in S^j_\sqcup$, $j \leq i$ that

$$m(\mathfrak{E}_1, \mathfrak{F}_1)^{\mathfrak{P}(\mathbb{K})}((A_{x_1}, B_{x_1}), \ldots, (A_{x_n}, B_{x_n}))$$
$$= (\{m(\mathfrak{G}, \mathfrak{H}) \in G \mid (\mathfrak{G}, \mathfrak{H}) \geq (\mathfrak{E}_1, \mathfrak{F}_1)\}, \{m(\mathfrak{G}, \mathfrak{H}) \in G \mid (\mathfrak{G}, \mathfrak{H}) \geq (\mathfrak{E}_1, \mathfrak{F}_1)\}')$$

and

$$j(\mathfrak{E}_2, \mathfrak{F}_2)^{\mathfrak{P}(\mathbb{K})}((A_{x_1}, B_{x_1}), \ldots, (A_{x_n}, B_{x_n}))$$
$$= (\{j(\mathfrak{G}, \mathfrak{H}) \in M_i \mid (\mathfrak{G}, \mathfrak{H}) \geq (\mathfrak{E}_2, \mathfrak{F}_2)\}', \{j(\mathfrak{G}, \mathfrak{H}) \in M_i \mid (\mathfrak{G}, \mathfrak{H}) \geq (\mathfrak{E}_2, \mathfrak{F}_2)\})$$

by induction over j:

For $j = 1$ we have

$$m(\mathfrak{E}_1, \mathfrak{F}_1)^{\mathfrak{P}(\mathbb{K})}((A_{x_1}, B_{x_1}), \ldots, (A_{x_n}, B_{x_n}))$$
$$= \bigsqcap\{(A_{x_j}, A'_{x_j}) \mid x_{j\sqcap} \in \mathfrak{E}_1\} \sqcap \neg \bigsqcup\{(A_{x_j}, A'_{x_j}) \mid x_{j\sqcap} \in \mathfrak{F}_1\}$$
$$= (\{\overline{m((\mathfrak{G}, \mathfrak{H})} \in G \mid \mathfrak{E}_1 \subseteq \mathfrak{G}\}, \{\overline{m((\mathfrak{G}, \mathfrak{H})} \in G \mid \mathfrak{E}_1 \subseteq \mathfrak{G}\}')$$
$$\sqcap \neg(\{\overline{m(\mathfrak{G}, \mathfrak{H})} \in G \mid \mathfrak{G} \cap \mathfrak{F}_1 \neq \emptyset\}, \{\overline{m(\mathfrak{G}, \mathfrak{H})} \in G \mid \mathfrak{G} \cap \mathfrak{F}_1 \neq \emptyset\}')$$
$$= (\{\overline{m((\mathfrak{G}, \mathfrak{H})} \in G \mid \mathfrak{E}_1 \subseteq \mathfrak{G}\}, \{\overline{m((\mathfrak{G}, \mathfrak{H})} \in G \mid \mathfrak{E}_1 \subseteq \mathfrak{G}\}')$$
$$\sqcap(\{\overline{m(\mathfrak{G}, \mathfrak{H})} \in G \mid \mathfrak{F}_1 \subseteq \mathfrak{H}\}, \{\overline{m(\mathfrak{G}, \mathfrak{H})} \in G \mid \mathfrak{F}_1 \subseteq \mathfrak{H}\}')$$
$$= (\{\overline{m(\mathfrak{G}, \mathfrak{H})} \in G \mid (\mathfrak{G}, \mathfrak{H}) \geq (\mathfrak{E}_1, \mathfrak{F}_1)\}, \{\overline{m(\mathfrak{G}, \mathfrak{H})} \in G \mid (\mathfrak{G}, \mathfrak{H}) \geq (\mathfrak{E}_1, \mathfrak{F}_1)\}').$$

Dually, we obtain

$$j(\mathfrak{E}_2, \mathfrak{F}_2)^{\mathfrak{P}(\mathbb{K})}((A_{x_1}, B_{x_1}), \ldots, (A_{x_n}, B_{x_n}))$$
$$= (\{j(\mathfrak{G}, \mathfrak{H}) \in M_i \mid (\mathfrak{G}, \mathfrak{H}) \geq (\mathfrak{E}_2, \mathfrak{F}_2)\}', \{j(\mathfrak{G}, \mathfrak{H}) \in M_i \mid (\mathfrak{G}, \mathfrak{H}) \geq (\mathfrak{E}_2, \mathfrak{F}_2)\}).$$

In order to conclude from j to $j + 1$ for $j + 1 \leq i$ we set for $(\mathfrak{E}_1, \mathfrak{F}_1) \in S^{j+1}_\sqcap$

$$\mathfrak{E}^*_1 := \mathfrak{E}_1 \cap \tilde{S}^j_\sqcap \text{ and } \mathfrak{F}^*_1 := \mathfrak{F}_1 \cap \tilde{S}^j_\sqcap.$$

Then

$$m(\mathfrak{E}_1, \mathfrak{F}_1) \approx m(\mathfrak{E}^*_1, \mathfrak{F}^*_1) \sqcap \bigsqcap(\mathfrak{E}_1 \setminus \mathfrak{E}^*_1) \sqcap \neg \bigsqcup(\mathfrak{F}_1 \setminus \mathfrak{F}^*_1)$$

and $(\mathfrak{E}_1 \setminus \mathfrak{E}^*_1) \cup (\mathfrak{F}_1 \setminus \mathfrak{F}^*_1) \subseteq \tilde{S}^{j+1}_\sqcap \setminus \tilde{S}^j_\sqcap$. For $j(\mathfrak{G}, \mathfrak{H})_\sqcap \in \tilde{S}^{j+1}_\sqcap \setminus \tilde{S}^j_\sqcap$ we have by assumption

$$\begin{aligned}&j(\mathfrak{G},\mathfrak{H})^{\mathfrak{P}(\mathbb{K})}((A_{x_1},B_{x_1}),\ldots,(A_{x_n},B_{x_n}))\\&\quad=(\{j(\mathfrak{I},\mathfrak{J})\in M_i\mid(\mathfrak{I},\mathfrak{J})\geq(\mathfrak{G},\mathfrak{H})\}',\{j(\mathfrak{I},\mathfrak{J})\in M_i\mid(\mathfrak{I},\mathfrak{J})\geq(\mathfrak{G},\mathfrak{H})\})\\&\quad=:(B',B).\end{aligned}$$

From $k(j(\mathfrak{G},\mathfrak{H}))<k(j(\mathfrak{G},\mathfrak{H})_\sqcap)\leq i$ follows

$$\begin{aligned}\overline{m(\mathfrak{I},\mathfrak{J})}\in B'&\Leftrightarrow\overline{m(\mathfrak{I},\mathfrak{J})}\sqsubseteq\overline{\bigsqcap B}=\overline{j(\mathfrak{G},\mathfrak{H})}\\&\Leftrightarrow\overline{m(\mathfrak{I},\mathfrak{J})}\sqsubseteq\overline{j(\mathfrak{G},\mathfrak{H})_\sqcap}\end{aligned}$$

and by Lemma 7 this is equivalent to $j(\mathfrak{G},\mathfrak{H})_\sqcap\in\mathfrak{I}$. Thus,

$$\begin{aligned}&m(\mathfrak{E}_1,\mathfrak{F}_1)^{\mathfrak{P}(\mathbb{K})}((A_{x_1},B_{x_1}),\ldots,(A_{x_n},B_{x_n}))\\&\quad=(m(\mathfrak{E}_1^*,\mathfrak{F}_1^*)\sqcap\bigsqcap(\mathfrak{E}_1\setminus\mathfrak{E}_1^*)\sqcap\neg\bigsqcup(\mathfrak{F}_1\setminus\mathfrak{F}_1^*))^{\mathfrak{P}(\mathbb{K})}((A_{x_1},B_{x_1}),\ldots,(A_{x_n},B_{x_n}))\\&\quad=(\{\overline{m(\mathfrak{G},\mathfrak{H})}\in G\mid(\mathfrak{G},\mathfrak{H})\geq(\mathfrak{E}_1^*,\mathfrak{F}_1^*)\},\{\overline{m(\mathfrak{G},\mathfrak{H})}\in G\mid(\mathfrak{G},\mathfrak{H})\geq(\mathfrak{E}_1^*,\mathfrak{F}_1^*)\}')\\&\qquad\sqcap\bigsqcap(\{\overline{m(\mathfrak{G},\mathfrak{H})}\in G\mid j(\mathfrak{I},\mathfrak{J})_\sqcap\in\mathfrak{G}\text{ for some }j(\mathfrak{I},\mathfrak{J})_\sqcap\in\mathfrak{E}_1\setminus\mathfrak{E}_1^*\},\\&\qquad\qquad\{\overline{m(\mathfrak{G},\mathfrak{H})}\in G\mid j(\mathfrak{I},\mathfrak{J})_\sqcap\in\mathfrak{G}\text{ for some }j(\mathfrak{I},\mathfrak{J})_\sqcap\in\mathfrak{E}_1\setminus\mathfrak{E}_1^*\}')\\&\qquad\sqcap\neg\bigsqcup(\{\overline{m(\mathfrak{G},\mathfrak{H})}\in G\mid j(\mathfrak{I},\mathfrak{J})_\sqcap\in\mathfrak{G}\text{ for some }j(\mathfrak{I},\mathfrak{J})_\sqcap\in\mathfrak{F}_1\setminus\mathfrak{F}_1^*\},\\&\qquad\qquad\overline{\{\overline{m(\mathfrak{G},\mathfrak{H})}\in G\mid j(\mathfrak{I},\mathfrak{J})_\sqcap\in\mathfrak{G}\text{ for some }j(\mathfrak{I},\mathfrak{J})_\sqcap\in\mathfrak{F}_1\setminus\mathfrak{F}_1^*\}'})\\&\quad=(\{\overline{m(\mathfrak{G},\mathfrak{H})}\in G\mid(\mathfrak{G},\mathfrak{H})\geq(\mathfrak{E}_1^*,\mathfrak{F}_1^*)\},\{\overline{m(\mathfrak{G},\mathfrak{H})}\in G\mid(\mathfrak{G},\mathfrak{H})\geq(\mathfrak{E}_1^*,\mathfrak{F}_1^*)\}')\\&\qquad\sqcap(\{\overline{m(\mathfrak{G},\mathfrak{H})}\in G\mid\mathfrak{E}_1\setminus\mathfrak{E}_1^*\subseteq\mathfrak{G}\},\{\overline{m(\mathfrak{G},\mathfrak{H})}\in G\mid\mathfrak{E}_1\setminus\mathfrak{E}_1^*\subseteq\mathfrak{G}\}')\\&\qquad\sqcap(\{\overline{m(\mathfrak{G},\mathfrak{H})}\in G\mid\mathfrak{F}_1\setminus\mathfrak{F}_1^*\subseteq\mathfrak{H}\},\{\overline{m(\mathfrak{G},\mathfrak{H})}\in G\mid\mathfrak{F}_1\setminus\mathfrak{F}_1^*\subseteq\mathfrak{H}\}')\\&\quad=(\{\overline{m(\mathfrak{G},\mathfrak{H})}\in G\mid(\mathfrak{G},\mathfrak{H})\geq(\mathfrak{E}_1,\mathfrak{F}_1)\},\{\overline{m(\mathfrak{G},\mathfrak{H})}\in G\mid(\mathfrak{G},\mathfrak{H})\geq(\mathfrak{E}_1,\mathfrak{F}_1)\}').\end{aligned}$$

For $j(\mathfrak{E}_2,\mathfrak{F}_2)^{\mathfrak{P}(\mathbb{K})}((A_{x_1},B_{x_1}),\ldots,(A_{x_n},B_{x_n}))$ a similar argument is used: Again, we set $\mathfrak{E}_2^*:=\mathfrak{E}_2\cap\tilde{S}^j_\sqcup$ and $\mathfrak{F}_2^*:=\mathfrak{F}_2\cap\tilde{S}^j_\sqcup$. For $m(\mathfrak{G},\mathfrak{H})_\sqcup\in\tilde{S}^{j+1}_\sqcup\setminus\tilde{S}^j_\sqcup$ we have by assumption

$$\begin{aligned}m(\mathfrak{G},\mathfrak{H})&=(\{m(\mathfrak{I},\mathfrak{J})\in G\mid(\mathfrak{I},\mathfrak{J})\geq(\mathfrak{G},\mathfrak{H}))\},\{m(\mathfrak{I},\mathfrak{J})\in G\mid(\mathfrak{I},\mathfrak{J})\geq(\mathfrak{G},\mathfrak{H}))\}')\\&=:(A,A').\end{aligned}$$

In order to conclude $j(\mathfrak{I},\mathfrak{J})\in A'\Leftrightarrow m(\mathfrak{G},\mathfrak{H})_\sqcup\in\mathfrak{J}$ we use that

$$\bigsqcup_{1\leq j\leq k}m(\mathfrak{A}\cup Y_j,\mathfrak{B}\cup(T\setminus Y_j)\approx m(\mathfrak{A},\mathfrak{B})\in G_i$$

and that $k(m(\mathfrak{G},\mathfrak{H}))=j\leq i$ implies

$$(\mathfrak{G},\mathfrak{H})\leq(\mathfrak{A}\cup Y_j,\mathfrak{B}\cup(T\setminus Y_j)\Leftrightarrow(\mathfrak{G},\mathfrak{H})\leq(\mathfrak{A},\mathfrak{B}).$$

Then we continue as in the case of $m(\mathfrak{E}_1,\mathfrak{F}_1)^{\mathfrak{P}(\mathbb{K})}((A_{x_1},B_{x_1}),\ldots,(A_{x_n},B_{x_n}))$ to obtain

$$\begin{aligned}&j(\mathfrak{E}_2,\mathfrak{F}_2)^{\mathfrak{P}(\mathbb{K})}((A_{x_1},B_{x_1}),\ldots,(A_{x_n},B_{x_n}))\\&\quad=(\{j(\mathfrak{G},\mathfrak{H})\in M_i\mid(\mathfrak{G},\mathfrak{H})\geq(\mathfrak{E}_2,\mathfrak{F}_2)\}',\{j(\mathfrak{G},\mathfrak{H})\in M_i\mid(\mathfrak{G},\mathfrak{H})\geq(\mathfrak{E}_2,\mathfrak{F}_2)\}).\end{aligned}$$

Finally, we show that $m(\mathfrak{A}\cup Y_j,\mathfrak{B}\cup(T\setminus Y_j)^{\mathfrak{P}(\mathbb{K})}((A_{x_1},B_{x_1}),\ldots,(A_{x_n},B_{x_n}))\neq(\emptyset,M_i)$: Obviously, $(\mathfrak{A}\cup Y_j,\mathfrak{B}\cup(T\setminus Y_j)\geq(\mathfrak{A},\mathfrak{B})$ and thus $\overline{m(\mathfrak{A}\cup Y_j,\mathfrak{B}\cup(T\setminus Y_j))}$ is in the extent of $m(\mathfrak{A},\mathfrak{B})^{\mathfrak{P}(\mathbb{K})}((A_{x_1},B_{x_1}),\ldots,(A_{x_n},B_{x_n}))$. For $j(\mathfrak{C}_l,\mathfrak{D}_l)_\sqcap\in T$

we have $j(\mathfrak{C}_l, \mathfrak{D}_l)_{\sqcap}^{\mathfrak{P}(\mathbb{K})}((A_{x_1}, B_{x_1}), \ldots, (A_{x_n}, B_{x_n})) = (\{\overline{j(\mathfrak{C}_l, \mathfrak{D}_l)}\}', \{\overline{j(\mathfrak{C}_l, \mathfrak{D}_l)}\})$ and by Lemma 7

$$\overline{m(\mathfrak{A} \cup Y_j, \mathfrak{B} \cup (T \setminus Y_j))} \sqsubseteq \overline{j(\mathfrak{C}_l, \mathfrak{D}_l)} \text{ if } j(\mathfrak{C}_l, \mathfrak{D}_l) \in Y_j$$

and

$$\overline{m(\mathfrak{A} \cup Y_j, \mathfrak{B} \cup (T \setminus Y_j))} \not\sqsubseteq \overline{j(\mathfrak{C}_l, \mathfrak{D}_l)} \text{ if } j(\mathfrak{C}_l, \mathfrak{D}_l) \notin Y_j.$$

Hence $\overline{m(\mathfrak{A} \cup Y_j, \mathfrak{B} \cup (T \setminus Y_j))}$ is in the extent of $m(\mathfrak{A} \cup Y_j, \mathfrak{B} \cup (T \setminus Y_j))_{\sqcap}^{\mathfrak{P}(\mathbb{K})}((A_{x_1}, B_{x_1}), \ldots, (A_{x_n}, B_{x_n}))$. □

The following examples show, how Theorem 6 can be used:

Example 4. Fig. 5 depicts the context $\mathbb{K}_1(\{x\})$. If we want to determine $S_{\sqcap}^2$ then Theorem 6 allows us to deal seperately with the $\sqcap$-sequents extending $(\{x_{\sqcap}\}, \emptyset)$ and the $\sqcap$-sequents extending $(\emptyset, \{x_{\sqcap}\})$.

For $(\{x_{\sqcap}\}, \emptyset)$, which corresponds to the first line of the context in Fig. 5 we obtain immediately from $\overline{x_{\sqcap}} \sqsubseteq \overline{x_{\sqcup}}$ that $m(\{x_{\sqcap}\}, \{x_{\sqcup}\}) = x_{\sqcap} \sqcap \neg(x_{\sqcup})_{\sqcap} \approx \bot$. Thus, we only need to check $m(\mathfrak{A}_1, \mathfrak{B}_1) := m(\{x_{\sqcap}, (x_{\sqcup})_{\sqcap}, (\lrcorner x_{\sqcup})_{\sqcap}\}, \emptyset)$ and $m(\mathfrak{A}_2, \mathfrak{B}_2) := m(\{x_{\sqcap}, (x_{\sqcup})_{\sqcap}\}, \{(\lrcorner x_{\sqcup})_{\sqcap}\})$. If both are in $S_{\sqcap}^2$ then, by Theorem 6, the context depicted in Fig. 6 contains a counterexample to $m(\mathfrak{A}_i, \mathfrak{B}_i) \approx \bot$. Indeed, with $A_x = \{1.1, 1.2\}$ and $B_x = \{a\}$ we obtain

$$m(\mathfrak{A}_1, \mathfrak{B}_1)_{\sqcap}^{\mathfrak{P}(\mathbb{K}_{1.1})}(A_x, B_x) = (\{1.1\}, \{a, b\})$$
$$m(\mathfrak{A}_2, \mathfrak{B}_2)_{\sqcap}^{\mathfrak{P}(\mathbb{K}_{1.1})}(A_x, B_x) = (\{1.2\}, \{a\}).$$

For $(\emptyset, \{x_{\sqcap}\})$, which corresponds to the second line of the context in Fig. 5, four possible $\sqcap$-sequents have to be checked. Again, if they are all different from $\bot$ then the context in Fig. 7 contains an example. Indeed, with $A_x = \{1\}$ and $B_x = \{a\}$ we find

$$m(\{(x_{\sqcup})_{\sqcap}, (\lrcorner x_{\sqcup})_{\sqcap}\}, \{x_{\sqcap}\})_{\sqcap}^{\mathfrak{P}(\mathbb{K}_{1.2})}(A_x, B_x) = (\{2\}, \{a, b\})$$
$$m(\{(x_{\sqcup})_{\sqcap}\}, \{x_{\sqcap}, \lrcorner x_{\sqcup})_{\sqcap}\})_{\sqcap}^{\mathfrak{P}(\mathbb{K}_{1.2})}(A_x, B_x) = (\{3\}, \{a\})$$
$$m(\{(\lrcorner x_{\sqcup})_{\sqcap}\}, \{x_{\sqcap}, (x_{\sqcup})_{\sqcap}\})_{\sqcap}^{\mathfrak{P}(\mathbb{K}_{1.2})}(A_x, B_x) = (\{4\}, \{b\})$$
$$m(\emptyset, \{x_{\sqcap}, (x_{\sqcup})_{\sqcap}, (\lrcorner x_{\sqcup})_{\sqcap}\})_{\sqcap}^{\mathfrak{P}(\mathbb{K}_{1.2})}(A_x, B_x) = (\{5\}, \emptyset).$$

Thus, these four $\sqcap$-sequents are in $S_{\sqcap}^2$. Duality yields

$$j(\{x_{\sqcup}\}, \{(x_{\sqcap})_{\sqcup}, (\neg x_{\sqcap})_{\sqcup}\}) \not\approx \top$$
$$j(\{x_{\sqcup}, (\neg x_{\sqcap})_{\sqcup}\}, \{(x_{\sqcap})_{\sqcup}\}) \not\approx \top$$
$$j(\{x_{\sqcup}, (x_{\sqcap})_{\sqcup}\}, \{(\neg x_{\sqcap})_{\sqcup}\}) \not\approx \top$$
$$j(\{x_{\sqcup}, (x_{\sqcap})_{\sqcup}, (\neg x_{\sqcap})_{\sqcup}\}, \emptyset) \not\approx \top.$$

Example 5. Fig. 8 depicts the context $\mathbb{K}_2(\{x\})$. Consider the terms $t_1 := (\neg x_{\sqcap}) \sqcap \neg(x_{\sqcup})_{\sqcap} \sqcap (\lrcorner(x_{\sqcup})_{\sqcap}$ (correspondig to the fourth line) and $t_2 := x_{\sqcup} \sqcup (x_{\sqcap})_{\sqcup} \sqcup \lrcorner(\neg(x_{\sqcap}))_{\sqcup}$ (correspondig to the second column). In order to determine if $t_1 \sqcap$

$\mathbb{K}_1(\{x\})$	$\overline{x_\sqcup}$	$\overline{\lrcorner x_\sqcup}$
$\overline{x_\sqcap}$	X	
$\overline{\neg x_\sqcap}$		

Fig. 5. $\mathbb{K}_1(\{x\})$

$\mathbb{K}_{1.1}$	a / $\overline{x_\sqcup}$	b / $\overline{\lrcorner x_\sqcup}$
1.1 / $\overline{x_\sqcap \sqcap (\lrcorner x_\sqcup)_\sqcap}$	X	X
1.2 / $\overline{x_\sqcap \sqcap \neg(\lrcorner x_\sqcup)_\sqcap}$	X	
2 / $\overline{\neg x_\sqcap}$		

Fig. 6. The first extension of $\mathbb{K}_1(\{x\})$

	a / $\overline{x_\sqcup}$	b / $\overline{\lrcorner x_\sqcup}$
1 / $\overline{x_\sqcap}$	X	
2 / $\overline{\neg x_\sqcap \sqcap (x_\sqcup)_\sqcap \sqcap(\lrcorner(x_\sqcup))_\sqcap}$	X	X
3 / $\overline{\neg x_\sqcap \sqcap (x_\sqcup)_\sqcap \sqcap\neg(\lrcorner(x_\sqcup))_\sqcap}$	X	
4 / $\overline{\neg x_\sqcap \sqcap \neg(x_\sqcup)_\sqcap \sqcap(\lrcorner(x_\sqcup))_\sqcap}$		X
5 / $\overline{\neg x_\sqcap \sqcap \neg(x_\sqcup)_\sqcap \sqcap\neg(\lrcorner(x_\sqcup))_\sqcap}$		

Fig. 7. The second extension of $\mathbb{K}_1(\{x\})$

$(t_2)_\sqcap \not\approx \bot$ and $t_1 \sqcap \neg(t_2)_\sqcap \not\approx \bot$, we modify $\mathbb{K}_2(\{x\})$ replacing object 4 with new objects 4.1 and 4.2. 4.1 represents then $t_1 \sqcap t_{2\sqcap}$ and 4.2 represents $t_1 \sqcap \neg t_{2\sqcap}$. This yields the context $\mathbb{K}_{2.1}$ as depicted in Fig. 9. We find

$$(t_1 \sqcap (t_2)_\sqcap)^{\underline{\mathfrak{P}}(\mathbb{K}_{2.1})}(\{1,2\},\{a,b\}) = (\emptyset, \{a,b,c,d,e,f\}) = \bot$$

and

$$(t_1 \sqcap \neg(t_2)_\sqcap)^{\underline{\mathfrak{P}}(\mathbb{K}_{2.1})}(\{1,2\},\{a,b\}) = (\{4.1\}, \{a,c,d,e,f\}) \neq \bot.$$

Then Theorem 6 yields $t_1 \sqcap (t_2)_\sqcap \approx \bot$.

In general, Theorem 6 allows to use the following algorithm to determine all $\sqcap$-sequents in $S_\sqcap^{m+1}$ extending a $\sqcap$-sequent $(\mathfrak{A}, \mathfrak{B}) \in S_\sqcap^m$ if $S_\sqcap^m$, $S_\sqcup^m$ and $\mathbb{K}_m(X) := (G_m, M_m, \sqsubseteq)$ are known.

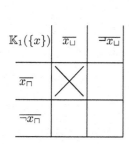

$\mathbb{K}_2(\{x\})$	a	b	c	d	e	f
1	X	X	X	X	X	X
2	X	X	X		X	
3	X	X	X	X	X	X
4	X		X	X	X	X
5	X	X	X	X	X	
6	X		X	X		

Fig. 8. The context $\mathbb{K}_2(\{x\})$. The abbreviated objects and attributes are explained below.

object	element of $\mathcal{D}(\{x\})$
1	$\overline{x_\sqcap \sqcap (x_\sqcup)_\sqcap \sqcap (\lrcorner(x_\sqcup))_\sqcap}$
2	$\overline{x_\sqcap \sqcap (x_\sqcup)_\sqcap \sqcap \neg(\lrcorner(x_\sqcup))_\sqcap}$
3	$\overline{\neg x_\sqcap \sqcap (x_\sqcup)_\sqcap \sqcap (\lrcorner(x_\sqcup))_\sqcap}$
4	$\overline{\neg x_\sqcap \sqcap \neg(x_\sqcup)_\sqcap \sqcap (\lrcorner(x_\sqcup))_\sqcap}$
5	$\overline{\neg x_\sqcap \sqcap (x_\sqcup)_\sqcap \sqcap \neg(\lrcorner(x_\sqcup))_\sqcap}$
6	$\overline{\neg x_\sqcap \sqcap \neg(x_\sqcup)_\sqcap \sqcap \neg(\lrcorner(x_\sqcup))_\sqcap}$

attribute	element of $\mathcal{D}(\{x\})$
a	$\overline{x_\sqcup \sqcup (x_\sqcap)_\sqcup \sqcup (\neg(x_\sqcap))_\sqcup}$
b	$\overline{x_\sqcup \sqcup (x_\sqcap)_\sqcup \sqcup \lrcorner(\neg(x_\sqcap))_\sqcup}$
c	$\overline{\lrcorner x_\sqcup \sqcup (x_\sqcap)_\sqcup \sqcup (\neg(x_\sqcap))_\sqcup}$
d	$\overline{\lrcorner x_\sqcup \sqcup \lrcorner(x_\sqcap)_\sqcup \sqcup (\neg(x_\sqcap))_\sqcup}$
e	$\overline{\lrcorner x_\sqcup \sqcup (x_\sqcap)_\sqcup \sqcup \lrcorner(\neg(x_\sqcap))_\sqcup}$
f	$\overline{\lrcorner x_\sqcup \sqcup \lrcorner(x_\sqcap)_\sqcup \sqcup \lrcorner(\neg(x_\sqcap))_\sqcup}$

Algorithm

1. Set $i = 1$, $A := \{j(\mathfrak{C},\mathfrak{D})_\sqcap \in \tilde{S}_\sqcap^{m+1} \mid m(\mathfrak{A},\mathfrak{B}) \not\sqsubseteq j(\mathfrak{C},\mathfrak{D})\}$ and $\mathcal{P}_1 := \mathfrak{P}(\mathfrak{P}(A))$. Then every $p \in \mathcal{P}_1$ with $p = \{Y_1,\dots,Y_k\}$ yields a set of possible extensions $(\mathfrak{A} \cup Y_1, \mathfrak{B} \cup (A \setminus Y_1)),\dots,(\mathfrak{A} \cup Y_k, \mathfrak{B} \cup (A \setminus Y_k))$ of $(\mathfrak{A},\mathfrak{B})$. Since $m(\mathfrak{A},\mathfrak{B} \cup A) \not\eqsim \bot$, we remove all $p \in \mathcal{P}_1$ with $\emptyset \notin p$ and set $T_0 := \{\emptyset\}$.
2. Choose a $p_i \in \mathcal{P}_i$ and set

$$G := (G_m \setminus \{\overline{m(\mathfrak{A},\mathfrak{B})}\}) \cup \{\overline{m(\mathfrak{A} \cup Y_j, \mathfrak{B} \cup (A \setminus Y_j))} \mid Y_j \in p_i\}.$$

   On $G \times M_m$ an incidence relation I is defined as

$$I := \{(g,m) \in G \times M_m \mid g \in G_m \text{ and } g \sqsubseteq m\}$$
$$\cup\{(\overline{m(\mathfrak{A} \cup Y_j, \mathfrak{B} \cup (A \setminus Y_j))}, \overline{j(\mathfrak{C},\mathfrak{D})}) \in G \times M_m \mid \overline{j(\mathfrak{C},\mathfrak{D})} \in Y_j\}.$$

   Set $\mathbb{K} := (G, M_m, I)$.
3. For $Y_k \in p \setminus T_{i-1}$ check if

$$m(\mathfrak{A} \cup Y_j, \mathfrak{B} \cup (A \setminus Y_j))^{\underline{\mathfrak{P}(\mathbb{K})}}((A_{x_1}, B_{x_1}),\dots,(A_{x_n}, B_{x_n})) \neq (\emptyset, M_m).$$

Set

$$T_i := T_{i-1} \cup \{Y_k \in p \mid m(\mathfrak{A} \cup Y_j, \mathfrak{B} \cup (A \setminus Y_j))^{\mathfrak{P}(\mathbb{K})}((A_{x_1}, B_{x_1}), \dots, (A_{x_n}, B_{x_n})) \neq (\emptyset, M_m)\}.$$

4. Set $P_{i+1} := \{q \in P_i \mid T_i \subseteq q \text{ and } q \neq p_i\}$.
5. If $P_{i+1} = \emptyset$ set $T := T_i$ and stop. Otherwise increase i and continue with 2).

When the algorithm stops, the set $\{(\mathfrak{A} \cup (\tilde{S}_\sqcap^{m+1} \setminus A) \cup Y_j, \mathfrak{B} \cup (A \setminus Y_j) \mid Y_j \in T\}$ is the set of all $\sqcap$-sequents in $S_\sqcap^{m+1}$ extending $(\mathfrak{A}, \mathfrak{B})$. Although in theory the set P_1 containing all possibilities we have to check may grow rapidly ($|P_1| = 2^{2^{|A|}-1} \leq 2^{2^{|S_\sqcap^m|}-1}$), the examples given above give reason to hope that the contexts $\mathbb{K}_m(X)$ have many crosses and therefore that $|A|$ is small compared to the total number of rows and columns of $\mathbb{K}_m(X)$.

4.2 Determination of $\sqsubseteq$

Once we know the object and the attribute set of $\mathbb{K}_{i+1}(X)$, the incidence relation needs to be found. A major part of this relation is obtained easily from:

1. If $(\mathfrak{A}, \mathfrak{B}) \in S_\sqcap^{i+1}$, $(\mathfrak{C}, \mathfrak{D}) \in S_\sqcup^{i+1}$ and $\overline{m(\mathfrak{A} \cap \tilde{S}_\sqcap^i, \mathfrak{B} \cap \tilde{S}_\sqcap^i)} \sqsubseteq \overline{j(\mathfrak{C} \cap S_\sqcup^i, \mathfrak{D} \cap S_\sqcup^i)}$ then $\overline{m(\mathfrak{A}, \mathfrak{B})} \sqsubseteq \overline{j(\mathfrak{C}, \mathfrak{D})}$, i.e. the incidence relation of $\mathbb{K}_i(X)$ has to be respected by $\mathbb{K}_{i+1}(X)$.
2. If $(\mathfrak{A}, \mathfrak{B}) \in S_\sqcap^{i+1}$, $(\mathfrak{C}_1, \mathfrak{D}_1) \in S_\sqcup^i$, $(\mathfrak{C}_2, \mathfrak{D}_2) \in S_\sqcup^{i+1}$ with $(\mathfrak{C}_2, \mathfrak{D}_2) \geq (\mathfrak{C}_1, \mathfrak{D}_1)$ and $(j(\mathfrak{C}_1, \mathfrak{D}_1))_\sqcap \in \mathfrak{A}$ then $\overline{m(\mathfrak{A}, \mathfrak{B})} \sqsubseteq \overline{j(\mathfrak{C}_2, \mathfrak{D}_2)}$.
   Dually, if $(\mathfrak{C}, \mathfrak{D}) \in S_\sqcup^{i+1}$, $(\mathfrak{A}_1, \mathfrak{B}_1) \in S_\sqcap^i$, $(\mathfrak{A}_2, \mathfrak{B}_2) \in S_\sqcap^{i+1}$ with $(\mathfrak{A}_2, \mathfrak{B}_2) \geq (\mathfrak{A}_1, \mathfrak{B}_1)$ and $(m(\mathfrak{A}_1, \mathfrak{B}_1))_\sqcup \in \mathfrak{D}$ then $\overline{m(\mathfrak{A}_2, \mathfrak{B}_2)} \sqsubseteq \overline{j(\mathfrak{C}, \mathfrak{D})}$.

$\mathbb{K}_{2.1}$	a	b	c	d	e	f
1	×	×	×	×	×	×
2	×	×	×		×	
3	×	×	×	×	×	×
4.1	×	×	×	×	×	×
4.2	×		×	×	×	×
5	×	×	×	×	×	
6	×		×	×		

Fig. 9. The extended context $\mathbb{K}_{2.1}$. The numbers and letters represent the same objects and attributes as in Fig. 8 except for 4.1 and 4.2, which are explained in Example 5.

We denote with $\tilde{I} \subseteq G_{i+1} \times M_{i+1}$ the relation between the objects and attributes of $\mathbb{K}_{i+1}(X)$ obtained from these rules.

A second helpful property of $\mathbb{K}_{i+1}(X)$ is its self-duality:

Lemma 8. *For a finite set X of variables and $i \in \mathbb{N}$ the context $\mathbb{K}_i(X)$ is isomorphic to its dual $\mathbb{K}_i(X)^d := (M_i, G_i, \sqsupseteq)$.*

Proof. Let $\alpha : \mathcal{T}(X) \to \mathcal{T}(X)$ map every term to its dual. Obviously (α, α) is an isomorphism between $\mathbb{K}_i(X)$ and $\mathbb{K}_i(X)^d$. □

We will call every context $\mathbb{K} := (G, M, I)$ satisfying $\mathbb{K} \cong \mathbb{K}^d$ *self-dual.*

Finally, Prop. 6 yields that $\mathbb{K}_{i+1}$ satisfies the following condition:

$$\text{(R)}\quad \begin{array}{l} \text{For } m(\mathfrak{A},\mathfrak{B}) \in S_\sqcap^{i+1},\ j(\mathfrak{C},\mathfrak{D}) \in S_\sqcup^{i+1}: \\ m(\mathfrak{A},\mathfrak{B})^{\underline{\mathfrak{P}}(\mathbb{K}_{i+1}(X))}((A_{x_1},B_{x_1}),\ldots,(A_{x_n},B_{x_n})) = (\{\overline{m(\mathfrak{A},\mathfrak{B})}\}, \{\overline{m(\mathfrak{A},\mathfrak{B})}\}') \\ \text{and} \\ j(\mathfrak{C},\mathfrak{D})^{\underline{\mathfrak{P}}(\mathbb{K}_{i+1}(X))}((A_{x_1},B_{x_1}),\ldots,(A_{x_n},B_{x_n})) = (\{\overline{j(\mathfrak{C},\mathfrak{D})}\}', \{\overline{j(\mathfrak{C},\mathfrak{D})}\}). \end{array}$$

Now the incidence relation of $\mathbb{K}_{i+1}(X)$ can be found as the smallest incidence relation extending $\tilde{I}$ and satisfying (R).

Proposition 8. *Let $\mathbb{K}_1 := (G_{i+1}, M_{i+1}, I_1), \ldots, \mathbb{K}_k := (G_{i+1}, M_{i+1}, I_k)$ be all self-dual contexts with an incidence relation extending $\tilde{I}$ and satisfying (R). Then $\bigcap_{j=1,\ldots,k} I_j$ is the incidence relation of $\mathbb{K}_{i+1}(X)$.*

Proof. We have already shown that the incidence relation of $\mathbb{K}_{i+1}(X)$ is equal to one of the I_j. Assume $(\overline{m(\mathfrak{A},\mathfrak{B})}, \overline{j(\mathfrak{C},\mathfrak{D})}) \notin I_j$ for some j. The Basic Theorem on Protoconcept Algebras yields

$$\begin{aligned} & m(\mathfrak{A},\mathfrak{B})^{\underline{\mathfrak{P}}(\mathbb{K}_j)}((A_{x_1},B_{x_1}),\ldots,(A_{x_n},B_{x_n})) \\ & \qquad \not\sqsubseteq j(\mathfrak{C},\mathfrak{D})^{\underline{\mathfrak{P}}(\mathbb{K}_j)}((A_{x_1},B_{x_1}),\ldots,(A_{x_n},B_{x_n})) \end{aligned}$$

and thus $\overline{m(\mathfrak{A},\mathfrak{B})} \not\sqsubseteq \overline{j(\mathfrak{C},\mathfrak{D})}$ in $\mathcal{D}(X)$. □

Thus a combination of the exploration algorithm given in Section 2, the algorithm in Section 4.1 and these results allow us to compute the contexts $\mathbb{K}_i(X)$ automatically: For a given set of variables $X := \{x_1, \ldots, x_n\}$, Proposition 7 yields immediately the sets $S_\sqcap^1$ and $S_\sqcup^1$ as well as the incidence relation of $\mathbb{K}_1(X) := (G_1, M_1, \sqsubseteq)$ computed in the first iteration of the exploration algorithm. In the following iterations of the exloration the algorithm given in Section 4.1 is used to determine the sets $S_\sqcap^{i+1}$ and $S_\sqcup^{i+1}$ in step 2. Proposition 8 yields the incidence relation of $\mathbb{K}_{i+1}(X)$ in step 3.

For two terms s, t over X with complexity $k(s), k(t) < m$ it follows from Theorem 4 that $s \approx t$ if and only if

$$s^{\mathfrak{P}(\mathbb{K}_m(X))}(((A_{x_1}, B_{x_1}), \ldots, (A_{x_n}, B_{x_n})))$$
$$= t^{\mathfrak{P}(\mathbb{K}_m(X))}(((A_{x_1}, B_{x_1}), \ldots, (A_{x_n}, B_{x_n})))$$

with

$$A_{x_j} := \{\overline{m(\mathfrak{A}, \mathfrak{B})} \in G_{i+1} \mid x_{j\sqcap} \in \mathfrak{A}\},$$
$$B_{x_j} := \{\overline{j(\mathfrak{C}, \mathfrak{D})} \in M_{i+1} \mid x_{j\sqcup} \in \mathfrak{D}\}.$$

Therefore, the combination of the exploration algorithm with the algorithm given in Section 4.1 and Proposition 8 yields a solution of the word problem for free double Boolean algebras.

5 Concluding Remarks

The solution of the word problem for free double Boolean algebras presented in this paper constructs, if possible, a counterexample to a term identity $s \sim t$ for terms over a set of variables X that depends only of the complexities $k(s)$ and $k(t)$ of s, t and $|X|$. This allows to use the counterexample, once constructed, for all terms with equal complexity over the same set of variables. If $\mathbb{K}_i(X)$ and the protoconcepts (A_{x_j}, B_{x_j}) assigned to the variables x_j are known, the terms may be evaluated rapidly in $\underline{\mathfrak{P}}(\mathbb{K}_i(X))$. This, and the development of a solution in the framework of double Boolean algebras, are the main differences to the solution intended in [HLSW00], where 2-sorted and 3-sorted approaches are used.

Work related to the investigation of the generation process in double Boolean algebras was published by M. Skorsky in 1989. In [Sk89] an earlier approach to introduce negations on formal concepts led to an investigation of compositions of Galois connections and dual Galois connections between ordered sets. For a context $\mathbb{K} := (G, M, I)$ we obtain a Galois connection $\mu : \mathfrak{P}(G) \to \mathfrak{P}(M)$, $\nu : \mathfrak{P}(M) \to \mathfrak{P}(G)$ as $\mu(A) := A'$ and $\nu(B) := B'$. In semiconcept algebras these maps correspond to $\mathfrak{a} \mapsto \mathfrak{a}_\sqcup$ for $\sqcap$-semiconcepts and $\mathfrak{b} \mapsto \mathfrak{b}_\sqcap$ for $\sqcup$-semiconcepts. With the complementation operation $X \mapsto X^c$ a dual Galois connection is obtained as $\gamma : \mathfrak{P}(G) \to \mathfrak{P}(M)$, $\gamma(A) := (\mu(A^c))^c$ and $\delta : \mathfrak{P}(M) \to \mathfrak{P}(G)$, $\delta(B) := (\nu(B^c))^c$. For $\sqcap$-semiconcepts, the map γ corresponds to $\mathfrak{a} \mapsto \lrcorner\neg\mathfrak{a}$, for $\sqcup$-semiconcepts, the map δ corresponds to $\mathfrak{b} \mapsto \neg\lrcorner\mathfrak{b}$. Thus compositions of these maps yield an example for the regular monoids studied in [Sk89].

References

[Bu00] P. Burmeister: ConImp – Ein Programm zur Fomalen Begriffsanalyse. In: G. Stumme, R. Wille (eds.): *Begriffliche Wissensverarbeitung: Methoden und Anwendungen.* Springer, Heidelberg 2000, 25–56.

[BS00] S. Burris, H. P. Sankappanavar: *A course in universal algebra.* Millenium Edition, 2000.
`<http://www.math.uwaterloo.ca/ snburris/htdocs/ualg.html>`.

[GW99] B. Ganter, R. Wille: *Formal Concept Analysis: Mathematical Foundations.* Springer, Heidelberg 1999.

[HLSW00] C. Herrmann, P. Luksch, M. Skorsky, R. Wille: Algebras of Semiconcepts and Double Boolean Algebras. *Contributions to General Algebra* **13**. Verlag Johannes Heyn, Klagenfurt 2000.

[Sk89] M. Skorsky: Regular Monoids Generated by two Galois Connections. *Semigroup Forum* **39**. Springer, Heidelberg 1989, 263 – 293.

[St97] G. Stumme: *Concept Exploration. Knowledge Acquisition in Conceptual Knowledge Systems.* Shaker, Aachen 1997.

[Vo02] B. Vormbrock: Kongruenzrelationen auf doppelt-booleschen Algebren. Diplomarbeit, FB Mathematik, TU Darmstadt 2002.

[Vo03] B. Vormbrock: Congruence Relations on Double Boolean Algebras. *Algebra Universalis* (submitted).

[Vo04] B. Vormbrock: A First Step Towards Protoconcept Exploration. In: P. Eklund (Ed.): *Concept Lattices.* LNAI **2961**. Springer-Verlag Heidelberg 2004, 208 –221.

[VW03] B. Vormbrock, R. Wille: Semiconcept and Protoconcept Algebras: The Basic Theorems. To appear in: B. Ganter, G. Stumme, R. Wille (Eds.): *Formal Concept Analysis. State of the Art.* LNAI **3626**. Springer, Heidelberg 2005.

[Wi82] R. Wille: Restructuring lattice theory: an approach based on hierarchies of concepts. In: I. Rival (ed.): *Ordered sets.* Reidel, Dordrecht – Boston 1982, 445–470.

[Wi00a] R. Wille: Boolean Concept Logic. In: B. Ganter, G. Mineau (Eds.): *Conceptual Structures: Logical, Linguistic and Computational Issues.* LNAI **1867**. Springer, Heidelberg 2000, 317 –331.

[Wi00b] R. Wille: Contextual Logic Summary. In: G. Stumme (Ed.): *Working with Conceptual Structures. Contributions to ICCS 2000.* Shaker, Aachen 2000, 265–276.

On the MacNeille Completion of Weakly Dicomplemented Lattices

Léonard Kwuida[1], Branimir Seselja[2], and Andreja Tepavčević[2]

[1] Universität Bern
Mathematisches Institut
Sidlerstrasse 5, CH-3012 Bern
kwuida@math.unibe.ch
[2] University of Novi Sad
Department of Mathematics and Informatics
Trg D. Obradovića 4
21000 Novi Sad

Abstract. The MacNeille completion of a poset $(P, \leq)$ is the smallest (up to isomorphism) complete poset containing $(P, \leq)$ that preserves existing joins and existing meets. It is wellknown that the MacNeille completion of a Boolean algebra is a Boolean algebra. It is also wellknown that the MacNeille completion of a distributive lattice is not always a distributive lattice (see [Fu44]). The MacNeille completion even seems to destroy many properties of the initial lattice (see [Ha93]). Weakly dicomplemented lattices are bounded lattices equipped with two unary operations satisfying the equations (1) to (3') of Theorem 3. They generalise Boolean algebras (see [Kw04]). The main result of this contribution states that *under chain conditions* the MacNeille completion of a weakly dicomplemented lattice is a weakly dicomplemented lattice. The needed definitions are given in subsections 1.2 and 1.3.

2000 Mathematics Subject Classification: 06B23.

Keywords and Phrases: MacNeille completion, weakly dicomplemneted lattices, Formal Concept Analysis.

1 Introduction

1.1 Motivation

Concept algebras are concept lattices enriched by a weak negation and a weak opposition. They should play for Boolean Concept Logic the rôle played by the powerset algebras for Classical Propositional Logic. The class of weakly dicomplemented lattices is a variety defined by some equations valid in all concept algebras. One important and still open problem in this topic is whether every weakly dicomplemented lattice can be embedded into a concept algebra of a suitable context (called concrete embedding problem in [Kw04, Section 1.4]).

S.O. Kuznetsov and S. Schmidt (Eds.): ICFCA 2007, LNAI 4390, pp. 271–280, 2007.

A promising step is the prime ideal theorem. Using this result on a weakly dicomplemented lattice $(L, \wedge, \vee, ^\triangle, ^\triangledown, 0, 1)$, a canonical context $\mathbb{K}_\triangledown^\triangle(L)$ has been constructed and a bounded lattice embedding $\varphi : L \to \mathfrak{B}(\mathbb{K}_\triangledown^\triangle(L))$ exhibited (see Subsection 1.4), that satisfies

$$\varphi(x^\triangledown) \leq \varphi(x)^\triangledown \leq \varphi(x)^\triangle \leq \varphi(x^\triangle)$$

where in $\mathcal{A}(\mathbb{K}_\triangledown^\triangle(L))$ the weak negation and weak opposition are also denoted by $^\triangle$ and $^\triangledown$ respectively. So the following question arises: does it make any difference if L is assumed to be a complete lattice? To answer this question we first examine the MacNeille completion $\tilde{L}$ of L, on which we extend the operations $^\triangle$ and $^\triangledown$. Of course $\tilde{L}$ embeds into $\mathfrak{B}(\mathbb{K}_\triangledown^\triangle(L)$. Is this a weakly dicomplemented lattice embedding of L into $\mathcal{A}(\mathbb{K}_\triangledown^\triangle(L))$? Our aim is to prove that the MacNeille completion $\tilde{L}$ of a weakly dicomplemented lattice L is a weakly dicomplemented lattice and that L embeds into $\mathcal{A}(\mathbb{K}_\triangledown^\triangle(L)$ iff $\tilde{L}$ embeds into $\mathcal{A}(\mathbb{K}_\triangledown^\triangle(L)$. Section 2 presents preliminary results for the first claim. The second claim is still not proved. Before that we recall some basic notions of Formal Concept Analysis in Subsection 1.2 and introduce weakly dicomplemented lattices in Subsection 1.3. The proofs of stated results can be found in [GW99] or [Kw04].

1.2 Formal Concept Analysis

Formal Concept Analysis is a mathematical field that aims to support human thinking. It has been introduced by Rudolf Wille in the early 80ies, and is based on the theory of lattices and ordered sets. It is started by formalizing the notions of "concept" and "concept hierarchy". The notion of *concept* is rather philosophical. A concept is considered to be determined by its extent and its intent. The extent consists of all entities belonging to the concept and the intent is the set of all common properties shared by all objects of the concept. The hierarchy on concept states that "a concept is more general if it contains more entities". For this purpose the following notions were adopted.

Definition 1. *A* **formal context** *is a triple (G, M, I) of sets such that $I \subseteq G \times M$. The members of G are called* **objects** *and those of M* **attributes**. *If $(g, m) \in I$ the object g is said to have m as an attribute. For $A \subseteq G$ and $B \subseteq M$, the* **derivation operation** $'$ *is defined by*

$$A' := \{m \in M \mid \forall g \in A \quad gIm\} \quad \text{and} \quad B' := \{g \in G \mid \forall m \in B \quad gIm\}.$$

A **formal concept** *of (G, M, I) is a pair (A, B) with $A \subseteq G$ and $B \subseteq M$ such that $A' = B$ and $B' = A$. We call A the* **extent** *and B the* **intent** *of the concept (A, B). The set of all formal concepts of (G, M, I) is denoted by $\mathfrak{B}(G, M, I)$. For concepts (A, B) and (C, D), we call (A, B) a* **subconcept** *of (C, D) provided that $A \subseteq C$ (which is equivalent to $D \subseteq B$). In this case, (C, D) is a* **superconcept** *of (A, B) and we write $(A, B) \leq (C, D)$.*

The pair $(', ')$ forms a Galois connection between the powersets of G and that of M. The basic theorem on concept lattices says that:

Theorem 1. *[GW99, Thm. 3] Let* (G,M,I) *be a formal context.* $\mathfrak{B}(G,M,I)$ *is a complete lattice in which infimum and supremum are given by:*

$$\bigwedge_{t\in T}(A_t,B_t)=\left(\bigcap_{t\in T}A_t,\left(\bigcup_{t\in T}B_t\right)''\right) \quad \text{and} \quad \bigvee_{t\in T}(A_t,B_t)=\left(\left(\bigcup_{t\in T}A_t\right)'',\bigcap_{t\in T}B_t\right).$$

A complete lattice L *is isomorphic to* $\mathfrak{B}(G,M,I)$ *iff there are mappings* $\tilde{\gamma}: G \to L$ *and* $\tilde{\mu}: M \to L$ *such that* $\tilde{\gamma}(G)$ *is supremum-dense in* L*,* $\tilde{\mu}(M)$ *is infimum-dense in* L *and for all* $g \in G$ *and* $m \in M$

$$gIm \iff \tilde{\gamma}(g) \le \tilde{\mu}(m).$$

In particular $L \cong \mathfrak{B}(L,L,\le)$.

$\mathfrak{B}(G,M,I)$ is called the **concept lattice** of the context (G,M,I). A particular case is the context $(P,P,\le)$ where $(P,\le)$ is a poset. Its concept lattice is (isomorphic to) the MacNeille completion of $(P,\le)$). In fact

Theorem 2. *[GW99, Thm. 4] For a poset* $(P,\le)$ *the map*

$$\begin{aligned}\varphi: P &\to \mathfrak{B}(P,P,\le)\\ x &\mapsto (\downarrow x, \uparrow x)\end{aligned}$$

is an order embedding of $(P,\le)$ *into* $\mathfrak{B}(P,P,\le)$ *preserving existing suprema and infima. If* ψ *is another embedding of* $(P,\le)$ *into a complete lattice* L*, then there is an order embedding* λ *of* $\mathfrak{B}(P,P,\le)$ *into* L *such that* $\psi = \lambda \circ \varphi$.

To formalize a negation on concepts, two unary operations are introduced: a weak negation and a weak opposition. We refer to [Wi00] for more details and similar operations encoding a negation.

1.3 Weakly Dicomplemented Lattices

Definition 2. *Let* $\mathbb{K} := (G,M,I)$ *be a formal context. For a formal concept* (A,B) *we define*

its **weak negation** *by* $(A,B)^{\triangle} := ((G\setminus A)'', (G\setminus A)')$

and its **weak opposition** *by* $(A,B)^{\triangledown} := ((M\setminus B)', (M\setminus B)'')$.

$\mathcal{A}(\mathbb{K}) := (\mathfrak{B}(\mathbb{K}); \wedge, \vee, ^{\triangle}, ^{\triangledown}, 0, 1)$ *is called the* **concept algebra** *of the context* $\mathbb{K}$*, where* $\wedge$ *and* $\vee$ *denote the supremum and infimum operations of the concept lattice.*

Rudolf Wille showed that

Theorem 3 ([Wi00]). *For each formal context* $\mathbb{K} := (G,M,I)$*, the following properties hold in its concept algebra* $\mathcal{A}(\mathbb{K})$*:*

(1) $x^{\triangle\triangle} \le x$,
(2) $x \le y \implies x^{\triangle} \ge y^{\triangle}$,
(3) $(x \wedge y) \vee (x \wedge y^{\triangle}) = x$,

(1') $x^{\triangledown\triangledown} \ge x$,
(2') $x \le y \implies x^{\triangledown} \ge y^{\triangledown}$,
(3') $(x \vee y) \wedge (x \vee y^{\triangledown}) = x$.

Since we are interested in the equational theory (if there is one) of concept algebras, we define abstract structures using the equations in Theorem 3. Then we try to find whether these are enough to describe concept algebras; i.e., if each equation valid in all concept algebras is also valid in these abstract structures and vice versa.

Definition 3. *A* **weakly dicomplemented lattice** *is an algebra* $(L; \wedge, \vee, ^{\triangle}, ^{\triangledown}, 0, 1)$ *of type* $(2, 2, 1, 1, 0, 0)$ *for which* $(L; \wedge, \vee, 0, 1)$ *is a bounded lattice satisfying properties* $(1), (2), (3), (1'), (2')$ *and* $(3')$ *of Theorem 3.* $x^{\triangle}$ *is called a weak complement and* $x^{\triangledown}$ *a dual weak complement of* x*. The pair* $(^{\triangle}, ^{\triangledown})$ *is called a* **weak dicomplementation**.

All concept lattices are complete lattices. If we expect each weakly dicomplemented lattice to be (isomorphic to) a concept algebra, we should add the completeness property in Definition 3. However, being a complete lattice cannot be expressed by means of equations.

1.4 Canonical Context

A context can be assigned canonically to each weakly dicomplemented lattice as follows. Let L be a weakly dicomplemented lattice. A subset X of L is called primary filter (resp. primary ideal) if for all $x \in L$ we have $x \in X$ or $x^{\triangle} \in X$ (resp. $x \in X$ or $x^{\triangledown} \in X$). The prime ideal theorem holds for weakly dicomplemented lattices namely,

> **for any filter F and any ideal I such that $F \cap I = \emptyset$ there is a primary filter G, a primary ideal J with $F \subseteq G$, $I \subseteq J$ such that** $G \cap J = \emptyset$. [Kw04, Lemma 2.2.1]

An immediate consequence is the existence of primary filters and primary ideals. We denote by $\mathcal{F}_{pr}(L)$ the set of primary filters of L and by $\mathcal{I}_{pr}(L)$ the set of primary ideals of L. A relation $\Delta \subseteq \mathcal{F}_{pr}(L) \times \mathcal{I}_{pr}(L)$ is defined by $F \Delta I :\iff F \cap I \neq \emptyset$. We call $(\mathcal{F}_{pr}(L), \mathcal{I}_{pr}(L), \Delta)$ the canonical context of L and denote it by $\mathbb{K}_{\triangledown}^{\triangle}(L)$. For $x \in L$, we set

$$\mathcal{F}_x := \{F \in \mathcal{F}_{pr}(L) \mid x \in F\} \quad \text{and} \quad \mathcal{I}_x := \{I \in \mathcal{I}_{pr}(L) \mid x \in I\}.$$

The map $\varphi\colon x \mapsto (\mathcal{F}_x, \mathcal{I}_x)$ is a bounded lattice embedding of L into $\mathfrak{B}(\mathbb{K}_{\triangledown}^{\triangle}(L))$ such that

$$\varphi(x^{\triangledown}) \leq \varphi(x)^{\triangledown} \leq \varphi(x)^{\triangle} \leq \varphi(x^{\triangle}). \qquad \text{[Kw04, Thm 2.2.4]}$$

Is φ an embedding of L into $\mathcal{A}(\mathbb{K}_{\triangledown}^{\triangle}(L))$? This question is still open and is known as the concrete representation problem for weakly dicomplemented lattices.

2 MacNeille Completion of Weakly Dicomplemented Lattices

In this section we prove that in the MacNeille completion of a weakly dicomplemented lattice L, unary operations can be naturally defined to extend the ones existing on L. The first observation towards this definition is the more general form of the De Morgan laws. For a subset X of L we set

$$X^{\triangle} := \{x^{\triangle} \mid x \in X\} \qquad \text{and} \qquad X^{\triangledown} := \{x^{\triangledown} \mid x \in X\}.$$

Lemma 1. *Let L be a weakly dicomplemented lattice and X a subset of L. We have*

$$(i) \qquad (\bigwedge X)^{\triangle} = \bigvee X^{\triangle}$$

and

$$(ii) \qquad (\bigvee X)^{\triangledown} = \bigwedge X^{\triangledown}$$

Proof. (*i*) Note that

$$(\bigwedge \emptyset)^{\triangle} = 1^{\triangle} = 0 = \bigvee \emptyset = \bigvee \emptyset^{\triangle}.$$

The equality trivially holds for $X = \emptyset$. We consider X to be a nonempty subset and set $X := \{x_i \mid i \in I\}$. We have

$$(\bigwedge X)^{\triangle} \geq x_i^{\triangle} \quad \text{for all} \quad i \in I, \quad \text{and hence} \quad (\bigwedge X)^{\triangle} \geq \bigvee X^{\triangle}.$$

Now let $t \geq x_i^{\triangle}$ for all $i \in I$. Consequently $t^{\triangle} \leq x_i^{\triangle\triangle} \leq x_i$ for all $i \in I$ holds as well as $t^{\triangle} \leq \bigwedge X$. Therefore $t \geq t^{\triangle\triangle} \geq (\bigwedge X)^{\triangle}$ holds. Thus $(\bigwedge X)^{\triangle}$ is the lowest upper bound of $X^{\triangle}$ i.e., $(\bigwedge X)^{\triangle} = \bigvee X^{\triangle}$. The equality in (*ii*) is proved dually. □

We consider L, a weakly dicomplemented lattice and $\mathfrak{B}(L, L, \leq)$, its MacNeille completion. Since L is $\bigvee$-dense and $\bigwedge$-dense in $\mathfrak{B}(L, L, \leq)$, we take advantage of Lemma 1 (*i*) to define the weak complement and of Lemma 1 (*ii*) to define the dual weak complement of each $x \in \mathfrak{B}(L, L, \leq)$. Before that, we have to prove that the definition does not depend on the representation. That is what Lemma 2 says.

Lemma 2. *Let $(L, \wedge, \vee, ^{\triangle}, ^{\triangledown}, 0, 1)$ be a weakly dicomplemented lattice. For any subset X of L we denote by $ub(X)$ (resp. $lb(X)$) the set of upper bounds (resp. the set of lower bounds) of X. The following statements hold:*

(*i*) *If $lb(X_1) = lb(X_2)$ then $ub(X_1^{\triangle}) = ub(X_2^{\triangle})$.*
(*ii*) *If $ub(X_1) = ub(X_2)$ then $lb(X_1^{\triangledown}) = lb(X_2^{\triangledown})$.*

Proof. To prove (*i*) we assume that the sets of lower bounds of X_1 and that of X_2 are equal. Let $s \in ub(X_1^{\triangle})$. We have $s \geq t^{\triangle}$ for all $t \in X_1$. Therefore, by (2) in Theorem 3, $s^{\triangle} \leq t^{\triangle\triangle} \leq t$ for all $t \in X_1$. Thus $s^{\triangle}$ belongs to $lb(X_1)$, which is by assumption the same as $lb(X_2)$. It follows that $s^{\triangle} \leq t$ for all $t \in X_2$, and hence $s^{\triangle\triangle} \geq t^{\triangle}$ for all $t \in X_2$. Therefore $t^{\triangle} \leq s^{\triangle\triangle} \leq s$ for all $t \in X_2$. This means that s belongs to $ub(X_2^{\triangle})$. Analogously we can prove that $ub(X_2)$ is contained in $ub(X_1)$ and get the equality.

(*ii*) can be established dually. □

Lemma 2 says on one side that if X_1 and X_2 are subsets of L such that the equality $\bigwedge X_1 = \bigwedge X_2$ holds in a completion, then the equality $\bigvee X_1^{\triangle} = \bigvee X_2^{\triangle}$ also holds, and on the other that if $\bigvee X_1 = \bigvee X_2$ then $\bigwedge X_1^{\triangledown} = \bigwedge X_2^{\triangledown}$.

In the sequel, let $(L, \wedge, \vee, {}^{\triangle}, {}^{\triangledown}, 0, 1)$ be a weakly dicomplemented lattice. We denote by $\tilde{L}$ the MacNeille completion of L. Since the binary operations $\wedge, \vee$ on L are restrictions of the binary operations on the complete lattice $\tilde{L}$, we denote the binary operations on $\tilde{L}$ by the same symbols. We consider the lattice L as a subset of $\tilde{L}$. Therefore, the top and the bottom elements of both lattices (which are the same) are denoted by the same symbols, 0 and 1. By a well known property, for each $x \in \tilde{L}$ there are subsets X and Y of L such that $x = \bigwedge X$ and $x = \bigvee Y$. Using this representation, we define unary operations on $\tilde{L}$:

$$x \mapsto x^{\blacktriangle} := \bigvee X^{\triangle} \quad \text{and} \quad x \mapsto x^{\blacktriangledown} := \bigwedge Y^{\triangledown}.$$

Lemma 3. *Let $(L, \wedge, \vee, {}^{\triangle}, {}^{\triangledown}, 0, 1)$ be a weakly dicomplemented lattice and $\tilde{L}$ its MacNeille completion. The unary operations ${}^{\blacktriangle}$ and ${}^{\blacktriangledown}$ are well defined, and the equations*

(1) $x^{\blacktriangle\blacktriangle} \leq x$, *(1')* $x^{\blacktriangledown\blacktriangledown} \geq x$,

(2) $x \leq y \implies x^{\blacktriangle} \geq y^{\blacktriangle}$, *(2')* $x \leq y \implies x^{\blacktriangledown} \geq y^{\blacktriangledown}$,

are satisfied.

Proof. The operations $x \mapsto x^{\blacktriangle}$ and $x \mapsto x^{\blacktriangledown}$ are well defined by Lemma 2. Therefore, the subsets X and Y can be chosen systematically. Note that $\tilde{L}$ can be identified with the concept lattice $\mathfrak{B}(L, L, \leq)$, X with the order filter of L generated by x and Y with the order ideal of L generated by x. We use this identification in the rest of the proof. We proceed as follows: we are going to prove the equations (1) and (2). The others are obtained dually. Let us start with (2). Let x_1 and x_2 be elements in L with $x_1 \leq x_2$. Then the order filter generated by x_1, say X_1 contains the order filter generated by x_2, say X_2. Therefore

$$x_1^{\blacktriangle} = \bigvee X_1^{\triangle} \geq \bigvee X_2^{\triangle} = x_2^{\blacktriangle},$$

and (2) is proved. Now let us prove (1). Using (2) we have

$$x^{\blacktriangle\blacktriangle} = (\bigvee X^{\triangle})^{\blacktriangle} \leq (t^{\triangle})^{\blacktriangle} \quad \text{for all } t \in X,$$

where X is the order filter generated by x. But

$$(t^{\triangle})^{\blacktriangle} = \bigvee (\uparrow t^{\triangle} \cap L)^{\triangle} = \bigvee \{a^{\triangle} \mid a \geq t^{\triangle},\ a \in L\}.$$

Therefore

$$(t^{\triangle})^{\blacktriangle} = \bigvee \{a^{\triangle} \mid a^{\triangle} \leq t^{\triangle\triangle},\ a \in L\} \leq t.$$

Thus $x^{\blacktriangle\blacktriangle} \leq t$ for all $t \in X$, and finally $x^{\blacktriangle\blacktriangle} \leq \bigwedge X = x$, achieving the proof of (1). □

From now on, we consider algebra $(\tilde{L}; \wedge, \vee, ^{\blacktriangle}, ^{\blacktriangledown}, 0, 1)$ of type $(2,2,1,1,0,0)$. The 3rd equation in the definition of the weakly dicomplemented lattice implies that $^{\triangle}$ is a dual semicomplementation (i.e., $x \vee x^{\triangle} = 1$ for all $x \in L$). Before we continue, we verify that it also holds in $\tilde{L}$. It will be useful for later need.

Lemma 4. *The equation $x \vee x^{\blacktriangle} = 1$ holds for all $x \in \tilde{L}$.*

Proof. In the isomorphism $\tilde{L} \cong \mathfrak{B}(L, L, \leq)$, the top element is represented by the concept $(L, \{1\})$. Now let $x := (Y, X) \in \tilde{L}$. According to the definition, $x^{\blacktriangle} = \bigvee X^{\triangle}$. The element $x^{\blacktriangle}$ can also be represented by $(\downarrow x^{\blacktriangle} \cap L, \uparrow x^{\blacktriangle} \cap L)$. To prove that $x \vee x^{\blacktriangle} = 1$, it is enough to prove

$$\uparrow x^{\blacktriangle} \cap \uparrow x \cap L = \{1\}.$$

Now let $b \in \uparrow x^{\blacktriangle} \cap \uparrow x \cap L$. We have $b \geq x^{\blacktriangle} = \bigvee X^{\triangle}$, and consequently

$$b \geq t^{\triangle}, \text{ for all } t \in X = \uparrow x \cap L.$$

Moreover, $b \geq x$ implies $b \in X$. Thus $b \geq b^{\triangle}$ and $1 = b \vee b^{\triangle} = b$. □

The intended result is immediate for MacNeille completions that are distributive (see [Er82]). In fact, in a distributive lattice, the equation $(x \wedge y) \vee (x \wedge y^{\triangle}) = x$ is equivalent to $y \vee y^{\triangle} = 1$. Therefore if $\tilde{L}$ is distributive, we can conclude from Lemma 3 and Lemma 4 that the operation $x \mapsto (x^{\blacktriangle}, x^{\blacktriangledown})$ defines a weak dicomplementation on $\tilde{L}$.

Corollary 1. *If the MacNeille completion $\tilde{L}$ of a weakly dicomplemented lattice L is distributive then $(\tilde{L}, \wedge, \vee, ^{\blacktriangle}, ^{\blacktriangledown}, 0, 1)$ is a weakly dicomplemented lattice.*

As the following theorem proves, the above result holds also in lattices with enough 1-prime and 0-prime elements. We call an element a 1-prime (resp. 0-prime) if $a < 1 = c \vee d$ implies $a \leq c$ or $a \leq d$ (resp. $a > 0 = c \wedge d$ implies $a \geq c$ or $a \geq d$) for all $c, d \in L$.

Theorem 4. *Let L be a weakly dicomplemented lattice such that the set $J_1(L)$ (resp. $M_0(L)$) of 1-prime (resp. 0-prime) elements is supremum (resp. infimum) dense in L. Then the MacNeille completion of L is a weakly dicomplemented lattice.*

Proof. To prove the theorem, we only need to prove that $(x \wedge y) \vee (x \wedge y^{\blacktriangle}) = x$ and its dual hold. Here the calculation is more tedious. We take advantage of the context representation. We are going to prove that in $\mathfrak{B}(J_1(L), M_0(L), \leq)$ the equation

$$(x \wedge y) \vee (x \wedge y^{\blacktriangle}) = x$$

holds. It is enough to prove that $(x \wedge y) \vee (x \wedge y^{\blacktriangle}) \geq x$. We set $x := (Y_1, X_1)$, $y := (Y_2, X_2)$ and $y^{\blacktriangle} := (Y_3, X_3)$. Note that there is no evidence to write $y^{\blacktriangle} = ((J_1(L) \setminus Y_2)'', (J_1(L) \setminus Y_2)')$. However $y^{\blacktriangle} = \bigvee X_2^{\triangle}$ and $(Y_2 \cup Y_3)'' = J_1(L)$. By Theorem 1 we have

$$x \wedge y = (Y_1 \cap Y_2, (X_1 \cup X_2)'') \quad \text{and } x \wedge y^{\blacktriangle} = (Y_1 \cap Y_3, (X_1 \cup X_3)'').$$

Hence,

$$(x \wedge y) \vee (x \wedge y^{\blacktriangle}) = \Big(((X_1 \cup X_2)'' \cap (X_1 \cup X_3)'')', (X_1 \cup X_2)'' \cap (X_1 \cup X_3)''\Big).$$

But

$$\begin{aligned}(X_1 \cup X_2)'' \cap (X_1 \cup X_3)'' &= ((X_1' \cap X_2') \cup (X_1' \cap X_3'))' \\ &= (X_1' \cap (X_2' \cup X_3'))' \\ &= (Y_1 \cap (Y_2 \cup Y_3))'\end{aligned}$$

and

$$(Y_1 \cap (Y_2 \cup Y_3))'' \subseteq Y_1'' \cap (Y_2 \cup Y_3)'' = Y_1 \quad \text{(see Lemma 3)}.$$

Now we want to prove that $(Y_1 \cap (Y_2 \cup Y_3))' \subseteq Y_1'$. Let $m \in M(L)$ such that $m \in (Y_1 \cap (Y_2 \cup Y_3))'$

$$\forall g \in J(L), \quad g \in Y_1 \cap (Y_2 \cup Y_3) \implies g \leq m. \qquad (*)$$

We should prove that $m \geq h$ for all $h \in Y_1$. Let $h \in Y_1$. We have $h \leq 1 = y \vee y^{\blacktriangle}$ (see Lemma 4). It follows that $h \leq y$ or $h \leq y^{\blacktriangle}$. Thus $h \in Y_2$ or $h \in Y_3$. Therefore $h \in Y_1 \cap (Y_2 \cup Y_3)$, and thus implies $h \leq m$. □

To capture a larger class of weakly dicomplemented lattices, 1-prime and 0-prime elements can be replaced by primary elements. An element $a \in L$ is called $\vee$-primary in L (resp. $\wedge$-primary) if for all $x \in L$ we have

$$a \not\leq x \implies a \leq x^{\triangle} \qquad (\text{ resp. } a \not\geq x \implies a \geq x^{\triangledown}).$$

In particular join-irreducible elements are $\vee$-primary, and meet-irreducible elements $\wedge$-primary.

Lemma 5. *Primary elements in L are also primary in $\tilde{L}$.*

Proof. Let $y := (Y_2, X_2) \in \tilde{L}$ and a a $\vee$-primary element in L such that $a \not\leq y$. We should prove that $a \leq y^{\blacktriangle}$. Note that in $\tilde{L}$ we have $\bigvee Y_2 = y = \bigwedge X_2$. Since $g \not\leq \bigwedge X_2$, there is some $m \in X_2 \subseteq L$ such that $g \not\leq m$. As a is $\vee$-primary it follows that $g \leq m^{\triangle}$. Therefore

$$a \leq \bigvee X_2^{\triangle} = y^{\blacktriangle}.$$ □

There are weakly dicomplemented lattices without irreducible elements that still have enough primary elements. For example the grid $L := \{0\} \oplus \mathbb{Z} \times \mathbb{Z} \oplus \{1\}$ has no join-irreducible and no meet irreducible elements (apart from 0 and 1). But the unique weakly dicomplemented lattice structure on L is trivial. Here all the elements are $\vee$-primary and $\wedge$-primary. Since the crucial step in the proof of the theorem is the choice of h, with $h \leq y$ or $h \leq y^{\blacktriangle}$ for all $y \in \tilde{L}$, we have

Corollary 2. *The MacNeille completion of a weakly dicomplemented lattice with enough primary elements is a weakly dicomplemented lattice.*

For the general case we might expect each weakly dicomplemented lattice to have enough $\vee$-primary and $\wedge$-primary elements. Unfortunately, this is not the case. A typical example is an atomless Boolean algebra. Note that an element a of a Boolean algebra is $\vee$-primary if and only if it is an atom. Otherwise there would be an element a_1 such that $a_1 < a$. But $a = (a \wedge a_1) \vee (a \wedge a_1')$ with $a \not\leq a \wedge a_1$ and $a \not\leq a \wedge a_1'$. It would follow that $a \leq (a \wedge a_1)' \wedge (a \wedge a_1')' = a'$, forcing a to be 0, a contradiction. However its MacNeille completion is a Boolean algebra, thus a weakly dicomplemented lattice.

To get the result in the general case, we need one more lemma:

Lemma 6. *Let $h \in L$ and $y \in \tilde{L}$ such that $h \not\leq y$ and $h \not\leq y^{\blacktriangle}$. Let G and M be subsets of L such that $\mathfrak{B}(G, M, \leq)$ is isomorphic to $\tilde{L}$. Then $\mathfrak{B}(G \setminus \{h\}, M, \leq)$ is isomorphic to $\tilde{L}$.*

Proof. Let (Y, X) be the concept corresponding to y in $(G, M, \leq)$. We know that $\bigvee Y = y = \bigwedge X$ and $y^{\blacktriangle} = \bigvee X^{\triangle}$. From $h \not\leq y^{\blacktriangle}$ we obtain $h \not\leq x^{\triangle}$ for all $x \in X$. In addition $h \not\leq y = \bigwedge X$ implies the existence of an element x_0 of X such that $h \not\leq x_0$. Therefore, $h \not\leq x_0$ and $h \not\leq x_0^{\triangle}$. But the 3rd axiom of weakly dicomplemented lattice gives $h = (h \wedge x_0) \vee (h \wedge x_0^{\triangle})$ with $h > h \wedge x_0$ and $h > h \wedge x_0^{\triangle}$. Therefore

$$h'' = \{g \in G \mid g \leq h \wedge x_0 \text{ or } g \leq h \wedge x_0^{\triangle}\}''$$

and

$$h \notin \{g \in G \mid g \leq h \wedge x_0 \text{ or } g \leq h \wedge x_0^{\triangle}\}.$$

Thus h can be removed from G and the concept lattice of the remaining context is still isomorphic to that of the initial one. □

Recall that a lattice satisfies the (descending and ascending) chain conditions if it has no infinite chains. We can now state the result in the general case.

Theorem 5. *The MacNeille completion of a weakly dicomplemented lattice satisfying the chain conditions is a weakly dicomplemented lattice.*

Proof. The proof is similar to that of Theorem 4. The change is to prove $(\star)$ replacing $J_1(L)$ with a suitable subset of L. By Lemma 6 such a subset always exists. In fact Lemma 6 states that Y_1 can be replaced by $\tilde{Y}_1$ with $\bigvee Y_1 = \bigvee \tilde{Y}_1$ such that $\tilde{Y}_1 \subseteq Y_2 \cup Y_3$. This gives the desired equality and concludes the proof. □

Unfortunately we cannot yet remove the chain conditions. The conjecture is that this result holds in general. If this is the case then the MacNeille completion of Boolean algebras can be deduced as a special case. In fact Boolean algebras are determined in the class of weakly dicomplemented lattices by the equation $x^{\triangle} = x^{\triangledown}$ [Kw04, Thm. 3.3.4]. Therefore it is enough to check this equality for $^{\blacktriangle}$ and $^{\blacktriangledown}$. The 3rd axiom of weakly dicomplemented lattices gives $x^{\blacktriangledown} \leq x^{\blacktriangle}$. Now let $y \in \tilde{L}$ represented by (Y, X). Then $\bigvee Y = y = \bigwedge X$. For all $g \in Y$ and $m \in X$, $g \leq m$ and by then $g^{\triangle} \geq m^{\triangle}$. Thus

$$y^{\blacktriangle} = \bigvee\{m^{\triangle} \mid m \in X\} \leq \bigwedge\{g^{\triangle} \mid g \in Y\} = \bigwedge\{g^{\triangledown} \mid g \in Y\} = y^{\blacktriangledown}.$$

This would prove that the completion of a Boolean algebra is again a Boolean algebra.

3 Conclusion

What is the consequence of Theorem 5 on the concrete representation problem of weakly dicomplemented lattices? First observe that for each $x \in L$, x "is in" $\tilde{L}$ and $x^{\blacktriangle} = x^{\triangle}$ as well as $x^{\blacktriangledown} = x^{\triangledown}$. Thus the $\varphi : x \mapsto (\downarrow x, \uparrow x)$ is a weakly dicomplemented lattice embedding of L into $\tilde{L}$. The conjecture is that "**L embeds into $\mathcal{A}(\mathbb{K}_{\triangledown}^{\triangle}(L))$ iff $\tilde{L}$ embeds into $\mathcal{A}(\mathbb{K}_{\triangledown}^{\triangle}(L))$**".

References

[Er82] M. Erne. *Distributivgesetze und Dedekind'sche Schnitte.* Abh. Braunschweig Wiss. Ges. 33, (1982) 117-145.

[GW99] B. Ganter & R. Wille. *Formal Concept Analysis. Mathematical Foundations* Springer Verlag 1999

[Kw04] L. Kwuida. *Dicomplemented Lattices. A Contextual Generalization of Boolean algebras* Shaker Verlag. 2004

[Fu44] N. Funayama. *On the completion by cuts of distributive lattices.* Proc. Imp. Acad. Tokyo **20** (1944), 1-2.

[Ha93] J. Harding. *Any lattice can be regularly embedded into the MacNeille completion of a distributive lattice.* Houston J. Math. **19**, (1993), No. **1**, 39-44.

[Wi00] R. Wille. *Boolean Concept Logic* in B. Ganter & G. W. Mineau (Eds.) *Conceptual Structures: Logical, Linguistic, and Computational Issues* LNCS **1867** Springer (2000), 317-331.

Polynomial Embeddings and Representations

Tim Becker

Institute for Medical Biometry, Informatics, and Epideomology, Germany

1 Introduction

The following paper activates polynomial methods for data analysis. We want to embed a given formal context into a polynomial context $(K^n, K[x_1, \ldots, x_n], \perp)$ in such a way that implications can be computed in the polynomial context, using the algebraic structure. The basic ideas of formal concept analysis that are needed here are sketched in [TB1].

The idea to describe given data algebraically has been investigated, for instance, in [V] where **linear embeddings** into a context of the form $(V, V^*, \perp)$ are treated. Here V is a finite-dimensional vector-space, V^* is its dual space and $a \perp \varphi$ holds if and only if $\varphi(a) = 0$.

One finds that polynomial solutions always exist, whereas empirical data in general does not allow for linear solutions. But nevertheless, polynomial solutions are not always satifactory because it is difficult to limit the (total) degree of the polynomials involved, and polynomial descriptions of a high (total) degree are not very useful for applications.

Unfortunately, if we work over an algebraically closed field, there is a large class of formal contexts which can only be embedded into a polynomial context in one variable. A problem of future research therefore could be to find suitable underlying fields or rings or even universal algebras to work with.

At the end of this paper we briefly touch polynomial representations, a related topic. As before, polynomial solutions always exist, but it is difficult to limit their total degree.

2 Polynomial Embeddings

We want to embed a given formal context $\mathbb{K}_1 := (G, M, I)$ into a context $\mathbb{K}_2$ in such a way that the closure B^{II} of a subset $B \subseteq M$ can be computed by intersecting the closure of the image of B with the image of M in $\mathbb{K}_2$. An embedding of this kind will be called an "intent-preserving embedding".

The idea is to gain a new description of the formal concepts of $\mathbb{K}_1$ and a new way to compute them, for instance, when $\mathbb{K}_2$ carries an algebraic structure.

Vogt, [V], has investigated "linear embeddings" into the context $(V, V^*, \perp)$, where V is a vector space, V^* is its dual space, and where $a \perp \varphi$ holds if and only if $\varphi(a) = 0$. He has shown that not every formal context is linearly embeddable.

So we will try to find embeddings into a polynomial context

$$(K^n, K[x_1, \ldots, x_n], \perp),$$

S.O. Kuznetsov and S. Schmidt (Eds.): ICFCA 2007, LNAI 4390, pp. 281–302, 2007.

where K is a field, $K[x_1, \ldots, x_n]$ is the polynomial ring in n variables over K and where $a \perp f$ holds if and only if $f(a) = 0$.

Definition 1. *A* **quasi-embedding** *of a formal context* $\mathbb{K}_1 := (H, N, J)$ *into a formal context* $\mathbb{K}_2 := (G, M, I)$ *is a pair* (α, β), *where* α *maps from* H *to* G, β *maps from* N *to* M *and where* hIn *holds if and only if* $\alpha(h)J\beta(n)$ *holds for all* $h \in H$ *and all* $n \in N$.

An **intent-preserving quasi-embedding** *of* $\mathbb{K}_1$ *into* $\mathbb{K}_2$ *is a quasi-embedding which satisfies for all* $D \subseteq N$ *the extra condition*

$$(1) \qquad \beta(D^{JJ}) = (\beta(D))^{II} \cap \beta(N).$$

A **linear embedding** *is an intent-preserving quasi-embedding into a linear context* $(V, V^*, \perp)$.

A **polynomial embedding** *is an intent-preserving quasi-embedding into a polynomial context*

$$(K^n, K[x_1, \ldots, x_n], \perp)$$

over a given field K.

Note that the mappings which constitute the quasi-embedding do not have to be injective by definition. However, since they are compatible with the relations, they will be injective on purified contexts.

The following lemma is taken from [V] and shows that one inclusion of equation (1) holds for arbitrary embeddings.

Lemma 1. *A quasi-embedding* (α, β) *from* (H, N, J) *into* (G, M, I) *is an intent-preserving quasi-embedding if and only if* $\beta(D^{JJ}) \subseteq (\beta(D))^{II} \cap \beta(N)$ *holds for all* $D \subseteq N$.

Proof. We show that $\beta(D^{JJ}) \supseteq (\beta(D))^{II} \cap \beta(N)$ holds for all $D \subseteq N$ and all quasi-embeddings (α, β) from (H, N, J) into (G, M, I).

For all $D \subseteq N$ the equations $\alpha(D^J) = (\beta(D))^I \cap \alpha(H)$ and $\beta(D^{JJ}) = ((\beta(D))^I \cap \alpha(H))^I \cap \beta(N)$ hold. Since $(\beta(N))^I \cap \alpha(H) \subseteq (\beta(N))^I$, we have $((\beta(D))^I \cap \alpha(H))^I \supseteq (\beta(D))^{II}$ and therefore the desired inclusion.

Lemma 2. *Let* (α, β) *be an intent-preserving quasi-embedding from* (H, N, J) *into* (G, M, I). *Then the map* ϕ_β *defined by*

$$\phi_\beta(C, D) := ((\beta(D))^I, (\beta(D))^{II})$$

is a $\bigwedge$-*preserving order embedding of* $\underline{\mathfrak{B}}(H, N, J)$ *into* $\underline{\mathfrak{B}}(G, M, I)$.

Proof. [V], p.78.

The lemma tells us that the concept lattice of the context to be embedded can be visualized within the concept lattice of the context which contains the embedding. In particular, infima of concepts of (G, M, I) can be read of from $\mathfrak{B}(H, N, J)$.

We want to find intent-preserving embeddings. To this purpose we will use the logic which is inherit in the set of attributes of a formal context. This logic can be desribed by **implications** between attributes. Intuitively speaking, a subset $B \subseteq M$ implies an attribute $m \in M$ if all the objects that are common to B are in relation with m as well. We write $B \longrightarrow m$. Thus $B \longrightarrow m$ holds if and only if $B^I \subseteq m^I$ and the latter condition is equivalent to $m \in B^{II}$.

An implication of the form $B \longrightarrow D$ with $D \subseteq M$ holds if $D \longrightarrow m$ holds for all $m \in D$. If (α, β) is an intent-preserving quasi-embedding from (G, M, I) to (H, N, J), we see that $B \longrightarrow m$ holds in (G, M, I) if and only if $\beta(B) \longrightarrow \beta(m)$ holds in (H, N, J): $\beta(B) \longrightarrow \beta(m) \iff \beta(m) \in (\beta(B))^{JJ} \iff \beta(m) \in (\beta(B))^{JJ} \cap \beta(M) \iff \beta(m) \in \beta(B^{II}) \iff m \in B^{II} \iff B \longrightarrow m$.

Let us have a look at the identical embedding below:

I	a	b	c	d
1	×	×		
2		×	×	
3			×	×

↓

I	a	b	c	d	e
1	×	×			×
2		×	×		
3			×	×	×

It is easily computed that this embedding is intent-preserving. Thus, we have a $\bigwedge$-preserving order embedding given by $\psi(C, D) = (\beta(D)^J, \beta(D)^{JJ})$, and implications between attributes can also be read of in the concept lattice of the bigger context.

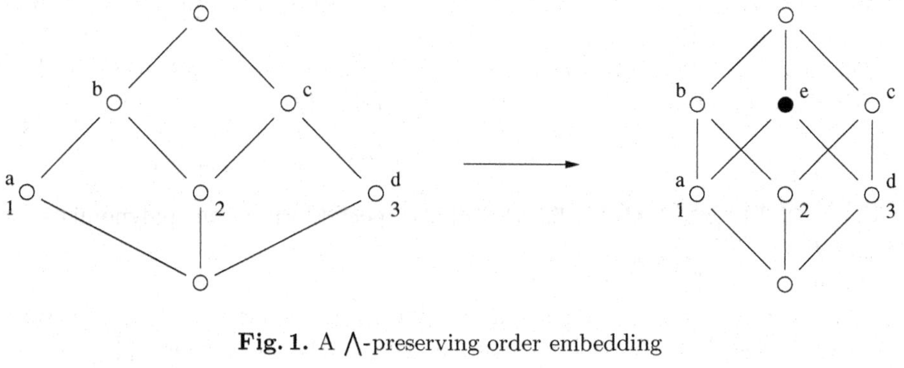

Fig. 1. A $\bigwedge$-preserving order embedding

Note that the smaller context is an intent-compatible[1] subcontext of the bigger context. Indeed, it has been shown in [V, p.81] that the identical embedding of

[1] A subcontext (H, N, J) of (G, M, I) is called **intent-compatible** if for all concepts (C, D) of (G, M, I) the pair $((D \cap N)^J, D \cap N)$ is a concept of (H, N, J).

a subcontext into a supercontext is an implication embedding if and only if it is intent-compatible.

We want to regard polynomial embeddings. To this purpose we need some results from algebraic geometry.

Theorem 1. *Let $\mathbb{K} := (K^n, K[x_1, \ldots, x_n], \perp)$ be a polynomial context over an algebraically closed field K.*
a) (Hilbert's Nullstellensatz) For any subset $A \subseteq K[x_1, \ldots, x_n]$, we have $A^{\perp\perp} = \sqrt{I}$ where I is the ideal generated by A and the **radical** *of I is defined as follows: $\sqrt{I} := \{f \in K[x_1, \ldots, x_n] \mid f^m \in I \, for \, some \, m \in \mathbb{N}\}$.*
b) Extents of $\mathbb{K}$ are called **algebraic varieties**. *Every finite set $V \subseteq K^n$ is an algebraic variety and finite unions of algebraic varieties are again algebraic varieties.*

Proof. For a **proof** see [K].

Example 1. Consider the polynomial context over $\mathbb{C}$ in two variables

$$(\mathbb{C}, \mathbb{C}[x, y], \perp).$$

For $I =< xy^2, yz^3 >$ we have

$$I^{\perp} = \{(a, 0, c) \mid a, c \in \mathbb{C}\} \cup \{(0, b, 0) \mid b \in \mathbb{C}\},$$

which means that $I^{\perp}$ is the union of the x, z-plane and the y-plane. $I^{\perp\perp}$ is equal to $< xy, yz >$.

Theorem 2. *Every finite context $\mathbb{K}$ allows for an intent-preserving quasi-embedding into $(\mathbb{C}, \mathbb{C}[x], \perp)$.*

Proof. We may assume that $\mathbb{K}$ is purified. Let $G = \{g_0, \ldots, g_n\}$ be the set of objects. Define $\alpha : G \longrightarrow \mathbb{C}$ by $\alpha(g_i) = i, i \in \mathbb{N}_0$ and $\beta(m) := \prod_{mIg_k}(x - k)$ for all $m \in M$. (Here the product is taken over all $k \in \{0, \ldots n\}$).

Obviously, (α, β) is an embedding. Let $D \subseteq M$. We only have to show that $\beta(D^{II})$ is a subset of $(\beta(D))^{\perp\perp} \cap \beta(M)$. If $\beta(D) = \{f_1, \ldots, f_t\}$ we have

$$(\beta(D))^{\perp\perp} = \sqrt{< (\beta(D)) >} = \sqrt{< gcd(\beta(D)) >} =< gcd(\beta(D)) >,$$

because $\mathbb{C}[x]$ is principal and because of the special form of our polynomials.

Let $D^I = \{g_{i_1}, \ldots, g_{i_p}\}$.

Then we have $\{i_1, \ldots, i_p\} = (\beta(D))^{\perp}$.

Each polynomial $\tilde{f} \in (\beta(D))^{\perp\perp}$ has the form $\tilde{f} = (x - i_1) \ldots (x - i_p)g$ for some polynomial $g \in \mathbb{C}[x]$. Since all polynomials from $\beta(D)$ are of the form $\prod_{s \in S}(x - s)$ where $S \subseteq \{0, 1, \ldots, n\}$ we know that $gcd(\beta(D)) = (x - i_1) \ldots (x - i_p)$, because (α, β) is an embedding. We conclude $\tilde{f} = (x - i_1) \ldots (x - i_p)g \in< gcd(\beta(D)) >= (\beta(D))^{\perp\perp}$.

Example 2. It has been shown in [V] that the interordinal scale $\mathbb{I}_3$ has no linear embedding into $(V, V^*, \perp)$ for any finite dimensional vector space V.

	≤ 1	≤ 2	≥ 2	≥ 3
1	×	×		
2		×	×	
3			×	×

According to the theorem we get a polynomial embedding (α, β) if we choose $\alpha = id_G$, $\beta(\leq 1) = x - 1$, $\beta(\leq 2) = (x-1)(x-2)$, $\beta(\geq 2) = (x-2)(x-3)$ and $\beta(\geq 3) = x - 3$.
Thus the image of $\mathbb{I}_3$ in $(\mathbb{R}, \mathbb{R}[x], \perp)$ has the following form:

	$x-1$	$(x-1)(x-2)$	$(x-2)(x-3)$	$x-3$
1	×	×		
2		×	×	
3			×	×

Lemmma 2 tells us that we can visualize the concept lattice of $\mathbb{I}_3$ within $\mathfrak{B}(\mathbb{K})$ via a $\bigwedge$-preserving order embedding ϕ_β defined by $\phi_\beta(C, D) = ((\beta(D))^\perp, (\beta(D))^{\perp\perp})$ for all concepts (C, D) of $\mathbb{I}_3$. We depict only the image of $\mathfrak{B}(\mathbb{I}_3)$ because $\mathfrak{B}(\mathbb{K})$ is infinite.

It should be mentioned, that there is a weakness in this method. In order to find out if an element m belongs to the closure B^{II} for a given $B \subseteq M$, one has to check wether $\beta(m)$ is a multiple of $gcd\,\{\beta(d) \mid d \in B\}$ or not.

This procedure can become rather complicated when the context to be embedded is big, because we can get polynomials of very high degree. So it is desirable to find better embeddings into polynomial contexts over several variables, limitting the (total) degree of the polynomials used. We will return to this problem in the course of this section.

Another problem is, that our embedding depends very much on the fact that the intents of $(\mathbb{C}, \mathbb{C}[x], \perp)$ are ideals and not just subrings. If we had an embedding into a context (T, R, J) where R is a ring and whose intents are exactly the

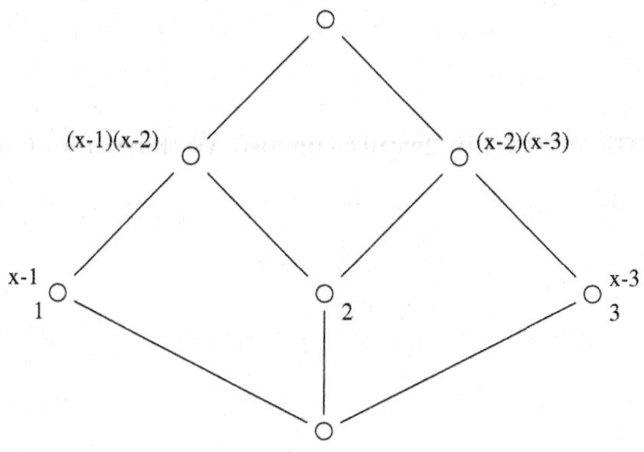

Fig. 2. $\phi_\beta(\mathfrak{B}(\mathbb{I}_3))$

subrings[2] of R, then for an element $m \in B^{II}$ we would derive $\beta(m)$ from the elements $\beta(b)$, $b \in B$, by repeated addition and multiplication. Hence we had an equation of the form $\beta(m) = \sum \pm\beta(b_i)\beta(b_j) \pm \beta(b_k)$, where $b_i, b_j, b_k \in B$ and no multiplication with arbitrary elements as in the case of polynomial embeddings would be involved.

Also note that for any other principal ring, we can get intent-preserving embeddings using the same method as in 2.

Consider for instance $(\mathbb{Z}, \mathbb{Z}, |)$, where $a|b$ means that a devides b. The formal concepts of this context are exactly the pairs of the form $(\{d \mid d\, devides\, a\}, I_a)$, $a \in \mathbb{Z}$. Here I_a is the ideal generated by a. We get an intent-preserving embedding as follows: Let p_i be the ith prime number and let $G = \{g_1, \dots, g_n\}$. Define (α, β) by $\alpha(g_i) = p_i$ and $\beta(m_j) = \prod_{g_k I m_j} p_k$. Then we can use the same proof as in 2, replacing $\perp$ by $|$, to show that we have an implication embedding.

As an example, we consider the following intent-preserving embedding into $(\mathbb{Z}, \mathbb{Z}, |)$:

I	m_1	m_2	m_3	m_4
g_1	×			
g_2	×	×	×	×
g_3		×		
g_4			×	

↓

|	6	15	21	3
2	×			
3	×	×	×	×
5		×		
7			×	

We can visualize $\psi(\mathfrak{B}(\mathbb{K}))$ as a $\bigwedge$-sublattice of the infinite lattice $\mathfrak{B}(\mathbb{Z}, \mathbb{Z}, |)$.

We can read of the implications between attributes. We have, for instance, $\beta(m_1), \beta(m_2) \longrightarrow \beta(m_3)$, since $\beta(m_3) = 21$ is a multiple of $gcd(\beta(m_1), \beta(m_2)) = 3$, and so $m_1, m_2 \longrightarrow m_3$.

As mentioned before, it is desirable to find a better embedding for a given formal context than the method in 2would suggest. We have a look at the following context $\mathbb{K}$:

I	a	b	c	d	e
1	×			×	×
2		×		×	
3			×		×
4	×	×	×	×	×

[2] If we consider R as an universal algebra, and if F is the free algebra over $\{x\}$ in the variety generated by R, then the context $(F \times F, R, \perp)$, with $(f, g) \perp r :\Longleftrightarrow f(r) = g(r)$, has this property. [TB2].

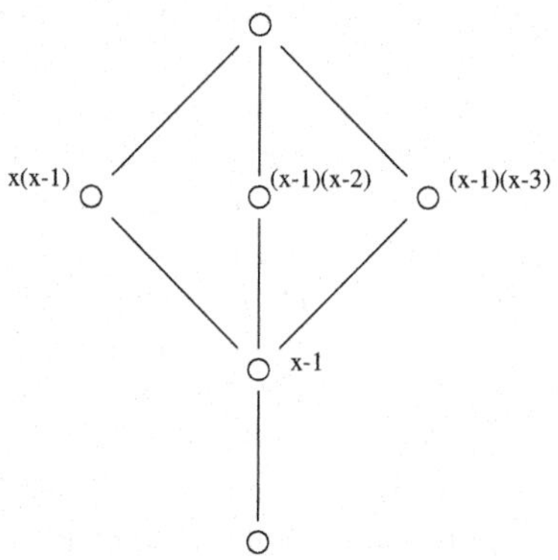

Fig. 3. $\psi(\mathfrak{B}(\mathbb{K}))$

The method from 2 yields the following embedding:

I	x^2-5x+4	x^2-6x+8	$x^2-7x+12$	$x^3-7x^2+14x-8$	$x^3-8x^2+19x-12$
1	×			×	×
2		×		×	
3			×		×
4	×	×	×	×	×

We see that the polynomials involved are already rather complicated, particularly those representing d and e. But there is a simpler embedding using two variables which lowers the degrees of the ọccuring polynomials:

I	x	y	$x+y$	xy	$(x+y)x$
$(0,1)$	×			×	×
$(1,0)$		×		×	
$(-1,1)$			×		×
$(0,0)$	×	×	×	×	×

One computes that this embedding is an intent-preserving embedding into

$$(\mathbb{C}^2, \mathbb{C}[x,y], \perp).$$

The interesting question is how we did find this embedding.

The attributes a, b are both not implied by any other single attribute. So it is natural to choose rather simple polynomials to represent them. Since they are different and since we want to involve at least two variables, we choose x and y respectively. We observe that the attributes c, d and e are implied by $\{a, b\}$. Consequently, we must choose elements from $\sqrt{<x,y>} = <x,y>$ to represent $\beta(c), \beta(d)$ and $\beta(e)$. The attribute c is not implied by a or b. So we must have $\beta(c) \in <x,y> \setminus [<x> \cup <y>]$. The simplest possibility fulfilling this requirement is $\beta(c) = x+y$. We observe that $\{x, x+y\}$ implies y and that $\{y, x+y\}$ implies x in $(\mathbb{C}^2, \mathbb{C}[x,y], \perp)$. Therefore we have to check if the implications $\{a,c\} \longrightarrow b$ and $\{b,c\} \longrightarrow a$ hold. Since this is true, we may choose $\beta(c) = x + y$. Since d is already implied both by a and b, we also have $d \in < x >$ and $d \in < y >$. Therefore, we check if we can choose the simplest possibilty fulfilling these requirements, $\beta(d) = xy$. Since d^I is the union of a^I and b^I and since we have $(fg)^\perp = f^\perp \cup g^\perp$ in $(\mathbb{C}^2, \mathbb{C}[x,y], \perp)$, it is actually possible to choose $\beta(d) = \beta(a)\beta(b) = xy$. As already mentioned, we must have $\beta(e) \in < x, y >$. Since $e^\perp$ is the union of $a^\perp$ and $c^\perp$, we choose $\beta(e) = (x+y)x$. We observe that all other implcations are fulfilled as well. For instance, $\{d,e\}$ implies a, but we also have $x \in \sqrt{< x+y, xy >}$. Indeed, $x^2 = x(x+y) - xy \in < x+y, xy >$.

The question arises how arbitrary formal contexts can be treated analogously. One idea is to write down systematically all implications that hold in the context to be embedded and to find polynomials for which exactly those implications are fulfilled in $\beta(M)$. We will show that such a systematic listing always exists and that it is sufficient to consider this list. We need some definitions to be able to introduce a "basis" of the set of implications.

Definition 2. *A subset $T \subseteq M$ respects an implication $B \longrightarrow D$ if $B \not\subseteq T$ or $D \subseteq T$. T respects a set of implications $\mathfrak{L}$ if it respects all implications from $\mathfrak{L}$.*

Lemma 3. *An implication $B \longrightarrow D$ holds in a context (G,M,I) if and only if each object intent g^I, $g \in G$, respects $B \longrightarrow D$.*

Proof. [GW, p.80]

Definition 3. *An implication $B \longrightarrow D$* **follows** *from a set $\mathfrak{L}$ of implications between attributes if each subset of M which respects $\mathfrak{L}$, respects $B \longrightarrow D$ as well.*

A family of implications is **closed** *if all the implications that follow from $\mathfrak{L}$ belongs already to $\mathfrak{L}$.*

$\mathfrak{L}$ is **complete** *if each implication of (G,M,I) follows from $\mathfrak{L}$.*

$\mathfrak{L}$ is **irredundant** *if no implication of $\mathfrak{L}$ follows from the other implications of $\mathfrak{L}$.*

Note that there are many implications which follow trivially from other implications. For instance, $A \longrightarrow B$ holds for all $B \subseteq A$ and if $A \longrightarrow B$ holds, we conclude that $A \longrightarrow C$ holds if $C \subseteq B$. In the same way, we have $\bigcup_{j \in J} A_j \longrightarrow \bigcup_{j \in J} B_j$ if $A_j \longrightarrow B_j$ holds for all $j \in J$. In particular, $A \longrightarrow C$ holds if $A \supseteq B$ and $B \longrightarrow C$ holds.

The following recursive definition is the decisive step in finding a basis of the implications of a formal context.

Definition 4. *Let (G,M,I) be a finite context. $P \subseteq M$ is called a* **pseudo-intent** *of (G,M,I) if $P \neq P^{II}$ and if for every pseudo-intent $Q \subseteq P$, $Q \neq P$, we have $Q^{II} \subseteq P$.*

Theorem 3. *The set of implications $\mathfrak{L} := \{P \longrightarrow P^{II} \mid P \text{ pseudo} - \text{intent}\}$ is irredundant and complete.*

Proof. [GW, p.85].

Of course, it is sufficient to write the implications in the form $P \longrightarrow P^{II} \setminus P$. The set of these implications is called **Duquenne-Guiges basis** of the implications between the attributes of (G, M, I). An algorithm to compute the Duquenne-Guiges basis can be found in [GW, p.85].

Theorem 3 and the following lemma will yield a method to find intent-preserving embeddings.

Lemma 4. *A quasi-embedding* $(\alpha, \beta) : (H, N, J) \longrightarrow (G, M, I)$ *is an intent-preserving quasi-embedding if and only if for all* $m \in M$ *and all* $D \subseteq M$ *the validity of* $D \longrightarrow m$ *in* (G, M, J) *is equivalent to the validity of* $\beta(D) \longrightarrow m$ *in* (G, M, I).

Proof. One implication has already been shown afore Theorem 1.

Therefore let $(\alpha, \beta) : (H, N, J) \longrightarrow (G, M, I)$ be a quasi-embedding with the equivalence of validities of implications and suppose that $\beta(D^{JJ}) \neq (\beta(D))^{II} \cap \beta(N)$. Using Lemma 1,we see that there is an element $\beta(n) \in (\beta(D))^{II} \cap \beta(N)$ with $n \notin D^{JJ}$. This means that the implication $D \longrightarrow n$ does not hold in (G, M, I). Thus $\beta(D) \longrightarrow \beta(n)$ does not hold in (G, M, I), hence $\beta(n) \notin (\beta(D))^{II}$. This is a contradiction to the choice of n and the lemma is proven.

Example 3. We wish to illustrate how we can use the Duquenne-Guiges basis and the lemma above to obtain intent-preserving quasi-embeddings.

In [GW, p.29] a relatively large context concerning the membership of countries of the third world to certain groups of countries can be found. Here we present a purified version of this context.

I	G77	Blockfrei	LLDC	MSAC	OPEC	AKP
Afghanistan	×	×	×	×		
Aegypten	×	×		×		
Angola	×	×				×
Algerien	×	×			×	
Bahamas	×					×
Botswana	×	×	×			×
Brasilien	×					
Burundi	×	×	×	×		×
Bolivien	×	×				
Malediven	×	×	×			
Birma	×		×	×		
Brunei						
Elfenbeinkueste	×	×		×		×
El Salvador	×			×		
Gabun	×	×			×	×
Haiti	×		×	×		×
Kirbati			×			×
Mongolei			×			

The abbreveations mean: LLDC:=Least Developed Countries, MSAC:= Most Seriously Affected Countries, AKP:=African, Karribean and Pacific Countries, G77:=Gruppe der 77.

In order to find a polynomial embedding of this contexts we compute its Duquenne-Guiges basis. We find the implications

$$\begin{gathered}\text{OPEC} \longrightarrow \text{BLOCKFREI, GRUPPE DER 77}\\ \text{MSAC} \longrightarrow \text{GRUPPE DER 77}\\ \text{BLOCKFREI} \longrightarrow \text{GRUPPE DER 77}\\ \text{G77, BLOCKFREI, MSAC, OPEC} \longrightarrow \text{LLDC, AKP}\\ \text{G77, BLOCKFREI, LLDC, OPEC} \longrightarrow \text{MSAC, AKP}\end{gathered}$$

The premisses of these implications can still be simplified. We observe that $\{MSAC, OPEC\}$ already implies LLDC and AKP and that $\{LLDC, OPEC\}$ already implies AKP and MSAC. Therefore it is enough to consider the Duquenne-Guige basis with minimal premisses.

$$\begin{gathered}\text{OPEC} \longrightarrow \text{BLOCKFREI, GRUPPE DER 77}\\ \text{MSAC} \longrightarrow \text{GRUPPE DER 77}\\ \text{BLOCKFREI} \longrightarrow \text{GRUPPE DER 77}\\ \text{MSAC, OPEC} \longrightarrow \text{LLDC, AKP}\\ \text{LLDC, OPEC} \longrightarrow \text{MSAC, AKP}\end{gathered}$$

Now suppose we have a quasi-embedding (α, β) into $(K^n, K[x_1, \ldots, x_n], \perp)$. Let $\beta(\text{OPEC}) = O$, $\beta(\text{GRUPPE DER 77}) = G$, $\beta(\text{BLOCKFREI}) = B$, $\beta(\text{MSAC}) = M$, $\beta(\text{LLDC}) = L$ and $\beta(\text{AKP}) = A$.

In order to obtain an intent-preserving embedding, the polynomials G, B, L, M, O and A have to satisfy the implications that correspond to the implications above, namely

$$\begin{gathered}O \longrightarrow B, G\\ M \longrightarrow G\\ B \longrightarrow G\\ M, O \longrightarrow L, A\\ L, O \longrightarrow M, A\ .\end{gathered}$$

If these implications hold in $(K^n, K[x_1, \ldots, x_n], \perp)$, then also all implications which follow from them will hold in $(K^n, K[x_1, \ldots, x_n], \perp)$. If we choose our polynomials carefully, i.e. if they are chosen such that no further implications between them hold, Lemma 4 tells us that we have found an intent-preserving embedding.

To determine our polynomials we start with those premisses which consist of a single element. In our case these are O, M and B and we chose x, y and z respectively to represent them.

Next we consider the conclusions of the one-element premisses. We observe that the premiss B is already implied by O. Therefore, we have to revise B.

Since B is implied by O, a power of B must be in the ideal generated by O. So we can choose B as a multiple of O, $B := xz$.

G is implied by all the one-element premisses and we can take G as a common multiple of x, y and xz. However, it is not sufficient to define $G := xyz$, because $\{G77\}^I$ is not equal to the union of $\{OPEC\}^I$, $\{MSAC\}^I$ and $\{BLOCKFREI\}^I$. So we choose $G = wxyz$, where w is a new variable. (Of course, we could choose a prime element different from x, y, z as well).

We continue with those premisses which consist of two elements. Now we start with premisses whose elements are already determined, in our case $\{M, O\}$, which implies L an A. Hence a power of L must be in $< x, y >$. Furthermore, L is not implied by M or O alone and we must have $L \in < x, y > \setminus [< x > \cup < y >]$, $L = f_1 x + h_1 y$ with suitable polynomials $f_1, h_1 \in K[w, x, y, z]$. We may not choose $L = x + y$, because in this case the implication $M, L \longrightarrow O$ would hold since $\{x + y, y\}$ implies x. But $MSAC, LLDC \longrightarrow OPEC$ is an implication that does not follow from our list and so $M, L \longrightarrow O$ must not hold, which means that at least one of the polynomials f_1 and h_1 must not be a constant.

However, the remaining combination $L, O \longrightarrow M$ is valid, hence

$$y \in \sqrt{< g_1 x + h_1 y, x >}$$

and we can choose $h_1 = 1$. Of course, we cannot chose an arbitrary polynomial f_1, f_1 still has to be determined. The same is true for all other polynomials which occur later.

Next we consider $M, O \longrightarrow A$. Since the implications $M, A \longrightarrow O$ and $O, A \longrightarrow M$ are not valid we must choose $A = f_2 x + g_2 y$, where both f_2 and g_2 are polynomials of degree ≥ 1.

Finally, we have to check the implication $L, O \longrightarrow M, A$. The implication $L, O \longrightarrow M$ has already been treated.

It remains to regard $L, O \longrightarrow A$. We have already found requirements for L, O and A. $\{f_1 x + y, x\}$ implies $A = f_2 x + g_2 y$ for any choice of f_1, f_2 and g_2. Thus $L, O \longrightarrow A$ will hold. Neither $O, A \longrightarrow L$ nor $L, A \longrightarrow O$ must hold. But if we choose our polynomials f_1, f_2 and g_2 carefully these requirements will be fulfilled.

We summarize that our reasoning suggests to look for an intent-preserving quasi-embedding with $O = x, M = y, B = xz, G = wxyz, L = f_1 x + y$ and $A = f_2 x + g_2 y$ where f_1, f_2 and g_2 have still to be determined. In view of Lemma 4it remains to find a quasi-embedding (α, β) into $(\mathbb{C}^4, \mathbb{C}[w, x, y, z], \perp)$ for which the elements of the image under β have the form above. (Note that it is not clear by now if such a quasi-embedding exists because 4 presupposes the existence of the embedding).

We will obtain a quasi-embedding if we determine all unknown numbers and polynomials in the table below. All letters representing coordinates are not equal to zero.

	I	$wxyz$	xz	f_1x+y	y	x	f_2x+g_2y
1)	$(0,1,0,0)$	×	×	×	×		
2)	$(0,a,0,0)$	×	×		×		
3)	$(0,b,c,0)$	×	×				×
4)	$(0,0,d,0)$	×	×			×	
5)	$(0,e,l,m)$	×					×
6)	$(0,n,p,0)$	×	×	×			×
7)	$(0,q,r,s)$	×					
8)	$(0,t,0,0)$	×	×	×	×		×
9)	$(0,a_1,a_2,0)$	×	×				
10)	$(0,b_1,b_2,0)$	×	×	×			
11)	$(0,c_1,0,c_2)$	×		×	×		
12)	(d_1,d_2,d_3,d_4)						
13)	$(0,e_1,0,0)$	×	×		×		×
14)	$(0,l_1,0,l_2)$	×			×		
15)	$(0,0,m_1,0)$	×	×			×	×
16)	$(0,n_1,0,n_2)$	×		×	×		×
17)	(p_1,p_2,p_3,p_4)			×			×
18)	(q_1,q_2,q_3,q_4)			×			

We try to determine the coordinates and the polynomials simultaneously. We first consider those points whose only non-zero coordinate is the x-coordinate. These points correspond to the countries which are members of BF but neither of $MSAC$ nor of $OPEC$. These are the objects 1), 2), 8) and 13) which corespond to $(0,1,0,0)$, $(0,a,0,0)$, $(0,t,0,0)$ and $(0,e_1,0,0)$ respectively. We read from the cross table that we must find polynomials $f_1^{(1)}$ and $f_2^{(1)}$ which fulfill the following equalities and inequalities:

1)	$f_1^{(1)}(0,1,0,0)=0$	$f_2^{(1)}(0,1,0,0)\neq 0$
2)	$f_1^{(1)}(0,a,0,0)\neq 0$	$f_2^{(1)}(0,a,0,0)\neq 0$
8)	$f_1^{(1)}(0,t,0,0)=0$	$f_2^{(1)}(0,1,0,0)=0$
13)	$f_1^{(1)}(0,e_1,0,0)\neq 0$	$f_2^{(1)}(0,e_1,0,0)=0$

Note that $f_1^{(1)}$ vanishes both on $(0,1,0,0)$ and $(0,t,0,0)$, which are different points as we see from the cross table. Hence the total degree of $f_1^{(1)}$ must be at least two since the degree of $f_1^{(1)}$ in x is already two. In the same way we conclude that the total degree of $f_2^{(1)}$ must be at least two as well.

We choose the following solution: $f_1^{(1)}=(x-1)(x-2)$, $f_2^{(1)}=(x-2)(x-3)$, $a=4$, $t=2$ and $e_1=3$.

Next we consider all points which involve only the y-coordinate, $(0,0,d,0)$ and $(0,0,m_1,0)$. This yields restrictions only for g_2.

$g_2^{(1)}(0,0,d,0) \neq 0$ and $g_2^{(1)}(0,0,m_1,0) = 0$.

We choose $g_2^{(1)} = y - 1$, $m_1 = 1$ and $d = 2$.

We proceed with those points whose w-coordinate and z-coordinate are zero - i.e. 3), 6), 9) and 10) - and obtain the following conditions:

3) $$f_1^{(2)}(0,b,c,0) + c \neq 0 \qquad f_2^{(2)}(0,b,c,0)b + (c-1)c + 0$$

6) $$f_1^{(2)}(0,n,p,0)n + p = 0 \qquad f_2^{(2)}(0,n,p,0)n + (p-1)p \neq 0$$

9) $$f_1^{(2)}(0,a_1,a_2,0)a_1 + a_2 \neq 0 \qquad f_2^{(2)}(0,a_1,a_2,0)a_1 + (a_2-1)a_2 \neq 0$$

10) $$f_1^{(2)}(0,b_1,b_2,0)b_1 + b_2 = 0$$
$$f_2^{(2)}(0,b_1,b_2,0)b_1 + (b_2-1)b_2 \neq 0$$

(Here we use the observation that it is possible to maintain $g_2^{(1)}$ when further points are involved). Of course, we demand that the polynomials $f_1^{(2)}$ and $f_2^{(2)}$ fulfill the equalities and inequalities settled so far as well.

By now, $f_1^{(1)}$ and $f_2^{(1)}$ are polynomials in x alone. To keep our polynomials as simple as possible we look for solutions of the form $f_1^{(2)} = f_1^{(1)} + \tilde{f}_1$ and $f_1^{(2)} = f_2^{(1)} + \tilde{f}_2$, where $\tilde{f}_1$ and $\tilde{f}_2$ are polynomials in y which vanish on zero. Therefore we obtain the following new conditions:

$$b(f_1^{(1)} + \tilde{f}_1)(0,b,c,0) + c \neq 0 \qquad b(f_2^{(1)} + \tilde{f}_2)(0,b,c,0) + (c-1)c + 0$$

$$n(f_1^{(1)} + \tilde{f}_1)(0,n,p,0) + p = 0 \qquad n(f_2^{(2)} + \tilde{f}_2)(0,n,p,0) + (p-1)p \neq 0$$

$$a_1(f_1^{(2)} + \tilde{f}_1)(0,a_1,a_2,0) + a_2 \neq 0$$
$$a_1(f_2^{(2)} + \tilde{f}_2)(0,a_1,a_2,0) + (a_2-1)a_2 \neq 0$$

$$b_1(f_1^{(2)} + \tilde{f}_1)(0,b_1,b_2,0) + b_2 = 0 \qquad b_1(f_2^{(2)} + \tilde{f}_2(0,b_1,b_2,0) + (b_2-1)b_2 \neq 0$$

After some easy computations we find that we can choose $\tilde{f}_1 = -y$ and $\tilde{f}_2 = -2y$, hence $f_1^{(2)} = (x-1)(x-2) - y$ and $f_2^{(2)} = (x-2)(x-3) - 2y$. Furthermore, we can choose $b = 2$, $c = 5$, $p = 1$, $a_1 = 2$, $a_2 = 1$, $b_1 = 1$ and $b_2 = 3$.

We continue with those points where only the w-coordinate is equal to zero, namely 5), 7), 11), 14) and 16). Again we try to re-use the solutions we have found so far. We look for polynomials $\hat{f}_1$ and $\hat{f}_2$ in z alone which vanish on zero, such that $f_1^{(3)} = f_1^{(2)} + \hat{f}_1$ and $f_2^{(3)} = f_2^{(2)} + \hat{f}_2$ hold. We obtain the conditions listed below.

5) $e(f_1^{(2)} + \hat{f}_1)(0,e,l,m) + l \neq 0 \qquad e(f_2^{(2)} + \hat{f}_2)(0,e,l,m) + l(l-1) = 0$

7) $q(f_1^{(2)} + \hat{f}_1)(0,q,r,s) + r \neq 0 \qquad q(f_2^{(2)} + \hat{f}_2)(0,q,r,s) + r(r-1) \neq 0$

11)
$$c_1(f_1^{(2)} + \hat{f}_1)(0,c_1,0,c_2) = 0$$
$$c_1(f_2^{(2)} + \hat{f}_2)(0,c_1,0,c_2) \neq 0$$

14)
$$l_1(f_1^{(2)} + \hat{f}_1)(0,l_1,0,l_2) \neq 0$$
$$l_1(f_2^{(2)} + \hat{f}_2)(0,l_1,0,l_2) \neq 0$$

16)
$$n_1(f_1^{(2)} + \hat{f}_1)(0,n_1,0,n_2) = 0$$
$$n_1(f_2^{(2)} + \hat{f}_2)(0,n_1,0,n_2) = 0$$

We start with the last condition because it is a simultaneous equality and therefore the most difficult to solve. We check if we can choose $\hat{f}_1 = z$. Hence $n_1((n_1-1)(n_2-1)+n_2) = 0$ must hold and we choose $n_1 = 4$ and $n_2 = -6$. Now we must interpolate $\hat{f}_2$ from the equation $4((4-2)(4-3)+\hat{f}_2(0,4,0,-6)) = 0$. We obtain $\hat{f}_2 = \frac{1}{3}$. Now we must check whether the remaining eqalities which arise from 5) and 11) can be fulfilled. The equation $c_1[(c_1-1)(c_1-2)+c_2] = 0$ can be solved by $c_1 = 5$ and $c_2 = 12$ and we observe that the inequality which stems from 11), i.e. $5c_1(f_2^{(2)} + \hat{f}_2)(0,c_1,0,c_2) \neq 0$, is fulfilled as well. A solution for the equation $e((x-2)(x-3)-2y+\frac{1}{3}z)(0,e,l,m) = 0$ is $e = 2$, $l = 1$, $m = 6$ and again we see that this solution fulfills the inequality arising from 5) as well. Now it is easy to solve the remaining inequalities, we can choose, for instance, $q = r = s = 1$ and $l_1 = l_2 = 1$.

We proceed with 17) and 18). As before we try to find polynomials in w which can be added to $f_1^{(3)}$ and $f_2^{(3)}$ respectively. We find that we can choose $f_1 := f_1^{(4)} = (x-1)(x-2) - y + z + w$ and $f_2 := f_2^{(4)} = (x-2)(x-3) - 2y + \frac{1}{3}z + \frac{1}{3}w$ and $p_1 = -1$, $p_2 = p_3 = p_4 = 1$, $q_1 = -1$, $q_2 = q_3 = q_4 = 2$.

Finally, we must consider the empty row 12). We observe that none of the polynomials representing groups of countries vanishes on $(1,1,1,1)$. We obtain the result shown below.

I	$wxyz$	xz	f_1x+y	y	x	f_2x+g_2y
(0,1,0,0)	×	×	×	×		
(0,4,0,0)	×	×		×		
(0,2,5,0)	×	×				×
(0,0,2,0)	×	×			×	
(0,2,1,6)	×					×
(0,1,1,0)	×	×	×			×
(0,1,1,1)	×					
(0,2,0,0)	×	×	×	×		×
(0,2,1,0)	×	×				
(0,1,3,0)	×	×	×			
(0,5,0,-12)	×		×	×		
(1,1,1,1)						
(0,3,0,0)	×	×		×		×
(0,1,0,1)	×			×		
(0,0,1,0)	×	×			×	×
(0,4,0,-6)	×		×	×		×
(-1,1,1,1)			×			×
(-1,2,2,2)			×			

with $f_1 = (x-1)(x-2) - y + z + w$, $f_2 = (x-2)(x-3) - 2y + \frac{1}{3}z + \frac{1}{3}w$ and $g_2 = y - 1$.

Let us return to the beginning of our example. We observe at least two critical points. First, the element G had to be revised, which is an undesirable effect. Moreover, at the end of our reasoning we were left with an implication, namely $L, O \longrightarrow A$, where the occuring polynomials were already determined or subject to certain restrictions. However, how can we be sure that the implication $L, O \longrightarrow A$ does not contradict the restrictions we have found so far?

To avoid revision or contradiction we need an ordering on the set of implications. The following lemma suggests to work "column-wise".

Lemma 5. *Let (G, M, I) be a formal context, let $N \subseteq M$ and let $(G, N, I \cap (G \times N))$ be the corresponding subcontext.*

Then the set of all implications of $(G, N, I \cap (G \times N))$ is equal to the set of all implications of (G, M, I) in which only elements from N occur both in the premiss and in the conclusion.

Example 4. We consider the following formal context $\mathbb{K} = (G, M, I)$.

I	a	b	c	d	e	f
1	×	×				
2		×				
3			×	×		
4				×		
5					×	×
6						×

We first consider a small subset $N \subseteq M$ and try to find an intent-preserving embedding of $(G, N, I \cap (G \times N))$ into $(\mathbb{C}^n, \mathbb{C}[x_1, \ldots, x_n], \perp)$. Starting from this embedding we try to find an intent-preserving embedding from $(G, N \cup \{m\}, I \cap (G \times (N \cup \{m\}))$ into $(\mathbb{C}^n, \mathbb{C}[x_1, \ldots, x_n], \perp)$. The lemma tells us that we can use the embedding we have found so far, we only have to determine $\beta(m)$ from the implications that include m either in the premiss or in the conclusion. The other implications do not depend on the choice of $\beta(m)$. It is reasonable to start with a maximal independent subset N of M, i.e. only trivial implications hold among the elements of N, and to choose different variables to represent them.

In our example we can choose $N = \{a, c\}$ and we define $\beta(a) = x$ and $\beta(c) = y$ to represent them. Hence we are looking for a polynomial embedding of $\mathbb{K}$ into $(\mathbb{C}^2, \mathbb{C}[x, y], \perp)$.

Obviously,

	x	y
$(0,1)$	×	
$(1,0)$		×

is an intent-preserving embedding of $(G, N, I \cap (G \times N))$ into $(\mathbb{C}^2, \mathbb{C}[x, y], \perp)$. The empty rows symbolize that the missing objects will be determined in one of the following steps. Now we consider $N_1 := N \cup \{e\}$ and $\mathbb{K}_1 := (G, N_1, I \cap (G \times N_1))$. We must check all implications involving e. We observe that $\{a, c\} \longrightarrow e$, $\{a, e\} \longrightarrow c$ and $\{c, e\} \longrightarrow a$ hold. Therefore we can choose $\beta(e) = x + y$ and

	x	y	$x+y$
$(0,1)$	×		
$(1,0)$		×	
$(-1,1)$			×

is an intent-preserving embedding.

We proceed with those elements that are implied by one-element premisses. Let $N_2 := N_1 \cup \{b\}$ and $\mathbb{K}_2 := (G, N_2, I \cap (G \times N_2))$.

	x	b	y	$x+y$
$(0,1)$	×	×		
		×		
$(1,0)$			×	
$(-1,1)$				×

Thus we are looking for a polynomial $b \in \mathbb{C}[x,y]$ which satisfies exactly the implications that follow from

$$\begin{gathered} x \longrightarrow b \\ (y, x+y \longrightarrow b) \\ y, b \longrightarrow x, (x+y) \\ b, x+y \longrightarrow x, y. \end{gathered}$$

The implications in brackets will already be fulfilled if all the other implications are fulfilled. Hence, we must have $b = xg$ for some polynomial g, $x \in \sqrt{< xg, y >}$ and $x, y \in < xg, x+y >$. We observe that $b := x(x+2y)$ fulfills these requirements.

We proceed with the following subcontext and try to determine d.

	x	$x(x+2y)$	y	d	$x+y$
$(0,1)$	×	×			
$(1,-\frac{1}{2})$		×			
$(1,0)$			×	×	
				×	
$(-1,1)$					×

If we translate the implications involving d into requirements in our polynomial context we obtain $d = yh$ for some polynomial h, $d \in \sqrt{< x+y, x(x+2y) >}$, $\{x,y\} \subseteq \sqrt{< d, x+y >}$, $\{x,y\} \subseteq \sqrt{< d, x(x+2y) >}$, and $y \in \sqrt{< d, x >}$. A solution to this conditions is $d := y(y+2x)$.

Finally, we consider the object f. We obtain the following conditions for f: $f \in < x+y >$, $f \in < x, y >$, $f \in \sqrt{< x(x+2y), y >}$, $f \in \sqrt{< x, y(y+2x) >}$, $f \in \sqrt{< x(x+2y), y(y+2x) >}$, $\{x,y\} \subseteq \sqrt{< f, y(y+2x) >}$, $\{x+y, x\} \subseteq \sqrt{< f, y >}$, $\{x,y\} \subseteq \sqrt{< f, x(x+2y) >}$ and $y \in \sqrt{< f, x >}$. A solution is $f := (x+y)(x+3y)$ and we obtain the following intent-preseving embedding of $\mathbb{K}$ into $(\mathbb{C}^2, \mathbb{C}[x,y], \perp)$:

	x	$x(x+2y)$	y	$y(y+2x)$	$x+y$	$(x+y)(x+3y)$
$(0,1)$	×	×				
$(1,-\frac{1}{2}$		×				
$(1,0)$			×	×		
$(-\frac{1}{2},1)$				×		
$(-1,1)$					×	×
$(3,-1)$						×

Unfortunately, there is a large class of formal contexts which do not allow for a polynomial embedding over $\mathbb{C}$ in more than one variable.

Lemma 6. *Let us consider the formal context* $\mathbb{K}$ *below:*

	a	b	c	d
1	×			
2		×		×
3		×	×	
4			×	×

The set $\{b, c, d\}$ *is independent. Therefore our contemplation elaborated afore suggests to look for a polynomial embedding* $(\alpha, \beta) : \mathbb{K} \longrightarrow (\mathbb{C}^3, \mathbb{C}[x, y, z], \perp)$ *with* $\beta(b) = x$, $\beta(c) = y$ *and* $\beta(d) = z$.
We observe that the following conditions must hold for $f := \beta(a)$*:*

$$\begin{aligned} x, y, z &\longrightarrow f \\ f, x &\longrightarrow y, z \\ f, y &\longrightarrow x, z \\ f, z &\longrightarrow x, y. \end{aligned}$$

Let us consider the implication $x, f \longrightarrow y, z$. *Thus the polynomial* $\tilde{f} := f(0, y, z) \in \mathbb{C}[y, z]$ *must vanish on* $(0, 0)$ *exclusively. However, there is no polynomial in* $\mathbb{C}[y, z]$ *that vanishes only on a single point. Hence we cannot find a polynomial embedding of the desired form. It is easily seen that there is no polynomial embedding over* $\mathbb{C}$ *in more that two variables at all. The only polynomial embedding over* $\mathbb{C}$ *is the embedding in one variable described in 2.*

We summarize that polynomial embeddings over $\mathbb{C}$ in more than one variable in general do not exist when there is proper premiss that consists of three or more elements.

The situation is somewhat different if we consider polynomial embeddings over $\mathbb{R}$. Here we have polynomials that vanish only on a single point. For instance, the only zero of $x^2 + y^2$ is the origin $(0, 0)$. Therefore $\{x^2 + y^2, z\}$ implies x and y. In the same way $\{x^2 + z^2, y\}$ implies x and z and $\{y^2 + z^2, x\}$ implies y and z. Hence, we should try to choose $f := x^2 + y^2 + z^2$ in order to find an implication-embedding into $(\mathbb{R}^3, \mathbb{R}[x, y, z], \perp)$.

The polynomials $x^2 + y^2 + z^2$, x, y and z actually satisfy exactly the desired implications, yet they cannot be used to define an implication embedding from

$\mathbb{K}$ into $(\mathbb{R}^3, \mathbb{R}[x, y, z], \perp)$. Since the object "1" is in relation with a, f must be in relation with $\alpha(1)$. Thus we must have $\alpha(1) = (0, 0, 0)$ because $(0, 0, 0)$ is the only zero of f. However, this implies that $\beta(b) = x$, $\beta(c) = y$ and $\beta(d) = z$ are in relation with $\alpha(1) = (0, 0, 0)$ as well, which contradicts the fact that none of the attribures b, c and d is in relation with "1". But we get an implication of the following context $\tilde{\mathbb{K}}$, whose concept lattice is isomorphic to that of $\mathbb{K}$.

	a	b	c	d
1	×	×	×	×
2		×		×
3		×	×	
4			×	×

↓

	$x^2 + y^2 + z^2$	x	y	z
$(0, 0, 0)$	×	×	×	×
$(0, 1, 0)$		×		×
$(0, 0, 1)$		×	×	
$(1, 0, 0)$			×	×

3 Polynomial Representations

In the final section we briefly touch the related topic of polynomial representations. Similarly to the case of polynomial embedding, this leads to a genarelized interpolation problem. In our case the points of interpolation are not given explicitly, but they are subject to certain ordinal conditions. We observe that it is very difficult to find optimal solutions, i.e. solutions which minimize the total degree of the polynomials that are involved. We take up the ideas of geometric representation of data and algebraic descriptions of dependencies of attributes, as they are treated in [W]. There empirical structures are represented by numerical relational structures, for instance by real n-dimensional vector spaces. This approach is settled within the framework of numerical measurement. The existence of measurements and the meaningfulness of manipulations of measurements are analyzed fundamentally in [Kr]. Following the spirit of the first section, we are looking for representations in affine space rather than in vector spaces, and we describe dependencies of attributes via polynomials and not via elements from a vector space or an ordered quasigroup.

We present our ideas within the setting elaborated in [W]. First, we define a formal model for (an important kind of) empirical data.

Definition 5. *A quadruple* $\mathbb{K} := (G, M, (W, \geq), I)$ *is called* **ordinal context**, *if* G *and* M *are sets,* $(W, \geq)$ *is an ordered set, and* I *is a ternary relation on* $G \times M \times W$*, such that for each* $g \in G$*,* $m \in M$*, there is exactly one element* $w \in W$ *with* $(g, m, w) \in I$*.*

The elements of G, M and W are called objects, attributes and attribute values respectively. We often write $m(g) = w$ if $(g, m, w) \in I$ and read 'the object g has the value w for the attribute m'.

The following example of a linear representation is taken from [W]. It shows data from Schiffmann and Falkenberg (1968), describing the amount of absorption for four different colour stimuli by eleven receptors in the retina of a goldfish.

Receptor	Violet 430	Blue 458	Blue 485	Blue-Green 498
1	147	153	89	57
2	153	154	110	75
3	145	152	125	100
4	99	101	122	140
5	46	85	103	127
6	73	78	85	121
7	14	2	46	52
8	44	65	77	73
9	87	59	58	52
10	60	27	23	24
11	0	0	40	39

The data on colour stimuli is represented in the figure on the next page.

The orders with respect to the stimuli 'violet' can be read of from the picture by projecting onto the x-axis, the order with respect to 'blue-green' by projecting onto the y-axis. The attribute 'blue 485' is represented by the system of parallel lines in the direction of the arrow labelled 'b485', the attribute 'blue 458' is represented by the other system of parallel lines.

We give a general definition for the situatuion we encountered here.

Definition 6. *An ordinal context* $\mathbb{K} := (G, M, (W, \geq), I)$ *with attributes* $m_1, \ldots, m_{n+l}$ *has a simultaneous linear representation in a n-dimensional real vector space with respect to* $m_1, \ldots, m_n$, *if there are real n- tuples* $\alpha_1^j, \ldots \alpha_n^j$, $j = 1, \ldots, l$, *and an injective mapping* $\varphi : G \longrightarrow \mathbb{R}^n$ *with*

$$m_s(g) \geq m_s(h) \Longleftrightarrow \pi(\varphi(g)) \geq \pi(\varphi(h)),\ s = 1, \ldots, n$$

$$m_{n+j}(g) \geq m_{n+j}(h) \Longleftrightarrow \sum_{s=1}^{n} \alpha_s^j \pi_s(\varphi(g)) \geq \sum_{s=1}^{n} \alpha_s^j \pi_s(\varphi(h)),\ j = 1, \ldots, l$$

for all $g, h \in G$. *(Here* π_s *denotes the projection to the s coordinate axis).*

Thus, the definition means that the objects of the given ordinal context are mapped into $\mathbb{R}^n$ such that the orders of n attributes are represented on the n coordinate axis and the remaining l orders are described by l different linear combinations, specified by the n-tuples $\alpha_1^j, \ldots, \alpha_n^j$, $j = 1, \ldots, l$.

One finds that an ordinal context has to satisfy strong conditions in order to be linear representable. Results concerning this question can be found in [W]. Usually, empirical data will not fulfill the requirements. Therefore we consider representations by more general mathematical objects, for instance by polynomials.

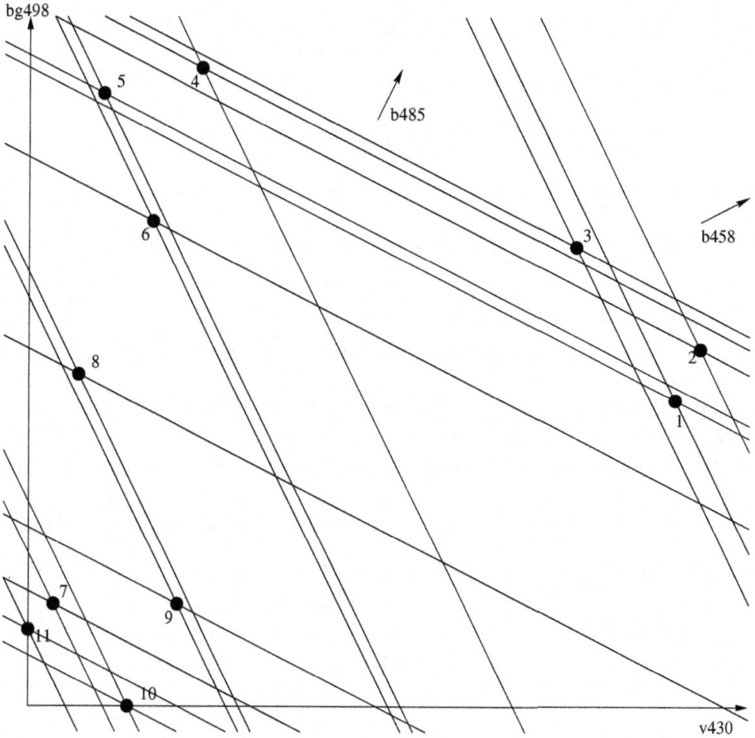

Fig. 4. A linear representation

Definition 7. *An ordinal context* $\mathbb{K} := (G, M, (W, \geq), I)$ *with attributes* $m_1, \ldots, m_{n+l}$ *has a simultaneous polynomial representation in the affine space* $\mathbb{R}^n$*, if there are real polynomials in* n *variables* f_j*,* $j = 1, \ldots, l$*, and an injective mapping* $\varphi : G \longrightarrow \mathbb{R}^n$*, such that*

$$m_s(g) \geq m_s(h) \iff \pi(\varphi(g)) \geq \pi(\varphi(h)),\ s = 1, \ldots, n$$
$$m_{n+j}(g) \geq m_{n+j}(h) \iff f_j(\varphi(g)) \geq f_j(\varphi(h)),\ j = 1, \ldots, l$$

holds for all $g, h \in G$.

Thus, polynomial representations are a generalization of linear representations, since every linear expression of the form $\sum_{s=1}^{n} \alpha_s^j x_s$ is a polynomial.

Theorem 4. *Let* $\mathbb{K} := (G, M, \mathbb{R}, I)$ *be a finite real-valued ordinal context with* $M = \{m_1, \ldots, m_n\}$.

Then $\mathbb{K}$ *is simultaneously representable in* $\mathbb{R}^2$.

Proof. We choose $\varphi : G \longrightarrow \mathbb{R}^2$ such that (1) is fulfilled, which is always possible. Now it is sufficient to find polynomials f_j, $j = 3, \ldots, n$ with $f_j(\varphi(g)) = m_j(g)$ for all $g \in G$. Therefore, the theorem follows from the lemma below.

Lemma 7. *Let $\boldsymbol{a}_1, \ldots, \boldsymbol{a}_m \in \mathbb{R}^n$ and $b_1, \ldots, b_m \in \mathbb{R}$ such that $b_1 = b_j$ if $\boldsymbol{a}_i = \boldsymbol{a}_j$ for $i, j \in \{1, \ldots, n\}$. Then there is a polynomial $f \in \mathbb{R}[x_1, \ldots, x_n]$ with*

$$(\star) \ f(\boldsymbol{a}_i) = b_i$$

for all $i \in \{1, \ldots, m\}$.

Proof. It is sufficient to find polynomials f_k, $k = 1, \ldots, m$ with

$$(\star\star) \ f_k(\boldsymbol{a}_i) = \delta_{ki}, \ i = 1, \ldots, m.$$

Then $f := \sum_{k=1}^{m} b_i f_k$ will yield the desired interpolation.

Consider $\{\boldsymbol{a}_1, \ldots, \boldsymbol{a}_{n-1}\}$. We have $\{\boldsymbol{a}_1, \ldots, \boldsymbol{a}_{n-1}\}^{\perp\perp} = \{\boldsymbol{a}_1, \ldots, \boldsymbol{a}_{n-1}\}$ and $\{\boldsymbol{a}_1, \ldots, \boldsymbol{a}_n\}^{\perp\perp} = \{\boldsymbol{a}_1, \ldots, \boldsymbol{a}_n\}$, because finite sets are always algebraic varieties. Since $\{\boldsymbol{a}_1, \ldots, \boldsymbol{a}_{n-1}\} \subseteq \{\boldsymbol{a}_1, \ldots, \boldsymbol{a}_n\}$, we have $\{\boldsymbol{a}_1, \ldots, \boldsymbol{a}_n\}^{\perp} \subseteq \{\boldsymbol{a}_1, \ldots, \boldsymbol{a}_{n-1}\}^{\perp}$. If we had $\{\boldsymbol{a}_1, \ldots, \boldsymbol{a}_n\}^{\perp} = \{\boldsymbol{a}_1, \ldots, \boldsymbol{a}_{n-1}\}^{\perp}$, we would conclude $\{\boldsymbol{a}_1, \ldots, \boldsymbol{a}_n\}^{\perp\perp} = \{\boldsymbol{a}_1, \ldots, \boldsymbol{a}_{n-1}\}^{\perp\perp}$, hence $\{\boldsymbol{a}_1, \ldots, \boldsymbol{a}_{n-1}\} = \{\boldsymbol{a}_1, \ldots, \boldsymbol{a}_n\}$ which is a contradiction. This means that there is a polynomial $\hat{f}_1 \in \{\boldsymbol{a}_1, \ldots, \boldsymbol{a}_{n-1}\}^{\perp} \setminus \{\boldsymbol{a}_1, \ldots, \boldsymbol{a}_n\}^{\perp}$, i.e. $f(\boldsymbol{a}_1) = \ldots = f(\boldsymbol{a}_{n-1}) = 0$, $f(\boldsymbol{a}_n) \neq 0$. If we devide $\hat{f}_1$ by $\hat{f}_1(\boldsymbol{a}_n)$ we obtain a polynomial f_1 which satisfies $(\star\star)$. By the same way of reasoning, we find polynomials f_k, $k = 2, \ldots, n$, which satisfy $(\star\star)$ and the proof is completed.

References

[GW] B. Ganter, R. Wille Formale Begriffsanalyse Mathematische Grundlagen, Springer-Verlag, Berlin Heidelberg 1996.

[K] E. Kunz Einführung in die kommutative Algebra und Algebraische Geometrie, Vieweg, Braunschweig Wiesbaden 1980.

[Kr] D.H. Krantz, R.D. Luce, R.Suppes and A.Tversky Foundations of Measurement, Academic Press New York and London 1971

[TB1] T. Becker Features of Interaction between Formal Concept Analysis and Algebraic Geometry, TU Darmstadt.

[TB2] T. Becker General Algebraic Geometry, TU Darmstadt.

[V] F. Vogt Bialgebraic Contexts, TU Darmstadt, Verlag Shaker, Aachen 1994.

[W] U. Wille Geometric Representation of Ordinal Contexts, Dissertation, Giessen 1995.

The Basic Theorem on Labelled Line Diagrams of Finite Concept Lattices

Rudolf Wille

Technische Universität Darmstadt, Fachbereich Mathematik,
Schloßgartenstr. 7, D–64289 Darmstadt
wille@mathematik.tu-darmstadt.de

Abstract. This paper offers a mathematical analysis of labelled line diagrams of finite concept lattices to gain a better understanding of those diagrams. The main result is the *Basic Theorem on Labelled Line Diagrams of Finite Concept Lattices*. This Theorem can be applied to justify, for instance, the training tool *"CAPESSIMUS - A Game of Conceiving Concepts"* which has been created to support the understanding and the drawing of appropriate line diagrams of finite concept lattices.

Contents

1 Introduction

For successfully applying *Formal Concept Analysis* in practice, well-readable diagrams of concept lattices are needed. Although the developed computer programs for drawing concept lattices are quite useful, they are still far from being satisfying, in particular, with respect to purpose support and adequate content representation. Human beings still have to take over the essential part of creating diagrammatic knowledge representations which are able to inspire, stimulate, and guide human thought. The purpose of this paper is to give some basic support for training humans in drawing *labelled line diagrams of finite concept lattices*.

It is important to realize that, in practice, labelled line diagrams of concept lattices have a *three-fold semantics*: a mathematical, a philosophical and a special purpose-oriented semantics. This triadic view is derived from the first level of *Peirce's classification of sciences* ([Pe92]; p.114) which lists the sciences in the order of abstractness: I. Mathematics II. Philosophy III. Special Sciences, where *Mathematics* is viewed as the most abstract science studying hypotheses exclusively and dealing only with potential realities, *Philosophy* is considered as the most abstract science dealing with actual phenomena and realities, while all

S.O. Kuznetsov and S. Schmidt (Eds.): ICFCA 2007, LNAI 4390, pp. 303–312, 2007.

other sciences are more concrete in dealing with special types of actual realities (see [GW06]; p.216).

Since, in practice, *labelled line diagrams of concept lattices* link mathematical, philosophical, and special thought, a well developed understanding of those diagrams is strongly desirable; in particular, users should be able to recognize whether a labelled line diagram represents a given concept lattice or not. This paper offers a mathematical analysis of labelled line diagrams to support the desired ability of understanding those diagrams. For that, *Section 2* recalls some basics of finite concept lattices and gives a proper proof for the Basic Theorem of Finite Concept Lattices, *Section 3* introduces a mathematization of line diagrams of finite bounded ordered sets, and *Section 4* formulates and proves the *Basic Theorem on Labelled Line Diagrams of Finite Concept Lattices.* In the final section, the Basic Theorem on Finite Labelled Line Diagrams is applied to support the conceptual training tool "*CAPESSIMUS - A Game of Conceiving Concepts*".

2 Basics of Finite Concept Lattices

Let us assume that the reader is familiar with the basic notions of Formal Concept Analysis as they are defined in [GW99]. From lattice theory the reader should particularly know that, in a finite lattice $\underline{L}$, each element of $\underline{L}$ is the supremum of $\vee$-*irreducible elements* and the infimum of $\wedge$-*irreducible elements* (see [DP02], p.55). In this paper, $J(\underline{L})$ denotes the set of all $\vee$-irreducible elements of $\underline{L}$ and $M(\underline{L})$ denotes the set of all $\wedge$-irreducible elements of $\underline{L}$. In a concept lattice $\underline{\mathfrak{B}}(\mathbb{K})$ of a finite context $\mathbb{K} := (G, M, I)$, each $\vee$-irreducible concept is of the form $\gamma g := (\{g\}'', \{g\}')$ for some $g \in G$ and each $\wedge$-irreducible concept is of the form $\mu m := (\{m\}', \{m\}'')$ for some $m \in M$, i.e., γG contains $J(\underline{\mathfrak{B}}(\mathbb{K}))$ and μM contains $M(\underline{\mathfrak{B}}(\mathbb{K}))$.

Now, we are prepared to formulate and to prove the finite case of the Basic Theorem on Concept Lattices (Part II) (cf. [GW99]; p.20):

Basic Theorem on Finite Concept Lattices (Part II). *A finite lattice $\underline{L}$ is isomorphic to the concept lattice $\underline{\mathfrak{B}}(\mathbb{K})$ of a finite context $\mathbb{K} := (G, M, I)$ if and only if there exist mappings $\tilde{\gamma} : G \to \underline{L}$ and $\tilde{\mu} : M \to \underline{L}$ such that*

1. *$\tilde{\gamma}G$ contains $J(\underline{L})$,*
2. *$\tilde{\mu}M$ contains $M(\underline{L})$,*
3. *$gIm \iff \tilde{\gamma}g \leq \tilde{\mu}m$ for $g \in G$ and $m \in M$.*

Proof: Let ξ be an isomorphism from $\underline{\mathfrak{B}}(\mathbb{K})$ onto $\underline{L}$. If, for some $(A, B) \in \mathfrak{B}(\mathbb{K})$, $\bigvee\{\gamma g \mid g \in A\}$ is a $\vee$-irreducible element of $\underline{\mathfrak{B}}(\mathbb{K})$, there must exist an object $h \in A$ with $\gamma h = \bigvee\{\gamma g \mid g \in A\}$. Since in a finite lattice every element is the supremum of $\vee$-irreducible elements, it follows that $J(\underline{\mathfrak{B}}(\mathbb{K})) \subseteq \gamma G$ and hence $J(\underline{L}) = \xi J(\underline{\mathfrak{B}}(\mathbb{K})) \subseteq \xi\gamma G$. Dually, we obtain $M(\underline{L}) \subseteq \xi\mu M$. For $g \in G$ and $m \in M$ we have the following equivalences: $gIm \iff \gamma g \leq \mu m \iff \xi\gamma g \leq \xi\mu m$. Thus, defining $\tilde{\gamma} := \xi\gamma$ and $\tilde{\mu} := \xi\mu$ yields the conditions 1, 2, and 3.

Conversely, let $\tilde{\gamma} : G \to \underline{L}$ and $\tilde{\mu} : M \to \underline{L}$ be mappings satisfying the conditions 1, 2, and 3. For every $(A, B) \in \mathfrak{B}(\mathbb{K})$ we have $\bigvee \tilde{\gamma}A \leq \bigwedge \tilde{\mu}B$ by condition 3. For each $x \in J(\underline{L})$ with $x \leq \bigwedge \tilde{\mu}B$ there exists a $g \in G$ with $x = \tilde{\gamma}g$ by condition 1; it follows that $g \in B' = A$ and that $\bigvee(\tilde{\gamma}A \cap J(\underline{L})) = \bigwedge \tilde{\mu}B$. Dually, for each $y \in M(\underline{L})$ with $y \geq \bigvee \tilde{\gamma}A$ there exists an $m \in M$ with $y = \tilde{\mu}m$ by condition 2; it follows that $m \in A' = B$ and that $\bigwedge(\tilde{\mu}B \cap M(\underline{L})) = \bigvee \tilde{\gamma}A$. Both cases together yield $\bigvee \tilde{\gamma}A = \bigvee(\tilde{\gamma}A \cap J(\underline{L})) = \bigwedge(\tilde{\mu}B \cap M(\underline{L})) = \bigwedge \tilde{\mu}B$. Now, we define $\xi : \mathfrak{B}(\mathbb{K}) \to \underline{L}$ by

$$\xi(A, B) := \bigvee \tilde{\gamma}A \,(= \bigwedge \tilde{\mu}B\,).$$

Evidently, the mapping ξ is order-preserving. Now, let $(A, B), (C, D) \in \mathfrak{B}(\mathbb{K})$ with $(A, B) \not\leq (C, D)$. Then there exists $g \in A$ and $m \in D$ with $(g, m) \notin I$ which yields $\tilde{\gamma}g \not\leq \tilde{\mu}m$ by condition 3. Thus, $\xi(A, B) = \bigvee \tilde{\gamma}A \not\leq \bigwedge \tilde{\mu}D = \xi(C, D)$. This proves that ξ is injective. Since $\bigvee(z] \cap J(\underline{L}) = z = \bigwedge[z) \cap M(\underline{L})$ for $z \in \underline{L}$ (see [DP02], p.55), there exists, by conditions 1 and 2, $A_z \subseteq G$ and $B_z \subseteq M$ with $\tilde{\gamma}A_z \subseteq J(\underline{L})$, $\tilde{\mu}B_z \subseteq M(\underline{L})$ and $\bigvee \tilde{\gamma}A_z = z = \bigwedge \tilde{\mu}B_z$; hence $\xi(A_z'', B_z'') = z$, i.e., ξ is also surjective and therefore an isomorphism from $\mathfrak{B}(\mathbb{K})$ onto $\underline{L}$.

3 Line Diagrams of Finite Bounded Ordered Sets

Let us recall that an *ordered set* is a pair $(O, \leq)$ consisting of a set O and a binary relation $\leq$ which is reflexive, antisymmetric, and transitive. An ordered set $(O, \leq)$ is said to be *bounded* if there exist elements 0 and 1 in O such that $0 \leq x \leq 1$ for all $x \in O$. In the following we consider only finite bounded ordered sets.

Mathematically, a *line diagram of a finite bounded ordered set* $\underline{O} := (O, \leq)$ can be defined as a quadruple $\mathbb{D}_\eta(\underline{O}) := (C_{\underline{O}}, S_{\underline{O}}, T_{\underline{O}}, \eta)$ formed by

- a set $C_{\underline{O}}$ of disjoint little circles of the same radius in the Euclidean plane $\mathbb{R}^2$,
- a set $S_{\underline{O}}$ of straight line segments in $\mathbb{R}^2$ having at most one point in common,
- a ternary relation $T_{\underline{O}} \subseteq C_{\underline{O}} \times S_{\underline{O}} \times C_{\underline{O}}$ which contains for each $s \in S_{\underline{O}}$ exactly one triple (c_1, s, c_2) indicating that the line segment s links up the circles c_1 and c_2 in $\mathbb{R}^2$ and that $c_1 <_2 c_2$ (i.e. for all points $p_i \in c_i$ with $i = 1, 2$, the second coordinate of p_1 is smaller than the second coordinate of p_2).
- a bijection $\eta : O \to C_{\underline{O}}$ which makes explicit that the covering pairs $o_1 \prec o_2$ in $\underline{O}$ are in one-to-one correspondence to the triples $(\eta(o_1), s, \eta(o_2))$ of $T_{\underline{O}}$ (consequently, $|\prec| = |T_{\underline{O}}|$).

The line diagrams $\mathbb{D}_\eta(\underline{O})$ and $\mathbb{D}_{\hat{\eta}}(\hat{\underline{O}})$ of finite bounded ordered sets $\underline{O} := (O, \leq)$ and $\hat{\underline{O}} := (\hat{O}, \leq)$ are called *isomorphic* if and only if there exist bijections $\zeta : C_{\underline{O}} \to C_{\hat{\underline{O}}}$ and $\sigma : S_{\underline{O}} \to S_{\hat{\underline{O}}}$ such that $(c_1, s, c_2) \in T_{\underline{O}} \iff (\zeta(c_1), \sigma(s), \zeta(c_2)) \in T_{\hat{\underline{O}}}$; the corresponding isomorphism is denoted by (ζ, σ).

Lemma 1. *Two finite bounded ordered sets are isomorphic if and only if they have isomorphic line diagrams.*

Proof: Let $\underline{O} := (O, \leq)$ and $\underline{\hat{O}} := (\hat{O}, \leq)$ be finite bounded ordered sets with corresponding line diagrams $\mathbb{D}_\eta(\underline{O})$ and $\mathbb{D}_{\hat{\eta}}(\underline{\hat{O}})$, respectively. Let θ be an isomorphim from $\underline{O}$ onto $\underline{\hat{O}}$. Then $\zeta := \hat{\eta}\theta(\eta^{-1})$ is a bijection from $C_{\underline{O}}$ onto $C_{\underline{\hat{O}}}$. For each $s \in S_{\underline{O}}$, there is a unique covering pair $o_1 \prec o_2$ in $\underline{O}$ with $(\eta(o_1), s, \eta(o_2)) \in T_{\underline{O}}$ and with $\theta(o_1) \prec \theta(o_2)$ in $\mathbb{D}_{\hat{\eta}}(\underline{\hat{O}})$; furthermore, there is a unique $\hat{s} \in S_{\underline{\hat{O}}}$ with $(\hat{\eta}\theta(o_1), \hat{s}, \hat{\eta}\theta(o_2)) \in T_{\underline{\hat{O}}}$. This shows that there is a bijection $\sigma : S_{\underline{O}} \to S_{\underline{\hat{O}}}$, defined by $\sigma(s) := \hat{s}$, such that $(c_1, s, c_2) \in T_{\underline{O}} \iff (\zeta(c_1), \sigma(s), \zeta(c_2)) \in T_{\underline{\hat{O}}}$. Hence $\mathbb{D}_\eta(\underline{O})$ and $\mathbb{D}_{\hat{\eta}}(\underline{\hat{O}})$ are isomorphic.

Conversely, let $\mathbb{D}_\eta(\underline{O})$ and $\mathbb{D}_{\hat{\eta}}(\underline{\hat{O}})$ be line diagrams of finite bounded ordered sets with bijections $\zeta : C_{\underline{O}} \to C_{\underline{\hat{O}}}$ and $\sigma : S_{\underline{O}} \to S_{\underline{\hat{O}}}$ such that $(c_1, s, c_2) \in T_{\underline{O}}$ $\iff (\zeta(c_1), \sigma(s), \zeta(c_2)) \in T_{\underline{\hat{O}}}$. Then a bijection $\theta : \underline{O} \to \underline{\hat{O}}$ can be defined by $\theta := (\hat{\eta}^{-1})\zeta\eta$. For each covering pair $o_1 \prec o_2$ in $\underline{O}$, there is a corresponding triple $(\eta(o_1), s, \eta(o_2))$ in $T_{\underline{O}}$ and hence a corresponding triple $(\zeta\eta(o_1), \sigma(s), \zeta\eta(o_2))$ in $T_{\underline{\hat{O}}}$ such that $(\hat{\eta}^{-1})\zeta\eta(o_1) \prec (\hat{\eta}^{-1})\zeta\eta(o_2)$, i.e. $\theta(o_1) \prec \theta(o_2)$; concluding backwards yields that $\theta(o_1) \prec \theta(o_2)$ implies $o_1 \prec o_2$. Since, in finite ordered sets, we have $o_1 \leq o_2 \iff o_1 = o_2$ or $o_1 \prec \cdots \prec o_2$, the bijection $\theta : \underline{O} \to \underline{\hat{O}}$ is an isomorphism.

4 The Basic Theorem on Finite Labelled Line Diagrams

A cross table which represents a finite context $\mathbb{K} := (G, M, I)$ contains the object names of the objects in G and the attribute names of the attributes in M. Since those names are understood as *proper names* (german: *Eigennamen*), there is a bijection ν mapping each object resp. attribute in $G\dot{\cup}M$ to its proper name, i.e., $\nu(G\dot{\cup}M) := \{\nu(x) \mid x \in G\dot{\cup}M\}$ is the set of all proper names of the objects and attributes in the context $\mathbb{K}$. A line diagram $\mathbb{D}_{\bar{\eta}}(\underline{\mathfrak{B}}(\mathbb{K}))$ together with the bijection ν is called a *$(\nu G, \nu M)$-labelled line diagram* denoted by $\mathbb{D}^\nu_{\bar{\eta}}(\underline{\mathfrak{B}}(\mathbb{K}))$. Analogously, for a finite bounded ordered set $\underline{O}$ and mappings $\check{\gamma} : G \to \underline{O}$ and $\check{\mu} : M \to \underline{O}$, a line diagram $\mathbb{D}_\eta(\underline{O})$ together with the introduced naming bijection ν on $G\dot{\cup}M$ is called a *$(\nu G, \nu M)$-labelled line diagram* denoted by $\mathbb{D}^\nu_\eta(\underline{O})$. In both types of labelled line diagrams, the object names νg are attached from below to the circle $\bar{\eta}(\gamma g)$ resp. $\eta(\check{\gamma} g)$ and the attribute names νm are attached from above to the circle $\bar{\eta}(\mu m)$ resp. $\eta(\check{\mu} m)$. A $(\nu G, \nu M)$-labelled line diagram $\mathbb{D}^\nu_\eta(\underline{O})$ is said to be *isomorphic* to a $(\nu G, \nu M)$-labelled line diagram $\mathbb{D}^\nu_{\bar{\eta}}(\underline{\mathfrak{B}}(\mathbb{K}))$ if there exist an isomorphism (ζ, σ) from the line diagram $\mathbb{D}_\eta(\underline{O})$ onto the line diagram $\mathbb{D}_{\bar{\eta}}(\underline{\mathfrak{B}}(\mathbb{K}))$ such that $\zeta\eta(\check{\gamma} g) = \bar{\eta}(\gamma g)$ for all $g \in G$ and $\zeta\eta(\check{\mu} m) = \bar{\eta}(\mu m)$ for all $m \in M$.

Lemma 2. *A finite bounded ordered set $\underline{O}$ is isomorphic to a finite concept lattice $\underline{\mathfrak{B}}(\mathbb{K})$ with $\mathbb{K} := (G, M, I)$ if and only if a corresponding $(\nu G, \nu M)$-labelled line diagram $\mathbb{D}^\nu_\eta(\underline{O})$ is isomorphic to a $(\nu G, \nu M)$-labelled line diagram $\mathbb{D}^\nu_{\bar{\eta}}(\underline{\mathfrak{B}}(\mathbb{K}))$.*

Proof: By Lemma 1, $(\zeta, \sigma) : \mathbb{D}_\eta(\underline{O}) \to \mathbb{D}_{\bar{\eta}}(\underline{\mathfrak{B}}(\mathbb{K}))$ is an isomorphism if and only if $(\bar{\eta}^{-1})\zeta\eta : \underline{O} \to \underline{\mathfrak{B}}(\mathbb{K})$ is an isomorphism. Additionally, $(\zeta, \sigma) : \mathbb{D}^\nu_\eta(\underline{O}) \to \mathbb{D}^\nu_{\bar{\eta}}(\underline{\mathfrak{B}}(\mathbb{K}))$ is an isomorphism if and only if $(\bar{\eta}^{-1})\zeta\eta : \underline{O} \to \underline{\mathfrak{B}}(\mathbb{K})$ is an isomorphism and $\zeta\eta(\check{\gamma}\nu^{-1}(\nu g)) = \bar{\eta}(\gamma\nu^{-1}(\nu g))$ for all $g \in G$ and $\zeta\eta(\check{\mu}\nu^{-1}(\nu m)) = \bar{\eta}(\mu\nu^{-1}(\nu m))$ for all $m \in M$.

Basic Theorem on Labelled Line Diagrams of Finite Concept Lattices. *Let $\underline{\mathfrak{B}}(\mathbb{K})$ be the concept lattice of a finite context $\mathbb{K} := (G, M, I)$ and let $\underline{O} := (O, \leq)$ be a finite bounded ordered set with mappings $\check{\gamma} : G \to \underline{O}$ and $\check{\mu} : M \to \underline{O}$. Then, a $(\nu G, \nu M)$-labelled line diagram $\mathbb{D}^\nu_\eta(\underline{O})$ of the ordered set $\underline{O}$ is isomorphic to a $(\nu G, \nu M)$-labelled line diagram $\mathbb{D}^\nu_{\bar{\eta}}(\underline{\mathfrak{B}}(\mathbb{K}))$ of the concept lattice $\underline{\mathfrak{B}}(\mathbb{K})$ if and only if, in $\mathbb{D}^\nu_\eta(\underline{O})$,*

1. *each circle, having exactly one line segment downwards, is labelled (from below) by at least one object name out of νG,*
2. *each circle, having exactly one line segment upwards, is labelled (from above) by at least one attribute name out of νM,*
3. *a circle labelled by an object name out of νG is linked up by an ascending chain of line segments to a circle labelled by an attribute name out of νM, or those labelled circles are identical, if and only if the named object has the named attribute,*
4. *there exists an injection $\zeta : C_{\underline{\mathfrak{B}}(\mathbb{K})} \to C_{\underline{O}}$ such that, for each circle $\bar{c}$ in the diagram $\mathbb{D}^\nu_{\bar{\eta}}(\underline{\mathfrak{B}}(\mathbb{K}))$, $\zeta(\bar{c})$ represents a minimal upper bound of $\{\check{\gamma}g \mid g \in G \text{ with } \gamma g \leq \bar{\eta}^{-1}\bar{c}\}$ which is also a maximal lower bound of $\{\check{\mu}m \mid m \in M \text{ with } \mu m \geq \bar{\eta}^{-1}\bar{c}\}$,*
5. *the number of all circles of $\mathbb{D}^\nu_\eta(\underline{O})$ equals the number of all circles of $\mathbb{D}^\nu_{\bar{\eta}}(\underline{\mathfrak{B}}(\mathbb{K}))$,*
6. *the number of all line segments of $\mathbb{D}^\nu_\eta(\underline{O})$ equals the number of all line segments of $\mathbb{D}^\nu_{\bar{\eta}}(\underline{\mathfrak{B}}(\mathbb{K}))$.*

Proof: Let us first assume that $\underline{O}$ is a finite lattice $\underline{L}$. Then, by the Basic Theorem on Finite Concept Lattices, $\underline{L}$ is isomorphic to $\underline{\mathfrak{B}}(\mathbb{K})$ if and only if there exist mappings $\tilde{\gamma} : G \to \underline{L}$ and $\tilde{\mu} : M \to \underline{L}$ such that 1. $\tilde{\gamma}G \supseteq J(\underline{L})$, 2. $\tilde{\mu}M \supseteq M(\underline{L})$, and 3. $gIm \iff \tilde{\gamma}g \leq \tilde{\mu}m$ for $g \in G$ and $m \in M$. Since, by Lemma 2, these conditions 1, 2, 3 are equivalent to the conditions 1, 2, 3 of the Theorem above, respectively, it follows from the Basic Theorem on Finite Concept Lattices that $\underline{L} \cong \underline{\mathfrak{B}}(\mathbb{K})$ if and only if the conditions 1, 2, 3 of the Theorem above are valid.

Now, let $\underline{O}$ be a finite bounded ordered set with a $(\nu G, \nu M)$-labelled line diagram $\mathbb{D}^\nu_\eta(\underline{O})$ satisfying the conditions 1 to 6 above. Condition 4 yields the existence of an injection $\zeta : C_{\underline{\mathfrak{B}}(\mathbb{K})} \to C_{\underline{O}}$ defined by $\zeta(\bar{c}) := c$ where the circle c represents a minimal upper bound of $\{\check{\gamma}g \mid g \in G \text{ with } \gamma g \leq \bar{\eta}^{-1}\bar{c}\}$ which is also a maximal lower bound of $\{\check{\mu}m \mid m \in M \text{ with } \mu m \geq \bar{\eta}^{-1}\bar{c}\}$. By condition 5, ζ is even a bijection. Condition 6 yields the existence of a bijection $\sigma : S_{\underline{\mathfrak{B}}(\mathbb{K})} \to S_{\underline{O}}$ such that $(\bar{c}_1, \bar{s}, \bar{c}_2) \in T_{\underline{\mathfrak{B}}(\mathbb{K})} \iff (\zeta(\bar{c}_1), \sigma(\bar{s}), \zeta(\bar{c}_2)) \in T_{\underline{O}}$. By Lemma 1, $\underline{O}$ and $\underline{\mathfrak{B}}(\mathbb{K})$ are isomorphic as ordered sets. Therefore $\underline{O}$ is a finite lattice for which the

	a	b	c	d
1	X	X		X
2	X	X	X	
3		X		
4	X			

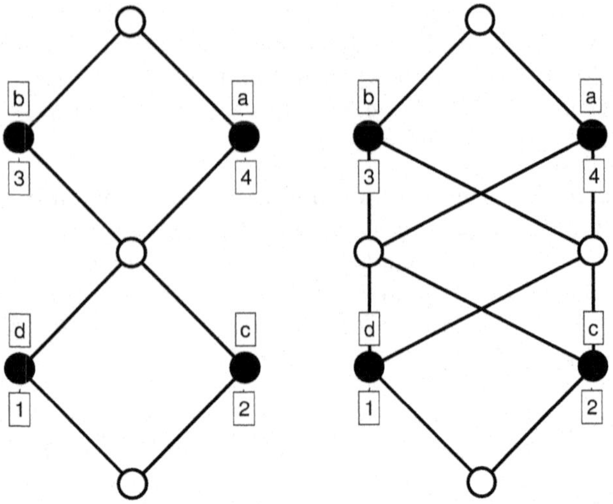

Fig. 1. A formal context, the labelled line diagram of its concept lattice, and a line diagram with one circle and four line segments more

	a	b	c	d	e	f
1	X	X	X	X	X	
2	X	X	X	X		X
3	X	X			X	
4	X	X				X
5	X		X			
6		X		X		

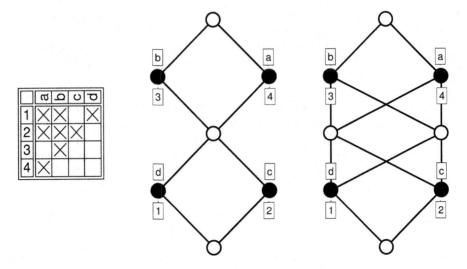

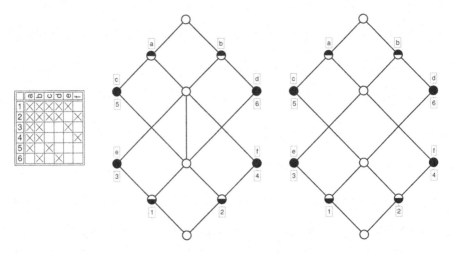

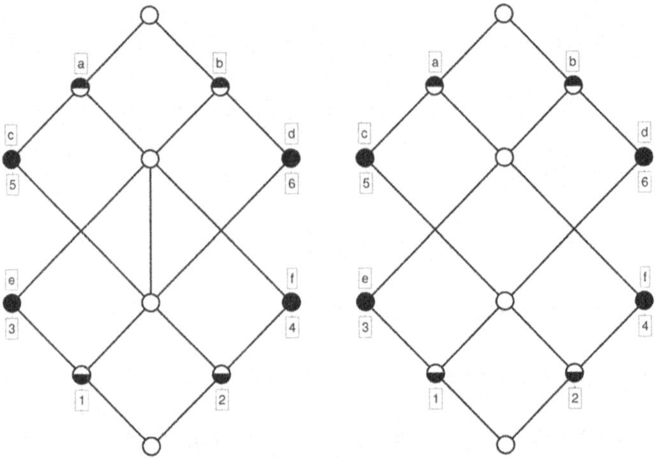

Fig. 2. A formal context, the labelled line diagram of its concept lattice, and a line diagram with the same number of circles, but with one line segment less

isomorphism of $\mathbb{D}^{\nu}_{\eta}(\underline{O})$ and $\mathbb{D}^{\nu}_{\bar{\eta}}(\underline{\mathfrak{B}}(\mathbb{K}))$ has already been proven in the preceding paragraph.

The conditions 5 and 6 are necessary for obtaining ζ and σ as bijections. That can be demonstrated by the examples in Fig.1 and Fig.2. The bounded ordered set on the right of Fig.1 satisfies the conditions 1 to 4, but not 5 and 6, and the bounded ordered set on the right of Fig.2 satisfies the conditions 1 to 5, but not 6.

5 CAPESSIMUS - A Game of Conceiving Concepts

CAPESSIMUS has been developed as a game of conceiving concepts and concept hierarchies. In particular, players of the game are supposed to learn and to understand the reading and drawing of labelled line diagrams of finite concept lattices. The basic task of the game is to complete incomplete labelled line diagrams. For each single task, the player obtains as *pre-information*

- a data table representing a formal context,
- a labelled line diagram of the concept lattice of that context in which all line segments are deleted except those line segments touching the lowest circle or the highest circle, and
- the number of line segments needed to complete the line diagram.[1]

Fig.3 shows a simple example of such pre-information. Since the "Canadian Tower" has the attribute "high", a line segment should be drawn between the circles labelled with "Canadian Tower" and labelled with "high". Similarly, line

	very high	high	broad	very broad
Canadian Tower	X	X		
Petronas Twin Tower	X	X	X	
Golden-Gate-Bridge		X	X	X
Pontchartrain-Bridge			X	X

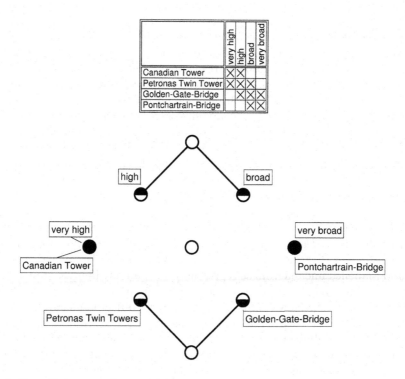

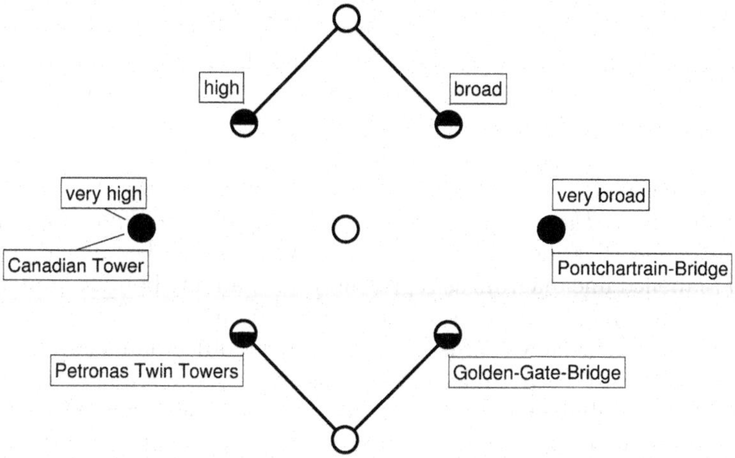

Fig. 3. Formal context about architectures and an incomplete line diagram of that context, which has to be completed by 8 line segments

[1] This number can, for instance, be obtained from the successor list in Burmeister's program *ConImp* [Bu00].

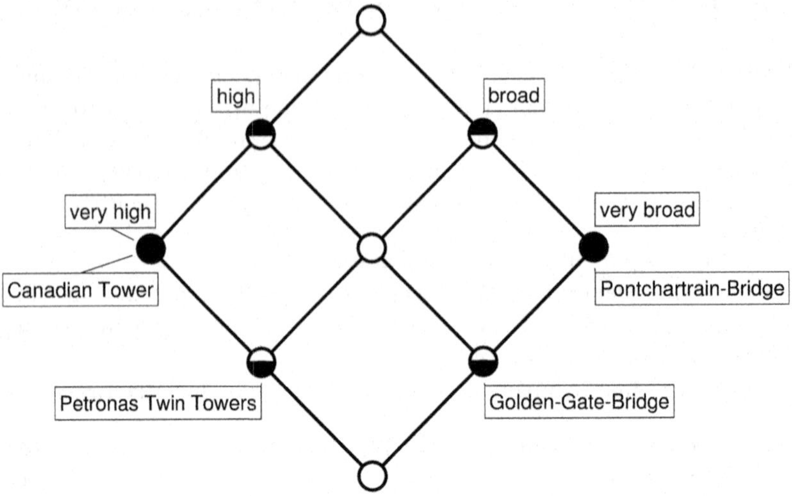

Fig. 4. Labelled line diagram of the concept lattice of the context in Fig.3

segments should be drawn between the circles labelled with "Petrona Twin Towers" and "very high", with "Golden-Gate-Bridge" and "very broad", and with "Pontchartrain-Bridge" and "broad", respectively. Furthermore, the circle of the "Petrona Twin Towers" should also be linked to the circle of the attribute "broad", and the circle of the "Golden-Gate-Bridge" should also be linked to the circle of the attribute "high". Since $\mathfrak{b}:=$ ({Pontchartrain-Bridge,Golden-Gate-Bridge},{high, broad}) is a formal concept of the given context, one has to add four line segments between the circle representing $\mathfrak{b}$ and the circle of the "Petrona Twin Towers", the "Golden-Gate-Bridge", the attribute "high", and the attribute "broad", respectively. In total, eight line segments have to be added to complete the incomplete line diagram, which yields the diagram shown in Fig.4.

The question is, of course, whether the completed labelled line diagram represents the concept lattice of the given context or not. An answer can be given by applying the *Basic Theorem on Labelled Line Diagrams of Finite Concept Lattices.* This means that the conditions 1 to 6 have to be checked in the presented $(\nu G, \nu M)$-labelled line diagram as representation of a finite bounded ordered set. In the completed diagram, *condition 1* is satisfied because each circle, having exactly one line segment downwards, is labelled by an object name, and *condition 2* is satisfied because each circle, having exactly one line segment upwards, is labelled by an attribute name. *Condition 3* can be easily checked by inspecting each cell of the context where a (non-)cross means that the corresponding object circle is (not) linked to the corresponding attribute circle. *Condition 5* is always satisfied because the pre-information yield that the circles are in one-to-one correspondence to the concepts of the concept lattice of the given context.

	C-major	d-minor	e-minor	F-major	G-major	a-minor
c	X			X		X
d		X			X	
e	X		X			X
f		X		X		
g	X		X		X	
a		X		X		X
b			X		X	

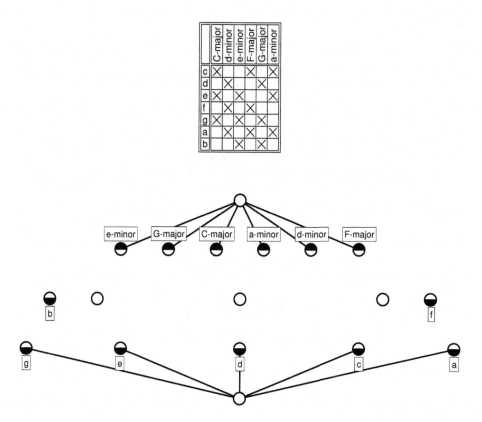

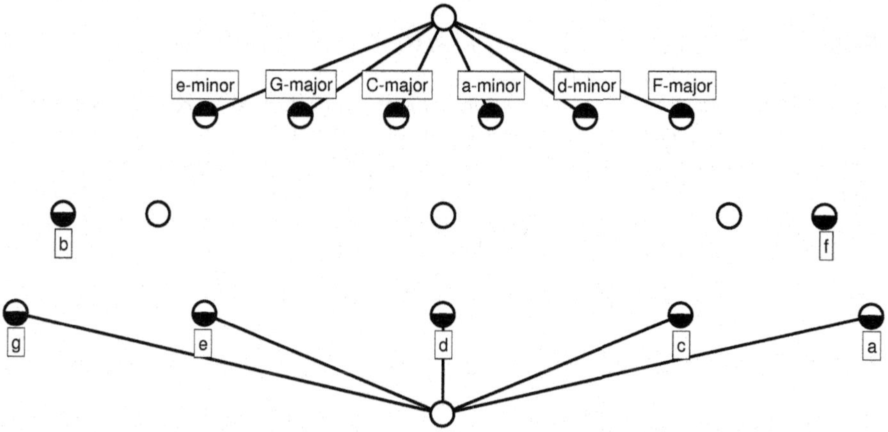

Fig. 5. Formal context about the major and minor 3-harmonies in C-major diatonicism and an incomplete labelled line diagram to be completed by 20 line segments

Condition 6 is valid if and only if the number of added line segments equals the number listed in the pre-information; for the example above this number is 8. Finally, we are able to check *condition 4* because the one-to-one correspondence of condition 5 can be viewed as the identity on the set of all circles of the presented line diagram and the one-to-one correspondence of condition 6 can be viewed as the identity on the set of all line segments of the presented line diagram. This together with the pre-information yields that condition 4 is always valid.

We can summerize that testing the *correctness of a labelled line diagram* in the game CAPESSIMUS consists of checking the conditions 1, 2, 3 and the proposed number of added line segments. First experiences with the game CAPESSIMUS have shown that it is quite attractive and instructive to practice the game. To taste this, the reader should try to work out the exercise in Fig.5 (the solution can be found in [Wi05], p.11).

References

[DP02] B. Davey, H. Priestley: *Introduction to Lattices and Order*. Second Edition. Cambridge University Press, Cambridge 2002.

[Bu00] P. Burmeister: ConImp - Ein Programm zur Formalen Begriffsanalyse. In: G. Stumme, R. Wille (Hrsg.): *Begriffliche Wissensverarbeitung: Methoden und Anwendungen*. Springer, 2000, 25–56.

[GW99] B. Ganter, R. Wille: *Formal Concept Analysis: mathematical foundations*. Springer, Heidelberg 1999.

[GW06] P. Gehring, R. Wille: Semantology: basic methods for knowledge representations. In: H. Schärfe, Pascal Hitzler, Peter Øhrstrøm (eds.): *Conceptual structures: inspiration and application*. LNAI **4068**. Springer, Heidelberg 2006, 215–228.

[HW06] M. Helmerich, R. Wille: *CAPESSIMUS - Ein Denkspiel zum Begreifen von Begriffen* (in preparation)

[Pe92] Ch. S. Peirce: *Reasoning and the logic of things*. Edited by K. L. Ketner; with an introduction by K. L. Ketner and H. Putnam. Havard University Press, Cambridge 1992.

[Wi05] R. Wille: Mathematik präsentieren, reflektieren, beurteilen. In: K. Lengnink, F. Siebel (Hrsg.): *Mathematik präsentieren, reflektieren, beurteilen*. Verlag Allgemeine Wissenschaft, Mühltal 2005, 3-19.

Bipartite Ferrers-Graphs and Planar Concept Lattices

Christian Zschalig

Institut für Algebra, TU Dresden, Germany
Christian.Zschalig@tu-dresden.de

Abstract. There exists a close relation between the Ferrers-dimension of a context and the order dimension of the appropriate concept lattice [4]. Based on this fact we will introduce *Ferrers-Graphs* on contexts and show how they characterize planar concept lattices.

1 Introduction

In this work we will show how to decide whether a concept lattice is planar out of their contexts. We will introduce *Ferrers-graphs* [6] for that purpose. They are indicating which elements of the cross table are not in the same *Ferrers-relation* [4]. In case of this graph being bipartite we introduce a *left-relation on the context* based on its vertex classes. This relation will be used to define a *left-relation on the concept lattice* [7]. Since this relation is a strict order, the lattice is planar.

2 Preliminaries

All contexts considered in this work will be finite and reduced.

2.1 Ferrers-Graphs

In this part we want to remind some basics about Ferrers-relations and the Ferrers-dimension. We will introduce the notion of Ferrers-graphs and state the conjecture that will be proven in the course of that work.

Definition 1. *[4] A* Ferrers-relation F *is a relation* $F \subseteq A \times B$ *with*

$$a_1 F b_1 \wedge a_2 F b_2 \quad \Longrightarrow \quad a_1 F b_2 \vee a_2 F b_1.$$

The Ferrers-dimension $fdim(\mathbb{K})$ *of a context* $\mathbb{K} = (G, M, I)$ *is the smallest number of Ferrers-Relations* $F_t \subseteq G \times M$, $t \in T$, *whose intersection is equal to* I, *i.e.* $I = \bigcap_{t \in T} F_t$.

In a cross table representing a context $\mathbb{K} = (G, M, I)$ we notice that I is a Ferrers-relation if and only if the configuration depicted on the right does not occur.

	m_1	m_2
g_1	×	
g_2		×

S.O. Kuznetsov and S. Schmidt (Eds.): ICFCA 2007, LNAI 4390, pp. 313–327, 2007.

The inverse $\overline{F}$ of a Ferrers-relation is again a Ferrers-relation. Hence this Ferrers-dimension of a context $\mathbb{K} = (G, M, I)$ is the smallest number of Ferrers-relations covering the empty cells of its cross table [4], i.e. $\overline{I} := (G \times M) \setminus I = \bigcup_{t \in T} F_t$. The following theorem gives a connection to the *order dimension* of a lattice.

Theorem 1. *[4] Let* $\mathbb{K}$ *be a context. Then* $fdim(\mathbb{K}) = dim(\mathfrak{B}(\mathbb{K}))$.

We know that a lattice is planar if and only if its order dimension is at most 2 (see Theorem 2). Hence the result already gives a nice characterization of contexts possessing planar concept lattices. Unfortunately the calculation of the Ferrers-dimension in general is $\mathcal{NP}$-complete [4].

Now we will introduce our notion of a *Ferrers-graph*. Its nodes are the empty cells of a context and its edges indicate which vertices can not belong to the same Ferrers-relation $\overline{F}$.

Definition 2. *[6] Let* $R \subseteq A \times B$ *be a relation. We define the* Ferrers-graph $\Gamma(R)$ *as follows:*

$$V(\Gamma(R)) := \overline{R} \qquad E(\Gamma(R)) := \{\{(a_1, b_2), (a_2, b_1)\} \mid (a_1, b_1), (a_2, b_2) \in R\}.$$

Let $\chi(\Gamma(I))$ is the chromatic number of $\Gamma(I)$. There is a conjecture claiming that

$$fdim(\mathbb{K}) = r \iff \chi(\Gamma(I)) = r$$

We will show in this work that this assertion holds for $r = 2$. It is easy to see that the first statement implies the second:

Lemma 1. *[5] For a context* $\mathbb{K} = (G, M, I)$ *the following implication holds:*

$$fdim(\mathbb{K}) = 2 \implies \Gamma(I) \text{ is bipartite.}$$

Proof. Since $fdim(\mathbb{K}) = 2$ there exist two Ferrers-relations F_1 and F_2 with $F_1 \cup F_2 = \overline{I} = V(\Gamma(I))$. Let (g_1, m_1) and (g_2, m_2) be elements of F_1. By Definition 1 we notice $g_1 \not I\, m_2$ or $g_2 \not I\, m_1$, i.e. $\{(g_1, m_1), (g_2, m_2)\} \notin E(\Gamma(I))$. Analogously we conclude that there exist no edges between elements of F_2. Hence $\Gamma(I)$ is bipartite with the vertex classes F_1 and $F_2 \setminus F_1$. □

2.2 Conjugate Orders

Conjugate orders are a powerful tool to characterize planar lattices and ordered sets. Here we need that notion for introducing left-relations which are more convenient for our purpose.

Definition 3. *[3] Let* $\underline{P} = (P, \leq)$ *be an ordered set. The incomparability relation in* $\underline{P}$ *is denoted by* $\|$.

1. *We call* L_c conjugate relation *if* $L_c \cup {L_c}^{-1} = \|$.
2. *We call* L_c conjugate order *if additionally* L_c *is a strict order.*

With the help of conjugate orders we can characterize planar lattices since we have:

Theorem 2. *[1,2,3] Let $\mathbb{K}$ be a context. Then the following are equivalent:*

1. $\underline{\mathfrak{B}}(\mathbb{K})$ *is planar.*
2. *There exists a conjugate order on* $\mathfrak{B}(\mathbb{K})$.
3. $dim(\underline{\mathfrak{B}}(\mathbb{K})) \leq 2$.

2.3 Left-Relations

This section will remind some properties of left-relations. They are closely related to conjugate orders in case they are (strict) orders. Since a left-relation is determined by a much smaller sorting relation acting only on some attributes instead of all lattice elements, it is of better use for our issue.

A *sorting relation* on a lattice is just the union of strict linear orders on incomparable $\bigwedge$-irreducibles sharing a common upper neighbour:

Definition 4. *[7] Let $\underline{\mathfrak{V}}$ be a finite lattice and M be the set of its $\bigwedge$-irreducible elements. A strict order $L_a \subseteq M \times M$ is called* sorting relation *if the following condition*[1] *holds for all elements $m, n \in M$:*

$$m^* = n^* \iff m\, L_a\, n \text{ or } n\, L_a\, m.$$

Based on the last definition we introduce *left-relations* on lattices. They extend a sorting relation to lattice elements below the appropriate $\bigwedge$-irreducibles.

Definition 5. *[7] Let $\underline{\mathfrak{V}}$ be a finite lattice with a given sorting relation L_a. For arbitrary lattice elements v and w, we define*

$$M(v,w) = \{(v',w') \subseteq M \times M \mid v \leq v', w \leq w', v \parallel w', w \parallel v'\}.$$

We define the relation $L \subseteq \mathfrak{V} \times \mathfrak{V}$ according to:

$$v\, L\, w :\iff \begin{cases} v\, L_a\, w, & v, w \in M, v^* = w^* \\ \exists (m,n) \in M(v,w) : m\, L\, n, & \text{else} \end{cases}$$

L is called left-relation *and $R := L^{-1}$ is called* right-relation *on the lattice $\underline{\mathfrak{V}}$.*

Proposition 1. *[7] Let L be a relation on a finite lattice $\underline{\mathfrak{V}}$. Then the below-mentioned statements are equivalent:*

1. *L is a conjugate order.*
2. *L is a left-relation and a strict order.*

Finally we want to remind the first planartiy condition. It is a necessary and sufficient condition for a lattice to be planar. It acts on incomparable $\bigwedge$-irreducibles only and will be needed later in the proof. An illustration is given in Figure 1.

[1] With m^* we denote the unique upperneighbour of an $\bigwedge$-irreducible m.

Definition 6. *[8] A conjugate relation R on a lattice $\underline{\mathfrak{V}}$ fulfills the* first planarity condition (FPC) *if*

$$m_i \, R \, m_k \, R \, m_j \implies m_k > (m_i \wedge m_j)$$

holds for all $\bigwedge$-irreducibles $m_i, m_k, m_j \in M$.

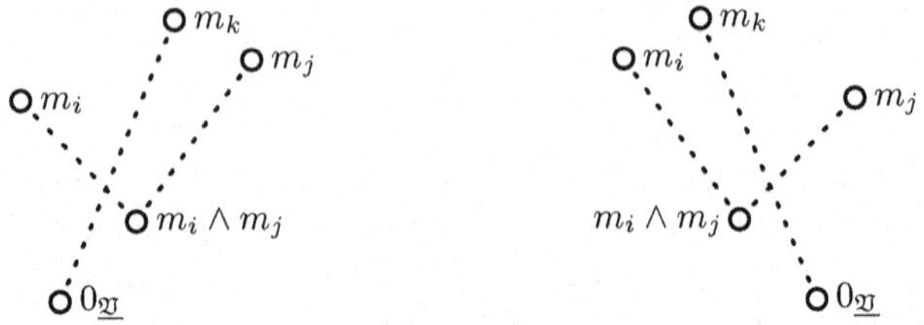

Fig. 1. When considering a diagram of a lattice, the necessity of the FPC is obvious for its planarity: If $m_i \, L \, m_k \, L \, m_j$ or $m_j \, L \, m_k \, L \, m_i$ holds then also $m_k > (m_i \wedge m_j)$. Otherwise every chain of diagram edges from m_k to the bottom element of the lattice intersects with a chain of edges from either m_i or m_j to $m_i \wedge m_j$ (see [7] and [8]).

Proposition 2. *[8] Let L be a left-relation on a lattice $\underline{\mathfrak{V}}$, then the following equivalence holds:*

$$L \text{ satisfies the FPC} \iff L \text{ is a conjugate order.}$$

3 Left-Relations on Contexts

Now we want to give a connection between the vertex classes F_1 and F_2 of the bipartite Ferrers-graph of a context $\mathbb{K}$ and a left relation on the concept lattice $\underline{\mathfrak{B}}(\mathbb{K})$. This is done by simply reinterpreting the vertices of the graph. Instead of "(g, m) is in vertex class F_1" we read "g is left of m", respectively for F_2 and right. In this way we define a left-relation $\mathfrak{L}$ on the context $\mathbb{K}$ which can be extended to a left-relation $\hat{L}$ on the concept lattice $\underline{\mathfrak{B}}(\mathbb{K})$.

Definition 7. *Let $\mathbb{K} = (G, M, I)$ be a context.*

1. *An object $g \in G$ and an attribute $m \in M$ are* incomparable (denoted by *$g \parallel_{\mathbb{K}} m$), if the respective concepts are, i.e. if the inequality $\gamma g \parallel \mu m$ holds.*
2. *The relation $\parallel_{\mathbb{K}} \subseteq G \times M$ is called* incomparability relation in $\mathbb{K}$.
3. *A relation $\mathfrak{L} \subseteq \parallel_{\mathbb{K}}$ is called* left-relation on $\mathbb{K}$.
4. *We denote $\mathfrak{R} := \parallel_{\mathbb{K}} \setminus \mathfrak{L}$.*

Two relations $\mathfrak{L}, \mathfrak{R} \subseteq \overline{I}$ on a context $\mathbb{K} = (G, M, I)$ induce a relation $\tilde{L}$ on $M \times M$ by

$$m \, \tilde{L} \, n :\iff \mu m \parallel \mu n \text{ and } (\exists g \in G : gIm, g \, \mathfrak{L} \, n \text{ or } g \, \mathfrak{R} \, m, gIn).$$

Let $\Gamma(I)$ be the bipartite Ferrers-graph of a context $\mathbb{K} = (G, M, I)$. Let the vertex classes be denoted by $\mathfrak{L}$ and $\mathfrak{R}$. Then $\mathfrak{L} \setminus \{(g, m) \mid \gamma g > \mu m\}$ is a left-relation on $\mathbb{K}$ and $\mathfrak{L} \cup \mathfrak{R} = \overline{I}$. If $\Gamma(I)$ consists of components $\Gamma_j(I), j \in J$ we partition the vertex classes $\mathfrak{L}$ and $\mathfrak{R}$ by $\mathfrak{L}_j := \mathfrak{L} \cap V(\Gamma_j(I))$ and $\mathfrak{R}_j := \mathfrak{R} \cap V(\Gamma_j(I))$. We introduce induced relations $\tilde{L}_j, j \in J$ by

$$m\ \tilde{L}_j\ n \iff \mu m \parallel \mu n \text{ and } (\exists g \in G : gIm, g\ \mathfrak{L}_j\ n \text{ or } g\ \mathfrak{R}_j\ m, gIn).$$

We observe that for a bipartite graph $\Gamma(I)$ and attribute concepts $\mu m \parallel \mu n$ the following equivalence holds:

$$\exists g \in G : gIm, g\ \mathfrak{L}_j\ n \iff \exists h \in G : h\ \mathfrak{R}_j\ m, hIn.$$

The induced relation $\tilde{L}$ is the key to our proof. We show in Lemma 2 that $\tilde{L}$ "almost" contains a sorting relation L_a. In Lemma 3 we prove that $\tilde{L}$ can be extended to a conjugate order (implying that $\underline{\mathfrak{B}}(\mathbb{K})$ planar) if it is transitive.

Lemma 2. *Let $\mathbb{K} = (G, M, I)$ be a reduced context and its Ferrers-graph $\Gamma(I)$ be bipartite with vertex classes $\mathfrak{L}$ and $\mathfrak{R}$. Let $\tilde{L}$ be the relation induced by $\mathfrak{L}$. Then $\tilde{L}$ is asymmetric and connex on pairs of incomparable attributes.*

Proof.

1. Let m and n be attributes fulfilling $m\ \tilde{L}\ n$ and $n\ \tilde{L}\ m$. In a respective cross table one of the cases depicted on the right occurs. The left contradicts the fact that $\mathfrak{L}$ is a vertex class of $\Gamma(I)$. The right one does as well since there exists an object $g \in G$ with $\gamma g \parallel \mu m$ and gIn (by definition of $\tilde{L}$ we know $\mu m \parallel \mu n$), both $g\ \mathfrak{L}\ m$ and $g\ \mathfrak{R}\ m$ are contradictions.

m	n
×	$\mathfrak{L}$
$\mathfrak{L}$	×

m	n
×	$\mathfrak{L}$
×	$\mathfrak{R}$

2. Connexity is obvious: Since $\mu m \parallel \mu n$ we find an object g with gIm and $g \not{I}\ n$, i.e. $(g, n) \in \mathfrak{L} \cup \mathfrak{R}$. □

Lemma 3. *Let $\mathbb{K} = (G, M, I)$ be a reduced context and its Ferrers-graph $\Gamma(I)$ be bipartite with vertex classes $\mathfrak{L}$ and $\mathfrak{R}$. Let $\tilde{L}$ be the relation induced by $\mathfrak{L}$. Then the following implication holds:*

$$\tilde{L} \text{ is transitive} \implies \underline{\mathfrak{B}}(\mathbb{K}) \text{ is planar.}$$

Proof. By Lemma 2 we know that $\tilde{L}$ is asymmetric and connex on incomparable pairs of attributes. Since $\tilde{L}$ is also transitive we conclude that $\tilde{L}$ is a strict order on attribute sets whose concepts possess the same upper neighbour. Hence there exists a sorting relation $L_a \subseteq \tilde{L}$. Let L be the (unique) left-relation induced by L_a.

1. We show $\tilde{L} \subseteq L$. Let $m_1\ \tilde{L}\ n_1$. According to Definition 5 we have to show that there exists a pair of attributes $(m_2, n_2) \in M(m_1, n_1)$ with $m_2\ \tilde{L}\ n_2$. Since $m_1\ \tilde{L}\ n_1$ there exists an object g fulfilling gIm_1 and $g\tilde{L}n_1$. From $m_1 \leq m_2$ we conclude gIm_2. Consider an object h with hIn_1 and $h \not{I}\ m_2$ (which exists because $\mu n_1 \parallel \mu m_2$). Then hIn_2 and since $\Gamma(I)$ is bipartite we conclude $h\ \mathfrak{R}\ m_2$, i.e. $m_2\ \tilde{L}\ n_2$.

	m_1	n_1	m_2	n_2
g	×	$\mathfrak{L}$	×	
h		×	$\mathfrak{R}$	×

2. We show that L is a strict order. We know with Lemma 2 and 1. that $m_1 \, L \, m_2 \iff m_1 \, \tilde{L} \, m_2$. By applying the FPC we need to prove only

$$m_1 \, \tilde{L} \, m_2 \, \tilde{L} \, m_3 \implies \mu m_2 > (\mu m_1 \wedge \mu m_3)$$

Let $m_1 \, \tilde{L} \, m_2 \, \tilde{L} \, m_3$. Hence there exist objects g_1 and g_2 fulfilling $g_1 I m_2$, $g_2 I m_2$, $g_1 \, \mathfrak{R} \, m_1$ and $g_2 \, \mathfrak{L} \, m_3$. Let g_3 be an object possessing m_1 and m_3. (If such an object does not exist, the infimum of m_1 and m_3 is the bottom element of $\underline{\mathfrak{B}}(\mathbb{K})$ and μm_2 trivially greater.)
Now $\gamma g_3 > \mu m_2$ contradicts our assertion $\mu m_1 \parallel \mu m_2$ and both $g_3 \, \mathfrak{L} \, m_2$ and $g_3 \, \mathfrak{R} \, m_2$ contradict the fact that $\mathfrak{L}$ and $\mathfrak{R}$ are vertex classes of the bipartite Graph $\Gamma(I)$. We conclude $g_3 I m_2$, i.e. $\mu m_2 > \mu m_1 \wedge \mu m_3$.

	m_1	m_2	m_3
g_1	$\mathfrak{R}$	×	
g_2		×	$\mathfrak{L}$
g_3	×	?	×

By the use of $\mathfrak{L}$ we constructed a left-relation L which is a strict order. With Propostion 1 and Theorem 2 we conclude that $\underline{\mathfrak{B}}(\mathbb{K})$ is planar. □

With the previous Lemmas 2 and 3 we notice that we are "nearly finished": From the vertex classes $\mathfrak{L}$ and $\mathfrak{R}$ of $\Gamma(I)$ we constructed via the relation $\tilde{L}$ a left-relation L which is a strict order meaning that the concept lattice $\underline{\mathfrak{B}}(\mathbb{K})$ is planar. The only requirement we had to meet was the transitivity of $\tilde{L}$. Unfortunately it is not possible to prove this assertion as straightforward as the others.

Let us consider the example in Figure 2. On the left you can see the context for the lattice M_3. We notice that its Ferrers-graph (depicted in the middle) is bipartite and consists of three components. However, the induced relation $\tilde{L}$ is not transitive. However, we can "flip" for instance the component in the middle to make $\tilde{L}$ transitive. This will be our strategy in the remaining of the proof:

If $\tilde{L}$ is not transitive, $\Gamma(I)$ consists of at least three components which can be "turned around smartly". This will keep $\Gamma(I)$ bipartite and the induced relation becomes transitive.

	m_1	m_2	m_3
g_1	×	$\mathfrak{L}$	$\mathfrak{R}$
g_2	$\mathfrak{R}$	×	$\mathfrak{L}$
g_3	$\mathfrak{L}$	$\mathfrak{R}$	×

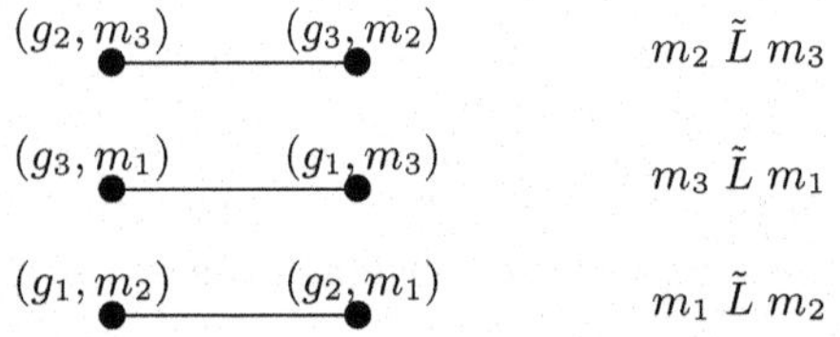

Fig. 2. A construction of a bipartite graph $\Gamma(I)$ with vertex classes $\mathfrak{L}$ on the left and $\mathfrak{R}$ on the right from a context $\mathbb{K}$. The induced relation $\tilde{L}$ is not transitive.

In the next section we will therefore observe how the components of $\Gamma(I)$ look like. In particular, we are interested which attributes are *connected*, i.e. we want to know, whether there is an edge sequence $(g_1, m_1)E \ldots E(g_2, m_2)$ in $\Gamma(I)$ for two attributes with incomparable concepts $\mu m_1 \parallel \mu m_2$ and appropriate objects g_1 and g_2.

4 The Components of Ferrers-Graphs

The next result gives a nice view on isolated vertices in $\Gamma(I)$. Additionally it explains the little ambiguity of the last section. There we used the symbols $\mathfrak{L}$ and $\mathfrak{R}$ both for the left-relation on a context and the vertex classes of $\Gamma(I)$. A pair (g,m) with $\gamma g > \mu m$ is included in $\overline{I}$, but not in $\|_{\mathbb{K}}$. The following lemma is stating that these are just the isolated vertices in $V(\Gamma(I))$, therefore we do not have to regard them in our subsequent considerations.

Proposition 3. *Let $\mathbb{K} = (G,M,I)$ be a reduced context and $\Gamma(I)$ its Ferrers-graph. A vertex (g,m) is isolated if and only if $\gamma g > \mu m$.*

Proof. $\Leftarrow$: Let $\gamma g > \mu m$. Let $h \in G$ be an object and $n \in M$ be an attribute s.t. hIm and gIn. We notice $\gamma h \leq \mu m < \gamma g \leq \mu n$, i.e. hIn. Hence, there is no edge between (g,m) and (h,n).
$\Rightarrow$: Since $(g,m) \in V(\Gamma)$ we know $g \not I m$.

1. If there is no object $h \in G$ with hIm then we conclude $\mu m = 0_{\underline{\mathfrak{V}}}$ and $\gamma g > \mu m$.
2. If there is no attribute $n \in M$ with gIn then we conclude $\gamma g = 1_{\underline{\mathfrak{V}}}$ and $\gamma g > \mu m$.
3. Otherwise the sets g' and m' are non-empty. All objects $h \in m'$ and all attributes $n \in g'$ meet the condition hIn, i.e. $\gamma h \leq \mu n$ since there is no edge between (g,m) and (h,n) in $\Gamma(I)$. We conclude $g'' \supseteq m' \uplus \{g\}$, i.e. $\mu m < \gamma g$.

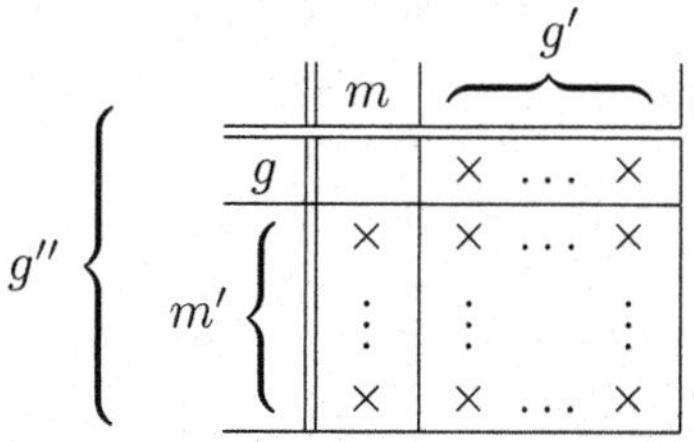

□

Now we want to observe which vertices of $\Gamma(I)$ are in the same component for a given context $\mathbb{K}$ n. It turns out that sequences of objects and attributes of the form

$$h_1 I n_1 I^{-1} h_2 I n_2 I^{-1} \dots I^{-1} n_r$$

play an important role for connectivity.

Definition 8. *Let $\mathbb{K} = (G,M,I)$ be a context and $[\underline{v}, \overline{v}]$ be an interval in $\underline{\mathfrak{B}}(\mathbb{K})$. A sequence $p = n_0, h_1, n_1, h_2, \dots, n_{r-1}, h_r, n_r$ of objects h_i and attributes n_i is called* connection of n_0 and n_r w.r.t. $[\underline{v}, \overline{v}]$ *if*

$$n_0 \notin Int(\overline{v}) \quad \wedge \quad \forall i \in \{1, \dots r\} : \; h_i I n_i, \; h_i I n_{i-1}, \; n_i \notin Int(\overline{v}), \; h_i \notin Ext(\underline{v})$$

If additionally the condition $h_i I n_j \implies j \in \{i, i-1\}$ holds for all $i \in \{1, \dots, r\}$ then p is called shortest connection.

It is obvious that every connection p between n_0 and n_r contains a shortest connection q. We can construct q in the following way: Start with $q := n_0$. If n_i is the last element of q, search for the highest index t with $h_t I n_i$ and add h_t to

q. If h_i is the last element of q, search for the highest index t with $h_i I n_t$ and add n_t to q. Stop after adding n_r.

A shortest connection between n_0 and n_r is represented by a subcontext as depicted on the right. The main diagonal and the secondary diagonal above are filled with crosses, all other cells are empty. It is easy to prove that every two of these empty cells are in the same component of $\Gamma(I)$ since a vertex $(h_i, n_j) \in V(\Gamma(I))$ with $i < j$ has the neighbours

	n_0	n_1	…	n_r
h_1	×	×	• … •	•
h_2	○	×	× •	•
⋮			⋱	⋮
h_{r-1}	○	○	×	•
h_r	○	○	×	×

$$N(h_i, n_j) = \{(h_j, n_{i-1}), (h_j, n_i), (h_{j+1}, n_{i-1}), (h_{j+1}, n_i)\} \cap V(\Gamma(I))$$

In particular, if $n_0 \tilde{L} n_r$ then we conclude $\begin{cases} h_i \mathfrak{R} n_j, & j < i, \\ h_i I n_j, & j \in \{i, i+1\}, \\ h_i \mathfrak{L} n_j, & j > i+1. \end{cases}$

Definition 9. *Let $\mathbb{K} = (G, M, I)$ be a context and $[\underline{v}, \overline{v}]$ be an interval in $\underline{\mathfrak{B}}(\mathbb{K})$.*

1. *An attribute m with $\mu m \in [\underline{v}, \overline{v}]$ is called* bound *if there exists an attribute n with $\mu n \notin [\underline{v}, \overline{v}]$ and a connection $p = m, \ldots, n$ in $[\underline{v}, \overline{v}]$.*
2. *Two attributes m, n with $\mu m, \mu n \in [\underline{v}, \overline{v}]$ are called* connected *if there exists a connection $p = m, \ldots, n$ in $[\underline{v}, \overline{v}]$.*
3. *We call the attributes m_1, m_2, m_3 (with pairwise incomparable concepts)* free triple *if none of them is bound and no two of them are connected in the interval $[\mu m_1 \wedge \mu m_2 \wedge \mu m_3, \mu m_1 \vee \mu m_2 \vee \mu m_3]$.*

In Figure 3 one can see how to imagine bounded and connected elements w.r.t. an interval $[\underline{v}, \overline{v}]$ and free triples in a diagram of a concept lattice.

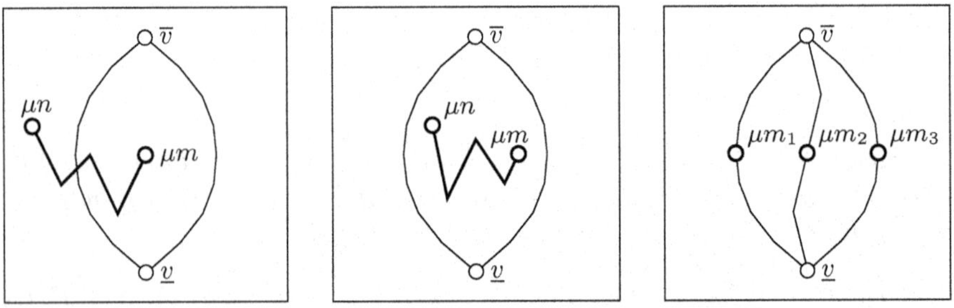

Fig. 3. Examples for Definition 9. In the left picture m is bound since there is a connection to an attribute n which is not in $[\underline{v}, \overline{v}]$. In the middle m and n are connected. On the right the three attributes m_1, m_2 and m_3 are a free triple. There are no edges between the three branches containing the appropriate attributs concepts.

Definition 10. *Let* $\mathbb{K} = (G, M, I)$ *be a context and* $[\underline{v}, \overline{v}]$ *be an interval in* $\underline{\mathfrak{B}}(\mathbb{K})$. *Let* $U(\underline{v}, \overline{v})$ *denote the set of attributes which are not bound in* $[\underline{v}, \overline{v}]$. *Let* $m \in U(\underline{v}, \overline{v})$. *Then we call the set*

$$U_m[\underline{v}, \overline{v}] := \{n \in M \mid m \text{ and } n \text{ are connected}\}$$

the m-component *of* $[\underline{v}, \overline{v}]$.

Obviously "connected" is an equivalence relation. Therefore the set of equivalence classes $\{U_m\}_{m \in M}$ is a partition of U.

Definition 11. *Let* $\mathbb{K} = (G, M, I)$ *be a context and* $\Gamma(I)$ *its Ferrers-graph. Let* m_1 *and* m_2 *be incomparable attributes. The* m_1, m_2-component of $\Gamma(I)$ *denoted by* $\Gamma(I)_{m_1,m_2}$ *is that component of* $\Gamma(I)$ *containing all edges between* m_1 *and* m_2, *i.e. all edges of the form* $\{(g_2, m_1), (g_1, m_2)\}$.

Remark 1. *1. Definition 11 is well defined since every two edges between* m_1 *and* m_2 *are in the same component of* $\Gamma(I)$. *Let* $(g_1, m_2)E(g_2, m_1)$ *and* $(g_3, m_2)E(g_4, m_1)$ *then* $g_2 I m_2$ *und* $g_3 I m_1$, *i.e. both edges are connected in* $\Gamma(I)$ *via the edge* $\{(g_2, m_1)(g_3, m_2)\}$.

2. We remind a basic result of lattice theory: Let $\underline{\mathfrak{B}}(\mathbb{K})$ *be a concept lattice and* $v \parallel w \in \mathfrak{B}(\mathbb{K})$ *then*

$$\exists g \in Ext(v) : \gamma g \parallel w \qquad \text{and} \qquad \exists m \in Int(v) : \mu m \parallel w$$

3. Let (m_1, m_2, m_3) *be a free triple then there exist objects* $g_1, g_2, g_3 \in G$ *with* $g_i I m_j \iff i = j$; *from* $g I m_i, g I m_j$ *we conclude* $g I m_k$, *otherwise* m_i *and* m_j *would be connected.*

4. Is there an object g *mit* $g I m_1$, $g I m_2$ *and* $g \not I m_3$ *then we conclude* $\Gamma_{m_1,m_3} = \Gamma_{m_2,m_3}$ *since we find objects* g_1 *and* g_2 *with* $g_1, g_2 \in m_3'$ *and* $g_1 \notin m_1'$, $g_2 \notin m_2'$, *i.e. a sequence of edges* $(g_1, m_1)E(g, m_3)E(g_2, m_2)$ *in* $\Gamma(I)$. *In particular,* $m_1 \tilde{L}_{13} m_3 \iff m_2 \tilde{L}_{13} m_3$.

	m_1	m_2	m_3
g	×	×	•
g_1	∘		×
g_2		∘	×

With the tools provided so far we can now treat the problem of transitivity. The next two lemmas claim that, for three attributes m_1, m_2 and m_3, the condition $m_1 \tilde{L} m_2 \tilde{L} m_3$ implies $m_1 \tilde{L} m_3$ if one of the attributes is bound or if two are connected. That means that only the free triples remain as "problematic cases".

Lemma 4. *Let* $\mathbb{K} = (G, M, I)$ *be a context and* $\Gamma(I)$ *its bipartite Ferrers-graph. Let* $m_1, m_2, m_3 \in M$ *be pairwise incomparable attributes, s.t.* m_1 *and* m_2 *are connected w.r.t.* $[\underline{v}, \overline{v}] := [\mu m_1 \wedge \mu m_2 \wedge \mu m_3, \mu m_1 \vee \mu m_2 \vee \mu m_3]$. *Let* $\tilde{L}_{ij}$ *be induced by* $\Gamma(I)_{m_i,m_j}$ $(i \neq j \in \{1, 2, 3\})$.

1. *If* m_1, m_3 *and* m_2, m_3 *are not connected in that interval then* $\tilde{L}_{13} = \tilde{L}_{23}$ *and* $m_1 \tilde{L}_{13} m_3 \iff m_2 \tilde{L}_{13} m_3$
2. *Otherwise one of the following cases occurs:* $m_1 \tilde{L}_{12} m_2 \implies m_1 \tilde{L}_{12} m_3$ *or* $m_1 \tilde{L}_{12} m_2 \implies m_3 \tilde{L}_{12} m_2$ *or* $m_1 \tilde{L}_{13} m_3 \iff m_2 \tilde{L}_{13} m_3$.

Both cases imply: $m_1 \tilde{L}_{12} m_2, m_2 \tilde{L}_{23} m_3 \implies m_1 \tilde{L}_{ij} m_3$, $\tilde{L}_{ij} \in \{\tilde{L}_{12}, \tilde{L}_{23}\}$.

Proof. Let $p = n_0, h_1, \ldots, h_r, n_r$ be a shortest connection between $n_0 := m_1$ and $n_r := m_2$. Let g_3 be an object possessing m_3 and not both m_1 and m_2.

1. Since m_1, m_3 and m_2, m_3 are not connected, we know that the conditions $h_s \nmid m_3$ and $g_3 \nmid n_s$ hold for all $s \in \{0, \ldots, r\}$. Then the following edge sequence is in $\Gamma(I)_{m_1,m_3}$.

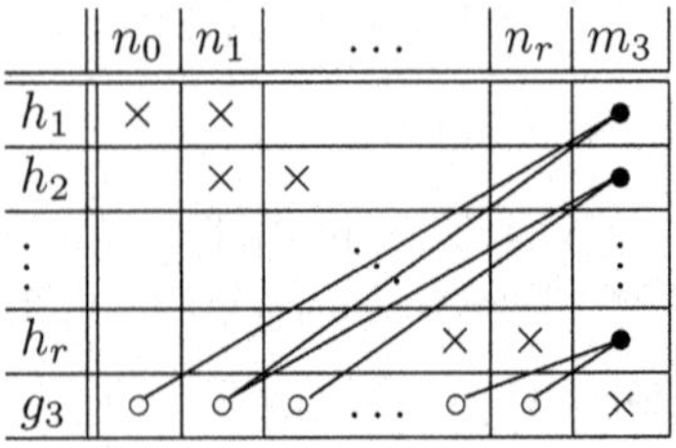

$$(g_3, n_0)E(h_1, m_3)E(g_3, n_1)E\ldots$$
$$E(h_r, m_3)E(g_3, n_r).$$

	n_0	n_1	$\ldots$	n_r	m_3
h_1	×	×			•
h_2		×	×		•
$\vdots$					$\vdots$
h_r				×	×	•
g_3	∘	∘	∘ $\ldots$ ∘	∘	×

We conclude $m_1 \tilde{L}_{13} m_3 \iff m_2 \tilde{L}_{13} m_3$ and $\tilde{L}_{13} = \tilde{L}_{23}$. Therefore we have $m_1 \tilde{L}_{12} m_2, m_2 \tilde{L}_{23} m_3 \implies m_1 \tilde{L}_{23} m_3$.

2. Let $p = m_1, \ldots, m_2$ be a connection as considered in 1. then we conclude in analogy $m_1 \tilde{L}_{13} m_3 \iff m_2 \tilde{L}_{13} m_3$.

Let $h_s I m_3$, $s < r$. Then $m_1 \tilde{L}_{12} m_2 \implies h_s \mathfrak{L}_{12} m_j \implies m_3 \tilde{L}_{12} m_2$.
Let $h_r I m_3$. Then $m_1 \tilde{L}_{12} m_2 \implies h_r \mathfrak{R}_{12} m_1 \implies m_1 \tilde{L}_{12} m_3$.
Let $g_3 I n_s, s < r - 1$. Then $m_1 \tilde{L}_{12} m_2 \implies h_r \mathfrak{R}_{12} n_s \implies g_3 \mathfrak{L}_{12} m_2 \implies m_3 \tilde{L}_{12} m_2$.
Let $g_3 I n_s, s > 1$. Then $m_1 \tilde{L}_{12} m_2 \implies h_1 \mathfrak{L}_{12} n_s \implies g_3 \mathfrak{R}_{12} m_i \implies m_1 \tilde{L}_{12} m_3$.
Let $g_3 I n_s$ and $r - 1 \leq s \leq 1$. We have either $r = 1$, i.e. $h_1 I m_2$ and hence $m_1 \tilde{L}_{13} m_3 \iff m_2 \tilde{L}_{13} m_3$ (see Remark 1 4.) or $r = 2$ and $s = 1$.

Since n_1 is part of the connection p and therefore $\mu n_1 \not\geq \overline{v}$ we conclude either $\mu n_1 \parallel \mu m_3$ (in this case we find another object $\tilde{g}_3$ possessing m_3 but not n_1 and can reduce this case to one of the above-stated) or $\mu n_1 \geq \mu m_3$ and w.l.o.g. $\mu n_1 \parallel \mu m_1$ (we find then objects $\tilde{g}_3$ possessing m_1 and not n_1 and $\hat{g}_3$ possessing n_1 and m_3, but not m_1 implying $m_1 \tilde{L}_{12} m_2 \implies h_2 \mathfrak{R}_{12} m_1 \implies \tilde{g}_3 \mathfrak{L}_{12} n_1 \implies \hat{g}_3 \mathfrak{R}_{12} m_1 \implies m_1 \tilde{L}_{12} m_3$). $\square$

Lemma 5. *Let $\mathbb{K} = (G, M, I)$ be a context and $\Gamma(I)$ its bipartite Ferrers-graph. Let $m_1, m_2, m_3 \in M$ be pairwise incomparable attributes, s.t. m_1 is bound and m_1, m_2 and m_1, m_3 are not connected in $[\underline{v}, \overline{v}] := [\mu m_1 \wedge \mu m_2 \wedge \mu m_3, \mu m_1 \vee \mu m_2 \vee \mu m_3]$. Let $\tilde{L}_{12}$ be induced by $\Gamma(I)_{m_1,m_2}$. Then $m_1 \tilde{L}_{12} m_2 \iff m_1 \tilde{L}_{12} m_3$*

Proof. Let $p = h_1, n_1 \ldots n_{r-1}, h_r$ be a shortest connection between $n_0 := m_1$ and n_r with $\mu n_r \notin [\underline{v}, \overline{v}]$. Let t be the biggest index with $\mu n_t \in [\underline{v}, \overline{v}]$. We show first that $n_t \tilde{L} m_2 \iff n_t \tilde{L} m_3$.

1. According to Definition 8 we know $h_{t+1} I n_t$ and $h_{t+1} I n_{t+1}$.
2. Since m_1, m_2 and m_1, m_3 are not connected we conclude $h_{t+1} \nmid m_2$ and $h_{t+1} \nmid m_3$.

3. Since m_1 is incomparable to m_2 and m_3, we find objects g_2 and g_3 meeting the conditions $g_2 I m_2, g_3 I m_3$ and $g_2, g_3 \not{I}\, m_1$. Since m_1, m_2 and m_1, m_3 are not connected we notice $g_2, g_3 \not{I}\, n_t, n_{t+1}$.
   (a) $n_{t+1} \notin Int(\underline{v})$:
      4. We find an object $h \in Ext(\underline{v})$ not inciding with n_{t+1} since $n_{t+1} \notin Int(\underline{v})$.

   Therefore the following edge sequence is in $\Gamma(I)$ (see Figure 4):

$$(g_1, n_t)E(h_{t+1}, m_1)E(h, n_{t+1})E(h_{t+1}, m_3)E(g_3, n_t).$$

   Since $\Gamma(I)$ is bipartite we conclude $n_t\ \tilde{L}\ m_2 \iff n_t\ \tilde{L}\ m_3$.

	n_t	m_2	m_3	n_{t+1}
h_{t+1}	$\times_1$	$\bullet_2$	$\bullet_2$	$\times_1$
g_2	$\circ_3$	$\times_3$		
g_3	$\circ_3$		$\times_3$	
h	$\times_4$	$\times_4$	$\times_4$	$\circ_4$

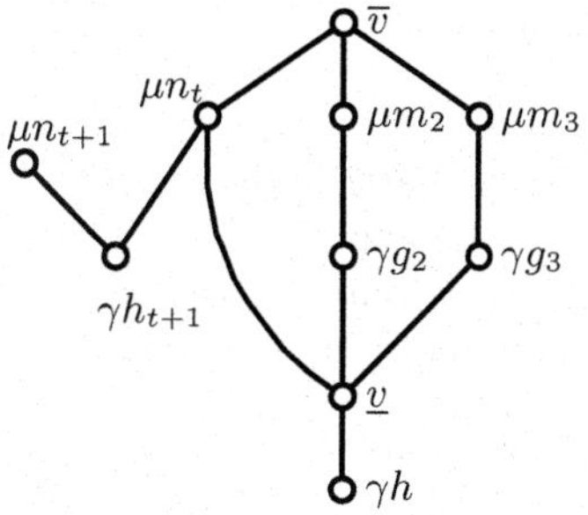

Fig. 4. An illustration for the proof above. The filled dots • symbolize $\mathfrak{L}$ and the circles ∘ symbolize $\mathfrak{R}$. The indices specify which part of the proof determines the appropriate symbol.

   (b) $n_{t+1} \in Int(\underline{v})$: With Definition 9 we conclude $\mu n_{t+1} \parallel \overline{v}$.
      5. According to Remark 1 2. we find an attribute $\tilde{m}$ with $\tilde{m} \in Int(\overline{v})$ and $\mu\tilde{m} \parallel \mu n_{t+1}$.
      6. Finally there exists an object h with hIn_{t+1} and $h \not{I}\, \tilde{m}$.

   Therefore we find the following edge sequence in $\Gamma(I)$ (see Figure 5):

$$(g_2, n_t)E(h_{t+1}, m_2)E(g_2, n_{t+1})E(h, \tilde{m})E(g_3, n_{t+1})E(h_{t+1}, m_3)E(g_3, n_t).$$

   Since $\Gamma(I)$ is bipartite we conclude $n_t\ \tilde{L}\ m_2 \iff n_t\ \tilde{L}\ m_3$.

Since n_t and m_1 are connected, we finally apply the first claim of Lemma 4 and conclude

$$m_1\ \tilde{L}_{12}\ m_2 \iff n_t\ \tilde{L}_{12}\ m_2 \iff n_t\ \tilde{L}_{12}\ m_3 \iff m_1\ \tilde{L}_{12}\ m_3. \qquad \square$$

Corollary 1. *Let* $\mathbb{K} = (G, M, I)$ *be a context and* $m_1, m_2, m_3 \in M$ *pairwise incomparable attributes. If* $m_1\ \tilde{L}\ m_2\ \tilde{L}\ m_3\ \tilde{L}\ m_1$ *then* (m_1, m_2, m_3) *is a free triple.*

Proof. : Let $\{i, j, k\} = \{1, 2, 3\}$.

	n_t	m_2	m_3	n_{t+1}	$\tilde{m}$
h_{t+1}	×1	•2	•2	×1	
g_2	○3	×3		○3	×5
g_3	○3		×3	○3	×5
h				×6	•6

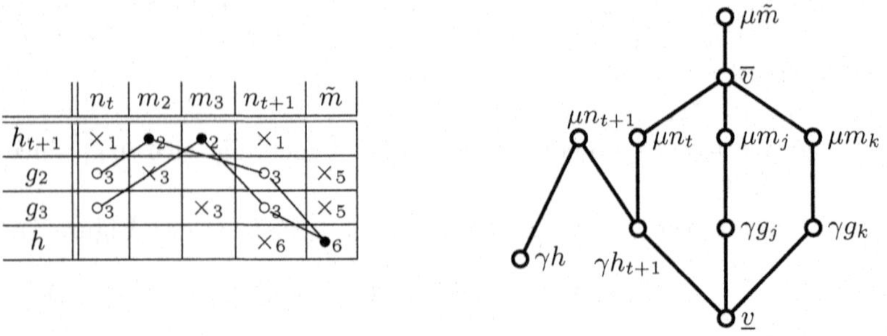

Fig. 5. An illustration for the proof above. The filled dots • symbolize $\mathfrak{L}$ and the circles ○ symbolize $\mathfrak{R}$. The indices specify which part of the proof determines the appropriate symbol.

1. No two attributes m_i, m_j are connected since $m_i \, \tilde{L} \, m_j \, \tilde{L} \, m_k \implies m_i \, \tilde{L} \, m_k$ (see Lemma 4).
2. No attribute m_i is bound since then $m_i \, \tilde{L} \, m_j \, \tilde{L} \, m_k \implies m_i \, \tilde{L} \, m_k$ (see Lemma 5).

By Definition 9 (m_1, m_2, m_3) is then a free triple. $\square$

In the remaining of this section we show that edges between elements of free triples (together with appropriate objects) are in different components of $\Gamma(I)$. This means that the relations induced by the vertex classes of components of $\Gamma(I)$ are transitive, i.e. strict orders.

Lemma 6. *Let* $\mathbb{K} = (G, M, I)$ *be a context. Then* (m_1, m_2, m_3) *is a free triple if and only if the components* Γ_{m_i, m_j} *($i \neq j \in \{1, 2, 3\}$) are pairwise disjoint.*

Proof. $\Rightarrow$: We suppose $\Gamma(I)_{m_i, m_j} = \Gamma(I)_{m_i, m_k}$. W.l.o.g. we find an edge sequence

$$(g_j, m_i)E(g_i, m_j)E(h_0, n_0)E(h_1, n_1)E \ldots E(h_r, n_r)E(g_i, m_k)E(g_k, m_i).$$

in $\Gamma(I)$. Since no attribute is bound this provides a connection $p = m_j, h_0, \ldots, m$ w.r.t. $[\mu m_1 \wedge \mu m_2 \wedge \mu m_3, \mu m_1 \vee \mu m_2 \vee \mu m_3]$ where $m \in \{m_i, m_k\}$. This contradicts our assertion that no two attributes are connected.
$\Leftarrow$: If (m_1, m_2, m_3) is not a free triple then two attributes are connected or one is bound. In both cases we find with Lemma 4 and Lemma 5 respectively that not all components $\Gamma(I)_{m_i, m_j}$ are disjoint.
$\square$

Proposition 4. *Let* $\mathbb{K} = (G, M, I)$ *be a context and* $\Gamma(I) = \uplus_{j \in J} \Gamma_j(I)$ *its bipartite Ferrers-graph. Let* $\tilde{L}_j$ *be the relation induced by the vertex classes* $\mathfrak{L}$ *and* $\mathfrak{R}$ *on the component* $\Gamma_j(I)$*. Then* $\tilde{L}_j$ *is a strict order.*

Proof. The relation $\tilde{L}_j$ is asymmetric since $\tilde{L}$ is (see Lemma 2). Let m_1, m_2, m_3 be attributes. We want to show $m_1 \, \tilde{L}_j \, m_2 \, \tilde{L}_j \, m_3 \implies m_1 \, \tilde{L}_j \, m_3$.

1. W.lo.g. let $\mu m_1 \leq \mu m_3$. We find objects g_1 with $g_1 I m_1, g_1 \, \mathfrak{L} \, m_2$ and g_2 with $g_2 I m_2, g_2 \, \mathfrak{L} \, m_3$. This contradicts $\Gamma(I)$ to be bipartite since $g_1 I m_3$, i.e. $(g_1, m_2) E (g_2, m_3)$.
2. Since we have $\Gamma(I)_{m_1,m_2} = \Gamma(I)_{m_2,m_3}$ we notice by applying Lemma 6 that (m_1, m_2, m_3) is not a free triple.

   If m_1, m_2 or m_2, m_3 are connected then we find $m_1 \, \tilde{L}_j \, m_3$ by applying Lemma 4. If only m_1, m_3 are connected we find with the first statement of Lemma 4 that $m_1 \, \tilde{L}_j \, m_2 \iff m_3 \, \tilde{L}_j \, m_2$ contradicting our precondition.

   If otherwise m_1 or m_3 is bound then we apply Lemma 5 and notice $m_1 \, \tilde{L}_j \, m_3$. If m_2 is bound we conclude $m_1 \, \tilde{L}_j \, m_2 \iff m_3 \, \tilde{L}_j \, m_2$ contradicting our precondition.

□

5 Bipartitions of Ferrers-Graphs and Their Induced Left-Relations

In this section we want to show how the components of $\Gamma(I)$ can be "turned around" s.t. the relation induced by the new vertex classes is transitive. At the beginning we remind a basic result of graph theory, namely that "turning around" components keeps a graph bipartite. See Figure 6 for an intuition of this fact.

Lemma 7. *Let $\Gamma = (V, E)$ be a bipartite graph with vertex classes X and Y and Γ_j, $j \in J$ its components. Let $X_j = X \cap V(\Gamma_j)$ and $Y_j = Y \cap V(\Gamma_j)$ be the vertex classes of the appropriate components Γ_j. Let $R_j \in \{X_j, Y_j\}$ for all $j \in J$. Then the sets $R = \bigcup_{j \in J} R_j$ and $V(\Gamma) \setminus R$ are a bipartition of Γ.*

Proof. Let $v \in R_a$ and $w \in R_b$. If $a = b$ then there is no edge $\{v, w\} \in E(\Gamma)$ since v, w are both either in X or Y. If $a \neq b$ then there is no edge between them since there is no edge between Γ_a and Γ_b.

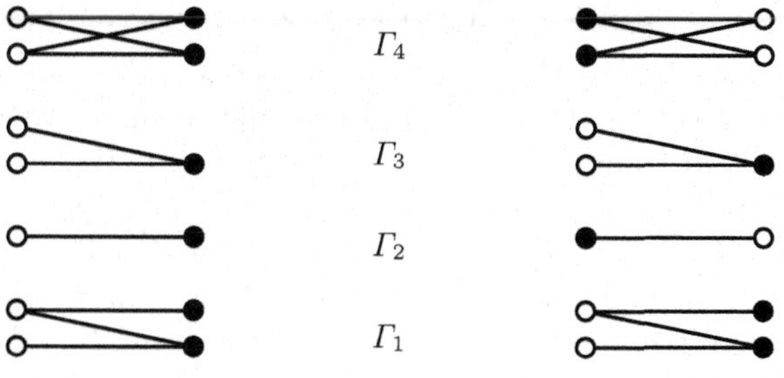

Fig. 6. The bipartite graph Γ consists of four components. "Turning around" some of them (here Γ_2 and Γ_4) supplies a new bipartition.

Lemma 8. *Let $\mathbb{K} = (G, M, I)$ be a context, $\Gamma(I)$ its bipartite Ferrers-graph and $[\underline{v}, \overline{v}]$ an interval in $\underline{\mathfrak{B}}(\mathbb{K})$. Let $m, n \in M$ and $U_m \neq U_n \subseteq U[\underline{v}, \overline{v}]$. The edge set E of $\Gamma(I)_{m,n}$ consists exactly of all edges of the form*

$$\{\{(g, m_1), (h, n_1)\} \mid m_1 \in U_m, n_1 \in U_n, \{(g, m_1), (h, n_1)\} \in E(\Gamma(I))\}.$$

Moreover, let $\tilde{L}_j$ be induced by $\Gamma(I)_{m,n}$ then

$$m \, \tilde{L}_j \, n \implies m_1 \, \tilde{L}_j \, n_1 \qquad \forall \, m_1 \in U_m, n_1 \in U_n.$$

Proof. Let $m_1 \in U_m$, $n_1 \in U_n$ and $\{(g, m_1), (h, n_1)\} \in E(\Gamma(I))$. Since there exists a connection $p = m_1, \tilde{h}_1, \tilde{n}_1, \ldots, \tilde{h}_r, m$ we find an edge sequence

$$(g, m_1)E(\tilde{h}_1, n_1)E(g, \tilde{n}_1)E \ldots E(\tilde{h}_r, n_1)E(g, m)$$

Since $(g, m) \in V(\Gamma(I)_{m,n})$ we know $\{(g, m_1), (h, n_1)\} \in E(\Gamma(I)_{m,n})$.

Let $\{(g_1, m_1), (h_1, n_1)\}$ in $E(\Gamma(I)_{m,n})$ then we find w.l.o.g an edge sequence

$$(g_1, m_1)E(h_1, n_1)E \ldots E(g, m)E(h, n).$$

W.l.o.g. this results in connections $p_1 = m_1, h_1, \ldots, m$ and $p_2 = n_1, \ldots, g, n$, i.e. $m_1 \in U_m$ and $n_1 \in U_n$. In this case we find

$$m \, \tilde{L}_j \, n \implies g \, \mathfrak{R}_j \, m \implies \ldots \implies h_1 \, \mathfrak{L}_j \, n_1 \implies g_1 \, \mathfrak{R}_j \, m_1 \implies m_1 \, \tilde{L}_j \, z n_1.$$

□

Lemma 9. *Let $\mathbb{K} = (G, M, I)$ be a context and $\Gamma(I)$ its bipartite Ferrers-graph. Let $\Gamma_j(I)$, $j \in J$ be the components of $\Gamma(I)$ and $\tilde{L}_j$ their respective induced relations. There exists a relation $\hat{L} = \biguplus_{j \in J} \hat{L}_j$ with $\hat{L}_j \in \{\tilde{L}_j, \tilde{L}_j^{-1}\}$ that is transitive.*

Proof. For each interval $[\underline{v}, \overline{v}] \subseteq \underline{\mathfrak{B}}(\mathbb{K})$ we introduce a linear order on its components:

$$U_{m_1} \leq U_{m_2} \leq \ldots \leq U_{m_t}.$$

Let $U_{m_i} \neq U_{m_j} \subseteq U[\underline{v}, \overline{v}]$, the relation $\tilde{L}_{ij}$ be induced by $\Gamma(I)_{m_i, m_j}$ and $m_i \, \tilde{L}_{ij} \, m_j$. We define

$$\hat{L}_{ij} := \begin{cases} \tilde{L}_{ij}, & i < j \\ \tilde{L}_{ij}^{-1}, & i > j. \end{cases}$$

According to Lemma 8 this setting is unambiguous w.r.t. the components of the interval $[\underline{v}, \overline{v}]$. We notice that $\tilde{L}_{ij}$ is uniquely determined since two attributes m_i and m_j are in different components $U_{m_1}, U_{m_2} \in [\underline{v}, \overline{v}]$ only if $\underline{v} = m_1 \wedge m_2$ and $\overline{v} = m_1 \vee m_2$. For all other relations $\tilde{L}_r$ induced by $\Gamma_r(I)$ we set $\hat{L}_r := \tilde{L}_r$.

Let $m_1 \, \hat{L}_{12} \, m_2 \, \hat{L}_{23} \, m_3$. Since $\Gamma(I)$ is bipartite we know $\mu m_1 \parallel \mu m_3$. In case of $\hat{L}_{13} \in \{\hat{L}_{12}, \hat{L}_{23}\}$ we conclude with Proposition 4 $m_1 \hat{L}_{13} m_3$. Otherwise (m_1, m_2, m_3) is a free triple by Lemma 6. The components U_{m_1}, U_{m_2} and U_{m_3} of $U[\underline{v}, \overline{v}]$ are pairwise disjoint. By applying the construction above we conclude $U_{m_1} \leq U_{m_2} \leq U_{m_3}$ and therefore $U_{m_1} \leq U_{m_3}$. This is, $m_1 \hat{L}_{13} m_3$. □

At this point, we provided everything to show that the above-stated conjecture is true.

Theorem 3. *Let $\mathbb{K} = (G, M, I)$ be a context and $\Gamma(I)$ its Ferrers-graph. Then the following conditions are equivalent.*

1. *$\Gamma(I)$ is bipartite*
2. *$\underline{\mathfrak{B}}(\mathbb{K})$ is planar.*

Proof.
1. $\implies$ 2.: Consider the relation $\hat{L}$ from Lemma 9. Let $\hat{\mathfrak{L}} := \bigcup_{j\in J} \hat{\mathfrak{L}}_j$ and $\hat{\mathfrak{R}} := \bigcup_{j\in J} \hat{\mathfrak{R}}_j$ with

$$\hat{\mathfrak{L}}_j := \begin{cases} \mathfrak{L}_j, & \hat{L}_j = \tilde{L}_j \\ \mathfrak{R}_j, & \hat{L}_j = \tilde{L}_j^{-1} \end{cases}.$$

$\hat{\mathfrak{L}}$ and $\hat{\mathfrak{R}}$ are a bipartition of $\Gamma(I)$ since $\mathfrak{L}_j = V_j(\Gamma(I)) \cap \mathfrak{L}$ (see Lemma 7) and $\hat{L}$ is its induced relation. The relation $\hat{L}$ is asymmetric and connex (Lemma 2) and transitive (Lemma 9). Therefore $\underline{\mathfrak{B}}(\mathbb{K})$ is planar (Lemma 3).
2. $\implies$ 1.: If $\underline{\mathfrak{B}}(\mathbb{K})$ is planar then its order dimension is at most two (Theorem 2). Therefore, the Ferrers-dimension is at most two (Theorem 1) and hence its Ferrers-graph bipartite (Lemma 1).

□

6 Conclusion

We could show in this work that planar concept lattices can be characterized in a simple way by their contexts. The complexity of an algorithm to decide whether $\Gamma(I)$ is bipartite, is $\mathcal{O}(E(\Gamma(I))) = \mathcal{O}((|G| \cdot |M|)^2)$. We still do not know how to develop an algorithm to find the conjugate order $\hat{L}$ out of $\Gamma(I)$.

It would be interesting to find estimations of the minimal number of edge crossings in non-planar concept lattices depending on the shape of the Ferrers-graph of the appropriate context.

References

1. K. A. Baker, P. Fishburn, F. S. Roberts: *Partial Orders of Dimension 2.* Networks, 2, 11-28, 1971.
2. G. Birkhoff: *Lattice Theory.* Amer. Math. Soc., Third Edition, 1967.
3. B. Dushnik, E.W. Miller: *Partially Ordered Sets.* Amer. J. Math. 63, 1941, pp. 600-610.
4. B. Ganter, R. Wille: *Formal Concept Analysis.* Springer, 1999.
5. Private communication with B. Ganter.
6. K. Reuter: *Removing Critical Pairs.* Preprint, no. 1241, TU Darmstadt, 1989.
7. C. Zschalig: *Planarity of Lattices - An approach based on attribute additivity.* Proc. of ICFCA 05, LNAI 3403, pp. 391-402, 2005.
8. C. Zschalig: *Characterizing Planar Lattices Using Left-relations* Proc. of ICFCA06, LNAI 3874, pp. 280-290, 2006.

Author Index

Printing: Mercedes-Druck, Berlin
Binding: Stein+Lehmann, Berlin

Lecture Notes in Artificial Intelligence (LNAI)

Vol. 4390: S.O. Kuznetsov, S. Schmidt (Eds.), Formal Concept Analysis. X, 329 pages. 2007.

Vol. 4384: T. Washio, K. Satoh, H. Takeda, A. Inokuchi (Eds.), New Frontiers in Artificial Intelligence. IX, 401 pages. 2007.

Vol. 4371: K. Inoue, K. Satoh, F. Toni (Eds.), Computational Logic in Multi-Agent Systems. X, 315 pages. 2007.

Vol. 4369: M. Umeda, A. Wolf, O. Bartenstein, U. Geske, D. Seipel, O. Takata (Eds.), Declarative Programming for Knowledge Management. X, 229 pages. 2006.

Vol. 4342: H. de Swart, E. Orłowska, G. Schmidt, M. Roubens (Eds.), Theory and Applications of Relational Structures as Knowledge Instruments II. X, 373 pages. 2006.

Vol. 4335: S.A. Brueckner, S. Hassas, M. Jelasity, D. Yamins (Eds.), Engineering Self-Organising Systems. XII, 212 pages. 2007.

Vol. 4334: B. Beckert, R. Hähnle, P.H. Schmitt (Eds.), Verification of Object-Oriented Software. XXIX, 658 pages. 2007.

Vol. 4333: U. Reimer, D. Karagiannis (Eds.), Practical Aspects of Knowledge Management. XII, 338 pages. 2006.

Vol. 4327: M. Baldoni, U. Endriss (Eds.), Declarative Agent Languages and Technologies IV. VIII, 257 pages. 2006.

Vol. 4314: C. Freksa, M. Kohlhase, K. Schill (Eds.), KI 2006: Advances in Artificial Intelligence. XII, 458 pages. 2007.

Vol. 4304: A. Sattar, B.-H. Kang (Eds.), AI 2006: Advances in Artificial Intelligence. XXVII, 1303 pages. 2006.

Vol. 4303: A. Hoffmann, B.-H. Kang, D. Richards, S. Tsumoto (Eds.), Advances in Knowledge Acquisition and Management. XI, 259 pages. 2006.

Vol. 4293: A. Gelbukh, C.A. Reyes-Garcia (Eds.), MICAI 2006: Advances in Artificial Intelligence. XXVIII, 1232 pages. 2006.

Vol. 4289: M. Ackermann, B. Berendt, M. Grobelnik, A. Hotho, D. Mladenič, G. Semeraro, M. Spiliopoulou, G. Stumme, V. Svátek, M. van Someren (Eds.), Semantics, Web and Mining. X, 197 pages. 2006.

Vol. 4285: Y. Matsumoto, R.W. Sproat, K.-F. Wong, M. Zhang (Eds.), Computer Processing of Oriental Languages. XVII, 544 pages. 2006.

Vol. 4274: Q. Huo, B. Ma, E.-S. Chng, H. Li (Eds.), Chinese Spoken Language Processing. XXIV, 805 pages. 2006.

Vol. 4265: L. Todorovski, N. Lavrač, K.P. Jantke (Eds.), Discovery Science. XIV, 384 pages. 2006.

Vol. 4264: J.L. Balcázar, P.M. Long, F. Stephan (Eds.), Algorithmic Learning Theory. XIII, 393 pages. 2006.

Vol. 4259: S. Greco, Y. Hata, S. Hirano, M. Inuiguchi, S. Miyamoto, H.S. Nguyen, R. Słowiński (Eds.), Rough Sets and Current Trends in Computing. XXII, 951 pages. 2006.

Vol. 4253: B. Gabrys, R.J. Howlett, L.C. Jain (Eds.), Knowledge-Based Intelligent Information and Engineering Systems, Part III. XXXII, 1301 pages. 2006.

Vol. 4252: B. Gabrys, R.J. Howlett, L.C. Jain (Eds.), Knowledge-Based Intelligent Information and Engineering Systems, Part II. XXXIII, 1335 pages. 2006.

Vol. 4251: B. Gabrys, R.J. Howlett, L.C. Jain (Eds.), Knowledge-Based Intelligent Information and Engineering Systems, Part I. LXVI, 1297 pages. 2006.

Vol. 4248: S. Staab, V. Svátek (Eds.), Managing Knowledge in a World of Networks. XIV, 400 pages. 2006.

Vol. 4246: M. Hermann, A. Voronkov (Eds.), Logic for Programming, Artificial Intelligence, and Reasoning. XIII, 588 pages. 2006.

Vol. 4223: L. Wang, L. Jiao, G. Shi, X. Li, J. Liu (Eds.), Fuzzy Systems and Knowledge Discovery. XXVIII, 1335 pages. 2006.

Vol. 4213: J. Fürnkranz, T. Scheffer, M. Spiliopoulou (Eds.), Knowledge Discovery in Databases: PKDD 2006. XXII, 660 pages. 2006.

Vol. 4212: J. Fürnkranz, T. Scheffer, M. Spiliopoulou (Eds.), Machine Learning: ECML 2006. XXIII, 851 pages. 2006.

Vol. 4211: P. Vogt, Y. Sugita, E. Tuci, C.L. Nehaniv (Eds.), Symbol Grounding and Beyond. VIII, 237 pages. 2006.

Vol. 4203: F. Esposito, Z.W. Raś, D. Malerba, G. Semeraro (Eds.), Foundations of Intelligent Systems. XVIII, 767 pages. 2006.

Vol. 4201: Y. Sakakibara, S. Kobayashi, K. Sato, T. Nishino, E. Tomita (Eds.), Grammatical Inference: Algorithms and Applications. XII, 359 pages. 2006.

Vol. 4200: I.F.C. Smith (Ed.), Intelligent Computing in Engineering and Architecture. XIII, 692 pages. 2006.

Vol. 4198: O. Nasraoui, O. Zaïane, M. Spiliopoulou, B. Mobasher, B. Masand, P.S. Yu (Eds.), Advances in Web Mining and Web Usage Analysis. IX, 177 pages. 2006.

Vol. 4196: K. Fischer, I.J. Timm, E. André, N. Zhong (Eds.), Multiagent System Technologies. X, 185 pages. 2006.

Vol. 4188: P. Sojka, I. Kopeček, K. Pala (Eds.), Text, Speech and Dialogue. XV, 721 pages. 2006.

Vol. 4183: J. Euzenat, J. Domingue (Eds.), Artificial Intelligence: Methodology, Systems, and Applications. XIII, 291 pages. 2006.

Vol. 4180: M. Kohlhase, OMDoc – An Open Markup Format for Mathematical Documents [version 1.2]. XIX, 428 pages. 2006.

Vol. 4177: R. Marín, E. Onaindía, A. Bugarín, J. Santos (Eds.), Current Topics in Artificial Intelligence. XV, 482 pages. 2006.

Vol. 4160: M. Fisher, W. van der Hoek, B. Konev, A. Lisitsa (Eds.), Logics in Artificial Intelligence. XII, 516 pages. 2006.

Vol. 4155: O. Stock, M. Schaerf (Eds.), Reasoning, Action and Interaction in AI Theories and Systems. XVIII, 343 pages. 2006.

Vol. 4149: M. Klusch, M. Rovatsos, T.R. Payne (Eds.), Cooperative Information Agents X. XII, 477 pages. 2006.

Vol. 4140: J.S. Sichman, H. Coelho, S.O. Rezende (Eds.), Advances in Artificial Intelligence - IBERAMIA-SBIA 2006. XXIII, 635 pages. 2006.

Vol. 4139: T. Salakoski, F. Ginter, S. Pyysalo, T. Pahikkala (Eds.), Advances in Natural Language Processing. XVI, 771 pages. 2006.

Vol. 4133: J. Gratch, M. Young, R. Aylett, D. Ballin, P. Olivier (Eds.), Intelligent Virtual Agents. XIV, 472 pages. 2006.

Vol. 4130: U. Furbach, N. Shankar (Eds.), Automated Reasoning. XV, 680 pages. 2006.

Vol. 4120: J. Calmet, T. Ida, D. Wang (Eds.), Artificial Intelligence and Symbolic Computation. XIII, 269 pages. 2006.

Vol. 4118: Z. Despotovic, S. Joseph, C. Sartori (Eds.), Agents and Peer-to-Peer Computing. XIV, 173 pages. 2006.

Vol. 4114: D.-S. Huang, K. Li, G.W. Irwin (Eds.), Computational Intelligence, Part II. XXVII, 1337 pages. 2006.

Vol. 4108: J.M. Borwein, W.M. Farmer (Eds.), Mathematical Knowledge Management. VIII, 295 pages. 2006.

Vol. 4106: T.R. Roth-Berghofer, M.H. Göker, H.A. Güvenir (Eds.), Advances in Case-Based Reasoning. XIV, 566 pages. 2006.

Vol. 4099: Q. Yang, G. Webb (Eds.), PRICAI 2006: Trends in Artificial Intelligence. XXVIII, 1263 pages. 2006.

Vol. 4095: S. Nolfi, G. Baldassarre, R. Calabretta, J.C.T. Hallam, D. Marocco, J.-A. Meyer, O. Miglino, D. Parisi (Eds.), From Animals to Animats 9. XV, 869 pages. 2006.

Vol. 4093: X. Li, O.R. Zaïane, Z. Li (Eds.), Advanced Data Mining and Applications. XXI, 1110 pages. 2006.

Vol. 4092: J. Lang, F. Lin, J. Wang (Eds.), Knowledge Science, Engineering and Management. XV, 664 pages. 2006.

Vol. 4088: Z.-Z. Shi, R. Sadananda (Eds.), Agent Computing and Multi-Agent Systems. XVII, 827 pages. 2006.

Vol. 4087: F. Schwenker, S. Marinai (Eds.), Artificial Neural Networks in Pattern Recognition. IX, 299 pages. 2006.

Vol. 4068: H. Schärfe, P. Hitzler, P. Øhrstrøm (Eds.), Conceptual Structures: Inspiration and Application. XI, 455 pages. 2006.

Vol. 4065: P. Perner (Ed.), Advances in Data Mining. XI, 592 pages. 2006.

Vol. 4062: G.-Y. Wang, J.F. Peters, A. Skowron, Y. Yao (Eds.), Rough Sets and Knowledge Technology. XX, 810 pages. 2006.

Vol. 4049: S. Parsons, N. Maudet, P. Moraitis, I. Rahwan (Eds.), Argumentation in Multi-Agent Systems. XIV, 313 pages. 2006.

Vol. 4048: L. Goble, J.-J.C.. Meyer (Eds.), Deontic Logic and Artificial Normative Systems. X, 273 pages. 2006.

Vol. 4045: D. Barker-Plummer, R. Cox, N. Swoboda (Eds.), Diagrammatic Representation and Inference. XII, 301 pages. 2006.

Vol. 4031: M. Ali, R. Dapoigny (Eds.), Advances in Applied Artificial Intelligence. XXIII, 1353 pages. 2006.

Vol. 4029: L. Rutkowski, R. Tadeusiewicz, L.A. Zadeh, J.M. Zurada (Eds.), Artificial Intelligence and Soft Computing – ICAISC 2006. XXI, 1235 pages. 2006.

Vol. 4027: H.L. Larsen, G. Pasi, D. Ortiz-Arroyo, T. Andreasen, H. Christiansen (Eds.), Flexible Query Answering Systems. XVIII, 714 pages. 2006.

Vol. 4021: E. André, L. Dybkjær, W. Minker, H. Neumann, M. Weber (Eds.), Perception and Interactive Technologies. XI, 217 pages. 2006.

Vol. 4020: A. Bredenfeld, A. Jacoff, I. Noda, Y. Takahashi (Eds.), RoboCup 2005: Robot Soccer World Cup IX. XVII, 727 pages. 2006.

Vol. 4013: L. Lamontagne, M. Marchand (Eds.), Advances in Artificial Intelligence. XIII, 564 pages. 2006.

Vol. 4012: T. Washio, A. Sakurai, K. Nakajima, H. Takeda, S. Tojo, M. Yokoo (Eds.), New Frontiers in Artificial Intelligence. XIII, 484 pages. 2006.

Vol. 4008: J.C. Augusto, C.D. Nugent (Eds.), Designing Smart Homes. XI, 183 pages. 2006.

Vol. 4005: G. Lugosi, H.U. Simon (Eds.), Learning Theory. XI, 656 pages. 2006.

Vol. 4002: A. Yli-Jyrä, L. Karttunen, J. Karhumäki (Eds.), Finite-State Methods and Natural Language Processing. XIV, 312 pages. 2006.

Vol. 3978: B. Hnich, M. Carlsson, F. Fages, F. Rossi (Eds.), Recent Advances in Constraints. VIII, 179 pages. 2006.

Vol. 3963: O. Dikenelli, M.-P. Gleizes, A. Ricci (Eds.), Engineering Societies in the Agents World VI. XII, 303 pages. 2006.

Vol. 3960: R. Vieira, P. Quaresma, M.d.G.V. Nunes, N.J. Mamede, C. Oliveira, M.C. Dias (Eds.), Computational Processing of the Portuguese Language. XII, 274 pages. 2006.

Vol. 3955: G. Antoniou, G. Potamias, C. Spyropoulos, D. Plexousakis (Eds.), Advances in Artificial Intelligence. XVII, 611 pages. 2006.